CONSTRUCTED SUBSURFACE WETLANDS
Case Study and Modeling

CONSTRUCTED SUBSURFACE WETLANDS
Case Study and Modeling

Abdel Razik Ahmed Zidan, PhD, and
Mohammed Ahmed Abdel Hady, PhD

Apple Academic Press Inc. | Apple Academic Press Inc.
3333 Mistwell Crescent | 9 Spinnaker Way
Oakville, ON L6L 0A2 Canada | Waretown, NJ 08758 USA

© 2018 by Apple Academic Press, Inc.
Exclusive worldwide distribution by CRC Press, a member of Taylor & Francis Group
No claim to original U.S. Government works
Typeset by Accent Premedia Group (www.accentpremedia.com)
Printed in the United States of America on acid-free paper
International Standard Book Number-13: 978-1-77188-463-1 (Hardcover)
International Standard Book Number-13: 978-1-315-36589-3 (eBook)

All rights reserved. No part of this work may be reprinted or reproduced or utilized in any form or by any electric, mechanical or other means, now known or hereafter invented, including photocopying and recording, or in any information storage or retrieval system, without permission in writing from the publisher or its distributor, except in the case of brief excerpts or quotations for use in reviews or critical articles.

This book contains information obtained from authentic and highly regarded sources. Reprinted material is quoted with permission and sources are indicated. Copyright for individual articles remains with the authors as indicated. A wide variety of references are listed. Reasonable efforts have been made to publish reliable data and information, but the authors, editors, and the publisher cannot assume responsibility for the validity of all materials or the consequences of their use. The authors, editors, and the publisher have attempted to trace the copyright holders of all material reproduced in this publication and apologize to copyright holders if permission to publish in this form has not been obtained. If any copyright material has not been acknowledged, please write and let us know so we may rectify in any future reprint.

Trademark Notice: Registered trademark of products or corporate names are used only for explanation and identification without intent to infringe.

Library and Archives Canada Cataloguing in Publication

Zidan, Abdel Razik Ahmed, author
Constructed subsurface wetlands : case study and modeling / Abdel Razik Ahmed Zidan, PhD, and Mohammed Ahmed Abdel Hady, PhD.

Includes bibliographical references and index.
Issued in print and electronic formats.
ISBN 978-1-77188-463-1 (hardcover).--ISBN 978-1-315-36589-3 (PDF)
1. Constructed wetlands--Case studies. I. Hady, Mohammed Ahmed Abdel, author II. Title.

TD756.5.Z53 2017   333.91'8   C2017-902582-1   C2017-902583-X

Library of Congress Cataloging-in-Publication Data

Names: Zidan, Abdel Razik Ahmed, author. | Hady, Mohammed Ahmed Abdel, author.
Title: Constructed subsurface wetlands : case study and modeling / Abdel
Razik Ahmed Zidan, PhD, and Mohammed Ahmed Abdel Hady, PhD.
Description: Toronto ; Waretown, NJ, USA : Apple Academic Press, 2017. |
Includes bibliographical references and index.
Identifiers: LCCN 2017017445 (print) | LCCN 2017019984 (ebook) | ISBN 9781315365893 (ebook) | ISBN 9781771884631 (hardcover : alk. paper)
Subjects: LCSH: Constructed wetlands--Case studies. | Constructed wetlands--Simulation methods.
Classification: LCC TD756.5 (ebook) | LCC TD756.5 .Z53 2017 (print) | DDC 628.1/68--dc23
LC record available at https://lccn.loc.gov/2017017445

Apple Academic Press also publishes its books in a variety of electronic formats. Some content that appears in print may not be available in electronic format. For information about Apple Academic Press products, visit our website at **www.appleacademicpress.com** and the CRC Press website at **www.crcpress.com**

CONTENTS

List of Abbreviations .... *vii*
List of Symbols .... *ix*
List of Chemical Compounds and Elements... *xv*
Acknowledgments.. *xvii*
Preface .... *xix*
About the Authors.. *xxi*

1. **Introduction** 1
2. **Literature Review**.. 7
3. **Field and Experimental Works**... 69
4. **Theoretical Approach** 111
5. **Field and Experimental Results**.. 129
6. **Set Up Stage Analysis**.. 181
7. **Steady Stage Analysis**.. 229
8. **ANNs Modeling and SPSS Analysis** 357
9. **Summary, Conclusions, and Recommendations** 515
   References *527*
   Appendices *539*
   Index.. *651*

LIST OF ABBREVIATIONS

AAS	Atomic Absorption Spectrometer
ANNs	Artificial Neural Networks
APHA	American Public Health Association
BOD	Biochemical Oxygen Demand
BP	Back Propagation
COD	Chemical Oxygen Demand
CWs	Constructed Wetlands
D.F.	Degree of Freedom
DFID	Department For International Development
DO	Dissolved Oxygen
E-JUST	Egypt-Japan University of Science and Technology
EPA	Environmental Protection Agency
FC	Fecal Coliforms
FG	Fine Gravel
FHL	First Hidden Layer
GBH	Gravel Bed Hydroponics
ICSBE	International Conference on Sustainable Built Environment
I/O	Input/Output
HC	Hydraulic Conductivity
HSSF	Horizontal Subsurface Flow
LE	Egyptian Pounds
LECA	Light Expanded Clay Aggregated
MFFNNs	Multi-Feed Forward Neural Networks
ML	Marquardt-Levenberg
MLP	Multi-Layer Perceptron
MPN	Most Probable Number
MRA	Multiple Regression Analysis
MSE	Mean Square Error
NAWQAM	National Water Quality and Availability Management
PCA	Principle Component Analysis
PFD	Plug Flow with Dispersion
PVC	Polyvinyl Chloride
RBFN	Radial Basis Function Network

RE	Removal Efficiency
RFCW	Reciprocating Flow Constructed Wetlands
RTM	Regional Technical Meeting
RZM	Root Zone Method
S.D.	Standard Deviation
SF	Surface Flow
SHL	Second Hidden Layer
SOP	Soluble Organic Phosphorus
SPSS	Stochastic Package for Social Since
SS	Suspended Solids
SSF	Subsurface Flow
TDS	Total Dissolved Solids
TC	Total Califorms
TIS	Tanks-In-Series
TKN	Total Kjeldahl Nitrogen
TN	Total Nitrogen
TOC	Total Organic Carbon
TP	Total Phosphorus
TSS	Total Suspended Solids
TVA	Tennessee Valley Authority
USA	United States of America
USAID	United States Agency for International Development
USBR	United States Bureau of Reclamation
USDA	United States Department of Agriculture
UV	Ultra Violet
VF	Vertical Flow
WLCS	Water Level Control System
WWTP	Wastewater Treatment Plant

LIST OF SYMBOLS

A	wetland top surface area, L^2
A_{active}	area of wetland containing water in active flow, L^2
A_c	cross-sectional area, L^2
A_p	surface area of sphere, L^2
A_s	specific surface area, L^{-1}
A_{six}	surface area of media (i = media type and x = distance from cell inlet), L^2
B	local bed bottom elevation, L
b_i	constant, D.L.
$B(x)$	elevation of bed bottom, L
C	pollutant concentration, M/L^3
C_{ANN}	artificial neural network output concentration. M/L^3
C_{Exp}	experimental measured output concentration. M/L^3
C_t	sorting coefficient, D.L.
C_{SPSS}	regression equation output concentration. M/L^3
$C(x)$	concentration at length (x), M/L^3
C^*	background concentration, M/L^3
C^*_{BOD}	background concentration for BOD pollutant, M/L^3
C^*_{COD}	background concentration for COD pollutant, M/L^3
C^*_{TSS}	background concentration for TSS pollutant, M/L^3
Cd	coefficient of discharge, D.L.
C_i	influent concentration, M/L^3
C_{i1}	influent concentration for run number 1, M/L^3
C_o	effluent concentration, M/L^3
C_{oix}	effluent concentration (i = media type and x = distance from cell inlet), M/L^3
d_1	first media bucket, at distance 2.5 m from inlet
d_2	second media bucket, at distance 5.5 m from inlet
d_3	third media bucket, at distance 7.5 m from inlet
d_{10}	effective grain diameter, L
d_{20}	grain diameter for which 20% of sample are finer than, L
d_{50}	median grain diameter, L

d_{60}	diameter through which no more than 60% of sample passes, L
d_p	particle diameter, L
Da	Damköhler number, $D.L.$
D_s	dispersion coefficient, L^2/T
E_n	ANN percentage error, $D.L.$
E_s	SPSS percentage error, $D.L.$
ET	evapotranspiration rate, L/T
g	acceleration of gravity, L/T^2
$G(x)$	elevation of bed surface, L
h	wetland water depth, L
h_i	water elevation at inlet, L
h_o	water elevation at outlet, L
$h(x)$	water depth, L
H	elevation of water surface, L
H_d	head over the notch, L
$H(x)$	elevation of water surface at distance x, L
I_o	intercept of line formed by d_{50} and d_{10} with grain size axis, L
J	removal load, M/L^2T
k	removal rate constant, L/T
K	half-saturation constant, M/L^3
k_{20}	first-order areal rate constant at 20°C, T^{-1}
K_c	media hydraulic conductivity, L/T
k_T	first-order areal rate constant at T°C, T^{-1}
k_v	volumetric rate constant, T^{-1}
L	wetland length, L
L_a	first layer, 16.7 cm height
L_b	first and second layers, 33.3 cm height
L_c	the three layers, 50 cm height
L_i	influent loads, M/L^2T
L_o	effluent loads, M/L^2T
L_r	load removal, M/L^2T
L_t	total film thickness, L
m	number of contact points, $D.L.$
m_i	specific inlet mass loading, M/L^2T
m_o	specific outlet mass loading, M/L^2T
M_i	inlet mass loading, M/T

List of Symbols

n	porosity, *D.L.*
n_1	porosity at time T_{o1}, *D.L.*
n_2	porosity at time T_{o2}, *D.L.*
n_{avg}	average porosity, *D.L.*
n_{cg}	porosity of course gravel. *D.L.*
n_g	porosity of gravel media, *D.L.*
n_i	porosity at time T_{oi}, *D.L.*
n_m	porosity of used media, *D.L.*
n_p	porosity of plastic media, *D.L.*
n_r	porosity of rubber media, *D.L.*
N	number of tanks in series, *D.L.*
N_1	number of neurons in the first hidden layers, *D.L.*
N_2	number of neurons in the second hidden layers, *D.L.*
N_i	number of input variables, *D.L.*
N_o	number of output variables, *D.L.*
pH	hydrogen ion, *D.L.*
P	apparent number of TIS, *D.L.*
P_e	Peclet number, *D.L.*
P_r	precipitation rate, *L/T*
q_{xi}	loading rate (x = distance from cell inlet and i = media type), *M/L*
Q	hydraulic loading rate, *L/T*
Q	discharge, L^3/T
Q_{act}	actual discharge, L^3/T
Q_{avg}	average discharge, L^3/T
Q_b	Bank loss rate, L^3/T
Q_c	Catchment's runoff rate, L^3/T
Q_{gw}	infiltration to groundwater, L^3/T
Q_i	inlet flow rate, L^3/T
Q_o	outlet flow rate, L^3/T
Q_{th}	theoretical discharge, L^3/T
R	water recovery fraction, *D.L.*
R^2	determination coefficient, *D.L.*
Re	Reynolds number, *D.L.*
RE	pollutant removal efficiency, *D.L.*
R_s	rate of sediment storage, *M/L.T.*
RE_{xg}	removal efficiency of gravel cell at distance x, *D.L.*
RE_{xp}	removal efficiency of plastic cell at distance x, *D.L.*

RE_{xr}	removal efficiency of rubber cell at distance x, D.L.
S	water slope, D.L.
S_1	sampling point at distance of 2 m from inlet (right)
S_2	sampling point at distance of 2 m from inlet (middle)
S_3	sampling point at distance of 2 m from inlet (left)
S_4	sampling point at distance of 5 m from inlet (right)
S_5	sampling point at distance of 5 m from inlet (middle)
S_6	sampling point at distance of 5 m from inlet (left)
S_7	sampling point at distance of 8 m from inlet (right)
S_8	sampling point at distance of 8 m from inlet (middle)
S_9	sampling point at distance of 8 m from inlet (left)
S_b	bottom slope, D.L.
S_e	effluent water samples
S_i	influent water samples
t	time, T
T	desired temperature, D.L.
T_i	inlet flow-based T_r, T
T_o	time from start of operation, T
T_{oi}	in-between time from start of operation, T
T_{o1}	time from start of operation for run number 1, T
T_{o2}	time from start of operation for run number 2, T
T_r	hydraulic retention time, T
T_s	time from start of sampling, T
u	average water velocity, L/T
u_s	superficial velocity, L/T
U	grain uniformity coefficient, D.L.
v	actual water velocity, L/T
v_{xi}	actual water velocity (x = distance from cell inlet and i = media type), L/T
V	water volume in wetlands, L^3
V_{active}	volume of wetland containing water in active flow, L^3
V_b	volume of buckets, L^3
V_d	the space volume between 6 inch pipe and bucket, L^3
V_{m1}	volume of coarse gravel, L^3
V_{m2}	volume of surface media, L^3
V_{m3}	volume of used media, L^3
V_p	volume of 6 inch pipes, L^3
V_s	volume of sphere, L^3

List of Symbols

V_u	total volume considering contact points, L^3
V_v	volume of media voids, L^3
V_w	volume of added water, L^3
V_{w2}	water volume at the first 2 m, L^3
V_{w5}	water volume at the first 5 m, L^3
V_{w8}	water volume at the first 8 m, L^3
V_{w10}	water volume in the wetland cell, L^3
V_{wx}	volume of water inside cell at distance x, L^3
W	wetland width, L
W_1	weight the dried filter paper, ML/T^2
W_2	weight the paper, ML/T^2
W_i	inlet weight of suspended solids, M/T
W_o	outlet weight of suspended solids, M/T
x	longitudinal distance, L
x_i	inlet distance, L
x_o	outlet distance, L
X	independent variable
y	transverse distance, L
Y	dependent variable
Z	constant
α_m	backing factor, $D.L.$
β	constant
θ	apex angle of V-notch, $D.L.$
θ_t	empirical temperature coefficient, $D.L.$
δ	bed depth, L
δ_b	thickness of the biofilm, L
δ_w	thickness of the stagnant boundary layer, L
μ	viscosity of water, M/LT
ρ	density of water, M/L^3
ω	constant depends on media type and discharge

LIST OF CHEMICAL COMPOUNDS AND ELEMENTS

Al	aluminum
Alkali	sodium hydroxide, NaOH
Azide	sodium azide, NAN_3
B	boron
C	carbon
$C_9H_{17}N_5S$	ametryn
Ca	calcium
$CaCl_2$	calcium chloride
$CaCO_3$	calcium carbonate
Cd	cadmium
Cr	chromium
Cu	copper
Fe	iron
$FeCl_3$	ferric chloride
H_2SO_4 reagent	sulfuric acid + silver sulphate, Ag_2SO_4
H_2SO_4	sulfuric acid
HgI_2	mercuric iodide
HNO_3	nitric acid
Iodide	sodium iodide, NaI
$K_2Cr_2O_7$	potassium di-chromate
K_2HPO_4	di-potassium hydrogen orthophosphate anhydrous
KH_2PO_4	potassium di-hydrogen orthophosphate
KI	potassium iodide
$MgSO_4$	magnesium sulphate
M	molar
Mn	manganese
$MnSO_4$	manganese(II)-sulfate-1-hydrate
N	nitrogen
Na_2HPO_4	di-sodium hydrogen orthophosphate anhydrous
$Na_2S_2O_3$	sodium thiosulfate
NaOH	sodium hydroxide
Nessler	HGI_2 + KI + NaOH

NH_3	ammonia
NH_4	ammonium
NH_4Cl	ammonium chloride
Ni	nickel
P	phosphorus
Pb	lead
Phosphate buffer	$KH_2PO_4 + K_2HPO_4 + Na_2HPO_4 + NH_4Cl$
PO_4	phosphate
Reagent 1	stannous chloride, $SnCl_2 \cdot 2H_2O$
Reagent 2	ammonium molybidate, $(NH_4)6\ MO_7O_{24} \cdot 4H_2O$
Zn	zinc

ACKNOWLEDGMENTS

The authors would like to thank Prof. Dr. Ahmed Ali Rashed, Head of Drainage Studies Department, Drainage Research Institute (NWRC), Ministry of Water Resources and Irrigation, Egypt, for his efforts. His cooperation, valuable comments, and help in the construction of subsurface wetlands are greatly appreciated. Thanks are due to Dr. Ahmed Yousef Hatata, faculty of engineering, Mansoura University, for his help in carrying out the ANNs Models.

The assistance given by the Dakahlia company for potable water and domestic sewage, Egypt, is acknowledged for the company's financial support, preparing the research site at Samaha wetland for constructing the subsurface wetlands, carrying out the experimental work, and analyzing of collected samples in the company laboratories. Thanks are due to the Ex-Chairman of the company, Lt. Gen. Eng. Ahmed Abdeen, as well as to the company team, Eng. Mohamed Ragab El-Zoghby and Dr. Mohamed Ibrahem.

PREFACE

The last few decades witnessed a sharp focus on environmental pollution and its impact on life and nature. Scientists and engineers have studied the water treatment effect of natural wetlands for many years, resulting in the development of constructed wetlands (CWs) for treating wastewater.

A constructed wetland is an engineered sequence of water bodies designed to filter and treat waterborne pollutants found in sewage, industrial effluent or storm water runoff. Constructed wetlands are used for wastewater treatment for gray water and can be incorporated into an ecological sanitation approach. They can be used after septic tank for primary treatment in order to separate the solids from liquid effluent. The application of CWs for wastewater treatment was recently introduced in Egypt. Studies on removal rates and using different bed media are, therefore, required in order to model the behavior and estimate removal constants for different contaminants in wastewater and the optimum bed media to use.

Constructed wetlands for wastewater treatment are an extensively used technology for the treatment of different types of wastewaters in the last decades. Sanitary service in the Egyptian rural areas lags far behind the potable water supply. Only urban centers and some larger rural villages have wastewater treatment facilities. Domestic wastewater of many rural areas is typically discharged directly or indirectly to drains, causing degradation of drainage water quality against its reuse plans. The concern for the deteriorating quality of the environment has been increasing in recent years, and there is a need to apply the simple technology with low cost and very minimum controls, which is CW.

The Irrigation and Hydraulics Department at the Faculty of Engineering, Mansoura University established the early milestone of research and investigation at the CW fields. Free surface CW drain water treatment facility in Lake Manzala, NE Egypt, was tested, evaluated and modeled in early 2000, followed by this intensive study about subsurface CW treating municipal wastewater of a crowded Nile Delta village that was carried out during 2010–2013. Several ministries, universities and research centers have followed these goals in several locations of the Delta and valley of the Nile River and its deserts fringes. Such efforts help in adopting the fundamental of CW to cope with the arid and semiarid climate conditions of Egypt and the surrounding Middle East countries.

As the CW growing knowledge base leads to proving how complex treatment wetlands are due to a variety of internal and external ecological cycles, the assumptions that simplify the analysis of treatment reactors can no longer be justified. Wetland design factors and performance continues to develop, as much effort is being applied, to understand both setup and steady cycles within CW operation, and much more effort is still required in such fields.

The main objective of this study is to evaluate the performance of horizontal subsurface flow constructed wetlands in treating domestic wastewater to the limit that can be safely discharged to agricultural drains. Two-step procedures were used for the preparation of this book. The first one was the construction of an experimental project using three media (rubber, gravel, and plastic) and the analysis of the treated water samples. Statistical analysis using a stochastic package for a social science (SPSS) program showed that a significant difference between these three media for the treatment of most pollutants. The second step was the design of artificial neural network models (ANNs) using the Matlab software to simulate some of the experimental data and to generate the parameters output concentration.

The ability of the ANN model to predict results close to the measured ones was demonstrated. The plastic media gave the best treatment performance, better than both gravel and rubber media by percentages that varied between 5.3 and 11.6% (more than gravel) and between 10.9 and 19.5% (more than rubber). The wetland systems appear to have an ability to deal with various pollutants with different concentrations and return the treated water to the standard limits.

The intent of this book is to represent current information and provide guidance on the construction, performance, operation, and maintenance of subsurface flow constructed wetlands of domestic and municipal wastewaters.

This book is based on the thesis of Dr. Mohammed Ahmed Abdel Hady for his doctor of philosophy at the faculty of engineering at Mansoura University, Egypt. This work was suggested and guided in close cooperation by the principal supervisor Prof. Dr. Abdel Razik Ahmed Zidan, Professor of Hydraulics. Co-supervisors were Prof. Dr. Mahmoud Mohamed El-Gamal, Professor of Irrigation Works Design, and Associate Prof. Ahmed Ali Rashed, Head of Drainage Studies Department, Drainage Research Institute (NWRC), Ministry of Water Resources and Irrigation, Egypt.

ABOUT THE AUTHORS

Abdel Razik Ahmed Zidan, PhD
Professor, Hydraulics and Water Resources, Mansoura University, Egypt

Prof. Abdel Razik Ahmed Zidan, PhD, is currently a professor of hydraulics and water resources at Mansoura University, Egypt. Dr. Zidan has supervised more than 25 MSc and PhD theses in the field of hydraulics, water resources, irrigation, and wetlands. He is a member of the permanent scientific committee of water resources for promotion to the professorship, Egypt; a peer reviewer of more than 15 international journals, and a peer reviewer of national and international projects presented to the Academy of Scientific Research, Egypt, in the field of water resources. Dr. Zidan is member of the editorial board of the *Journal of Earth Science and Engineering* and Associate Editor-in-Chief of the *International Water Technology Journal* (IWTJ). He is the chairperson of the scientific committee of the International Water Technology Conferences. Dr. Zidan was an external examiner for PhD theses at Delft University of Technology, the Netherlands, and for MSc and PhD theses presented to Egyptian Japanese University of Science and Technology (E-JUST). He is an author of several books, such as *Breakwaters Effect on Wave Energy Dissipation* and *Numerical Modeling of Rivers Using Turbulence Models*. He was a visiting professor at Salford University, UK, and Delft University of Technology, and also worked as a consultant engineer at the Dakahlia company for potable water and domestic sewage, Egypt. Prof. Zidan holds BSc and PhD degrees in civil engineering, obtained respectively from Cairo University, Egypt, and Strathclyde University, UK.

Mohammed Ahmed Abdel Hady, PhD
Assistant Professor, Irrigation and Hydraulics Department, Mansoura University, Egypt

Mohammed Ahmed Abdel Hady, PhD, is currently an assistant professor of irrigation and drainage engineering at the Irrigation and Hydraulics Department at Mansoura University, Egypt. Dr. Abdel Hady earned his doctor of philosophy in subsurface constructed wetlands. He holds BSc, MSc, and PhD degrees in civil engineering from Mansoura University, Egypt.

CHAPTER 1

INTRODUCTION

CONTENTS

1.1 Introduction 1
1.2 Goals and Objectives 2
1.3 Scope of the Study 3
1.4 Layout of the Book 3
Appendices 4
Keywords 5

1.1 INTRODUCTION

Every community needs to treat its wastewater because of the serious health problems it can cause. Although this may seem obvious, untreated wastewater is still the root cause of much environmental damage and human illness, misery, and death around the world. Egypt, like most of the developing countries is facing an increase of the generation of wastes and of accompanying problems with the disposal of these wastes. Total untreated municipal wastewater in Egypt is about 10.7 million m^3/d (1.3 mm^3/d for urban zone and 9.4 mm^3/d for rural zone) so, it is clear that a wastewater treatment problem concentrated in rural areas (El-Zoghby, 2010).

Conventional wastewater treatment plants involve large capital investments and operating costs, and could be economically unsustainable for small or medium communities. So constructed wetland systems offer a suitable social economic alternative to the classic wastewater treatment plants for small and medium rural communities areas. The horizontal subsurface flow (HSSF) wetlands are biological porous media systems, colonized with emergent plants, which use the physical, chemical, and biological processes to remove wastewater pollutants. The focus is on the standard water quality variables as biochemical oxygen demand (BOD), chemical oxygen demand (COD), total suspended solids (TSS), in addition to ammonia (NH_3), phosphate (PO_4), dissolved oxygen (DO), fecal coliforms (FC), and concerned selected elements of heavy metals.

1.2 GOALS AND OBJECTIVES

The main goals of the present book are: (*i*) developing a physical model that can be used as a planning tool for the design and investigate the performance of the HSSF constructed wetlands using three bed substrates, since natural gravel, shredded tires pieces, and plastic pipes pieces differ in their porosity then their purification efficiency may be different to choose the effective one for treating domestic wastewater in a village represents the Egyptian rural areas, and (*ii*) the design and training of artificial neural networks (ANNs) and statistical models to represent the field data of the study. In order to achieve these goals, several specific objectives have to be formulated:

- Planning and constructing HSSF treatment cells using three different media.
- Determining the concentrations of the selected parameters to study their variation with distance and discharge.
- Examining the removal rates for each of the wetland cells.
- Determining the major role of media type on pollutants removal performance during operation.
- Measuring the treatment media porosity and computing the media surface area.
- Calculating pollutants removal rate constants applying the currently used plug and mixed flow first order empirical design equations for HSSF constructed wetlands in Egypt treating primary treated municipal wastewater.
- Evaluating the proposed ANNs models for treatment of some water parameters.
- Establishing linear and nonlinear regression models to represent a simple treatment tool of water parameters using SPSS statistical software.
- Using statistical analysis to differentiate between the three media under study.

Introduction

1.3 SCOPE OF THE STUDY

The experimental work was carried out using four main intervals. The first was the construction period of physical model which took about 18 months. The second was the preparation period which took about one month after the complete of project construction. The third was the set up stage, continued about five months after the preparation time. Finally, the fourth interval was the steady stage which took about six months. The following items were involved in the experimental work to study the performance evaluation for horizontal subsurface flow constructed treatment wetland systems:

- Cells set up adoption including adjustment of water depth, discharge, and monitoring of plants and harvesting.
- Hydraulic loading rate, porosity for used media, and media surface area.
- Water quality sampling and hydraulic retention time.

Three artificial neural networks programs were designed, the first for set up stage (Program No. 1) which represents the three pollutants in this stage (BOD, COD, and TSS). For steady stage, which consists of 12 parameters, the second program was designed to represent BOD, COD, and TSS (Program No. 2); and the third one was for the residual parameters (Program No. 3). The field data for set up and steady stages were statistically modeled by using stochastic package for social science (SPSS) program version 17 for windows.

1.4 LAYOUT OF THE BOOK

The structure of the book includes nine chapters and five appendices and is given as follows:

Chapter 1: Presents a general introduction, objectives and scope of the study for which this book was prepared.

Chapter 2: Provides the background information on models used to predict constructed wetlands performance, various wetland types, and wetland components. The function and values as well as wetland hydraulics are presented. Also, applications of neural networks; and statistical modeling and comparisons in constructed wetlands are given.

Chapter 3: Encompasses a complete description of the experimental facilities, planting in wetland cells, field sampling. The different methodologies and measuring devices are also given.

Chapter 4: Presents the hydraulic representation of the physical model, removal rate constants, and a brief description for ANNs. The SPSS software is used to deduce a set of regression equations to represent the experimental data. Also contains the preparation of the data. Finally, the calibration and validation processes for the obtained models.

Chapter 5: Contains the experimental results represented by tabular forms and graphs. The chapter also includes the computation of removal efficiencies for different pollutants and the values of measured media porosities.

Chapter 6: Displays the performance evaluation for treatment of the studied pollutants through the three bed media and its variation with length, loading rate, hydraulic retention time, and the time progression through set up stage.

Chapter 7: Gives the performance evaluation for studied pollutants treatment through the three bed media and its variation with length, loading rate, hydraulic retention time, and discharge through steady stage.

Chapter 8: Exhibits both applications of the artificial neural networks and SPSS models, which simulates the performance of the set up and steady stages in horizontal subsurface flow constructed wetlands. A comparison study was performed between these models outputs and the corresponding experimental results. Also, statistical analysis was performed to differentiate between the used media and models.

Chapter 9: Highlights summary of the book outcomes, conclusions, and recommendations for future research.

APPENDICES

Appendix I: Lists tables of the porosity.
Appendix II: Shows results of water samples analysis.
Appendix III: Gives the ANNs programs and results.
Appendix IV: Exhibits the tables of the statistical analysis.
Appendix V: Presents a glossary of terms.

KEYWORDS

- Egypt
- horizontal subsurface flow (HSSF)
- media surface area
- plastic pipes
- shredded tires
- wastewater treatment

CHAPTER 2

LITERATURE REVIEW

CONTENTS

2.1 Introduction 7
2.2 Definition of Wetlands 8
2.3 Types of Wetlands 9
2.4 Wetlands in Egypt 13
2.5 Wetlands Functions and Values 15
2.6 Hydrology of HSSF Wetlands 17
2.7 Hydraulics of HSSF Wetlands 20
2.8 Main Components of SSF Constructed Wetlands 26
2.9 Representing Treatment Performance 38
2.10 Models for Constructed Wetlands 42
2.11 Water Parameters in Wetlands 47
2.12 Nutrient Cycles in Constructed Wetlands 53
2.13 Wetland Advantages and Disadvantages 58
2.14 Implementation of Constructed Wetlands 61
2.15 Applications of Neural Networks in Wetlands 64
2.16 Statistical Modeling and Comparison in Wetlands 66
Keywords 67

2.1 INTRODUCTION

Wastewater treatment is a problem that has faced man ever since he discovered that discharging his wastes into surface water can lead to many additional environmental problems. Constructed wetlands (CWs) are techniques aim to improve water quality and reduce the harmful effect of effluent (Sarafraz et al., 2009). The export of conventional sewage technology to developing countries has often been unsuccessful due to complex operating requirements and expensive maintenance procedures (Butler and Williams, 1997). Constructed wetlands are considered a technical, economical and environmental sustainable solution for wastewater treatment in small communities since they are efficient diverse with pollutants removal (Araújo et al., 2008; Chen et al., 2008).

Constructed wetlands are an innovating technology around the world, which developed over the past few decades, primarily by wetland scientists in Europe and the United States (Czech, 2005). Wetlands can effectively treat domestic, industrial, and agricultural wastes; acid mine drainage; contaminated groundwater; and other polluted waters (Hodgson et al., 2004; Gearheart, 2006; Islam et al., 2009; Powell et al., 2009). Wetlands filter out pollutants by purifying the water through physical (setting, sedimentation, filtration), physical-chemical (adsorption, oxidation, reduction, precipitation), and biochemical processes (biochemical degradation, nitrification, denitrification, decomposition, and plant uptake) (Baskar et al., 2009).

Wetland can transform many of the common pollutants into harmless effects, because they have a higher rate of biological activity than most ecosystems. These pollutant transformations can be obtained for the relatively low cost of earthwork, piping, pumping, and a few structures (Fraser et al., 2004). From the modeling viewpoint, constructed wetlands are definitely more complex than conventional treatment processes because the diffusive flow and the large number of processes involved in pollution reduction. For these reasons, many authors have pointed out that the removal efficiency of constructed wetlands is not easily predictable, and being highly influenced by the hydraulic or environmental conditions (Marsili-Libelli and Checchi, 2005).

2.2 DEFINITION OF WETLANDS

Wetlands are not easily defined, especially for legal purposes, because they have a considerable range of hydrologic condition, uplands and deep water system, and because of their great variation in size, location, species, and human influence. Wetland definitions often include three main components (Mitsch and Gosselink, 1993):

- Wetlands are distinguished by the presence of water.
- Wetlands often have unique soil conditions.
- Wetlands support vegetation adapted to the wet conditions.

Marsh, bog, swamp, slough, and fen, these are just few names commonly applied to the class of natural habitats called "wetlands." Broadly defined, wetlands are land areas that have a transition habitat between dry land and deep water environment, they support plants specially adapted to grow in alternating wet and dry conditions (Moore, 1993).

Constructed wetland is a treatment system engineered to mimic the physical, chemical, and biological purification processes of a natural wetland. Natural and constructed wetlands can reduce total suspended solids chemical and biochemical oxygen demand, nutrients, metals, pathogens, and toxic organics from inflowing water by a variety of processes including sedimentation, filtration, precipitation, microbial metabolism (both aerobic and anaerobic), and plant uptake (Hanko and Cundiff, 1997). The main difference between natural and constructed wetland systems is that the constructed wetland enables wastewater treatment designs to be based on specific effluent quality goals (Wynn et al., 1997).

2.3 TYPES OF WETLANDS

Wetlands are categorized into two main groups; natural and constructed systems. Constructed wetlands differ from natural wetlands in several ways, as they (Kadlec and Wallace, 2008):

- remain constant in size.
- are not directly connected with groundwater.
- accommodate greater volumes of sediment.
- more quickly develop the desired diversity of plants.

2.3.1 NATURAL WETLANDS

Natural treatment wetlands rely on renewable, naturally occurring energies, including solar radiation; the kinetic energy of wind; the chemical-free energy of rainwater, surface water, and groundwater; and storage of potential energy in biomass and soils (Kadlec and Knight, 1996). Natural wetlands are providing many other benefits, include: food and habitat for wildlife; flood protection; and opportunities for recreation and esthetic appreciation. Many of these benefits have been realized by projects across the countries that involve the use of constructed wetlands in wastewater treatment (U.S. EPA, 1993a).

2.3.2 CONSTRUCTED WETLANDS

Constructed wetlands are man-made systems designed to imitate the functions of natural wetland systems. There are two fundamental types of constructed wetlands; surface and subsurface flow systems (U.S. EPA, 1993b). Subsurface flow constructed wetlands are designed to maintain the water flow below the upper surface of the media, thus minimizing human and ecological exposure. In surface flow constructed wetlands, water flows primarily horizontal and above ground. Both types of wetlands treatment systems typically are constructed in basins with a natural or constructed barrier to limit seepage (Crites et al., 2006; Tanner and Sukias, 2002).

Various types of constructed wetlands may be combined in order to achieve higher treatment performances, especially for nitrogen and pathogen removal (hybrid systems) (Abidi et al., 2009; Vymazal, 2010; Hoffmann et al., 2011). In these systems, the advantages and disadvantages of used wetland systems can be combined to complement each other (Borkar and Mahatme, 2011). The basic classification of constructed wetlands system for treating wastewater is shown in Figure 2.1 (Nikolić et al., 2009). Hawkins (2002) stated some factors for choice which wetland system is suitable for use, as: influent wastewater characteristics, pollutant removal goals, amount and timing of wastewater flows, land area availability, suitable plants, and media cost.

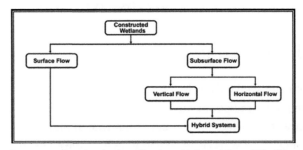

FIGURE 2.1   Constructed wetlands classification. (Adapted from Nikolić et al., 2009)

Literature Review

2.3.2.1 Surface Flow Systems

Surface flow (SF) wetlands contain emergent aquatic vegetation in a shallow bed. The water is exposed to the atmosphere as it slowly flows through the wetlands (Masi and Martinuzzi, 2007). The SF wetlands are commonly used to treat municipal wastewater, storm water run off, and mine drainage waters (Nilsson et al., 2012). The components in a typical SF wetland are shown in Figure 2.2.

2.3.2.2 Subsurface Flow Systems

Subsurface flow (SSF) wetlands consist of shallow basins with a seepage barrier and inlet and outlet structures. The bed is filled with porous media and vegetation is planted in the media. The term SSF constructed wetlands have also been called gravel bed wetlands, microbial rock-reed filters, root zone method (RZM), and vegetative submerged bed systems (Brown and Reed, 1994; Crites et al., 2006; Farooqi et al., 2008). The direction of the water flow provides the names of the two most known designs for the SSF systems as: horizontal flow and vertical flow systems as shown in Figures 2.3 and 2.4, respectively (Kadlec and Wallace, 2008).

The SSF wetlands demonstrate higher rates of pollutant removal per unit area of land than SF wetlands. Therefore subsurface flow wetlands can be smaller while achieving the same level of contaminants removal. The emergent vegetation, mostly bulrush, reeds, and sometimes cattails, supplies oxygen to the media and allows biological growth to accumulate on its roots. Bacteria and beneficial fungi live in the media as biofilm attached to its

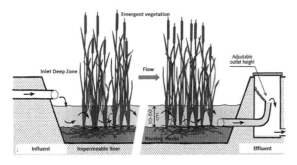

FIGURE 2.2 Basic elements of surface flow wetlands.

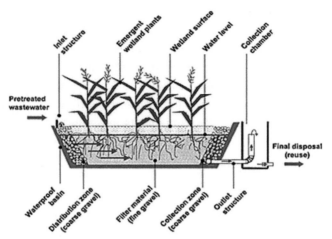

FIGURE 2.3 A schematic of horizontal subsurface flow wetlands.

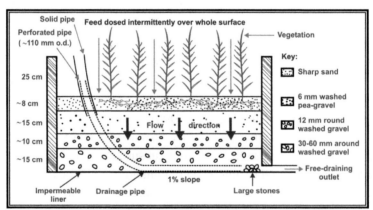

FIGURE 2.4 Typical arrangement of vertical flow wetlands. (Copyright 2007. From Kadlec, R.H.; Wallace, S.D. Treatment Wetlands. Second Edition. Reproduced by permission of Taylor and Francis Group, LLC, a division of Informa plc.)

particles. Flow is maintained by either a sloping bottom and/or an adjustable outlet structure which allows the water level to be lowered at the end of the bed, producing the pressure head required to overcome flow resistance through the media (Halverson, 2004).

Horizontal subsurface flow (HSSF) constructed wetlands have successfully been used for treatment various types of wastewaters such as municipal, domestic, animal wastewater, mine waters, industrial and agricultural wastes, leachate and remediation, urban storm water and field run off (Vymazal and Kröpfelová, 2009). Al-Omari and Fayyad (2003) demonstrated the capacity of HSSF constructed wetlands to significantly reduce organic carbon and particulate matter from wastewater, as measured by biological oxygen demand (BOD), chemical oxygen demand (COD), total organic carbon (TOC), and total suspended solids (TSS). The HSSF wetlands have been less successful for the removal of nitrogen and phosphorous than conventional treatment systems (Mitchell and McNevin, 2001).

Bed depth for horizontal subsurface flow constructed wetlands is generally between 0.6 to 0.9 m. Typical flow depths vary from 0.5 to 0.8 m. Typical hydraulic loading rates are between 2 and 20 cm/d, which correspond to a wetland of 50 m^2 to 500 m^2 per 10 m^3/d of flow (Kadlec and Knight, 1996). Vertical flow (VF) systems are constructed so that water moves uniformly down or up through the media. Water is generally poured on the surface to help maintain aerobic conditions in the media and to provide hydraulic head to encourage flow, and is collected at the bottom of the bed by drainage pipes. An important factor is that the bed has to be completely drained so the air could refill the bed (Nilsson et al., 2012). Vertical flow systems are more common in mining applications (treating wastewater from mining processes).

Lavrova and Koumanova (2007) studied constructed vertical flow wetland system for polishing of aerobically treated wastewater. It was established that the values of the treated water characteristics were significantly decreased for comparatively short time accompanied with odor elimination. Recirculation vertical flow constructed wetlands are modified type of VF wetlands which sometimes defined as vegetated recirculation gravel filters. They treat wastewater by passing sewage through the constructed wetland where it is filtered through the gravel media in the bottom layer and then recirculation back around the roots and rhizomes several times for more filtration and treatment before it finally discharges to the soil absorption area (Garcia-Perez et al., 2007).

Single-stage constructed wetlands cannot achieve high removal of total nitrogen due to their inability to provide both aerobic and anaerobic conditions at the same time. Vertical flow CWs remove successfully ammonia-N but very limited denitrification takes place in these systems. On the other hand, horizontal flow CWs provide good conditions for denitrification but the ability of these systems to nitrify ammonia is very limited. Therefore, various types of constructed wetlands may be combined (hybrid systems) with each other (Vymazal, 2007).

2.4 WETLANDS IN EGYPT

In Egypt, there are some important natural wetlands on the Mediterranean and Red coasts in addition to the Nile delta lakes, and some lagoons and salt-marshes in different sites. Recently, the constructed wetlands projects had been conducted and working successfully in treatment of variety of wastewaters.

2.4.1 NATURAL WETLANDS IN EGYPT

The most natural wetlands in Egypt are (RTM, 2009):
- Nile delta lakes (El-Burullus, El-Manzala, and El-Bardawil).
- Small lakes (Idku and Maryut).
- Nile River and its islands.
- Lake Nasser and Toshka depression.
- Suez Canal and its associated lakes.
- Red sea (Gulfs of Aqaba and Suez).
- Siwa oasis and Qattara depressions.
- Closed basins (Wadi El-Natrun, Quaroun, and Wadi El-Rayan).
- Matrouh closed lagoons.

2.4.2 CONSTRUCTED WETLANDS IN EGYPT

The constructed wetland technologies have become a good treatment tool in developing countries. In Egypt, there are seven projects for treating wastewater. A brief description for these projects is giving as follows (Awad and Saleh, 2001; NAWQAM, 2002):
- Abu-Attwa plant (SSF wetlands used for domestic sewage and planted by reed in Ismailia governorate, the plant consists of six treatment beds) is not working now due to stop of maintenance.
- The project at 10th of Ramadan city (gravel bed hydroponics with reeds received a mixture of wastewater from a wide range of industries in Sharkia governorate) is not working now.
- Drainage canals in Fayoum (passive water quality system, which proposed to be situated close to the source of pollution production, this strategy would facilitate the reuse of drain water) is currently working successfully.
- Samaha plant (full scale horizontal subsurface flow wetlands, treating domestic wastewater and consists of eight wetland cells in Dakahlia governorate) is our case study.
- Lake Manzala project (engineered wetlands for treatment of Bahr El Baqar drain having an ability to reduce a variety of pollutants in its water, the wetland system consists of ten surface flow treatment cells)

Literature Review

(Rashed et al., 2000) is working perfectly and one of the most successful project in Egypt.
- Edfina drain in Edfina city (on-stream remediation, the drain designed as a sedimentation pond followed by 4 surface wetland cells and lasted by open water reach. This design gives a good treatment performance matching the allowable limits of law 48) (Rashed and Abdel-Rashid, 2008) is the good model for in-stream treatment.
- El Salam canal (microcosm horizontal and vertical flow constructed wetlands for improving mixed irrigation water, nonland consumer) (El Refaie et al., 2004). Many similar projects were constructed based on this type.

Tennessee Valley Authority patented a treatment technology is known as reciprocating flow constructed wetland (RFCW) for enhanced nutrients reduction through increasing oxygen content in the treatment processes. In Egypt, a pilot-scale RFCW was built near Lake Manzala to treat a portion of Bahr El-Baqar drain wastewater that carries drainage water contains domestic and industrial pollution loads. The wetland system consists of one sedimentation basin followed by two subsurface flow cells working in fill-empty sequence mechanism. Rashed (2007) evaluated the treatment performance of reciprocating wetland for different pollutants. Results showed that RFCW managed to polish contaminants of drainage influent in a safe manner for the agricultural purpose or dumping it into Lake Manzala without negative impacts on its ecosystem.

2.5 WETLANDS FUNCTIONS AND VALUES

During last decades the multiple functions and values of wetlands have been recognized (Brix, 1994). Many people use the terms purposes (functions) and values interchangeably when discussing wetlands, even through functions and values are different. Functions are properties that a wetland naturally provides; values are the properties which are beneficial to humans. Functions of wetlands are the physical, chemical, and biological processes occurring in and making up an ecosystem (Smith et al., 1995). Wetland functions include pollution interception, floodwater storage, coastal protection, groundwater recharge, carbon fixation and carbon dioxide balance, fish and wildlife habitat, and biological productivity (U.S. EPA, 2001; Hanson et al., 2008).

The removal effects take place during passage of wastewater through wetland soil; they are based on various complex processes within the media, plants and microorganisms (Thiyagarajan et al., 2006) as:

- Settling of suspended particulate matter.
- Filtration, and chemical precipitation and transformations.
- Adsorption and ion exchange.
- Breakdown and uptake of pollutants and nutrients.
- Perdition and natural die-off of pathogens.

Values of wetlands are an estimate, usually of subjective importance. Wetlands values may be derived from outputs that can be consumed directly, such as food, or timber; indirect uses which arise from the functions occurring within the ecosystem, such as an improvement of water quality, and flood control; possible future direct outputs or indirect uses (Fisher and Acreman, 2004; Snow et al., 2009).

Treated water may be destined for one of three primary receivers' surface water, groundwater, or irrigation (reuse). There are often stringent specifications of quality that must be met to allow discharges to these recipients, and they are quite different (Kadlec and Wallace, 2008). Borges et al. (2009) studied the use of subsurface flow constructed wetlands for the mitigation of pesticide runoff. It was concluded that CWs are capable of mitigating water contaminated with ametryn ($C_9H_{17}N_5S$, is an herbicide used to control broadleaf and grass weeds in fields, U.S. EPA, 2005).

Ferro et al. (2002) built a pilot scale subsurface flow wetlands system to test whether a full scale CWs could be used to treat recovered groundwater contaminated with petroleum hydrocarbons. This study showed that SSF wetlands can treat this type of pollutant. El-Khateeb and El-Gohary (2003) investigated two treatment systems (conventional and wetland) consisting of an up-flow anaerobic sludge blanket reactor followed by either SSF or SF wetlands. From this study it can be concluded that the use of HSSF wetland is a promising technology for domestic wastewater reclamation and reuse in arid and semiarid areas. Murray-Gulde et al. (2003) used hybrid-constructed wetlands to produce treated waters from oil fields (i.e., waters that have been in contact with oil in site). The produced water can be reused for irrigation purposes or recharged to groundwater supplies.

2.6 HYDROLOGY OF HSSF WETLANDS

Hydrology conditions influence the soils and nutrients, which in turn influence the character of the system. The proper hydrologic conditions will give the potential chemical and biological elements necessary for a properly functioning wetland to exist (Mitsch and Gosselink, 1993). Wetland hydrology includes the following subsection items.

2.6.1 HYDRAULIC LOADING RATE

The hydraulic loading rate is defined as the rainfall equivalent of whatever flow is under consideration. In HSSF wetlands, the wetted area is usually known with good accuracy, because of berms or other confining features (Kadlec and Knight, 1996). The defining equation may be written as follows:

$$q = \frac{Q}{A} \tag{2.1}$$

where: q = hydraulic loading rate, m/d; Q = average flow through the wetlands, m^3/d; A = wetland top surface area, m^2.

Some wetlands are operated with intermittent feed. Under these circumstances, the term hydraulic loading rate refers to the time average flow rate. The loading rate during a feed portion of a cycle is the instantaneous hydraulic loading rate, which is also called the hydraulic application rate (Kadlec and Wallace, 2008).

2.6.2 MEAN WATER DEPTH

The success or failure of a treatment wetland depends upon creating and maintaining correct water depths and flows. Constructed treatment wetlands, typically have some form of outlet water level control structure. The notation and variables are illustrated in Figure 2.5 (Kadlec and Wallace, 2008). The mean water depth is denoted by:

$$h = H - B \tag{2.2}$$

where: h = water depth, m; H = local water elevation, m; B = local bed bottom elevation, m.

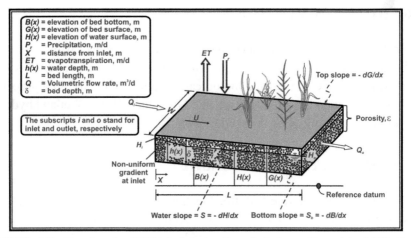

FIGURE 2.5 Notation for HSSF bed hydraulic calculations. (Copyright 2007. From Kadlec, R.H.; Wallace, S.D. *Treatment Wetlands*. Second Edition. Reproduced by permission of Taylor and Francis Group, LLC, a division of Informa plc.)

2.6.3 HYDRAULIC RETENTION TIME

Actual wetland hydraulic retention time (T_r) is a measurement of how long on average, the water is in contact with the wetland media and is defined as the wetland water volume divided by the flow rate (Myers and Jackson, 2001):

$$T_r = \frac{V_{active}}{Q} = \frac{nhA_{active}}{Q} \qquad (2.3)$$

where: V_{active} = volume of wetland containing water in active flow, m³; n = porosity (volume occupied by water), % expressed as decimal; A_{active} = area of wetland containing water in active flow, m².

Wetlands routinely experience water gains (precipitation) and losses (evapotranspiration and seepage), so that outflows differ from inflows. If there is net gain, the water accelerates; if there is net loss, the water slows. Chazarenc et al. (2003) found that evapotranspiration played a major role in summer by improving retention time, this may be written as:

$$T_r = T_i \left(\frac{R}{R-} \right) \qquad (2.4)$$

$$R = \frac{Q_o}{Q_i} \qquad (2.5)$$

Literature Review

where: T_r = actual retention time, d; T_i = inlet flow-based T_r, d; R = water recovery fraction; Q_o = outlet flow rate, m³/d; Q_i = inlet flow rate, m³/d.

If hydraulic calculations are done on average flow rate, this approximation is good within 4% as long as (0.5 < R < 2.0).

2.6.4 WATER VELOCITY IN WETLAND CELL

The actual water velocity (v) which would be measured within the wetland may be given as:

$$v = \frac{Q}{nhW} \tag{2.6}$$

where: v = actual water velocity, m/d; nhW = open area perpendicular to flow, m²; h = average depth of water in wetland, m; W = width of wetland cell, m.

The superficial water velocity (u_s) is the empty wetland velocity and may be written as:

$$u_s = \frac{Q}{hW} \tag{2.7}$$

where: u_s = superficial water velocity, m/d; hW = total wetland area perpendicular to flow, m².

For SF wetlands, difference between u_s and v is very small, because SF porosity is nearly unity (typically around 0.95). However, there is a large difference for HSSF systems because of the bed media porosity was typically around 0.35–0.40 for gravel media (Kadlec and Wallace, 2008).

2.6.5 OVERALL WATER MASS BALANCES

Transfer of water to and from surface and subsurface flow constructed wetlands follow the same pattern for the natural wetland, Figure 2.6. The dynamic overall water budget for a wetland may be defined as (Kadlec and Knight, 1996):

$$\frac{dV}{dt} = Q_i - Q_o + Q_c - Q_b - Q_{gw} + (P_r - ET) \times A \tag{2.8}$$

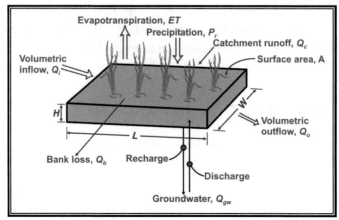

FIGURE 2.6 Components of wetland water budget. (Reprinted from Kadlec, R.H.; Knight, R.L. *Treatment Wetlands*. CRC Lewis Publisher, New York, © 1995 with permission from Taylor & Francis.)

where: V = water storage (volume) in wetlands, m³; t = time, d; Q_i = input flow rate, m³/d; Q_o = output flow rate, m³/d; Q_c = catchment's runoff rate, m³/d; Q_b = bank loss rate, m³/d; Q_{gw} = infiltration to groundwater, m³/d; P_r = precipitation rate, m/d; ET = evapotranspiration rate, m/d; A = wetland top surface area, m².

2.7 HYDRAULICS OF HSSF WETLANDS

Horizontal subsurface flow constructed wetlands are widely used for wastewater treatment but there's still a lack of information on flow characteristics changes throughout the porous bed over time. The idea of flowing water through a planted bed of porous media seems simple enough; yet numerous difficulties have arisen in practice. Hydraulics refers to the physical mechanisms used to convey the water through the wetland (Albuquerque and Bandeiras, 2007).

2.7.1 FLOW IN POROUS MEDIA

In the United States, the most common media for HSSF constructed wetlands is gravel, but sand and soil have been used in Europe. Darcy's law, as defined by Eq. (2.9), could provide a reasonable approximation of the flow through the HSSF wetlands under the following conditions (Chervek, 2005; Crites et al., 2006):

Literature Review

- Small to moderate sized gravel is used as the media.
- The system is properly constructed to minimize short circuiting.
- The gains and losses of water are recognized.
- The system is designed to depend on a minimal hydraulic gradient.

$$A_c = \frac{Q}{K_c S} \quad (2.9)$$

where: A_c = cross-sectional area perpendicular to the flow, m²; Q = average flow through the wetland, m³/d; K_c = hydraulic conductivity, m/d; S = hydraulic gradient (S = 0.005 for flat bottom), m/m.

The particle Reynolds number may be defined as:

$$\text{Re} = \frac{d_p \times \rho \times u_s}{(1-n)\mu} \quad (2.10)$$

where: d_p = particle diameter, m; ρ = density of water, kg/m³; μ = viscosity of water, kg/m/d.

The resistance to flow in the HSSF wetland is caused primarily by the gravel media. Over the longer term, the spread of plant roots in the bed and the accumulation of nondegradable residues in the gravel pore spaces will also add resistance. The energy required to overcome this resistance is provided by the differential head between the water surface at the inlet and the outlet of the treatment wetland (Brix, 2003; Crites et al., 2006).

The HSSF wetlands operate in thin sheet flow, with a free upper surface. Flows may be averaged over the vertical (thin) dimension, for the case of the upper surface exposed to the atmosphere, this yield the two-dimensional Dupuit-Forcheimer equation (Kadlec and Knight, 1996):

$$\frac{\partial(nH)}{dt} = \frac{\partial}{\partial x}(K_c H \frac{\partial H}{\partial x}) + \frac{\partial}{\partial y}(K_c H \frac{\partial H}{\partial y}) + P_r - ET \quad (2.11)$$

where: H = elevation of the free water surface, m; x = longitudinal distance, m; y = transverse distance, m.

It is important to note that this equation embodies the assumption that the driving force for flow is a tilt to the water surface ($\partial H/\partial x$).

Most operational subsurface flow wetlands in the U.S. have a treatment zone and operating water depth of 0.6 m. A few, in warm climates, operate with a depth of 0.3 m. The shallow depth enhances the oxygen transfer but requires a greater surface area and the system is at greater risk of freezing in cold climates. The bed of 0.6 m deep also requires special operation to induce desirable root penetration to the bottom of the bed (Crites et al., 2006).

2.7.2 HYDRAULIC CONDUCTIVITY OF POROUS MEDIA

Hydraulic conductivity (HC) depends on the characteristics of the media mainly (Kadlec and Wallace, 2008):

- Mean particle diameter and shape.
- Variance of the particle size distribution.
- Porosity of the bed.
- Arrangement of the particles.

Ergun equation is widely accepted for random packing of uniform spheres and is given by (Kadlec and Wallace, 2008):

$$-\frac{dH}{dx} = \left(\frac{150(1-n)^2 \mu}{\rho g n^3 d_p^2}\right) u_s + \left(\frac{1.75(1-n)}{g n^3 d_p}\right) u_s^2 \qquad (2.12)$$

where: H = elevation of water surface, m; g = acceleration of gravity, m/d^2.

Equation (2.12) works for spheres of single size; but gravel bed wetlands do not use such media. This equation is applied to HSSF systems and found that Ergun-predicted depths were about 10 cm larger than the actual depth. The presence of a particle size distribution lowers the hydraulic conductivity. This occurs because small particles have a disproportionately large amount of surface area, which causes drag on the water, and because the small particles can fit in the spaces between the larger particles.

Odong (2007) presented the following empirical formulae for computing the value of hydraulic conductivity according to particle size analysis as follow:

1. Hazen Formula

$$K_c = \frac{\rho \times g}{\mu} \times 6 \times 10^{-4} \left[1 + 10(n - 0.26)\right] d_{10}^2 \qquad (2.13)$$

Hazen formula was originally developed for determination of hydraulic conductivity of uniformly graded sand but it is also useful for fine sand to gravel range, provided the sediment has a uniformity coefficient less than 5 and effective grain size between 0.1 and 3.0 mm.

2. Kozeny-Carman Formula

$$K_c = \frac{\rho \times g}{\mu} \times 8.3 \times 10^{-3} \left(\frac{n^3}{(1-n)^2} \right) d_{10}^2 \qquad (2.14)$$

The Kozeny-Carman equation is one of the most widely accepted and used formula for computing the hydraulic conductivity as a function of the characteristics of the soil medium. It is not appropriate for either soil with effective size above 3 mm or for clayey soils.

3. Breyer Formula

$$K_c = \frac{\rho \times g}{\mu} \times 6 \times 10^{-4} \log \frac{500}{U} d_{10}^2 \qquad (2.15)$$

$$U = \left(\frac{d_{60}}{d_{10}} \right) \qquad (2.16)$$

The porosity (n) may be derived from the empirical relationship with the coefficient of grain uniformity (U) as follows:

$$n = 0.255 \left(1 + 0.83^U \right) \qquad (2.17)$$

This method does not consider porosity and therefore, porosity function takes value one. Breyer formula is often considered most useful for materials with heterogeneous distributions and poorly sorted grains with uniformity coefficient between 1 and 20, and effective grain size between 0.06 and 0.6 mm.

4. Slitcher Formula

$$K_c = \frac{\rho \times g}{\mu} \times 1 \times 10^{-2} n^{3.287} d_{10}^2 \qquad (2.18)$$

This formula is the most applicable for size between 0.01 and 5 mm.

5. Terzaghi Formula

$$K_c = \frac{\rho \times g}{\mu} C_t \left(\frac{n - 0.13}{\sqrt[3]{1-n}} \right)^2 d_{10}^2 \qquad (2.19)$$

This formula is used for large grain sand.

6. The USBR Formula

$$K_c = \frac{\rho \times g}{\mu} \times 4.8 \times 10^{-4} d_{20}^{0.3} \times d_{20}^2 \qquad (2.20)$$

The U.S. Bureau of Reclamation (USBR) formula calculates hydraulic conductivity from the effective grain size (d_{20}), and does not depend on porosity; hence porosity function is a unity. The formula is most suitable for medium grain sand with uniformity coefficient less than 5.

7. Alyamani and Sen Formula

The equation considers both sediment grain sizes d_{10} and d_{50} as well as the sorting characteristics.

$$K_c = 1300 \left[I_o + 0.025(d_{50} - d_{10}) \right]^2 \qquad (2.21)$$

where: C_t = sorting coefficient ($0.0061 < C_t < 0.0107$); I_o = intercept of line formed by d_{50} and d_{10} with grain size axis, mm; d_{10} = effective grain diameter, mm; d_{20} = grain diameter for which 20% of sample are finer than, mm; d_{50} = median grain diameter, mm; d_{60} = diameter through which no more than 60% of sample passes, mm.

2.7.3 CLOGGING OF HSSF BED MEDIA

The HSSF bed will not maintain the clean-bed hydraulic conductivity once the system is placed into operation. The greatest reductions in hydraulic conductivity occur at the wetland inlet. Bed clogging has created the majority of operational problems for HSSF wetlands around the world. Although the processes of bed clogging may be still being quantified, there appear to be two distinct sets of mechanisms that contribute to the problem:

Short-term clogging that reduces hydraulic conductivity over the first year of operation. These appear to be related primarily to the development of plant root networks and microbial biomat formation occurs primarily in the inlet region of the wetland. The nonuniform distribution of roots and biomat along the length of the cell, this result in a nonuniform distribution of hydraulic conductivity throughout the bed, as shown schematically in Figure 2.7, short-term clogging mechanism is basically due to (Kadlec and Wallace, 2008):

- Development of plant root networks that occupy pore volume within the wetland bed.
- Loading of organic matter (both suspended and dissolved) that stimulates the growth of microbial biofilm on the bed media. This biofilm entraps both organic and inorganic solids, forming a biomat. This biomat varies depending on the nature of the waste being treated. Biomat formation is greatest at the inlet end of the wetland where the organic loading is the highest. The loss of pore volume due to biomat formation reduces the hydraulic conductivity in this inlet zone. Organic matter is removed as wastewater flows through the wetland, resulting in declining biomat growth. At the outlet, where only small quantities of organic matter are available to microbes, biomat formation is negligible.

Long-term clogging that gradually reduces hydraulic conductivity. These appear to be primarily related to deposition of inert (mineral) suspended solids, accumulation of refractory organic material, and formation of insoluble chemical precipitates. Reasons of long-term clogging mechanism might be due to (Kadlec and Wallace, 2007):

- Deposition of inert suspended solids (minerals) in the inlet region of the wetland bed: Solids deposition can occur for a variety of reasons, beginning with the placement of the media. Unwashed media will

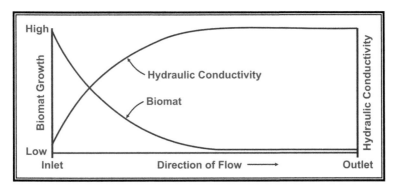

FIGURE 2.7 Relationship between HC and biomat formation. (Copyright 2007. From Kadlec, R.H.; Wallace, S.D. *Treatment Wetlands*. Second Edition. Reproduced by permission of Taylor and Francis Group, LLC, a division of Informa plc.)

carry a load of fine dust. Mud on the wheels of vehicles can add to the dirt supply during placement.
- Accumulation of refractory organic material in the inlet zone of the wetland bed: Due to the low flow velocities that occur within HSSF wetland beds, influent total suspended solids (TSS) will settle and deposit within the inlet region of the wetland bed. As pore volume is occupied by suspended solids, the hydraulic conductivity is reduced accordingly.
- Deposition of chemical precipitates in the wetland bed: Chemical reactions within HSSF wetlands can result in the formation of insoluble chemical precipitates. These precipitates can also block pore spaces within the wetland bed and have the same effect in reducing hydraulic conductivity.

The combined effect of short-term and long-term bed clogging mechanisms is to produce a drastic reduction in the hydraulic conductivity of the inlet zone of treatment wetland bed (Kadlec and Wallace, 2008). A combination of clogging and inappropriate design has produced overland flow in many existing HSSF system. Flooding is usually confined to the inlet region of the bed. Overland flow carries the excess water until the hydraulic conductivity and flow gradient over the remaining travel distance are sufficient to permit the flow to be carried below ground (Kadlec and Wallace, 2008).

Duarte et al. (2010) presented a synthesis of data obtained through an extensive survey performed in 20 Portuguese HSSF wetlands utilities. The survey showed that only 30% of systems had been operating without any problems and clogging of beds, identified as the main problem of this kind of wastewater treatment system. Subsurface flow wetlands use at least the equivalent of primary treatment as the preliminary treatment prior to the wetland component. The purpose of this treatment is to reduce the concentration of easily degraded organic solids that otherwise would accumulate in the entry zone of the wetland system and result in clogging, possible odors, and adverse impacts on the plants in the entry zone (Crites et al., 2006).

2.8 MAIN COMPONENTS OF SSF CONSTRUCTED WETLANDS

Tennessee valley authority (TVA) recommended the use of constructed wetlands in wastewater treatment for their simplicity and efficiency. The TVA lists the following main components for this type of treatment system (Botch and Light, 1994; McKenzie, 2004):

- Water column (water flowing in or above the surface of bed).
- Beds with various rates of hydraulic conductivity.
- Plants adapted to water-saturated anaerobic beds.

Other important components of wetlands, such as: invertebrates (mostly insect larvae and worms), and microorganisms (most importantly bacteria). Processes controlling contaminant retention in a constructed wetland sediment may be abiotic (physical and chemical) or biotic (microbial and phytological) and are often interrelated.

2.8.1 WATER AND SOURCES OF POLLUTANTS

Constructed wetlands may be used to treat water from different sources. Wastewater pollution sources (Table 2.1) are organized into the following main categories: municipal, industrial, storm-waters, and agricultural wastewater (Kadlec and Knight, 1996).

Municipal wastewater consists of a combination of domestic wastes, originating in households, offices, and public restrooms, and lesser contributions from many commercial and small industrial sources. The use of constructed wetlands (both horizontal and vertical flows) to treat

TABLE 2.1 Wastewater Sources Categorization, Frequency, and Origin

Category	Frequency	Origin	Description
Municipal	Continuous Flows	Primarily residential and commercial	Feces, urine, paper, plastics soaps, grease, and industrial chemicals
Industrial	Continuous to Intermittent	Primarily processes for industrial and leachates	Dilute to concentrated solution of biodegradable and nondegradable chemicals
Agricultural	Continuous to Intermittent	Intensive practices for agricultural as milking or washing barns	Dilute to concentrated mixture of biodegradable compounds
Storm-waters	Intermittent Flows	Runoff from urban, suburban, and rural areas	Dilute mixtures of mineral and organic solids, dissolved salts, and nutrients

(Reprinted from Kadlec, R.H.; Knight, R.L. *Treatment Wetlands*. CRC Lewis Publisher, New York, © 1995 with permission from Taylor & Francis.)

domestic wastewater has gained global attention in recent years (Steer et al., 2005; Wiessner et al., 2005; Zurita et al., 2009). Major pollutants in domestic wastewater are typically characterized by organic matter, suspended particulate matter, pathogenic organisms, and the nutrients nitrogen and phosphorus (Davison, 2001).

Domestic wastewater solids include a mixture made of about 33% soaps and solids, 20% urine, 18% ground food wastes, 16% feces, and 7% paper (Kadlec and Knight, 1996). The remaining 5% of the solids in domestic wastewater originate in the water supply itself. Three typical compositions of untreated domestic wastewater are shown in Table 2.2 (Pescod, 1992; Davis and Cornwell, 1998).

Commercial inputs to municipal wastewater include wastewater generated by car washes, restaurants, photofinishing shops, and laundries. Industrial wastewaters, with or without pretreatment, frequently are a component of municipal wastewater flows (Kadlec and Knight, 1996).

Urbanc-Bercic (1994) established a pilot-scale constructed reed bed system to investigate the potential of this low-technology approach to the treatment of contaminated municipal waste dump leachate which is proven to be a suitable approach. Horizontal subsurface flow wetland systems have been shown to provide poor performance when treating high strength wastewater, especially during winter operation in cold climates. This poor performance is likely due to the onset of limiting anaerobic conditions and the slowing of microbial processes under cold temperatures as winter progresses. In order to overcome these limitations, Pendleton et al. (2005) presented multicell

TABLE 2.2 Typical Composition of Untreated Domestic Wastewater

Wastewater Characteristics (mg/l)	Weak	Medium	Strong
Chloride	30	50	100
Alkalinity as (CaCO$_3$)	50	100	200
Total Phosphorus (TP)	5	10	20
Total Organic Carbon (TOC)	75	150	300
Total Dissolved Solids (TDS)	200	500	1000
Total Kjeldahl Nitrogen (TKN)	20	40	80
Total Suspended Solids (TSS)	100	200	350
Chemical Oxygen Demand (COD)	250	500	1000
Biochemical Oxygen Demand (BOD)	100	200	300

Source: After Pescod (1992); and Davis and Cornwell (1998).

subsurface vertical flow wetland bio-filter systems for leachate treatment which improved the performance.

The rural wastewater organic load rate is higher than the urban one (Abidi et al., (2009). Elmitwalli et al. (2002) showed that the domestic wastewater of the Egyptian rural-areas is relatively concentrated with a COD as high as 1100 mg/l, which is almost two times of that in the urban areas. This high level mainly is due to discharge of cow manure in served areas with gravity sewers. El-Gammal (2012) Studied the characterization of the wastewater in rural areas based on data collection from two villages in the Upper Egypt, one village from Lower Egypt, one village from Eastern of Delta, three villages from the Middle of Delta, and one village from Western of Delta. From This study, the following results might be written as:

- The pH values for the raw domestic wastewater range from 7.1 to 7.6 with an average value of 7.3 which is normal and close to the neutral water.
- The TDS values of the raw domestic wastewater vary between 980 to 2600 mg/l with an average value of 1600 mg/l. The average TDS of raw wastewater of the rural areas of the Nile Delta is 1050 mg/l while in the upper and lower region of the Nile valley is 2420 mg/l. This wide variation in the TDS values may be explained by the variations in the family traditions and habits in the food system in the rural areas.
- The BOD values of the raw domestic wastewater range from 160 to 760 mg/l with an average value of 450 mg/l. The data show a big variation especially in the lower region of the Nile valley from 150 to 500 mg/l with an average value of 340 mg/l.
- The COD values of the raw domestic wastewater vary between 260 and 1650 mg/l with an average value of 770 mg/l. The large variation in the results from the high variation in COD in middle of Nile Delta rural areas where some local industrials discharge directly to the sewerage networks like chemical and metal industries. The COD values in the middle Nile Delta rural areas range from 400 to 1650 mg/l with an average value of 860 mg/l. The average of COD value in the upper and lower region of the Nile Delta valley is 550 mg/l.
- The TSS values of the rural domestic wastewater range from 150 to 1200 mg/l with an average value of 360 mg/l. The data show a big variation from one region to another in the rural areas with no clear trend. The mean TSS in the upper and lower regions of the Nile Valley is 700 mg/l, while in the Nile Delta region is 400 mg/l.
- The TN values range from 30 to 70 mg/l with an average value of 50 mg/l. The high values come from discharging the manure of the

animals in farmer's households to the sewerage network. The TN values vary from one region to another based on the tradition and habits of farmers in the rural areas.

- The values of NH_3 range from 26 to 36 mg/l with an average value of 30 mg/l. The NH_3 values vary from one region to another, with no clear trend from the southern region of Nile Valley to the Nile Delta region.
- The TP values range from 2 to 15 mg/l with an average value 7 mg/l. The data show a big variation from one region to another. The TP values are higher in the Nile Delta region than those in the lower and upper region of Nile Valley due to the overuse of the detergents.
- The FC values vary from 1.3×10^6 to 24×10^8 with an average value of 2×10^8 MPN/100 ml. The measured values show a large variation in all regions of the rural areas with no clear trend. The variability of data is due to the nature and method of the measurements depending on personal counting.

The final composition of raw wastewater depends on the source and its characteristics. In the case of mixed municipal wastewater this depends on the types and numbers of industrial units and the characteristics of the residential communities. The composition of typical raw wastewater for some countries is given in Table 2.3 (Hussain et al., 2002).

2.8.2 CONSTRUCTED WETLAND MEDIA

To treat wastewater effectively, subsurface flow constructed wetland systems must contain the right kinds of media and the system must be maintained regularly. The following items are discussing in details, the media functions, size, characteristics, choice, configuration, and protection.

TABLE 2.3 Composition of Raw Wastewater for Some Countries

Parameter	India	USA	France	Morocco	Egypt
BOD	196–280	110–400	100–400	45	160–760
COD	Not available	250–1000	300–1000	200	260–1650
TSS	200–985	100–350	150–500	160	150–1200
TKN	28.5–73	20–85	30–100	29	30–70
TP	Not available	4–15	1–25	4–5	2–15

Source: After Hussain et al. (2002); and El-Gammal (2012).
Notice: All units in mg/l.

2.8.2.1 Functions of the Media

The media in a constructed wetland provide a path through which wastewater can flow and surfaces on which microorganisms can live. As the wastewater passes through the pores between the media particles, the microorganisms living there feed on the waste materials, removing them from the water. Another function of the media is to support the plants growing in the wetland (Vrhovšek et al., 1996). The wastewater also is treated by other processes as it flows through the media, including filtration, sedimentation, nitrification, denitrification, adsorption, and assimilation (Vrhovšek et al., 1996). The definitions of media types and different physicochemical interactions are listed in wetland glossary in Appendix V.

2.8.2.2 Media Particle Size

The size of the media particles greatly affects the system's ability to function. Table 2.4 gives a comparison of common media in SSF wetlands systems. Small particles have smaller pores, so they can filter smaller particles from wastewater and also have more surface area, where treatment occurs but clogging more than large particles. If the particles are too big, the effluent does not have enough contact with the biofilm on the media

TABLE 2.4 Comparison of Used Media in Constructed Wetlands System

Particle Size	Advantages	Disadvantages
Clean sand	Readily available	Pore size too small
		Easily clogged as microbes grow
Gravel (1.0 cm)	Readily available	Pore size too small
		Easily clogged as microbes grow
Gravel (1.6–1.90 cm)	Ideal size	Costs more because of limited availability
	Good porosity	Good barrier to odors
Concrete gravel (1.25–5.1 cm)	Ideal bulk porosity	Non-graded (most small particles removed)
	Readily available	Fewer large openings (variability of media sizes)
Gravel > 5.1 cm	Good permeability	Less surface area for microorganisms
	Lower clogging	Large openings may allow mosquitoes to enter
Chipped tires	Relatively inexpensive	Appearance, low esthetic value, lightweight
		Steel rusts and discolors effluent

Source: After Lesikar et al. (2005).

surfaces for proper treatment. Media with large particles have large openings which allow for a good air exchange and consequently prevent odors to exit (Lesikar et al., 2005).

2.8.2.3 Characteristics of Media

Various types of media are used in subsurface flow constructed wetlands. When choosing fill media, consider the essential properties required for reactive media as (Amos and Younger, 2003):

- Bulk porosity: should not be less than 30%, which allows effluent to flow through the cell, yet provides adequate contact time.
- Pore size: individual pore spaces must be large to resist clogging.
- Stability: media must not breakdown over time.
- Media size: particles must be small enough for treatment yet large enough for adequate water flow.
- Surface area: enough surface area for attaching biofilm.
- Uniformity: all particles should be about the same size.
- Permeability: large enough for water to flow through media.
- Availability and cost: must be available and cheap.
- Aesthetics: do not eyesore.

Cooke and Rowe (1999) evaluated the porosity and specific surface of porous media coated with an accumulating film geometrically to take account of limitations to film growth imposed by contact between particles and the film as it is developed around the particles, by representing the media with regularly packed spheres, and assuming a uniform film thickness. The unit volume of solid spheres and their coating film, taking contact points between spheres into account might be written as:

$$V_u = \left(\frac{\pi \times d_p + 2\pi \times L_t}{6}\right)^3 - m\frac{\pi}{3}L_t^2\left[3\left(\frac{d_p}{2} + L_t\right) - L_t\right] \quad (2.22)$$

The first term on the right side of this equation is the total volume of a solid sphere and film coating assuming uniform coverage and the second term is the sum of the volumes of each spherical cap of film where accumulation cannot occur because of contact with neighboring spheres. The variation of porosity (n) and specific surface area (A_s) with film thickness is given by Taylor et al. (1990) cited in Cooke and Rowe (1999):

$$n = 1 - \frac{\pi}{\alpha_m} \left[\frac{2-m}{12} \left(\frac{2L_t}{d_p} \right)^3 + \frac{4-m}{8} \left(\frac{2L_t}{d_p} \right)^2 + \frac{1}{2} \left(\frac{2L_t}{d_p} \right) + \frac{1}{6} \right] \quad (2.23)$$

$$A_s = \frac{\pi}{\alpha_m \times d_p} \left[\frac{2-m}{2} \left(\frac{2L_t}{d_p} \right)^2 + \frac{4-m}{2} \left(\frac{2L_t}{d_p} \right) + 1 \right] \quad (2.24)$$

where: V_u = total volume considering contact points, m³; d_p = diameter of porous media particles, m; L_t = total film thickness, m; m = number of contact points; α_m = packing factor; n = porosity, % expressed as decimal; A_s = specific surface of porous media, m²/m³.

2.8.2.4 Media Configuration

Wetland beds can contain two or more types of media in different layers. Generally, the larger media particles are placed on the bottom and different types of media, generally consisting of smaller particles, are placed on the top. There are several reasons to layer the media, including safety, esthetics, effluent treatment and water flow. Another consideration for media configuration is to avoid the tendency of system to clog near the inlet. There, more solids are present in the effluent, and the biofilm building is greater. Placing larger media at the inlet of the system will reduce the risk of clogging and distribute the wastewater across the inlet (Lesikar et al., 2005).

2.8.2.5 Types of Wetland Media

The gravel-bed hydroponics system (GBH) is an inclined gravel filled channel lined with an impermeable membrane and planted with helophytes such as common reed. Such system has considerable potential for the secondary and tertiary treatment of sewage effluent (Butler et al., 1991). Butler and Williams (1997) suggested that GBH beds have application for industrial wastes but may require a longer residence times or further treatment stages.

Pant et al. (2001) investigated the use of three types of bed media (Lockport dolomite, Queenston shale, and Fonthill sand) on phosphorus (P) sorption characteristics in SSF wetlands. They concluded that the use of Fonthill sand as bed media could be better to remove P from sewage waste.

Manios et al. (2003) used four different growing media (top soil, gravel, river sand, and mature sewage sludge compost) to determine the best substrate for TSS removal. Eight units were constructed, two for each media. One bed for each pair was planted with Cattails and the other unplanted. Primary treated domestic wastewater was continuously fed to the beds for move than six months. All eight beds performed very well, the best performance was achieved by the gravel beds. There was no significant effect in the removal of TSS of planted and unplanted wetlands.

Collaço and Roston (2006) investigated the use of shredded tires as a medium for SSF constructed wetlands for treating domestic wastewater (COD, TSS, ammonia, pH) with aquatic macrophytes from *Typha* species. The results indicated a potential use of shredded tires to substitute the conventional media used for subsurface flow constructed wetlands. Sirianuntapiboon et al. (2006) investigated the effect of varying soil-to-sand ratios on domestic wastewater treating efficiency using CWs planted with cattail. Treatment efficiency was evaluated for BOD, COD, TSS, TP, and total kjeldahl nitrogen (TKN). The results indicated that the pollutants reduction was corresponding to a longer retention time (3 days) and with a bed media containing a soil-to-sand ratio of 75:25. Also, the highest growth rate of cattail was found under the shortest T_r (0.75 day).

Calheiros et al. (2008) investigated the performance of HSSF wetlands planted with *Typhalatifolia* treating tannery wastewater under long-term operation. Two expanded clay aggregates (Filtralite MR3-8-FMR and Filtralite NR3-8-FNR) and fine gravel (FG) were used as substrate for the wetland units plus one unit with FMR was left as an unplanted. The systems were subjected to three q(6, 8, and 18 cm/d), and to periods of interruption in the feed. The planted units with FNR and FMR achieved significantly higher BOD and COD removal when compared to FG and to the unplanted units. The systems proved to be tolerant to high organic loadings and to interruptions in feed suggesting this technology as a viable option for the biological treatment of tannery wastewater.

Albuquerque et al. (2009) tested two bed media (gravel and Filtralite) in HSSF wetlands in order to evaluate the removal rates of ammonia and nitrate for different types of wastewater (acetate-based and domestic). They obtained that Filtralite enables a quick development of biofilm with the capability of removal of organic matter, ammonia and nitrate at high rates with respect to gravel media. Chen et al. (2009) tested the removal efficiencies of three heavy metals (Zinc, Copper, and Lead) using SSF wetlands. Two substrates were used in wetland cells (coke and gravel). The results suggested that the removal rates of tested heavy metals were influenced by the choice of media to some extent.

Cordesius and Hedström (2009) investigated the use of two types of bed media (gravel and plastic pieces) on treating domestic wastewater. The treatment parameters were BOD, COD, TSS, DO, nitrate, NH_3, NH_4, TN, PO_4, TP, and fecal coliforms (E-coli/thermotolerant). Their analyzes showed a little difference in treatment efficiency between these two media. Sarafraz et al. (2009) investigated the effective use of gravel beds mixed with 10% zeolite in the performance of horizontal subsurface flow constructed wetlands. They concluded that this system was efficient in removing zinc (Zn), lead (Pb) and cadmium (Cd) from agricultural wastewater. Also they mentioned that, the wetland had a strong potential for the reduction of total phosphorus (TP). The characteristics of the media type selected in this system (zeolite), containing higher amounts of calcium (Ca), aluminum (Al) and iron (Fe) oxides was inferred to be a factor causing such high removal of phosphorus by adsorption.

Noor et al. (2010) evaluated the performance of vegetated HSSF constructed wetlands for the removal of ammoniac nitrogen, total reactive phosphorus, and soluble reactive phosphorus from landfill leachate. Four reactors were used (granite without vegetation, granite, gravel, and sand and 67.5 L of charcoal); the last three reactors were planted with Cattails. Their results indicated that the last reactor was the best for the removal efficiency of the above parameters. Shuib et al. (2011) studied the performance of HSSF wetlands using two media (zeolite and gravel). Two similar laboratory scale wetland units were constructed and operated for approximately 10 months. These units were planted with *Phragmitesaustralis* and *Scirpusmaritimus* and the system was subjected to two hydraulic retention times (4 and 3 days). Based on the 39 weeks of operation, the CW unit with natural zeolite achieved significantly higher removal for COD, ammonium, total nitrogen, total phosphorus, and TSS at the two used T_r than CW unit with gravel.

2.8.3 PLANT COMMUNITIES AND THEIR FUNCTIONS

Plants are the most prominent feature of treatment wetlands (Liu, 2002). Wetland plants have physiological adaptations that allow growth in low oxygen soils. Plant growth rates are an indicator of the quality of the water and how well a well wetland operates. Plant height and stem density are common ways for measuring plant growth (Myers and Jackson, 2001).

A wide variety of aquatic plants have been used in wetland systems, namely: common reed (*Phragmites*), cattail (*Typha*), rush (*Juncus*), bulrush (*Scirpus*), giant cutgrass (*Zizaniopsis*), duck potato arrowhead (*Sagittaria*), and southern wild rice (*Zizania*) (USDA, 2002). Some conditions might be verified for selecting plant such as:

- It is prudent to consider only native plants that grow locally.
- It should be active vegetative colonizers with spreading rhizome.
- It should have considerable biomass and stem densities.
- Sometimes a combination of species is selected that would provide coverage over various water depths.

Mbuligwe (2005) studied the treatment of dye-rich wastewater in wetland systems (vegetated units and unplanted unit) concluding that wetland plants play a significant role in the treatment processes. Fraser et al. (2004) demonstrated the importance of macrophytes to reduce nutrient concentration encountered by single-family domestic SSF wetlands. Most systems are planted with the common reed, but some systems include other species of wetland plants. It was believed that the growth of roots and rhizomes of the reeds would increase and stabilize the hydraulic conductivity. The major roles of macrophytes in constructed treatment wetlands are summarized by Brix (1997) in Table 2.5.

Oxygen released by the roots of the helophytes as a result of diffusive and/or convective gas transport processes from the atmosphere through the plant tissues into the root system is believed to play an important role in the supply of oxygen to the microorganisms in the rhizosphere (Brisson and Chazarenc, 2009; Wieβner et al., 2005). Wetland vegetation serves to leak oxygen to the root zone, shade out algae and provide substrate for microbes (Jensen, 2001). Major plant adaptations in HSSF constructed wetlands are shown in Figure 2.8 (Kadlec and Wallace, 2008).

Aquatic plants of wetland act as the filter, nutrient up take organisms to provide a substrate for microbiota (algae, bacteria, fungi, protests) and to provide a carbon source for denitrification process (Sirianuntapiboon and

Literature Review

TABLE 2.5 Summary of the Major Roles of Macrophytes in Constructed Wetlands

Property	Role in Treatment Process
Aerial plant tissue	• Storage of nutrients – Aesthetic pleasing appearance of system • Influence on microclimate – insulation during winter • Light attenuation – reduced growth of phytoplankton • Reduced wind velocity – reduced risk of resuspension
Plant tissue in water	• Filtering effect – Stores of nutrients • Provide surface area for attached biofilm • Excretion of photosynthetic – increases aerobic degradation • Reduce velocity-increase sedimentation, reduces resuspension
Roots and rhizomes	• Uptake of nutrients – Maintain hydraulic pathways in media • Stabilizing the sediment surface – less erosion • Release of oxygen – increase degradation and nitrification • Prevents the medium from clogging in vertical flow systems

Source: After Brix (1997).

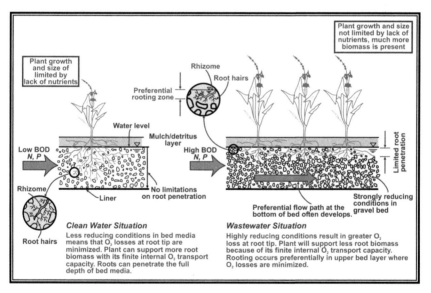

FIGURE 2.8 Major plant adaptations in HSSF wetlands. (Copyright 2007. From Kadlec, R.H.; Wallace, S.D. *Treatment Wetlands*. Second Edition. Reproduced by permission of Taylor and Francis Group, LLC, a division of Informa plc.)

Jitvimolnimit, 2007). Grismer et al. (2011) illustrated the main role of plant in the treating process by increasing the removal efficiency in the effluent of planted versus unplanted constructed wetland systems. Vegetation was confirmed to enhance the removal rate of all wetland types for all kinds of pollutants. However, the species of the vegetation does not significantly influence the removal rates (Xing, 2012).

2.9 REPRESENTING TREATMENT PERFORMANCE

Wetland performance can change from year-to-year due to changes in vegetative communities, organic loadings, or weather conditions.

2.9.1 CHEMICAL MASS BALANCES

The percent concentration reduction is often used as (Kadlec and Wallace, 2008):

$$\% \text{Concentration reduction} = 100 \times \left(\frac{C_i - C_o}{C_i} \right) \quad (2.25)$$

The inlet mass loading rates may be written as:

$$M_i = Q_i C_i \quad (2.26)$$

$$m_i = \frac{M_i}{A} = \frac{Q_i C_i}{A} = q_i C_i \quad (2.27)$$

where: m_i = specific inlet mass loading, g/m²/d; M_i = inlet mass loading, g/d. The mass removal rate (JA) may be defined as follows:

$$JA = (Q_i C_i - Q_o C_o) \quad (2.28)$$

This equation represents the average amount of pollutant that gets stored, destroyed or transformed. This single-number measures the treatment wetlands performance. The percentage of mass removal may be represented as:

$$\% \text{Mass removal} = 100 \times \left(\frac{Q_i C_i - Q_o C_o}{Q_i C_i} \right) = 100 \times \left(\frac{m_i - m_o}{m_i} \right) \quad (2.29)$$

where: m_o = specific outlet mass loading, g/m²/d.

2.9.2 IMPORTANT PROCESSES IN POLLUTANTS REMOVAL

A large number of constructed wetland processes may contribute to the reduction of any given pollutant. Many wetland reactions are microbial mediated,

Literature Review

which the result of activity of bacteria or other microorganisms. Very few such organisms are found free-floating in water; rather, the great majority is attached to solid surface. Often, the numbers are sufficient to form relatively thick coatings on immersed surfaces (Kadlec and Wallace, 2007).

Transfer of a chemical from water to immersed solid surfaces is the first step in the overall microbial removal mechanism. Those surfaces contain the biofilms responsible for microbial processing, as well as the binding sites for sorption processes. Roots are the locus for nutrient and chemical uptake by the macrophytes, and these are accessed by diffusion and transpiration flows. Dissolved materials must move from the bulk of the water to the vicinity of the solid surface, then diffuse through a stagnant water layer to the surface, and penetrate the biofilms while undergoing chemical transformations as shown in Figure 2.9 (Kadlec and Knight, 1996).

Various processes in wetland create product gases that are released from the wetland environment to the atmosphere, such as ammonia, hydrogen sulfide, nitrous oxide, and methane. Wetlands also take in atmospheric carbon dioxide for photosynthesis and expel it from respiratory processes. Sunlight can degrade or convert many waterborne substances. Many microorganisms, including pathogenic bacteria and viruses, can be killed by ultraviolet radiation. The effectiveness is presumptively determined by the radiation dose rate as well as the concentration of organisms. Plants take up nutrients to sustain their metabolism (Kadlec and Wallace, 2008).

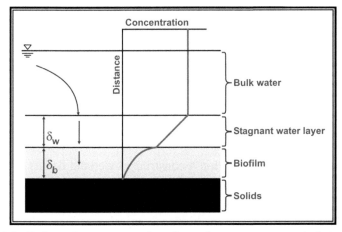

FIGURE 2.9 Pathway for movement of a pollutant from water across a deviation layer and into biofilm. (Reprinted from Kadlec, R.H.; Knight, R.L. *Treatment Wetlands*. CRC Lewis Publisher, New York, © 1995 with permission from Taylor & Francis.

The primary abiotic processes that are responsible for removing wastewater contaminants in a constructed wetland include the following (Halverson, 2004; Hammer, 1989):

- Settling and sedimentation, which achieve efficient removal of particulate matter and suspended solids.
- Sorption-including adsorption and absorption, chemical process occurring on surfaces of plants, media, sediment, and litter.
- Chemical oxidation/reduction/precipitation-conversion of metals in the influent stream, through contact of the water with media and litter, to an insoluble solid form that settles out.
- Volatilization, which occurs when compounds with significant vapor pressures partition to the gaseous state.

Biotic processes like biodegradation and plant uptake are also responsible for contaminant removal, in addition to the abiotic processes as follows (Davis, 1995a; and Halverson, 2004):

- Phytoaccumulation: accumulation of inorganic elements in plants.
- Phytostabilization: the ability to sequester inorganic compounds in plant roots.
- Aerobic/anaerobic biodegradation-metabolic processes of microorganisms, which play a significant role in removing organic compounds in wetlands.
- Phytodegradation: plants produce enzymes to break down the organic and inorganic contaminants that enter into the plant during transpiration.
- Rhizodegfadation: plants provide exudates that enhance microbial degradation of organic compounds.
- Phytovoiatilization/evapotranspiration: uptake and transpiration of volatile compounds through the leaves.

2.9.3 MODELS OF INTERNAL HYDRAULICS

The removal of pollutants within a constructed wetland occurs through the diverse range of interaction between the sediments, substrate, microorganisms, litter, plants, the atmosphere, and the wastewater as it moves through system. Many of the important biogeochemical reactions rely on contact time between wastewater constituents and microorganisms and the associated substrate, whereas wastewater velocity can be an important determining factor for other pollutant removal processes.

Any short-circuiting or dead zones that occur within the constructed wetlands will, consequently, have an effect on contact time as well as flow velocities and, therefore, impact on treatment efficiencies. Non-ideal flow patterns can have very large effects upon the removal performance of pollutants in wetland treatment systems. Figure 2.10 shows a sample of various models used to represent wetlands internal hydraulics.

There are many candidate models that may be used in constructed wetlands, which typically involve tanks in series, perfectly mixed units and plug flow sections. It is clear from numerous studies that treatment wetlands are neither plug flow nor well-mixed. The tanks-in-series (TIS) model captures the important one to represent flow. The TIS model requires two parameters: the number of tanks (N), and the actual retention time. As the model networks increase in complexity, the parallel path and finite stage models might be used but with adding more calibration parameters (Kadlec and Wallace, 2008).

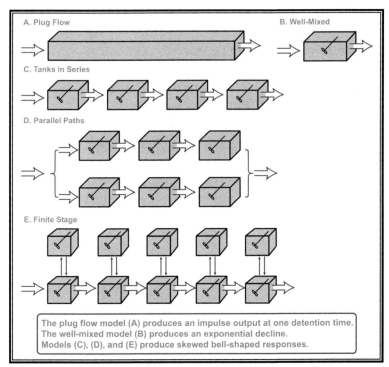

FIGURE 2.10 Various wetland tracer responses. (Copyright 2007. From Kadlec, R.H.; Wallace, S.D. *Treatment Wetlands*. Second Edition. Reproduced by permission of Taylor and Francis Group, LLC, a division of Informa plc.)

2.10 MODELS FOR CONSTRUCTED WETLANDS

This section reviews some models for horizontal subsurface flow constructed treatment wetland systems, starting with simple models like rules of thumb and regression equations then, the well known reaction rate models (zero and first-order models, Monod-type equations, tanks in series and relaxed TIS concentration models, and plug flow with dispersion model).

2.10.1 RULES OF THUMB

From an engineering point of view, rules of thumb are the fastest but also the roughest design methods. As an example, some of these rules for HSSF constructed wetlands are summarized in Table 2.6. Since, they are based on observations from a wide range of systems, climatic conditions and wastewater types, these rules of thumb show a large variation system (Rousseau et al., 2004).

2.10.2 REGRESSION EQUATIONS

Considering the fact that, the majority of the investigations on treatment wetlands have mainly been focused on input-output (I/O) data rather than on internal processes data. Constructed wetlands have long been seen as "black boxes" where wastewater enters and treated water leaves the system (Langergraber, 2008).

However, these black box "models" lump a complex system like a constructed treatment wetland into only two or three parameters, which is clearly an oversimplification. Important factors such as climate, bed material, bed

TABLE 2.6 Rule of Thumb Design for HSSF Constructed Wetlands

Criterion	Range of Values	
	Wood (1995)	**Kadlecand Knight (1996)**
Hydraulic retention time (d)	2–7	2–4
Max. BOD loading rate (kg/ha.d)	75	Not given
Hydraulic loading rate (cm/d)	2–30	8–30
Areal requirement (ha/m^3/d)	0.001–0.007	Not given

Source: After Rousseau et al. (2004).

Literature Review

design (length, width, and depth), etc. are neglected, leading to a wide variety of regression equations and thus a large uncertainty in the design. A literature overview of regression equations for BOD, COD, and TSS are presented in Table 2.7 (Rousseau et al., 2004).

2.10.3 REACTION RATE MODELS

The following part discusses the common reaction rate models, which represent the effluent concentration for the horizontal subsurface flow constructed wetlands.

2.10.3.1 Zero Order Model

The simplest quantitative model for contaminant reduction is a constant rate of removal, termed zero-order removal, because it does not depend upon how much of the contaminant is present at a given location. The zero-order removal rate is given by (Kadlec and Wallace, 2008):

TABLE 2.7 Regression Equations for HSSF Wetlands According to Different Authors

Type	Regression Equation	Input Range	Output Range	q Range
BOD[A]	$C_o = (0.11 C_i) + 1.87$	$1.0 < C_i < 330$	$1.0 < C_o < 50$	$0.8 < q < 22$
BOD[B]	$C_o = (0.33 C_i) + 1.40$	$1.0 < C_i < 57$	$1.0 < C_o < 36$	$1.9 < q < 11.4$
BOD[C]	$C_o = (0.099 C_i) + 3.24$	$5.8 < C_i < 328$	$1.3 < C_o < 51$	$0.6 < q < 14.2$
BOD[D]	$L_r = (0.653 L_i) + 0.292$	$4.0 < L_i < 145$	$4.0 < L_r < 88$	Not Given
BOD[E]	$L_o = (0.145 L_i) - 0.06$	$6.0 < L_i < 76$	$0.30 < L_o < 11$	Not Given
BOD[C]	$L_o = (0.13 L_i) + 0.27$	$2.6 < L_i < 99.6$	$0.32 < L_o < 21.7$	$0.6 < q < 14.2$
COD[E]	$L_o = (0.17 L_i) + 5.78$	$15 < L_i < 180$	$3.0 < L_o < 41$	Not Given
TSS[A]	$C_o = (0.09 C_i) + 4.7$	$0.0 < C_i < 330$	$0.0 < C_o < 60$	$0.8 < q < 22$
TSS[B]	$C_o = (0.063 C_i) + 7.8$	$0.1 < C_i < 253$	$0.1 < C_o < 60$	$1.9 < q < 44.2$
TSS[C]	$C_o = (0.021 C_i) + 9.17$	$13 < C_i < 179$	$1.7 < C_o < 30$	$0.6 < q < 14.2$
TSS[C]	$C_o = 0.76 C_i^{0.706}$	$8.0 < C_i < 595$	$2.0 < C_o < 58$	Not Given
TSS[E]	$L_o = (0.048 L_i) + 1.76$	$3.0 < L_i < 78$	$0.9 < L_o < 6.3$	Not Given
TSS[C]	$L_o = (0.083 L_i) + 1.18$	$3.7 < L_i < 123$	$0.45 < L_o < 15.4$	$0.6 < q < 14.2$

Source: After Brix (1994b); Knight et al. (1993b); Reed and Brown (1995); Vymazal (1998b); Rousseau et al. (2004); Vymazal (1998c).

L_i = influent loads, kg/ha/d; L_o = effluent loads, kg/ha/d; L_r = load removal, kg/ha/d; q = hydraulic loading rate, cm/d.

$$J = \text{cons}\tan t \qquad (2.30)$$

where: J = removal per unit area, or removed load, g/m²/d.

2.10.3.2 First Order Model

In case of constant conditions (e.g., inlet flow rate and pollutants concentration) and an ideal plug-flow behavior, the first-order equations predict an exponential profile between inlet and outlet concentrations (Nitisoravut and Klomjek, 2005):

$$C_o = C_i \times e^{(-k/q)} \qquad (2.31)$$

$$\frac{C_o - C^*}{C_i - C^*} = e^{(-nhk_v/q)} \qquad (2.32)$$

$$k = n \times h \times k_v \qquad (2.33)$$

where: C_o = effluent concentration, mg/l; C_i = influent concentration, mg/l; C^* = background concentration, mg/l; k = removal rate constant, m/d; k_v = volumetric removal rate constant, d⁻¹.

Rate constant (k) calculated using Eq. (2.31) resulted in a range of values from 0.01 to 0.5 for coliforms removal (Leonard, 2000). The influence of temperature is commonly modeled via an Arrhenius equation (2.34) (El-Hamouri et al., 2007):

$$k_T = k_{20}\theta_t^{(T-20)} \qquad (2.34)$$

where: k_T = first-order areal rate constant at T °C, d⁻¹; k_{20} = first-order areal rate constant at 20 °C, d⁻¹; θ_t = empirical temperature coefficient; T = desired temperature, °C.

According to Kadlec and Knight (1996) removal rate of BOD and TSS in treatment wetlands is generally found to be independent of temperature ($\theta_t = 1.0$). Table 2.8 presents an overview of first-order rate constants for HSSF constructed treatment wetlands.

The background concentration (C^*) for BOD, COD, and TSS may be represented (if $C_o < C^*$ use $C^* = 0.0$) as follows (Zurita et al., 2009):

Literature Review

TABLE 2.8 Rate Constants for HSSF Wetlands According to Different Authors

Type	Reference	K (m/d)	k_v (d⁻¹)	Remarks
BOD	Reed and Brown (1995)		1.10	k_{20} with $\theta_t = 1.06$
	Wood (1995)		1.35	$n = 0.39$ course sand (20 °C)
	Kadlec (1997)	0.49		$C^* > 3.0$ mg/l and $\theta_t = 1.0$ (20 °C)
	Brix (1994)	0.16		$C^* > 3.0$ mg/l soil based
	Cooper et al. (1996)	0.06		$C^* > 0.0$ mg/l secondary wetland
TSS	Kadlec and Knight (1996)	2.74		k_{20} with $\theta_t = 1.0$ and $C^* > 7.0$ mg/l
	Kadlec (1997)	8.22		k_{20} with $\theta_t = 1.0$ and $C^* > 7.0$ mg/l

Source: After Rousseau et al. (2004).

$$C^*_{BOD} = C^*_{COD} = 3.5 + 0.053 C_i \quad (2.35)$$

$$C^*_{TSS} = 7.8 + 0.063 C_i \quad (2.36)$$

2.10.3.3 Monod-Type Model

Monod-type kinetics is operating well in biological systems, where degradation rates are limited by pollutant availability at relative low concentration but would reach saturation at high concentrations. The Monod model, which interpolates between zero and first-order limits may be written as (Frazer-Williams, 2010):

$$J = k \left(\frac{C}{K + C} \right) \quad (2.37)$$

where: C = pollutant concentration, g/m³; K = half-saturation constant, g/m³.

2.10.3.4 Tanks in Series Model (TIS)

Water passes through assumed number N-tanks in series (TIS) and losses contaminant in each one ($N = 2:8$ by trail and error to fit the data). For longitudinal concentration profiles and case of no water losses or gains the result is as follows (Liu, 2002):

$$\frac{C_o - C^*}{C_i - C^*} = \left(1 + \frac{k_v \times T_r}{N} \right)^{-N} \quad (2.38)$$

where: T_r = hydraulic retention time, day; N = number of tanks in series.

Note that there are two reaction parameters in this model: the rate constant (k) and the hydraulic parameter (N). Equation (2.37) represents the reduction of a single compound on transit through a treatment system. However, much contaminates are, in fact, mixtures. It is clear that the individual component of such mixtures may be removed at different rates, and that there is a corresponding difference in removal rate constants (Kadlec and Wallace, 2008).

2.10.3.5 Relaxed TIS Concentration Model

It has been noted that observed weathering behavior in real wetland situations may be represented by the TIS model (Eq. 2.37), wherein the parameter values are relaxed to become fitting parameters. The relaxed TIS concentration model (P-k-C* model) is, therefore, defined to be (Kadlec and Wallace, 2008):

$$\frac{C-C^*}{C_i-C^*} = \frac{1}{(1+k/Pq)^P} = \frac{1}{(1+k_v\tau/P)^P} \tag{2.39}$$

where: k = modified first-order areal rate constant, m/d; k_v = modified first-order volumetric rate constant, d^{-1}; P = apparent number of TIS.

2.10.3.6 Plug Flow with Dispersion Model (PFD)

The first-order concentration reduction produced by the PFD model is well known as (Kadlec and Wallace, 2008):

$$\frac{C-C^*}{C_i-C^*} = \frac{4b\exp\left(\dfrac{Pe}{2}\right)}{(1+b)^2\exp\left(\dfrac{bPe}{2}\right) - (1-b)^2\exp\left(\dfrac{-bPe}{2}\right)} \tag{2.40}$$

$$b = \sqrt{1+4\frac{Da}{Pe}} \tag{2.41}$$

$$Da = \frac{k \times T_r}{h} \tag{2.42}$$

Literature Review

$$Pe = \frac{uL}{D_s} \tag{2.43}$$

where: Pe = Peclet number; Da = Damköhler number; D_s = dispersion coefficient, m²/d; u = average water velocity, m/d; L = wetland length, m.

Note that there are two reaction parameters in this model: the dispersion coefficient (D_s) and the rate constant (k), which is not constant but depends on factors such as loading rate, inlet concentrations, etc. (Rousseau et al., 2001). For instance, the PFD model forecasts the concentration profile through the wetland to be given by:

$$\frac{C(x)}{C_i} = \frac{2\exp\left(\frac{xPe}{2}\right)\left[(1+b)\exp\left(\frac{+bPe(1-x)}{2}\right) - (1-b)\exp\left(\frac{-bPe(1-x)}{2}\right)\right]}{(1+b)^2 \exp\left(\frac{bPe}{2}\right) - (1-b)^2 \exp\left(\frac{-bPe}{2}\right)} \tag{2.44}$$

where: $C(x)$ = concentration at length x, g/m³; x = distance in flow direction, m.

2.11 WATER PARAMETERS IN WETLANDS

Five of the most widely fluctuating and important abiotic factors are dissolved oxygen, hydrogen ion concentration, suspended solids, chemical oxygen demand, and biochemical oxygen demand. In addition to oxidation-reduction potential, wetland water temperature, dissolved solids, electric conductivity, and metals were presented. The mechanisms for constructed wetlands treatment are presented in Table 2.9 (Kayombo et al., 2004; Kiracofe, 2000).

2.11.1 DISSOLVED OXYGEN

Available oxygen in constructed wetlands is an important factor in the degradation of organic matter and transformation of ammonium-nitrogen, both of which are oxygen limiting processes (Zhang et al., 2010).

TABLE 2.9 Treatment Mechanisms in Constructed Wetlands

Pollutant	Treatment Mechanisms in Constructed Wetlands
BOD	Filtration and microbial degradation (aerobic and anaerobic)
TSS	Sedimentation, filtration, and aerobic or anaerobic microbial degradation
Phosphorus	Sedimentation, adsorption, precipitation, plant and microbial uptake
Nitrogen	Nitrification, ammonification, plant uptake, volatilization, and denitrification
Pathogens	UV degradation, filtration, predation, adsorption, and natural die-off
Heavy metals	Sedimentation, adsorption, plant uptake
Pesticides	Adsorption, Volatilization, photolysis, and biotic/abiotic degradation

Source: After Kayombo et al. (2004); Kiracofe (2000).

In wetland environments, dissolved oxygen (DO) is derived from both physical and biological processes. The DO may be photo-synthetically derived from aquatic plants, attached algae, and photo-plankton. Emergent aquatic plants are able to transport atmospheric oxygen to the root zone. Air-to-water diffusion is another source of DO, providing limited transport of atmospheric oxygen at the air-water interface. In most instances, these combined sources of oxygen in conventional constructed wetlands are not sufficient to meet the aerobic respiratory demands of organically enriched wetlands (Kadlec and Wallace, 2008).

Nitrification and oxidative consumption of organic compounds and BOD are dependent on dissolved oxygen. The DO is of interest in treatment wetlands for two principal reasons: it is an important participant in some pollutant removal mechanisms, and it is a regulatory parameter for discharges to surface waters. In many permits in the United States, a minimum DO of 5 mg/l is specified. The DO is depleted to meet wetland oxygen requirements in four major categories: sediment/litter oxygen demand, respiration requirements, dissolved carbonaceous BOD, and dissolved nitrogenous oxygen demand (Kadlec and Wallace, 2008).

Chan et al. (2004) gave a number of features to be incorporated into the design of subsurface flow constructed wetlands in an effort to enhance oxygen transfer. They include firstly, a pair of parallel wetlands that can be operated in a cycle "draw and fill" scheme, allowing filling of one wetland while draining the other (reciprocating wetland), and secondary a vertical filter with coarse gravel at the front section of each wetland cell. These modifications can significantly improve biological oxygen demand and nitrogen removal, therefore reducing the T_r of the system.

2.11.2 HYDROGEN ION CONCENTRATION

Hydrogen ion concentration (pH) influences many biochemical transformations. The pH is a term used to express the acidity or alkalinity of a solution. It influences the partitioning of ionized and un-ionized forms of carbonates and ammonia, and controls the solubility of gases, such as ammonia, and solids, such as calcite. Hydrogen ions are active in cation exchange processes with wetland sediments and soils, and determine the extent of metal binding (Kadlec and Wallace, 2008).

Healthy aquatic systems can function only within a limited pH range. As a consequence, surface water discharge permits frequently require 6.5 < pH < 9.0. Many treatment bacteria are not able to exist outside the range 4.0 < pH < 9.5. Denitrifiers operate best in the range 6.5 < pH < 7.5, and nitrifiers prefer pH = 7.2 and higher (Kadlec and Knight, 1996). Barton and Karathansis (1999) added a considerable amount of alkalinity from limestone dissolution within the wetland media for acid mine drainage.

2.11.3 TOTAL SUSPENDED SOLIDS

Low water velocities, coupled with the presence of plant litter (SF) or sand/gravel media (SSF), promote settling and interception of solid materials. This transfer of suspended solids from the water to the wetland sediment bed has important consequences for the quality of the water. Many pollutants are associated with the incoming suspended matter, such as metals and organic chemicals, which partition strongly to suspended matter (Kadlec and Wallace, 2008).

Most of the solids present within a HSSF wetland bed are an accumulation of microbial biofilms, intercepted particulate matter, and plant-root networks. This accumulated material, collectively called a biomat, occurs either as material attached to the bed media and plant roots or as colloidal material within the media pores (Kadlec and Wallace, 2008).

Suspended solids (SS) are generally efficiently removed in both types of constructed wetlands. The inlet concentrations of SS vary from low level for systems constructed for tertiary treatment, to moderate or high levels for systems constructed for secondary treatment (Brix, 1994). Suspended solids absorb heat from sunlight, which increases water temperature and subsequently decreases levels of DO "warmer water holds less oxygen than cooler water" (Kadlec and Wallace, 2008).

2.11.4 CHEMICAL OXYGEN DEMAND

Chemical oxygen demand (COD) is the equivalent quantity of oxygen used to oxidize the organic matter in a wastewater (Myers and Jackson, 2001). Vymazal and Kröpfelová (2009) made a survey of more than 400 SSF constructed wetlands from 36 countries around the world revealing that the highest removal efficiencies for BOD and COD were achieved in systems treating municipal wastewater, while the lowest efficiency was recorded for landfill leachate. This is caused by the fact that municipal wastewaters contain predominantly labile organics while landfill leachate contains often-recalcitrant organics which are difficult to degrade.

2.11.5 BIOCHEMICAL OXYGEN DEMAND

The BOD is a chemical procedure for determining the amount of DO needed by aerobic biological organisms in a body of water to break down organic material present in a given water sample at certain temperature over a specific time period. It is most commonly expressed in milligrams of oxygen consumed per liter of sample during 5 days of incubation at 20°C. A high BOD indicates the presence of a large number of microorganisms, which suggests a high level of pollution (Kopec, 2007).

The mean rate constants for BOD removal, K_{BOD}, are approximately 5 times higher in subsurface flow constructed wetlands than in surface flow systems. This probably partly reflects the greater surface area available for attached microbial growth in subsurface flow systems (Brix, 1994). Attiogbe et al. (1999) established empirical correlations between BOD and COD of effluents from industrial wastes.

2.11.6 OXIDATION-REDUCTION POTENTIAL

Oxidation-reduction is a chemical reaction in which electrons are transferred from a donor to an acceptor. The electron donor loses electrons and increases its oxidation number or is oxidized; the acceptor gains electrons

Literature Review

and decreases its oxidation number or is reduced. Considering a reaction in which *ne* electrons are transferred (Kadlec and Wallace, 2008):

$$Oxidation + ne^- \Leftrightarrow Reduction \qquad (2.45)$$

Constructed wetland systems for the treatment of domestic sewage usually cause the removal of ammonia due to nitrification and also the removal of nitrate and nitrite owing to denitrification. Moreover, organic carbon can be removed by both aerobic microbial mineralization and anaerobic microbial methane formation. Sulphate reduction and sulfide oxidation are characteristic processes in the rhizosphere of CWs (Wießner et al., 2005).

Processes with iron as electron acceptor: Iron reduction contributes only a maximum of 0.1% and 0.2% of the total removed acetate in nitrate rich and sulfate rich environment, respectively. Therefore, it is assumed that these processes play a minor role when treating domestic wastewater. However, they can be necessary for the treatment of industrial and mining wastewaters. Processes with hydrogen as electron donor: It is assumed that hydrogen occurs only as intermediate product and is rapidly consumed. Therefore, it is further assumed that processes with hydrogen as electron donor (SO_4 reduction with H_2, etc.) and H_2 volatilization is not considered (Langergraber et al., 2009).

Enhanced root penetration, attributed to shallow bed or intermittent feeding resulted in significantly higher oxidation-reduction values and thus a more oxidized treatment environment (Wynn et al., 1997).

2.11.7 WATER TEMPERATURE

The water temperature in treatment wetlands is of interest for several reasons (Kadlec and Wallace, 2008):

- Temperature modifies the rate of several biological processes.
- Temperature is sometimes a regulated water quality parameter.
- Water temperature is a prime determinant of evaporative loss.
- Cold-climate wetland systems have to remain functional in subfreezing conditions.

Water temperature is a critical parameter to the operation of a wetland. Most wetland plants cannot sustain growth in waters with temperatures greater than 38°C. At low temperatures, microbial activity slows down and plants become senescent (Myers and Jackson, 2001).

Mink et al. (2002) stated that a wetland to be effective, it must provide adequate nutrient treatment through the four seasons. This has been a problem in northern climates, especially during the winter season. Allen et al. (2002) studied the effect of temperature and wetland plant species on wastewater treatment and root zone oxidation. It resulted that subsurface flow wetlands can be effective in cold climates and suggested that plant species selection may be especially important to optimizing subsurface flow wetland performance in cold climate.

Mander and Jenssen (2003) suggested that, in cold regions, the treatment wetland must have a septic tank and an aerobic pretreatment step of gravel bed prior to reed bed discharge and finally to an infiltration gravel bed.

2.11.8 ELECTRIC CONDUCTIVITY AND TOTAL DISSOLVED SOLIDS

Conductivity in water is a numerical expression of the ability of an aqueous solution to carry an electric current. This ability depends up on the presence of ions, their total concentration, mobility, valence and temperature. High value of conductivity means the presence of excess minerals and dissolved matter in the water. Total dissolved solids (TDS) are an indication of ionic strength and an indirect measure of the salt content in the water (Myers and Jackson, 2001).

2.11.9 METALS REMOVAL

Heavy metal contamination is one of the most serious environment problems throughout the world (Yeh et al., 2009). Metal removal processes occurring in wetlands involve a series of mechanisms (Halverson, 2004; Kuschk et al., 2005; Marchand et al., 2010):

- The suspended solids and adsorbed metals are easily filtered and retained in wetlands.
- Oxidation and hydrolysis generally the main processes responsible for removal of Aluminum, Iron, and Manganese.
- Formation of carbonates which play an important role in the initial trapping of metals such as Copper, Manganese, and Nickel.
- Formation of insoluble sulfides metals such as Silver, Cadmium, Mercury, Arsenic, Copper, Lead, and Zinc form highly insoluble sulfide compounds.

- Binding to Iron and Manganese oxides which has been effective for Arsenic, Copper, Iron, Manganese, Nickel, Cobalt, Lead, Uranium, and Zinc.
- Reduction to nonmobile forms by bacterial activity effective for Chromium, Copper, Selenium and Uranium.
- Uptake by plants, algae, and bacterial.

Mitsch and Wise (1998) used the HSSF constructed wetlands to treat the acid mine drainage. They concluded that, the wetland has been fairly effective in the treatment of iron while it has been less successful in the control of acidity and related parameters. The extent of metal removal depends on dissolved metal concentrations, dissolved oxygen content, pH, and retention time of the water in the wetland.

Kamarudzaman et al. (2011) compared the efficiency of both horizontal and vertical SSF constructed wetlands in the removal of heavy metals (Iron and Manganese) in landfill leachate. They summarized that HSSF had higher removal rates of heavy metals than VSSF system.

2.12 NUTRIENT CYCLES IN CONSTRUCTED WETLANDS

Nutrients such as Nitrogen (N) and Phosphorous (P) are necessary for plant and animal growth, but their usefulness has a plateau after which all excess is potentially detrimental to the environment (Kadlec and Knight, 1996).

Nutrient removal and transformation processes in subsurface flow constructed wetlands include microbial conversion, decomposition, plant uptake, sedimentation, volatilization, and adsorption-fixation reactions. Aquatic plants enhance nutrient removal through biomass accumulation, fixation of inorganic and organic particulates and where ammonium-N is present, the creation of an oxidized rhizosphere (Huett et al., 2005).

2.12.1 CARBON CYCLE

The carbon cycle in wetlands is dominated by wetland's plant life. Wetland plants follow a cycle of growth and nutrient uptake, death, and lastly, decomposition and nutrient release. The microbial decomposition is described as a two-stage process. The first stage is an initial rapid weight loss over the first 30 to 60 days that is caused by the biological utilization of starches and sugars. The second one is an exponential decomposition with the biological release of additional nutrients (Kadlec and Wallace, 2008). Figure 2.11 shows the carbon cycle in constructed wetlands (Kayranli et al., 2010).

2.12.2 NITROGEN CYCLE

The major removal mechanism for nitrogen in constructed wetlands is nitrification-denitrification. The major nitrification problem that can occur is a lack of available oxygen for the Nitrosomones and Nitrobacter to survive (Leonard and Swanson, 2001). The oxygen required for nitrification is delivered either directly from the atmosphere through the water or sediment surface, or by leakage from plant roots. Nitrogen is also taken up by the plants and incorporated into the biomass (Brix, 1994). Simplified nitrogen transformations and associated oxygen consuming steps in SSF constructed wetlands are presented in Figure 2.12 (Tanner and Kadlec, 2003).

Kuschk et al. (2003) evaluated the annual course of nitrogen removal in a stable operating subsurface horizontal flow constructed wetland in a moderate climate. They stated that denitrification was nearly complete in midsummer and was clearly restricted at seasonal temperatures below 15°C. Mayo and Bigambo (2005) studied the nitrogen transformation in horizontal subsurface flow constructed wetlands. They found that the major pathways leading to permanent removal of nitrogen in HSSF wetlands system are denitrification, plant uptake and net sedimentation.

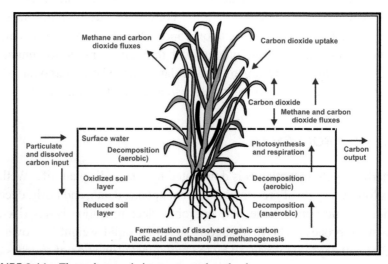

FIGURE 2.11 The carbon cycle in constructed wetlands.

Literature Review

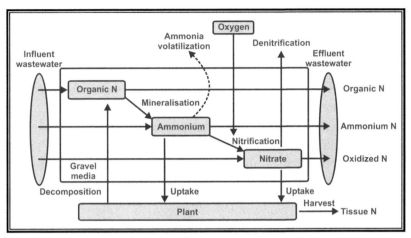

FIGURE 2.12 Nitrogen transformations in SSF wetlands. (Source: Tanner and Kadlec, 2003)

Wieβner et al. (1999) carried out a pilot-scale, subsurface flow constructed wetland treatment experiments on lignite pyrolysis wastewater which had been stored for a long time in an effluent pond in order to evaluate the removal efficiencies of NH_4 and organic components of a large molecular size. The results demonstrated the possibility of removing large amounts of ammonium contained in lignite pyrolysis wastewater in the rhizosphere of a HSSF constructed wetland with a relatively long T_r. Plant uptake could account for less than 10% of nitrogen removal; and bacterial denitrification seemed to be the dominated process for removing nitrogen within a wetland (Lin et al., 2000).

2.12.3 PHOSPHOROUS CYCLE

Removal of phosphorus in all types of constructed wetlands is low unless special substrates with high sorption capacity are used (Vymazal, 2007, 2010). Brix et al. (2001) investigated adding different materials to the media for sustainable phosphorus removal in subsurface flow constructed wetlands. Particularly calcite and crushed marble were found to have high phosphorous binding capacities.

Drizo et al. (1999) provided selected criteria for substrates that would enhance phosphate removal from wastewater in a subsurface horizontal flow constructed wetland system. Measured properties of seven substrates (bauxite, shale, burnt oil shale, limestone, zeolite, light expanded clay aggregated

LECA, and fly ash) were: pH, cation exchange capacity, hydraulic conductivity, porosity, specific surface area, particle size distribution and phosphate adsorption capacity. On the basis of these measurements it was concluded that, from the seven materials examined, shale had the best combination of properties as a substrate for constructed wetland systems, followed by fly ash, bauxite, limestone, and LECA.

Bubba et al. (2003) studied the P-adsorption capacities of 13 Danish sands for use as media in subsurface flow constructed reed beds by short-term experiments. They obtained that P-binding energy constants were not significantly related to the physicochemical properties of the sands. Figure 2.13 shows the phosphorus cycle in horizontal subsurface flow constructed wetlands (Lee, 1999).

Arias et al. (2005) stated the general strategies for P-removal in subsurface flow constructed wetlands as:

- The harvesting of emergent vegetation.
- The construction of the wetland beds using material with high Ca content to enhance P adsorption and/or precipitation.
- The establishment of external filters in order to retain P before final discharge.
- The implementation of independent chemical dosing systems to precipitate P.

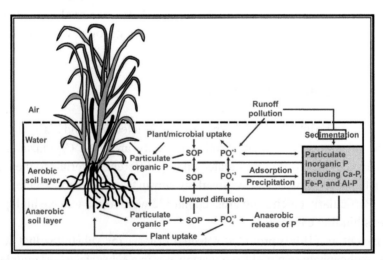

FIGURE 2.13 Phosphorus transformations in constructed wetlands. (Source: Lee R. E., 1999)

2.12.4 BACTERIA IN WETLANDS

Many nutrient transformations in wetlands are due to microbial metabolism and are directly related to microbial growth. Most treatments in wetlands are due to heterotrophic and autotrophic bacteria (Wynn and Liehr, 2001).

Microbial growth rate is determined by the availability of electron donors and acceptors, the amount of C and N, and environmental conditions (temperature, pH, space, etc.). While heterotrophs are responsible for ammonification, nitrification is inhibited when the DO concentrations drop below 2 mg/l. Conversely, the rate of denitrification is reduced in the presence of oxygen. Optimal conditions for bacteria growth are generally reported as being between pH of 6 and 9, and temperatures ranging from 15°C and 40°C. Growth of microbes still occurs outside of these ranges but the rates are reduced (Kadlec and Wallace, 2008).

Masi and Martinuzzi (2007) presented a comparison of the design and the performances of hybrid systems located in a medium scale. The hybrid constructed wetland (horizontal + vertical) SSF system, realized for the treatment of wastewater. This system was found proper for removal of pathogens and all the monitored chemical parameter (BOD, COD, TSS, NH_4, TN, and TP).

Dissolved biodegradable material is removed from the wastewater by decomposing microorganisms which are living on the exposed surfaces of the aquatic plants and soils. Decomposers such as bacteria, fungi, and actinomycetes are active in any wetland by breaking down this dissolved and particulate organic material to carbon dioxide and water. This active decomposition in the wetland produces final effluents with a characteristic low dissolved oxygen level with low pH in the water. The effluent from a constructed wetland usually has a low BOD as a result of this high level of decomposition (Kadlec and Wallace, 2008).

Ragusa et al. (2004) studied the microbial communities in sediments and within biofilms to study the development and activity of biofilms within laboratory bioreactors used to treat primary settled sewage. The results indicated that the biomass can take upwards of 100 days to be stabilized during operation of SSF wetlands and that the relative ratios of biomass components remain relatively constant during biofilm growth.

Vacca et al. (2005) investigated the effect of plants and filter materials on bacteria removal in six pilot-scale constructed wetland systems treating domestic wastewater: two vertical sand filters, two vertical expanded clay filters and two horizontal sand filters (each planted and unplanted). They showed that the microbial community in wetland system is strongly influenced by the filtration process, filter material, and plants.

Mburu et al. (2008) studied the performance of HSSF wetlands in removal of bacterial pathogens and TSS from domestic wastewater in tropical climates. The wetland received a continuous feed of settled sewage from primary facultative pond. They concluded that this type of wetlands can effectively remove fecal coliforms and total suspended solids in pretreated domestic wastewater under the tropics conditions.

Reinoso et al. (2008) studied combined wetland systems formed by facultative pond (FP), SF and SSF wetlands for about ten months for removal of bacteria and pathogenic parasites from wastewater. They concluded that, the type of wetland had a clear effect on the decay rate of the microorganisms being FP more efficient than SF and SSF in bacterial removal and SSF more efficient than SF and FP in protozoan parasites and coliphages removal with the exception of Giardia and Streptococci in SF.

Karathanasis et al. (2003) studied the vegetation effects on fecal bacteria (fecal coliforms and fecal streptococci) removal in 12 wetland systems treating domestic wastewater. The wetlands were monoculture systems planted with cattails and fescue; polyculture systems planted with a variety of flowering plants; and unplanted systems. They found that, no significant difference in the average yearly removal of fecal bacteria between systems, with the vegetated systems performing best during warmer months and the unplanted systems performing best during winter.

2.13 WETLAND ADVANTAGES AND DISADVANTAGES

Constructed wetlands systems have many benefits compared to conventional treatment techniques as they (Halverson, 2004; Thiyagarajan et al., 2006; Ulrich et al., 2005):

- Provide low cost, energy efficient treatment.
- Requires less skill operation.
- Tolerate fluctuations in flow and pollutant concentrations.
- Are capable of treating mixed contaminants.
- Have lower air and water emissions and secondary wastes.
- Provide habitat for plant and wildlife.
- Provide recreation and educational opportunities.
- Superior in removing microorganisms.

Literature Review

Also there are some limitations to use constructed wetland systems in treating wastewater as (Hawkins, 2002; Michele et al., 1998; Wynn et al., 1997):

- Wetlands generally require larger land areas than conventional wastewater systems, so adequate land must be available.
- The treatment wetland must have a pretreatment system.
- Wetlands may be relatively slow to provide treatment compared to more conventional treatment technologies.
- Performance may be less consistent than in conventional treatment system in cold climate weather.
- Surges in flow may temporarily reduce treatment effectiveness.
- Wetlands require a base flow of water for vegetation.
- They may provide breeding grounds for mosquitoes and pests.
- Long-term maintenance may be required.
- The biological components are sensitive to toxic chemicals.
- Wetlands are not a reliable treatment method for ammonia removal especially during the winter months.
- Anaerobic conditions might produce disagreeable odors.
- Wetlands require sapling management especially during start up operation.

The horizontal subsurface flow wetland treatment systems operate more efficiently in tropical regions (Rani et al., 2011). Reed bed type constructed wetlands are good for removing chemical and biochemical oxygen demands, and total suspended solids. But these treatment systems are not so effective for removing nitrogen and phosphorus compounds (McEntee, 2006).

2.13.1 SURFACE FLOW WETLANDS

The advantages and disadvantages of surface flow constructed wetlands for treating wastewater are summarized in Table 2.10 (Nelson et al., 2003).

2.13.2 SUBSURFACE FLOW WETLANDS

Table 2.11 presented the advantages and disadvantages of subsurface flow constructed wetlands for the wastewater treatment systems (Davis, 1995b).

Grove and Stein (2005) concluded that batch loaded constructed wetland systems can be expected to remove large quantities of three classes of nonhalogenated polar organic solvents from municipal wastewater. The

TABLE 2.10 Advantages and Disadvantages of Surface Flow Wetland

Advantages of SF Wetlands	Disadvantages of SF Wetlands
Less expensive to construct and operate and simpler to design than SSF wetlands and conventional treatment methods.	Lower rates of pollutant removal per unit of land area than SSF wetlands, thus they require more land to achieve a particular level of treatment.
Can be used for treating higher suspended solids wastewaters.	Requires more land than conventional treatment methods.
More operating data in the united states than for SSF wetlands.	Risk of ecological or human exposure to surface flowing wastewaters.
Offer greater flow control than SSF wetlands.	May be slower to provide treatment than conventional treatment.
Offer more diverse wildlife habitat.	Odors and insects may be a problem due to the free water surface.
Provides habitat for plants.	Particle resuspension due to wind, wave, or animal activity.

Source: After Nelson et al. (2003).

TABLE 2.11 Advantages and Disadvantages of Subsurface Flow Wetlands

Advantages of SSF Wetlands	Disadvantages of SSF Wetlands
Higher rates of pollutant removal per unit of land than FWS, thus they require less land to achieve particular level of treatment.	Requires more land than conventional treatment methods.
Lower total life-time and capital costs than conventional treatment systems. More accessible for maintenance because there is no standing water.	Variable treatment efficiencies due to the effects of season and weather. May be slower to provide treatment than conventional treatment.
Less expensive to operate than SF system. Provides habitat for plants and wildlife.	More expensive to construct than SF wetlands.
Minimal ecological risk due to absence of an exposure pathway. Odors and insects not a problem because the water level is below the media surface.	Waters containing high suspended solids may cause plugging. Sensitivity to high ammonia levels.
Media provides more surface area for bacteria biofilm growth than SF wetlands, resulting in increase treatment effectiveness.	Materials may not be readily available.

Source: After Davis (1995b).

three classes are a ketone, aliphatic and an aromatic represented by acetone, 1-butanol and tetrahydrofuran, respectively. Yamagiwa and Ong (2007) concluded that the up flow-constructed wetland with supplementary aeration is highly promising for on-site industrial wastewater treatment system.

2.14 IMPLEMENTATION OF CONSTRUCTED WETLANDS

The objectives of this section is to provide brief information on planning, design, construction, and operation and maintenance of horizontal subsurface flow wetlands for the purpose of water quality improvement (Kadlec and Wallace, 2008).

2.14.1 BASIC PRINCIPALS OF WETLAND PLANNING

It is recognized that SSF wetlands are not stand-alone treatment device but rather form part of an overall treatment process (Hoddinott, 2006). For HSSF wetlands primary treatment at a minimum is required to remove settle-able and floating solids prior to the wetland bed (Karu et al., 2000). Lee (1999) gave some ranges of wetlands design parameters (Table 2.12) in comparison with natural wetland systems. He summarized some basic principles of wetlands construction as:

TABLE 2.12 Range of CWs Design Parameters in Comparison With Natural Systems

Design Parameters	FWS Constructed Wetlands	SSF Constructed Wetlands	Natural Wetland Systems
Area (ha/1000 m^3/d)	2 to 4	1.2 to 1.7	5 to 10
q(cm/d)	2.5 to 5	5.8 to 8.3	1 to 2
Max. Depth (cm)	50	Below Media Surface	Depend on plant
Bed Thickness (cm)	Not available	30 to 90	Not available
Min. T_r(d)	5 to 10	5 to 10	14
Min. Aspect Ratio	1 to 2	2 to 5	1 to 4
Min. Pretreatment	Primary: Secondary	Primary	Primary: Secondary
Configuration of Cell	Parallel or Series	Parallel	Series
Distribution	Perforated Pipe	Large Gravel at Inlet	Perforated Pipe
Max. BOD (kg/ha/d)	100 to 110	80 to 120	4
Max. TSS (kg/ha/d)	Up to 150	Not available	Not available

Source: After Lee (1999).

- Design the system for minimum maintenance.
- The system has to use natural energies.
- Natural basin shape and uniform depths have to be maintained.
- Considering the surrounding lands and future land-use changes.
- Hydrologic conditions are paramount.
- Giving the system time to develop.
- Avoiding the construction of wetland on highly permeable soils.

2.14.2 DESIGN OF HSSF WETLANDS

Melton (2005) summarized the steps of sizing horizontal subsurface flow constructed wetlands as follows:

- Determining influent BOD concentration, average hydraulic load, and desired BOD concentration for the effluent.
- Selecting water level depth and fill medium size and type.
- Calculating the media porosity and select a length to width ratio.
- Determining the surface area required using Eq. (2.46).
- The removal of TSS in SSF wetlands has to be correlated to the hydraulic loading rate ($0.4 < q < 75$ cm/d, $22 < C_i < 118$ mg/l, $3 < C_o < 23$ mg/l) by Eq. (2.47) (Crites et al., 2006).
- Using Darcy's, Eq. (2.9) to determine if hydraulic conductivity of the medium is adequate.

$$A = \frac{Q(\ln C_i - \ln C_o)}{k_T \times h \times n} \tag{2.46}$$

$$C_o = C_i(0.1058 + 0.0011 \times q) \tag{2.47}$$

where: A = surface area of wetland, m²; Q = design flow rate, m³/d; C_i = influent concentration, mg/l; C_o = effluent concentration, mg/l; k_T = temperature dependent rate factor (k_T = 1.104 at T = 20°C), d⁻¹; h = average wastewater depth, m; q = hydraulic loading rate, cm/d.

In Eq. (2.46), the C represents the BOD concentrations but in Eq. (2.47) the C represents the TSS concentrations. The following Eq. (2.48) used in conceptual design to determine the maximum allowable flow to a specified media and geometry:

$$Q < \frac{K_c}{2}\left(\frac{h_i^2 - h_o^2}{L/W}\right) \tag{2.48}$$

where: K_c = hydraulic conductivity, m/d; L/W = aspect ratio.

The allowable head loss $(h_i - h_o)$ would typically be restricted to about 5 cm. The outlet depth would be set about 10 cm below the bed surface using water level control to allow for head loss and moderate rain events without flooding.

2.14.3 CONSTRUCTION OF HSSF WETLANDS

The higher length to width ratio of the wetland (aspect ratio) minimize short circuiting, and allows the wetland to closely perform plug flow hydraulics (Anurita, 2005). The design goal for influent flow in HSSF wetlands is the ability to distribute the flow as uniformity as possible across the section of the wetland bed, using V-notch weirs. For HSSF wetlands, flow collection is as simple as a perforated pipe extended across the width of the wetland cell (Kadlec and Wallace, 2008). The selection of a plant species in generally a function of two factors:

- The degree of rhizome spread and hence the ability to achieve plant canopy and crowd out unwanted invasive species.
- The development of more below ground root biomass and depth of root penetration.

2.14.4 OPERATION AND MAINTENANCE

Constructed wetlands are indeed characterized by low operation and maintenance costs and their operation requires only manpower (Toscano et al., 2009). Monitoring and adjustment of flows, water levels, water quality and biological parameters are the principal day-to-day activities required to achieve successful performance of these low-technology treatment systems.

Repair of pumps, levees, and water control structures; vegetation management; pest control; and removal of accumulated mineral solids, typically must be attended to less frequent intervals (Kadlec and Wallace, 2008). Three periods in the early life of a treatment wetland system are identified:

- Start-up phase: The start-up test for an individual flow-way begins when the previously mentioned samples demonstrate, over a four-week period, a noticeable reduction in the target pollutant.
- Stabilization phase: Performance during this period is not expected to be optimal because the wetland is not yet fully developed. The stabilization test is met when the long-term criteria for the facility are met.
- Routine operations phase: During the routine operations phase, the

wetland is deemed in compliance if the permit effluent limitation is being achieved.

1. Operation

Some operational requirements include (Kadlec and Wallace, 2008):

- Maintaining water levels in the wetland cells as appropriate for the vegetation, and below the surface of wetland media.
- Controlling flow into wetland in accordance with water budget.
- Monitoring treatment performance, collecting samples, and measuring flow rates into and out of the wetland regularly.
- Determining treatment efficiencies for use in adjusting wetland application rates.

2. Maintenance

Maintenance of a constructed wetland includes actions taken to prevent deterioration of the wetland components and to repair any damage as the inspect of (Kadlec and Wallace, 2008):

- inlet and outlet structures daily for plugging and damage.
- vegetation throughout the growing season and harvesting the plants periodically.
- pumps and piping systems as needed.
- dump the sedimentation basins.
- lining material for wetland cells.
- media clogging, in this case big quantity of fresh water is pumped to the system in a reverse direction.

2.15 APPLICATIONS OF NEURAL NETWORKS IN WETLANDS

Tomenko et al. (2007) compared between multiple regression analysis (MRA) and two artificial neural networks (ANNs); and multilayer perceptron (MLP) and radial basis function network (RBFN) in terms of their accuracy and efficiency when applied to prediction of BOD concentration at effluent and intermediate points of SSF wetlands. The data used in this work were obtained from various hydraulic and BOD loading of pilot units located in India and comprised of 91 patterns. MRA as well as ANN models were found to provide an efficient and robust tool in predicting constructed wetland performance. MLP and RBFN produced the most accurate results indicating strong potential for modeling of wastewater treatment processes.

Akratos et al. (2008) presented a model which can be used in the design of HSSF constructed wetlands. This model was developed based on experimental data from five pilot scale CW units, used in conjunction with ANNs. The CWs were operated for two years period under four different retention times. For the proper selection of the parameters entering the neural network, a principal component analysis (PCA) was performed first. From the PCA and model results, it was appeared that the main parameters affecting BOD removal were porous media porosity, wastewater temperature and retention time, and a set of other parameters which include the meteorological ones (barometric pressure, rainfall, wind speed, solar radiation and humidity). Two ANNs models were examined: the first one included only the three main parameters selected from the PCA, and the second included, in addition, the meteorological parameters. The first ANN predicted BOD removal rather satisfactorily and the second model examined resulted in even better predictions. From the predictions of the ANNs, a hyperbolic design equation, which combines zero and first order kinetics, was produced to predict BOD removal. The results of the ANNs and of the models design equations were compared with the available data from the literature, and showed a rather satisfactory performance. COD removal was found to be strongly correlated to BOD removal. An equation for COD removal prediction was produced.

Also, Akratos et al. (2009) examined if ANNs can predict nitrogen removal with the same above models entering parameters. The first model could predict TN removal rather satisfactorily ($R^2 = 0.53$), and the second model resulted in even better predictions ($R^2 = 0.69$) and gave a design equation for TN removal prediction.

Yalcuk (2012) developed an artificial neural network model to represent phenol removal in vertical and horizontal constructed wetlands. For this aim, a pilot scale horizontal-flow (planted and unplanted) and three vertical-flow (planted and unplanted) reactors made of PVC were designed. In these reactor systems two wetland plants, *Typhalatifolia* and *Cyperusalternatifolius*, and different bedding media (sand, zeolite, thin zeolite, and pebble) were used. A feed-forward network was used and fed with two subsets of operational data. The training procedure for effluent phenol concentrations from different wetlands was quite successful; a perfect match was obtained between measured and calculated concentrations.

2.16 STATISTICAL MODELING AND COMPARISON IN WETLANDS

Mays and Edwards (2001) compared between constructed and natural wetlands for accumulated metalsin sediments and plants. Results indicated that removal efficiencies of most metals (Mn, Zn, Cu, Ni, B, and Cr) were greater in constructed than in natural wetlands.

Ouellet-Plamondon et al. (2006) tested the contribution of artificial aeration on pollutant removal in summer and winter, with a combination of planted, unplanted, aerated and non-aerated wetland treating a fresh water fish farm effluent. Artificial aeration slightly enhanced TSS removal in all seasons regardless of treatment. In winter, the reduction in COD removal in nonaerated compared to summer was totally compensated for in aerated, in both planted and unplanted units. Artificial aeration improved summer and winter TKN removal efficiency for unplanted units, but the additional aeration did not fully compensate the absence of plants; they suggested that the role of macrophytes goes beyond the sole addition of oxygen in the rhizosphere. Artificial aeration also improved TKN removal in planted units, but to a lower extent than for unplanted units.

Zhang et al. (2010) studied the effect of limited artificial aeration on domestic wastewater treatment in CWs, four pilot-scales HSSF were operated from October 2006 to September 2007. The four units included aerated and planted, planted, aerated, and constructed wetland, and all the units have the identical dimensions of 3 m length, 0.7 m width and 1 m depth. The automated aeration was activated when the oxygen concentrations in the units were lower than 0.2 mg/l and ceased when the oxygen concentrations in the CWs were higher than 0.6 mg/l. More stable alkaline pH values and effective pollutant removal were found in aeration units than that in the nonaeration units. There were no significant differences(using SPSS) in TP removal between the aeration units and nonaeration units.

Mowjood et al. (2010) compared between the performance of vertical subsurface flow and horizontal subsurface flow wetlands at laboratory scale at tropical conditions and evaluated the effect of loading rate on treatment capacity of wastewater parameters (BOD, TSS, NO_3, PO_4, NH_4, FC and TC). Six wetland models of size 1.4 m length, 0.5 m width, and 0.5 m depth were constructed and arranged: two models as VSSF system with plants, two models as HSSF system with plants, one model as a VSSF without plants, and one model as a HSSF without plants. The used plant was Cattail with gravel media (10 to 20 mm diameter) and synthetic wastewater with average concentrations of 29.51, 3.22, 15.14, and 6.78 mg/l for BOD, NO_3, NH_3, and PO_4, respectively with Fecal and total coliforms equal to 495,120 and 915,500 counts/100 ml. The loading rate was increased from 2.5–25 cm/day at 12 days interval during 2.5 months period. Sampling was carried out from both influent and effluents of each wetland system after 12 days of constant flow rate and wastewater quality parameters were measured. Results demonstrated that VSSF systems perform better treatment than horizontal systems, but the treatment performance declined with the increasing loading rate in all six wetland models.

Babatunde et al. (2011) used dewatered alum sludge as a main substrate in a pilot on-site CW system treating agricultural wastewater for 11 months. Treatment performance was evaluated and SPSS software was used to establish correlations between water quality variables. Results showed that removal rates (loading rate multiplied by the difference in concentration between the influent and effluent, $g/m^2.d$) of 5:249 for BOD, 36:502 for COD, and 3:15 for PO_4 were achieved.

Ewemoje and Sangodoyin (2011) developed a pilot scale HSSF constructed wetland for treatment of primary lagoon effluent. The SPSS software was used for statistical analysis between three vegetated cells and one non-vegetated cell. Cells would be packed with 5–10 mm diameter pea gravel while 20–50 mm diameter pea gravel was used at the inlet and outlet areas to prevent clogging. Sampling was twice a month for a period of 2 years.

KEYWORDS

- **Egypt**
- **HSSF wetlands**
- **wetland functions and values**
- **wetland hydraulics**
- **wetland types**
- **wetlands definitions**
- **wetlands modeling**

CHAPTER 3

FIELD AND EXPERIMENTAL WORKS

CONTENTS

3.1	Introduction	69
3.2	Description of Samaha WWTP	70
3.3	Physical Model for HSSF Constructed Wetlands	74
3.4	Measuring Arrangements	82
3.5	Experimental Procedures	93
3.6	Reed Bed Establishment	94
3.7	Water Sampling	98
3.8	Water Quality Analysis	100
Keywords		109

3.1 INTRODUCTION

Generally, physical model supported by experimental and field data gives reliable principles for the specialists. Since HSSF constructed wetlands supported by uncommon biofilm carriers are relatively new to unreliable field data, hence experimental work has found a scope in the present project. Also, the aim of this project is to improve the treatment efficiency of the constructed wetland built in Samaha village.

The experiment was carried out on a field scale subsurface flow constructed wetland system exists in Samaha wastewater treatment plant (WWTP). Some modifications have been conducted on Samaha WWTP to serve different purposes in the present research work. A comparative study among the used media; rubber, gravel, and plastic to select the suitable one which gives the best efficiency in HSSF wetlands treatment system. The HSSF wetland cells were established and tested for two stages; the first one is the set up stage (the system was in the start of operational state) and the other is the steady stage conditions (the system reached to the stability state of biofilm growth, plants maturation and uniform flow).

This chapter presents the general description of Samaha wastewater treatment system, the different parts of the physical model, the measuring devices, and the arrangements for experiment requirements, as well as properties of the used media. The chapter also describes the experimental procedures, the plants cultivation process, the water sampling, the used laboratory tests and measuring instruments, the modeling of experimental data, and finally equations used for hydraulic and performance evaluation.

3.2 DESCRIPTION OF SAMAHA WWTP

The subsurface flow wetlands system for treating wastewater is located at Samaha village in Aga district of Dakahlia governorate, about 100 km northeast of Cairo as illustrated in Plate 3.1. Samaha is a small scale Egyptian village that has about 7000 inhabitants, Plate 3.2. There is a drinking water infrastructure provided through pipelines and groundwater hand pumps. There were no sewage treatment facilities before the project construction.

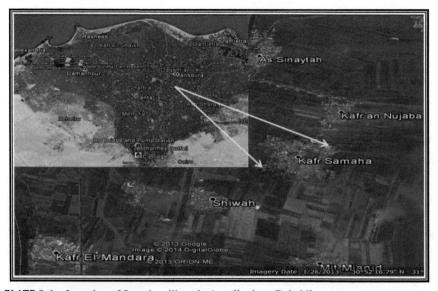

PLATE 3.1 Location of Samaha village in Aga district – Dakahlia governorate.

Field and Experimental Works 71

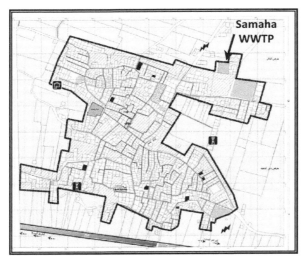

PLATE 3.2 Location of Samaha wastewater treatment plant.

The only sanitation facilities available were some private septic tanks with small storage capacities. Those tanks provided only primary treatment, and at some times the overflow seeped and infiltrated into the shallow water table causing contamination of the groundwater (NAWQAM, 2002).

To solve the sewage waste problems of the village, an environmental friendly solution was proposed and designed to cope with the future population levels, expected to reach about 10,000 capita by the year 2020. The project consists of the following components, Figure 3.1:

- Group of settling tanks (each house in the village has its own sewage collection tank; this tank acts as a settling facility to minimize the suspended solids of the sewage effluent and to provide anaerobic treatment for organic loads).
- Collecting network (PVC pipes with a number of concrete manholes, which collect the primary treated wastes at nine collecting tanks with an average size of 1.5 m wide × 6.0 m long and has the capability to collect a discharge of 12 h; sediments and sludge are collected regularly every 3 to 4 months from the collecting tanks).
- Alum adding unit (at the pump station, small adding Alum unit was considered to significantly reduce the suspended solids and total coliforms bacteria).
- Pump station (two submersible pumps work on an average discharge of 20 l/s for each, and has a big sump with enough capacity to collect

72   Constructed Subsurface Wetlands

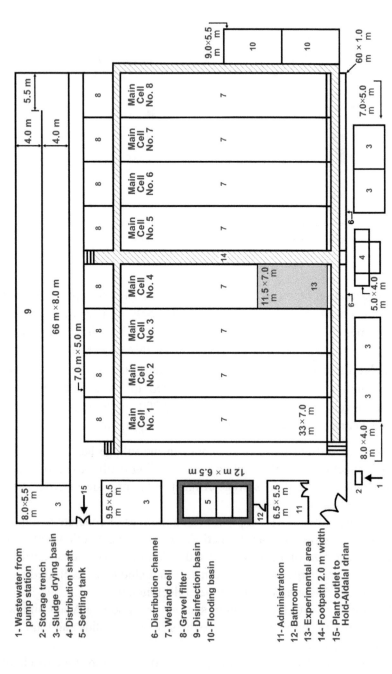

FIGURE 3.1 General layout of Samaha Wastewater treatment plant and the experimental area.

one day sewage discharges to feed the wetlands with the primary treated sewage).
- Subsurface wetland cells (a distribution channel 60 m length feeding eight gravel bed 33 × 7 × 0.75 m are located 400 m away from the pump station and consists of reinforced concrete walls lined with 350 micron plastic sheets and coated with 10 cm plain concrete; the gravel bed consists of 0.65 m thick regular medium size gravel of the type used in concrete construction works and cultivated with rhizomes root plant as papyrus; the retention time of each cell is about 24 h).
- Eight vertical gravel filters (each one 7.0 m × 5.0 m and contains 50 cm coarse gravel at the bottom covered with 20 cm fine gravel and 90 cm medium sands were located above; and the retention time is about 4 to 6 h).
- One disinfection basin (concrete tank 66 m long × 8 m wide × 1.5 m deep is located downstream the vertical gravel filters and is divided into two lanes to double the water pass as a tool for more aeration, it depends only on the sunlight, and the retention time is about 24 h then its water connected to outfall pipe to a private drain).

The average performance of Samaha wastewater is illustrated in Table 3.1 for years 2008, 2009, and 2010. The data in this table is the average concentrations of influent to wetland treatment cells and effluent from it. Description of main parameters and cost (in year of construction, 2001) for the project is presented as follows:

- The design capacity is about 1000 m³/day, while the actual discharge is about 1104 m³/day, USAID (2010).
- Total construction cost is about 600,000 Egyptian Pounds (LE), while the cost of a conventional system for the same treatment size is 3,000,000 LE for secondary treatment.

TABLE 3.1 Water Quality Parameters for Samaha Wastewater Treatment Plant

Water Characteristics	Year 2008 Influent	Year 2008 Effluent	Year 2009 Influent	Year 2009 Effluent	Year 2010 Influent	Year 2010 Effluent
pH	7.42	7.84	7.92	7.72	6.81	7.26
DO, mg/l	0.0	4.4	0.0	4.2	0.0	4.0
TSS, mg/l	345	34	206	39	180	46
COD, mg/l	477	57	334	69	295	75
BOD, mg/l	312	31	201	44	178	53
Ammonia, mg/l	27.3	10.8	39.14	15.36	22.6	9.78

- Area of the project is 4200 m² and area of a conventional system of the same treatment size is 12,600 m² for secondary treatment.
- Monthly operation and maintenance costs are less than 1000 pounds and the cost is 16,000 LE for conventional wastewater treatment plant.
- Manpower is only required for operation and maintenance, two full time technicians and one part time engineer.
- The plant produces bulrushes papers as a product (El-Zoghby, 2010).

3.3 PHYSICAL MODEL FOR HSSF CONSTRUCTED WETLANDS

With the increase of the entering wastewater discharge to Samaha plant than the design capacity; it causes deterioration in the plant treatment performance. The discharge of poorly treated wastewater to Hod-Aldalal drain created negative environmental impacts, which deteriorated its good quality water. Cooperation between Dakahlia Potable Water and Sanitary Drainage Company and Faculty of Engineering, El-Mansoura University, has been conducted to find some applicable solutions for raising the plant pollutants removal efficiencies. To improve treatment efficiency, a physical model was built in the first third of wetland cell number 4 in Samaha plant for examining the three different media for treatment.

The model inlet water was placed after the settling tanks of the original plant. So, the experimental cells of the model received primary treatment wastewater of Samaha plant. This section mainly deals with the establishment of wetland treatment cells, inlet and outlet zones, and the used media.

3.3.1 WETLAND TREATMENT CELLS

Experiments were carried out in three similar parallel field-scale horizontal subsurface flow constructed wetlands. All the units have a rectangular shape with identical dimensions of 10 m × 2.0 m × 0.65 m (length × width × depth). The aspect ratio is 5:1 which was taken as Samaha wetland cells. The wetland cells were built from bricks and lined several times with impermeable fabricated liner material in order to avoid seepage between cells and infiltration into groundwater to prevent groundwater contamination.

Wetland cells have a small bed slope which decreases runoff velocity. A decreased runoff velocity allows media particles and adsorbed pollutants to settle out. The experimental setup for treatment wetlands and the corresponding arrangements are illustrated in Figure 3.2. Plate 3.3 shows different stages of

Field and Experimental Works

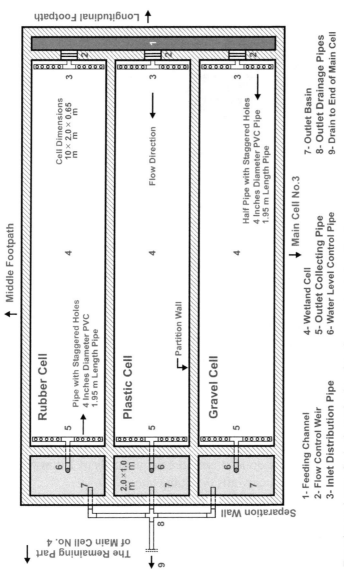

FIGURE 3.2 Experimental setup for horizontal subsurface flow constructed wetlands.

(A) Initial Stage

PLATE 3.3 Different stages for cells construction. (A) Initial Stage, (B) Transitional Stage, (C) Final Stage 1.

cells construction. Basin number four was chosen for two reasons, first it is close to the middle foot-path that helped in carrying construction materials and equipments to the experimental location, and the second one is to facilitate sampling. All works were conducted according to plan, budget, and engineering drawings.

The selected dimensions for treatment cells were underwent to the following conditions:

- *Cell depth*: the available main cell depth after lining was 65 cm, so the media depth was taken 50 cm plus 10 cm cover and 5 cm was taken as a freeboard. This is matching with the common used depths in horizontal subsurface flow wetlands in moderate climate regions also suitable for plant root penetration.
- *Cell width*: as the three selected media were chosen for this study (shredded tires as rubber media, graded gravel, and small pieces of hollow electricity corrugated pipes as plastic media), so the main basin width (7.0 m) was divided into three equal cells of 2.0 m wide separated by two partition walls 0.5 m thickness.
- *Cell length*: Samaha main cells have aspect ratio approximately equal to 5:1. The three treatment cells have the same ratio with length and width equals 10 m and 2 m, respectively.

3.3.2 INLET ZONE CONFIGURATIONS

The cell inlet purpose is to spread and regulate the wastewater evenly across the bed for effective treatment. As illustrated in Plates 3.4 and 3.5, inlet control structure consists of three main parts as:

Field and Experimental Works 77

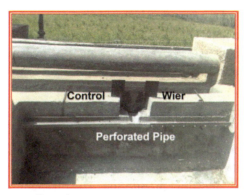

PLATE 3.4 Inlet zone configurations.

PLATE 3.5 Inlet zone media.

- Flow control weir, which receives wastewater from the main distributing channel of Samaha treatment plant. This control weir discharges at the mid width of each wetland cell. The selected notch shape is suitable for the small discharges of the experiment.
- The inlet distribution pipe is 100 mm diameter, fabricated from polyvinyl chloride (PVC). Its length is 1.95 m (half pipe with staggered drilled holes 1.0 cm diameter). The inlet pipe acts as a homogenous inflow distribution tool along the cell width. In addition, this pipe enhances aeration for the inlet wastewater. The lower part of this pipe was on the top surface of the treatment bed media.
- Large size inlet gravel (about 40 to 60 mm diameter) is used in the front section of wetland cell in order to limit the potential of clogging and to facilitate the inlet stream. The upper width of coarse gravel is taken as 1.0 m and the lower one is 1.5 m to cope with gravel angle of internal friction.

Plate 3.5 shows the separation between the inlet gravel zone and the cell media zone by the plastic screen (3 mm thickness and holes 1.0 cm diameter) to keep cell dimensions. Also, this plate illustrates the sampling holes and porosity buckets.

3.3.3 OUTLET ZONE CONFIGURATIONS

The outlet zone controls the depth of water in the wetland cell; collects the effluent water without creating dead zones in the wetlands; and provides access for sampling and flow monitoring. The different components of the outlet zone are as follows:

- Outlet collecting pipe is 100 mm diameter with staggered drilled holes (1.0 cm diameter) located at the end bottom of the treatment cells. Perforated pipe consists of a T-junction which is placed in the middle part and connected to two pipes carrying the flow to a water level control system, as illustrated in Plate 3.6. A gravel collecting outlet zone similar to the inlet zone is placed at the cell end to regulate the flow of water from the basin through the end perforated pipe.
- The three experimental cells were provided with water level control system (WLCS), which consists of a movable elbow attached to PVC, pipe (100 mm diameter) located in the outlet basin for controlling water level

PLATE 3.6 Outlet collecting pipe.

Field and Experimental Works 79

(the wastewater flows approximately at full bed depth from the start till it reaches the outlet at the beds bottom). Plate 3.7 shows the components of water level control system. This WLCS was the most proper tool to adjust water surface in the wetland cells (practical choice) compared with using flexible pipe or exchangeable pipe of different lengths.

- Outlet basin receives effluent water from WLCS to outlet collector pipes (2.0 m length, 1.0 m width, and 0.70 m depth). The outlet basin plays a double role by collecting effluent water samples and helping WLCS to adjust water depth of treatment cells.
- Outlet drainage pipes (three pipes from outlet basins connected by two elbows on left and right basins and by cross-junction at the

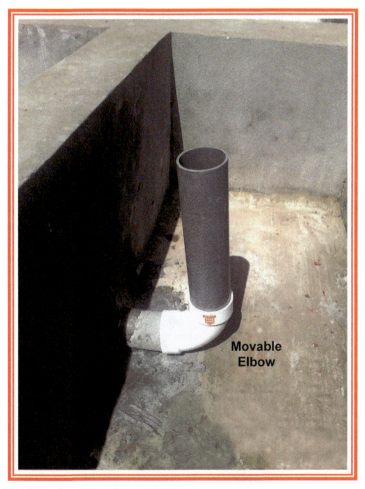

PLATE 3.7 Water level control arrangement.

middle one), which drains water to the end of main cell of Samaha WWTP through gravity pipes (22 m length). All used pipes and connections are 100 mm diameter, Plate 3.8.

3.3.4 TREATMENT MEDIA

Three types of attached biofilm carrier were used as a supportive medium for subsurface flow constructed treatment wetlands. The first selected media was rubber made from shredded tires, Plate 3.9. The destination of used tires has been delineated as a great environmental problem, as these are not degrade the environment, and thus can not be disposed in landfills and end up accumulating in rivers and public designations or burned releasing contaminated gases into the atmosphere.

The rubber media was obtained and chipped into small pieces (about 30:60 mm length, 25:55 mm width, and 5:15 mm thickness). The second studied media was made of corrugated hollow plastic pipes 50 mm length and 19 mm diameter (plastic media, Plate 3.10). The main problem for the use of plastic or rubber media was the floating of plastic or shredded tires pieces while emerging with wastewater, damaging the fixation of cell plants roots. To overcome these problems, plastic screen was placed on the media surface and covered with 10 cm coarse gravel.

PLATE 3.8 Details of outlet collector pipes.

Field and Experimental Works 81

PLATE 3.9 Sample of used shredded tires.

PLATE 3.10 Plastic media.

New plastic pipes were used in this study to save time. Waste plastic pipes are available in large quantities in construction companies could be recycled as media for reducing the cost. Finally, the natural gravel was used as the third bed media (washed gravel). The supportive media was stratified by coarse gravel (40:60 mm diameter) layer at the bottom, medium gravel (20:40 mm diameter) layer in-between, and fine gravel (less than 20 mm diameter) layer at the top. Each layer had a thickness of 16.7 cm. Plate 3.11 shows the natural graded gravel.

(A) Coarse Gravel (B) Medium Gravel (C) Fine Gravel

PLATE 3.11 Samples of the used gravel. (A) Coarse Gravel; (B) Medium Gravel; (C) Fine Gravel.

The gravel cell also covered with 10 cm coarse gravel for symmetrical shape of the other two treatment cells. Figure 3.3 illustrates three columns from media cells (the bed media after inlet zone coarse gravel).

3.4 MEASURING ARRANGEMENTS

Three important measurements were required for this experimental work, discharge, porosity, and media surface area. Some arrangements were made

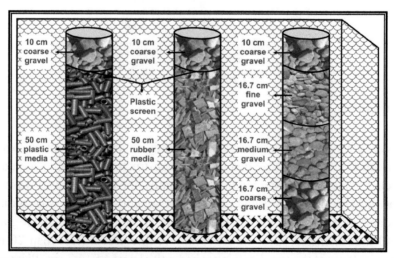

FIGURE 3.3 Three columns from media part of treatment cells.

Field and Experimental Works

in design to measure these variables. In addition to boring holes distributed over the treatment cells for sampling, and water depth control system as discussed before, Plates 3.4, 3.5, and 3.7, respectively.

3.4.1 INFLOW MEASUREMENT

The system should be initially designed for the average design flow and the impact of peak flows evaluated (Crites et al., 2006). To obtain the average discharge suitable for the available area of each treatment cell, the following assumptions were considered:

- Realizing an average retention time of 24 h as in Samaha treatment cell.
- The media porosity is about 0.5 then the water volume will be 5.0 m^3 (of 10 m^3 net treatment media volume) and the mean inflow rate amounted to 5.0 m^3/day.

The discharge of each cell was measured by 300 V-shaped sharp edge notch which is suitable in case of small discharge to get an acceptable head over the apex of the notch. This notch is shown in Plate 3.12. The notches were made of steel, 6 mm thickness and manufactured in the Faculty's workshop and painted several times to resist rust.

PLATE 3.12 V-notch for measuring flow rate.

The notch consists of two parts, one is a stationary part and the other is a movable one to control the flow rate. Figure 3.4 illustrates these two parts and their dimensions. Vertical scale with accuracy of 1 mm was fixed along the moving part for measuring the water head. Also a rubber sealing was fixed on this part to be watertight. The notch equation for computing discharge may be written as the follows (Anurita, 2005):

$$Q_{act} = \frac{8}{15} Cd \sqrt{2g} \tan\frac{\theta}{2} H_d^{5/2} \tag{3.1}$$

If $g = 981$ cm/s² and $\theta = 300$ then:

$$Q_{act} = 6.33\, Cd\, H_d^{5/2} \tag{3.2}$$

$$Q_{th} = 6.33\, H_d^{5/2} \tag{3.3}$$

where: Q_{act} = actual discharge, cm³/s; Q_{th} = theoretical discharge, cm³/s; Cd = coefficient of discharge, unit less; g = acceleration due to gravity, cm/s²; θ = apex angle of V-notch, unit less; H_d = head over the notch, cm.

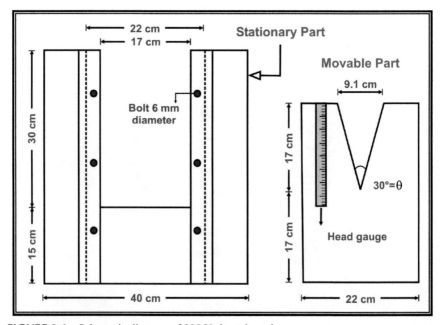

FIGURE 3.4 Schematic diagram of 300 V-shaped notch.

Field and Experimental Works 85

The notches were calibrated at the irrigation and hydraulics laboratory Faculty of Engineering El-Mansoura University, to compute the coefficient of discharge (Cd). For the calibration process the following steps are carried out:

- The movable part of notch is fixed at the end section of the laboratory flume (12 m length, 0.4 m width, and 0.4 m height) of a recirculating type, Plate 3.13.
- The flume regulated valve is opened to give the required flow.
- After few minutes of flume operation, a steady flow of water over the notch is established.
- The actual discharge (Q_{act}) is measured by means of measuring tank and stopwatch.
- The head over the notch is measured with a fixed head gauge.
- Applying this head in Eq. (3.3), the theoretical discharge (Q_{th}) is obtained.

The above steps are repeated 5 times (5 runs) for each notch with different regulated valve openings (i.e., different actual discharges). Then the values of actual and theoretical discharges are graphically plotted to obtain Cd values. The results of these runs are summarized in Table 3.2 for the used notches.

Figures 3.5–3.7 give the values of Cd for gravel, plastic, and rubber cell notches, respectively. From these figures the final notch equations for different wetland cells may be written as:

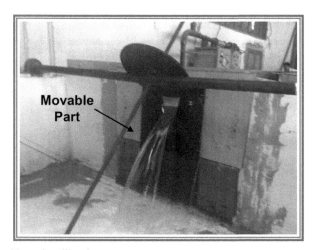

PLATE 3.13 V-notch calibration process.

TABLE 3.2 Results of Calibration for the Three V-Notches at Different Heads

Rubber Cell Notch			Plastic Cell Notch			Gravel Cell Notch		
Q_{th} (cm³/s)	Head (cm)	Q_{act} (cm³/s)	Q_{th} (cm³/s)	Head (cm)	Q_{act} (cm³/s)	Q_{th} (cm³/s)	Head (cm)	Q_{act} (cm³/s)
1808	9.6	1045	1715	9.4	1064	1333	8.5	788
2371	10.7	1472	2598	11.1	1548	2778	11.4	1748
2483	10.9	1567	3359	12.3	2137	2964	11.7	1853
3567	12.6	2375	3711	12.8	2375	3158	12.0	2261
5244	14.7	3895	4318	13.6	2973	3857	13.0	2793

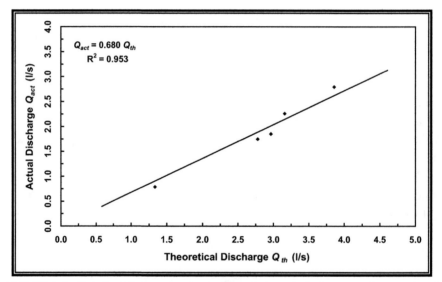

FIGURE 3.5 Relationship between the actual and theoretical discharges for gravel cell notch.

$$Q_{act}(gravel\ cell) = 4.30 H_d^{5/2} \quad (3.4)$$

$$Q_{act}(plastic\ cell) = 4.11 H_d^{5/2} \quad (3.5)$$

$$Q_{act}(rubber\ cell) = 4.37 H_d^{5/2} \quad (3.6)$$

Field and Experimental Works

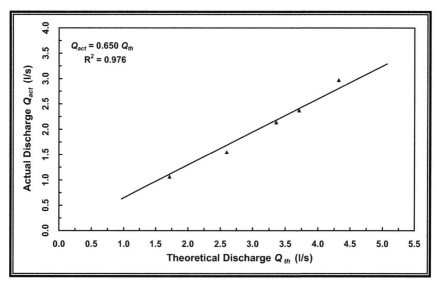

FIGURE 3.6 Relationship between the actual and theoretical discharges for plastic cell notch.

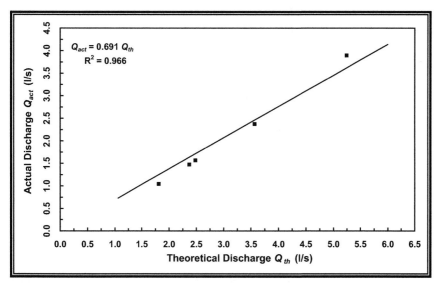

FIGURE 3.7 Relationship between the actual and theoretical discharges for rubber cell notch.

3.4.2 WATER DISCHARGES ALONG SET-UP AND STEADY STAGES

Through the set up stage, one discharge is applied for the three-wetland cells as this stage is an initial operation period. For steady stage (steady operation period) five decreasing discharges are used for cells. Each discharge is used for five consequence weeks (cycle discharge).

3.4.3 MEDIA POROSITY MEASUREMENT

In this study, an innovating method was adopted to measure the field media porosity. A porosity measuring apparatus was designed. A long-term porosity examination was carried out for the three treatment cells. The arrangements of this method could be explained as in the following points:

- Each cell was provided with three perforated cylindrical buckets with a solid base and side holes area smaller than the media size. These buckets were fabricated from steel (3 mm thickness) and painted several times to prevent rust.
- The cylindrical buckets were put in another bigger one. The outer bucket was centered and fixed by the surrounding media. These arrangements were distributed along the required locations in treatment cells, Plate 3.5.
- Inside each wetland cell, three porosity devices were located at distances of 2.5, 5.5, and 7.5 m from the inlet weir.
- The inner buckets were filled with bed media at the same gradation, and sequence found in the original cell.
- Three PVC pipes 15.24 cm diameter have been fixed in the cells side walls as illustrated in Plate 3.14 to receive the media buckets lifted from treatment cells. The short one is used to measure the porosity of the first lower layer. The pipe has a small drainage hole at 16.7 cm from bottom. The second pipe with drainage hole at 33.3 cm from bottom is used to measure the porosity of both first and second layers. The tall pipe measures the porosity of the three layers (the pipe has a small drainage hole at 50 cm from bottom).

The diameters of the nine cylindrical buckets used inside the treatment cells are presented in Table 3.3. Before operation of HSSF constructed wetlands system, the initial porosity of clean pretreatment media were measured by the following steps, Figure 3.8:

- Compute volumes of PVC pipes as (V_p = area of 15.24 cm pipe × height

Field and Experimental Works 89

PLATE 3.14 Pipes of media porosity measuring apparatus.

TABLE 3.3 Diameters of Perforated Media Buckets

Type of media	Diameter of media bucket at a distance		
	2.5 m from inlet	5.5 m from inlet	7.5 m from inlet
Rubber	11.30 cm	10.30 cm	10.80 cm
Plastic	11.00 cm	10.75 cm	10.50 cm
Gravel	10.50 cm	11.10 cm	11.00 cm

of drainage hole) (the first layer $V_{p1} = 3040.85$ cm^3, both the first and second layers $V_{p2} = 6079.88$ cm^3, and the three layers $V_{p3} = 9120.73$ cm^3).
- Calculate the volume of all buckets as (V_b = area of media bucket × height of layer) and Table 3.4 presents these values.
- The space volume between a fixed pipe and a media bucket is considered as $V_d = V_p - V_b$ (cm^3) at various layers and distances.
- Each cylindrical bucket was put in the short fixed pipe then water was added till drainage hole edge (V_w). The same steps are repeated with the second and tall fixed pipes. The added water is given by a 1000 cm^3 scaled bottle with accuracy of 10 cm^3.
- Calculate the volume of media voids as $V_v = V_w - V_d$ (cm^3).
- Finally, the porosity is obtained as $n = V_v/V_b$.

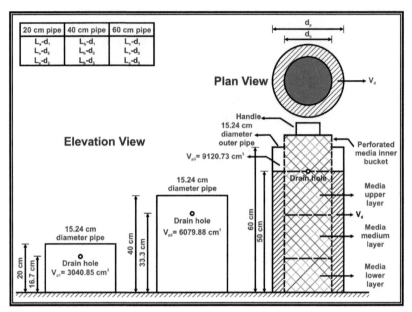

FIGURE 3.8 Sketch of porosity measuring apparatus.

TABLE 3.4 Volume of Media Buckets at Different Layers and Distances

Media Code	Volume of media bucket (V_b, cm³)		
	Rubber Media	**Gravel Media**	**Plastic Media**
L_a-d_1	1671.79	1443.46	1584.20
L_a-d_2	1388.99	1613.14	1513.01
L_a-d_3	1527.12	1584.20	1443.46
L_b-d_1	3342.58	2886.05	3167.46
L_b-d_2	2777.15	3225.31	3025.12
L_b-d_3	3053.32	3167.46	2886.05
L_c-d_1	5014.38	4329.51	4751.66
L_c-d_2	4166.15	4838.45	4538.13
L_c-d_3	4580.44	4751.66	4329.51

L_a = first layer, 16.7 cm height; L_b = first and second layers, 33.3 cm height; L_c = the three layers, 50 cm height; d_1 = first media bucket, at distance 2.5 m from inlet; d_2 = second media bucket, at distance 5.5 m from inlet; d_3 = third media bucket, at distance 7.5 m from inlet.

Eight media porosity runs were performed during the period from 5/9/2009 till 10/4/2010. Each run gave four values of porosity (coarse gravel, gravel, rubber, plastic media). The experiment was stopped when the difference between porosity results were small along time. This indicates

Field and Experimental Works

the end of experiment set up stage and the start of the steady conditions as the biofilm media growth reaches maturation stage. The detailed porosity calculations are tabulated in Appendix I.

3.4.4 MEDIA SURFACE AREA ESTIMATION

The surface area of the bed media in horizontal subsurface flow constructed wetlands is an important parameter in treatment functions. This surface area is controlling the volume of attached biofilm. These biofilms have a significant role in bio-degradation of pollutants such as BOD, COD, TSS, ammonia, phosphate, and fecal coliforms. But the surface area is related to many factors such as number of contact points; packing factor; the uniformity of biofilm thickness; size, shape, and gradation of media particles. Other factors such as the porosity of bed; the plant roots; as well as the dynamic of sediment accumulation at media voids affect media surface area. Approximate methods were applied for computing the surface area of each media as given in the following subsections.

3.4.4.1 Gravel Media

Assuming that, the one cubic meter of gravel will be divided into 12 equal parts according to the smallest and largest gravel sizes found in a random sample (four divisions for each of the three gravel sizes); each part will be computed as spheres according to its diameter as in Table 3.5. For each diameter, the surface area, the volume, and the specific surface area (surface area per unit volume) were calculated as given in the following equations (Taylor et al., 1990 cited in Cooke and Rowe, 1999):

$$A_p = \pi \times d_p^2 \tag{3.7}$$

$$V_s = \frac{1}{6} \pi \times d_p^3 \tag{3.8}$$

$$A_s = \frac{6(1-n_i)}{d_p} \tag{3.9}$$

where: A_p = surface area of sphere, m²; d_p = diameter of sphere, m; V_s = volume of sphere, m³; A_s = specific surface area, m⁻¹; n_i = initial porosity.

Values of specific surface area calculated in Table 3.5 are based on initial porosity equal to 0.431 as in Appendix (I). For 1.0 m³ of gravel media, the

TABLE 3.5 Calculations of Specific Surface Area For Gravel Media (1.0 m³ volume)

Type	Fine Gravel				Median Gravel				Coarse Gravel			
d, cm	0.5	1.0	1.5	2.0	2.5	3.0	3.5	4.0	4.5	5.0	5.5	6.0
A_s, m²/m³	683	341	228	171	137	114	98	85	76	68	62	57
Mean, m²/m³	1423/4 = 356				434/4 = 109				263/4 = 66			

corresponding surface area may be equal to $(356 + 109 + 66)/3 = 177$ m²/m³. Also, for 1.0 m³ of coarse gravel, the corresponding surface area may be calculated as $263/4 = 66$ m²/m³. Some of the ignored parameters may increase the specific surface area and the others may decrease it, assuming that the combined effect of these parameters will be in equilibrium and vanish each other.

3.4.4.2 Rubber Media

The shape of shredded tires pieces (rubber media) will be considered after practical measurements as a parallelepiped having dimensions of 30:60 mm length, 25:55 mm width, and 5:15 mm thickness. The average dimensions will be taken as 45 × 40 × 10 mm. To compute the rubber specific surface area, the following procedures were considered:

- A sample volume of 1300 cm³ was taken and number of rubber pieces was counted as it contained 32 pieces.
- One cubic meter of rubber media will contain ($10^6 \times 32/1300$) about 24615 pieces.
- Area of one piece = $2(4.5 \times 4.0 + 4.5 \times 1.0 + 4.0 \times 1.0) = 53$ cm².
- Finally, for 1.0 m³ of rubber media, the corresponding surface area may be equal to 130 m²/m³ ($24615 \times 53/10,000$).

3.4.4.3 Plastic Media

The plastic media has cylindrical shapes with an outer diameter of 19 mm. The thickness of each piece is about 0.5 mm. The total number of experimental plastic rounds was 312. Each round has a length of 40 m with a total length of 12,480 m. This length was cut to pieces of approximately 50 mm length. The specific surface area is obtained as follows:

- A tested volume of 134,900 cm³ contained 4600 pieces of plastic.
- One cubic meter of plastic media will contain ($10^6 \times 4600/134,900$) about 34100 pieces.

- One corrugated plastic piece was cut longitudinally and stretched. The surface area for the two faces was about 83 cm² (circumference = 5.97 cm, stretched length = 7.0 cm).
- Finally, for 1.0 m³ of plastic media, the corresponding surface area was about 283 m²/m³ (34100 × 83/10,000).

3.5 EXPERIMENTAL PROCEDURES

The planning and implementation of this physical model were started in March 2008 till august 2009. Several experimental preparations and procedures were carried out to construct the HSSF wetland system. Some of these managements were performed in outside and College workshops and the others in Samaha plant. A brief explain for these construction stages is mentioned as given in the following subsections.

3.5.1 WETLAND CELLS CONSTRUCTION

The work was started by building the separation wall inside the main cell after removing old media. The wetland cells and the outlet basins were built and lined several times with cement mortar and chemical adhesives to prevent seepage problems between cells. Many tools were prepared and fixed at the three cells to complete the experimental facilities. These tools included the following:

- Manufactured and painted notches. These notches were fixed at the center of each upstream side of wetland cell at the designed level.
- PVC perforated pipes for inlet and outlet zones.
- The water level control systems were prepared and established.
- Drainage pipes and their junctions located behind the separation wall till the end of main basin.
- Porosity buckets and sampling holes manufacturing, painting, and placing at the specific locations inside wetland cells.
- Preparation specific pipes for measuring the porosity at the side wall between cells.
- Suitable plastic screens (used for separation purposes between inlet and media zones, also between rubber and plastic media and the upper coarse gravel layer).
- Three wooden foot paths for plastic cell accessibility.

3.5.2 MEDIA ARRANGEMENT

In the inlet and outlet of the three treatment cells, 1.5 m^3 of coarse gravel was put to regulate wastewater flow. The coarse gravel in these zones was covered with plastic screen in order to keep the media position as designed. The graded gravel media was set in wetland cell as illustrated in Plate 3.15. Plastic and rubber media were put in the second and the third basins, as shown in Plates 3.16 and 3.17.

Plastic and rubber media were covered by wide plastic screens. A layer of gravel (10 cm depth) was then laid above these media to avoid floating, as shown in Plates 3.18 and 3.19. Also, 10 cm of gravel was placed on the top of the tire chips and plastic pipes to hold plants, as well as for safety. Gravel basin was covered also with the same upper layer for esthetic reasons, Plate 3.20.

3.6 REED BED ESTABLISHMENT

The treatment wetland system began to operate at the beginning of September 2009. The system was allowed to stabilize for about one month by flowing wastewater. After this stabilization period, the wetland cells were planted. The chosen plant species was *Phragmites Australis* (an emergent treatment

PLATE 3.15 Arrangement of gravel media.

Field and Experimental Works 95

PLATE 3.16 Plastic media arrangement.

PLATE 3.17 Rubber cell media.

PLATE 3.18 Covering plastic media with plastic screens.

PLATE 3.19 Covering rubber media with plastic screens.

plants that have high growth rates and can easily colonized), known popularly by common reeds, due to the availability of this plant in the surrounding areas of the experimental work.

In the preparation of saplings, the upper of the plant was removed leaving only about 0.30 m from the stem and rhizome of the plant, and were distributed in 3 lines, with 12 saplings each, aiming to achieve a stable population

Field and Experimental Works 97

PLATE 3.20   Final surface for the wetlands project.

of 36 saplings in each wetland bed (area = 20 m^2). Every sapling consists approximately of 10 rods. Plate 3.21 shows the beds soon after planting.

All planted basins were monitored, so that they are harvested (cut the above water stems) regularly every two to three months according to the growing seasons. Any invasive plants like ordinary grass were uprooted and removed immediately. Planting started with a density of 18 rods per square meter and was transplanted manually. After 2:3 months, roots were well developed and spread over the bed to a density nearly 200 rods per square meter. Plate 3.22 illustrates the state of cells after three months from planting, through this period the treatment performance was exactly improved.

PLATE 3.21   The reed bed establishment (initial stage – 5/10/2009).

PLATE 3.22 Reed bed after three months from planting (final stage – 5/1/2010).

3.7 WATER SAMPLING

This experimental work was performed at two main stages, the first stage was set-up operation (sampling period one) and the second one was steady-state conditions (sampling period two). Water samples were taken gently (with no turbulence) by 50 ml pipette. Every sample bottle was rinsed once with sample wastewater, filled to the top and capped. Bottled samples were, labeled and stored in an ice-cooled box and transported directly to the central laboratory, where analyzes are achieved for the required parameters.

3.7.1 SET-UP SAMPLING

After the cells cultivation, monitoring of the system had begun and continued for about 5 months for BOD, COD, and TSS parameters. The water samples were obtained bi-weekly from sampling holes in the system. Nine sampling well points (S_1 to S_9) were provided along each bed to collect water samples in addition to one inlet and three outlet samples.

Figure 3.9 presents schematic plan for wetland cell showing sampling points during set up stage. Water samples were collected manually in 500 ml cleaned polyethylene bottles. A total of 372 water samples were collected during this stage divided into 12 runs. It was noticed that, the variation in the transverse direction was small and the average value for the three sample points may be taken.

Field and Experimental Works

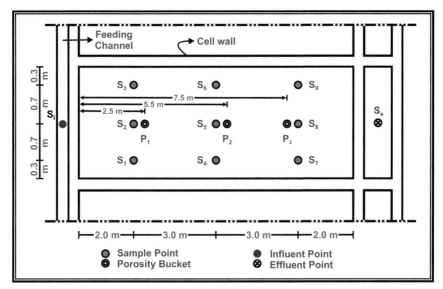

FIGURE 3.9 Plan view of water sampling and porosity measuring locations (set-up stage).

3.7.2 STEADY STAGE SAMPLING

Five water samples were collected during each sampling cycle. Sampling cycles were repeated 5 times (changing the flow rate) during the period from mid-April until mid-October, 2010. A total of 500 water samples were collected during this stage divided into 25 weeks.

The collected water samples were analyzed for BOD, COD, TSS (13 samples were taken, influent, 3 intermediate from each cell S_2, S_5, S_8, and 3 effluent). Influent and effluent samples were measured also for ammonia, phosphate, fecal coliforms, and some heavy metals. Effluent samples were obtained for dissolved oxygen. Water quality parameters such as pH and water temperature were recorded while taking water samples through the two stages.

For bacteriological (fecal coliforms) evaluation, the sample was 100 ml stored in sterilized bottle with sodium thiosulfate. Dissolved oxygen samples were collected in 300 ml glass bottles adding the required reagents in the field. For the heavy metals, the sample was 200 ml put in polyethylene bottle having HNO_3. For the remaining water quality parameters, samples were collected in 500 ml cleaned polyethylene bottles. Figure 3.10 presents schematic plan for wetland cell showing sampling points during steady stage.

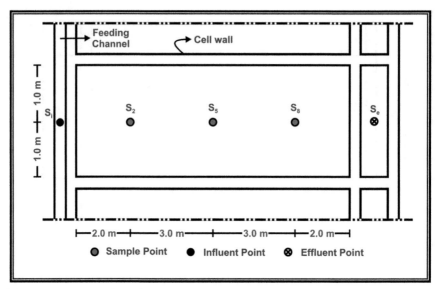

FIGURE 3.10 Plan view of water sampling points (steady stage).

Table 3.6 summarizes the wastewater samples number and locations for the two sampling periods. While sample handling and preservation are illustrated in Table 3.7.

3.8 WATER QUALITY ANALYSIS

The collected water samples were taken periodically from field to central laboratory of potable water and sanitary drainage company, Dakahlia Governorate. All analyzes were done according to standard methods (APHA, 1998). This section deals with the water analysis in brief, and the instruments used to provide these tests.

3.8.1 HYDROGEN ION

Measurement of pH is one of the most important tests in all phases of water and wastewater treatment. Aquatic organisms are sensitive to pH change, and biological treatment requires either pH control or monitoring, the role of pH in water chemistry is also associated with corrosively, alkalinity, and

Field and Experimental Works

TABLE 3.6 Wastewater Samples Number and Locations and Total Run Number

Stage Type	Water Quality Parameters	Samples Location in Treatment Wetland Cells	Samples Number	Run No.	Total
Set-up State	BOD COD TSS	1 – Influent + 3 – Effluent 9 – (S_1-S_9 Gravel Cell) 9 – (S_1-S_9 Rubber Cell) 9 – (S_1-S_9 Plastic Cell)	31	12	372
Steady-State	BOD – COD TSS – pH Temp.	1 – Influent + 3 – Effluent 3 – (S_2, S_5, S_8 Gravel Cell) 3 – (S_2, S_5, S_8 Rubber Cell) 3 – (S_2, S_5, S_8 Plastic Cell)	13	25	325
	Ammonia Phosphate Fecal Coliforms Heavy Metals	1 – Influent 3 – Effluent	4	25	100
	Dissolved Oxygen	3 – Effluent	3	25	75

TABLE 3.7 Samples Handling and Preservation

Pollutants Type	Container	Preservation
Heavy Metals	Polyethylene	Add HNO_3 to pH < 2.0 – Refrigerate
Fecal Coliforms	Sterilized Bottle	Refrigerate
Dissolved Oxygen	Glass Bottle	Acidification – Titration (Winkler Method)
TSS – BOD	Polyethylene	Refrigerate
Hydrogen Ion	Polyethylene	Analyze Immediately
Water Temperature	Polyethylene	Analyze in Site
Ammonia – Phosphate – COD	Polyethylene	Add H_2SO_4 to pH < 2.0 – Refrigerate

water softening or hardness, Plate 3.23 shows the pH meter (Cordesius and Hedstrom, 2009). The hydrogen ion test procedure can be summarized as:

- Calibrating the meter by dipping electrode in buffer solutions, (pH = 7.0 and 4.0).
- Stirring sample gently to insure homogeneity, then immerse electrode after washing by distilled water and record the reading.

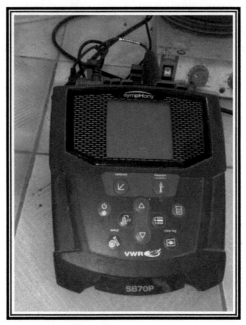

PLATE 3.23 pH meter, model SB70P, Europe.

3.8.2 DISSOLVED OXYGEN

The DO level in natural and wastewater depends on physical, chemical and biochemical activities. Analysis of dissolved oxygen is a key test for water pollution and wastewater treatment process control (Kadlec and Wallace, 2009). The DO test procedure can be arranged as follows:

- In 300 ml bottle, add 1.0 ml $MnSO_4$ solution and 1.0 ml alkali-oidide-azide reagent, then restopper and mix (in field).
- Add 1.0 ml concentrated H_2SO_4, restopper and mix by inverting several times until dissolution is complete.
- Measure 200 ml of the original sample and titrate it against 0.025 M $Na_2S_2O_3$ solution to a pale straw color and add a few drops of starch solution and continue titration till disappearance of blue.
- For titration of 200 ml sample, 1.0 ml 0.025 M $Na_2S_2O_3$ = 1.0 mg DO/liter.

$MnSO_4$ = *mangan (II) – sulfate – 1 – hydrat*
Alkali = *sodium hydroxide, NaOH*

Iodide = *sodium iodide, NaI*
Azide = *sodium azide, NaN_3*
H_2SO_4 = *sulfuric acid*
$Na_2S_2O_3$ = *sodium thiosulfate*
M = *molar*

3.8.3 BIOCHEMICAL OXYGEN DEMAND

The BOD indicates the amount of biodegradable substances in wastewater, and is widely used and recognized as an important parameter in wastewater treatment processes. It is a measure of the oxygen demand of microorganisms, when degrading organic matter in wastewater, at 200°C (Kopec, 2007). The BOD was calculated by measuring the DO content of a sample before and after 5 days incubation period. The BOD test procedure can be written as:

- Add 1.0 ml ($CaCl_2$-$FeCl_3$-$MgSO_4$-Phosphate buffer Solution)/1.0 L distilled water and aerated for 10:15 min.
- Add this solution to 5.0 ml sample (inlet) or 10 ml sample (outlet) in 300 ml bottle. This is due to the high organic matter in inlet samples. The samples at 2.0 m from notch are considered as inlet one, whereas the samples at 5.0 m and 8.0 m are taken as the outlet one.
- Prepare blank with distilled water. Incubate all the bottles for five days in the incubator at 20°C, Plate 3.24.
- After 5 days measure the dissolved oxygen concentration in the sample and blank bottles then calculate BOD from the following equation:

$$BOD\ (mg/L) = \frac{(DO_{blank} - DO_{sample}) \times 300}{Volume\ of\ sample} \quad (3.10)$$

$CaCL_2$ = *calcium chloride*
$FeCl_3$ = *ferric chloride*
$MgSO_4$ = *magnesium sulfate*
Phosphate buffer = *KH_2PO_4 + K_2HPO_4 + Na_2HPO_4 + NH_4Cl*
KH_2PO_4 = *potassium di hydrogen orthophosphate*
K_2HPO_4 = *di-potassium hydrogen orthophosphate unhydrous*
Na_2HPO_4 = *di-sodium hydrogen orthophosphate unhydrous*
NH_4Cl = *ammonium chloride*

PLATE 3.24   Incubator for BOD; model FC/FC111, Germany.

3.8.4   CHEMICAL OXYGEN DEMAND

The COD describes the total content of organic matter in a sample, and is given by the amount of organic material that can be oxidized by a strong chemical oxidant at high temperature, in a strongly acidic solution (Cordesius and Hedstrom, 2009). The COD is measured by the change of chromium from the hexavalent (VI) state to the trivalent (III) state, both of these chromium species are colored and absorb in the visible region of the spectrum (APHA, 1998). The COD test procedure can be mentioned as:

- Add 2.5 ml outlet sample or (1.0 ml inlet sample + 1.5 ml distilled water).
- Prepare blank with 2.5 ml distilled water.
- Add to all samples 1.5 ml $K_2Cr_2O_7$ and 3.5 ml H_2SO_4 reagent.
- Put in heater at 150°C for 2 h, Plate 3.24.
- After digestion, cool to room temperature.
- Use blank to calibrate the photometer, Plate 3.25.
- Record COD values for samples.

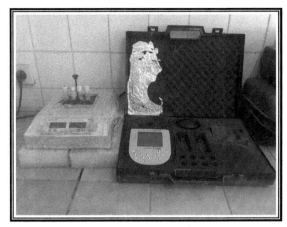

PLATE 3.25 Photometer, model 7100 and heater, model HB-1 (made in England).

$K_2Cr_2O_7$ = *potassium dichromate*
H_2SO_4 reagent = *sulfuric acid + silver sulfate(Ag_2SO_4)*

3.8.5 TOTAL SUSPENDED SOLIDS

Total suspended solids are solids in water that can be removed from the water sample by filtration. For adequate filtration to occur, the hydraulic conductivity of the bed must be large enough to allow the wastewater to contact the media (Kiracofe, 2000). High concentrations of total suspended solids entering the bed adsorption areas over time can cause it to clog which could lead to short circuiting, and failure of the system (Kopec, 2007). A well-mixed sample is filtered through a weighted standard glass-fiber filter and the residue retained on the filter is dried to a constant weight at 105°C. The increase in weight of the filter represents the total suspended solids (APHA, 1998). The TSS test procedure can be summarized as:

- Dry a filter paper in oven, Plate 3.26 at 105°C for 1.0 hr.
- Weight the dried filter paper W_1, in gm.
- Filter 100 ml sample by suction pump, Plate 3.27.
- Dry again in oven for 1.0 hr.
- Weight the paper W_2 (gm) then calculate TSS from the following equation:

$$TSS\ (mg/L) = (W_2 - W_1) \times 10000 \qquad (3.11)$$

PLATE 3.26 Drying oven, model UE400, Germany.

PLATE 3.27 Suction pump.

3.8.6 AMMONIA EXISTS

Ammonia occurs naturally in the environment as part of the nitrogen cycle. Ammonia is a decomposition product from urea and protein. High ammonia levels kill juvenile fish and other aquatic organisms (Wynn et al., 1997). The ammonia test procedure can be concluded as:

- Add to 10 ml sample, 3.0 drops nessler reagent.
- Wait 10 min and measure at 420 nm (nanometer) by using spectro photometer, Plate 3.28.

Nessler = $HgI_2 + KI + NaOH$
HgI_2 = *mercuric iodide*
KI = *potassium iodide*
NaOH = *sodium hydroxide*

PLATE 3.28 Spectro-photometer; model 4802 UV/VIS Double, USA.

3.8.7 PHOSPHATE

Phosphates are contributed to sewage by body wastes and food residues. Phosphate can be removed from the water column in wetlands by adsorption, sedimentation, or biological uptake. However, phosphate removal efficiency in wetlands is variable because phosphate can also be rereleased from wetland sediment to water body under certain conditions (Song et al., 2007). The phosphate test procedure can be summarized as:

- Add to 10 ml sample 3.0 drops reagent 1 + 3.0 drops reagent 2.
- Wait for 10 min and measure at 690 nm by using spectro photometer, see Plate 3.29.

Reagent 1 = *Stannous chloride, $SnCl_2.2H_2O$*
Reagent 2 = *Ammonium molybidate, $(NH_4)6\ MO_7O_{24}.4H_2O$*

3.8.8 FECAL COLIFORMS

There are 16 species of total coliforms (TC) found in soils, plants and in animal and human waste. A subgroup of coliforms, called fecal coliforms bacteria, is different from the total coliforms group because they can grow at higher temperatures and are found only in the fecal waste of warm-blooded animals. There are six species of fecal coliforms bacteria found in animal and human wastes. *Escherichia coli* (*E. coli*) is one type of the six species of fecal coliforms bacteria (Kopec, 2007). The fecal coliforms test procedure can be concluded as:

PLATE 3.29 Atomic absorption spectrometer; model ASS vario, Germany.

- Prepare dilutions of 10^{-1}, 10^{-2}, 10^{-3}, 10^{-4}, and 10^{-5} ml of the original sample (choose the best three dilutions series).
- Inoculate 15 tubes (each tube has 10 ml A-1 medium) by 1.0 ml diluted sample (5 tubes are used per dilution).
- Incubate at 35°C ±0.5°C for 3.0 hr then transfer to 44.5°C ±0.2°C for 21 hr.
- Count the number of positive tubes (gas production) from each set of tubes (each set consists of 5 tubes).
- Use the MPN index table (most probable number), to obtain the MPN/100 ml, APHA (1998).

3.8.9 WATER TEMPERATURE

Temperature exerts a strong influence on the rate of chemical and biological processes in wetland systems, including BOD decomposition, nitrification and denitrification. The rate of chemical reactions and biological activity increases with increasing temperature. Metabolism and growth of microorganisms are affected by this but only up to a certain level, after which the rate becomes lower. (Kayombo et al., 2004). Water temperature was measured during sampling periods using standard (0.0 to 100°C) thermometer accurate to 0.1°C.

3.8.10 HEAVY METALS

Metals are introduced into aquatic systems as a result of human activities. Although some metals such as Manganese (Mn) and Iron (Fe) are essential micronutrients, others such as Zinc (Zn), Cadmium (Cd), and Lead (Pb) are not required even in small amounts by any organisms. Virtually all metals, including the micronutrients are toxic to aquatic organisms as well as humans if exposure levels are sufficiently high.

Five heavy metals (Zn, Fe, Mn, Pb, and Cd) concentrations are measured by Atomic Absorption Spectrometer (AAS) (Plate 3.29). Influent and effluents at the end of treatment cells samples were analyzed for steady stage cycles. Table 3.8 illustrates the elements location and group in the periodic table in addition to the atomic number, the atomic weight, and the common valences for the five studied heavy metals elements (Periodic Table).

TABLE 3.8 Characteristics of the Studied Heavy Metals Elements (Periodic Table)

Element	Location	Group	Atomic Number	Atomic Weight	Valence
Zn	First	IIB	30	65.38	2
Fe	First	VIII	26	55.85	2 & 3
Mn	First	VIIB	25	54.94	2 & 7
Pb	Fifth	IVA	82	207.19	2 & 4
Cd	Second	IIB	48	112.41	2

KEYWORDS

- hollow plastic media
- media porosity
- rubber media
- samaha waste water treatment plant
- sampling
- treatment media
- water quality

CHAPTER 4

THEORETICAL APPROACH

CONTENTS

4.1	Introduction	111
4.2	Hydraulic Representation of Physical Model	112
4.3	Removal Rate Constants	116
4.4	Artificial Neural Networks Modeling	118
4.5	Set-Up Stage Data	122
4.6	Steady Stage Data	125
4.7	Statistical Modeling Applying SPSS	126
4.8	Calibration and Validation Processes for Models	127
4.9	Comparison Between Field, ANN, and SPSS Results	127
Keywords		128

4.1 INTRODUCTION

Two systems for modeling the input experimental data to estimate the effluent pollutant concentration will be followed in this study. The first one is the artificial neural networks (ANNs) and the second system is the statistical analysis using stochastic package for social science (SPSS) software. This chapter presents the hydraulic representation of the physical model, removal rate constants using plug and mixed flow models, and gives a brief description for the neural networks types and the training algorithms methods. The SPSS software is utilized to deduce a set of linear and nonlinear regression equations to represent the experimental data.

The chapter also contains the preparation of the experimental data for set up and steady stages and the methods used to increase the number of patterns that used in the ANNs and SPSS models. Finally, the calibration and validation processes for the obtained models and the equations that used for comparison between the experimental results and both the ANNs and SPSS model outputs are introduced.

4.2 HYDRAULIC REPRESENTATION OF PHYSICAL MODEL

This section introduces the hydraulic parameters used for the construction and computations of the physical model of the subsurface flow wetlands for both set up and steady stages.

4.2.1 HYDRAULIC LOADING RATE

The hydraulic loading rate for wetland cells (based on the input discharge, assuming no loss or gain of water, and surface area) is computed as follows:

$$q = \frac{Q_i}{A} \tag{4.1}$$

where: q = loading rate, m/d; Q_i = input discharge, m³/d; A = surface area of treatment cell, m².

4.2.2 HYDRAULIC RETENTION TIME

The hydraulic retention time is a basic factor for evaluating the treatment performance of any wetland system. The actual retention time at any distance from the cell inlet is written as:

$$T_r = \frac{V_{wx}}{Q_i} \tag{4.2}$$

where: T_r = hydraulic retention time, d; V_{wx} = volume of water inside cell at distance x, m³.

In this study two times are taken into consideration the first one is the time from start of operation, T_o (used in porosity measuring and its zero value means the porosity of used media before wastewater flow in the wetland cells), and the second time is the time from start of sampling, T_s (used in analysis of pollutants in set up stage, and its zero value means the first day that the water samples are taken from field and begins after 56 days from the start of operation).

Theoretical Approach

The water volume at any distance inside the wetland cell is calculated as the media volume multiplied by the media porosity at the specified time from start of sampling given by the following equation:

$$V_{wx} = V_{cg} \times n_{cg} + V_m \times n_m \tag{4.3}$$

where: V_{cg} = volume of coarse gravel, m³; n_{cg} = porosity of coarse gravel; V_m = volume of media, m³; n_m = porosity of used media.

As the porosity is measured at different times from start of operation, then for computing the porosity in-between these times (T_{oi}), the linear interpolation process is approximately considered for small difference between every two successive values. The T_{oi} value lies between two times from start of operation (T_{o1} and T_{o2}) and the porosity (n_i) also lies between two porosities (n_1 and n_2), Figure 4.1, and can be calculated by the following equation:

$$n_i = n_1 - (n_1 - n_2) \times \left(\frac{T_{oi} - T_{o1}}{T_{o2} - T_{o1}} \right) \tag{4.4}$$

where: n_i = porosity at time T_{oi}; n_1 and n_2 = porosity at time T_{o1} and T_{o2}.

4.2.3 POLLUTANT REMOVAL EFFICIENCY

For all pollutants in the two stages the removal efficiency is calculated as a percentage that measures the treatment rate through the media cells. This removal efficiency can be written as follows:

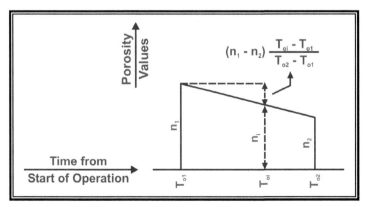

FIGURE 4.1 Schematic diagram for interpolation process for computing the porosity in-between times.

$$RE = \frac{(C_i - C_o)}{C_i} \times 100 \qquad (4.5)$$

where: RE = removal efficiency, %; C_i = influent concentrations; C_o = effluent concentrations.

4.2.4 ACTUAL WATER VELOCITY

This parameter computes the water velocity through the open perpendicular area which is affected by the change in the media porosity. The actual water velocity is given by the following equation:

$$v = \frac{Q}{nhW} \qquad (4.6)$$

where: v = actual water velocity, m/d; n = media porosity; h = average depth of water, m; W = width of wetland cell, m.

4.2.5 COMPARISONS BETWEEN MEDIA REMOVAL EFFICIENCIES

The following equations are used for computing the average difference in removal efficiencies between the three media through set up and steady stages to make comparisons between these media in treatment performance.

4.2.5.1 Average Difference (Set up Stage)

Equations (4.7) and (4.8) give the average difference between pollutant removal efficiency through 12 runs of the set up stage for plastic cell and both gravel and rubber cells. Whereas Eq. (4.9) gives this average difference between gravel and rubber cells.

$$\text{Average Difference (Plastic \& Gravel)} = \frac{\sum (RE_{xp} - RE_{xg})}{\text{No. of Runs}} \qquad (4.7)$$

$$\text{Average Difference (Plastic \& Rubber)} = \frac{\sum (RE_{xp} - RE_{xr})}{\text{No. of Runs}} \qquad (4.8)$$

Theoretical Approach

$$\text{Average Difference (Gravel \& Rubber)} = \frac{\sum\left(RE_{xg} - RE_{xr}\right)}{\text{No. of Runs}} \quad (4.9)$$

where: RE_{xp} = removal efficiency of plastic cell at distance x, %; RE_{xg} = removal efficiency of gravel cell at distance x, %; RE_{xr} = removal efficiency of rubber cell at distance x, %.

4.2.5.2 Average Difference (Steady Stage)

Equations (4.10) and (4.11) give the average difference of pollutant removal efficiency, through 5 discharge cycles between plastic cell and both gravel and rubber cells, whereas Eq. (4.12) gives this average difference between gravel and rubber.

$$\text{Average Difference (Plastic \& Gravel)} = \frac{\sum\left(RE_{xp} - RE_{xg}\right)}{\text{No. of Cycles}} \quad (4.10)$$

$$\text{Average Difference (Plastic \& Rubber)} = \frac{\sum\left(RE_{xp} - RE_{xr}\right)}{\text{No. of Cycles}} \quad (4.11)$$

$$\text{Average Difference (Gravel \& Rubber)} = \frac{\sum\left(RE_{xg} - RE_{xr}\right)}{\text{No. of Cycles}} \quad (4.12)$$

4.2.6 INLET AND OUTLET WEIGHTS OF SUSPENDED SOLIDS

Equations (4.13) and (4.14) give the weights of inlet and outlet suspended solids. The rate of sediment storage in the wetland cells through the steady stage is presented by Eq. (4.15).

$$W_i = Q \times C_i \quad (4.13)$$

$$W_o = Q \times C_o \quad (4.14)$$

$$R_s = \frac{W_i - W_o}{x_o - x_i} \quad (4.15)$$

where: W_i = weight of inlet suspended solids, gm/d; W_o = weight of outlet suspended solids, gm/d; R_s = rate of sediment storage, gm/m/d; C_i = inlet TSS concentrations, gm/m³; C_o = outlet TSS concentrations, gm/m³; x_i = inlet distance, m; x_o = outlet distance, m.

4.3 REMOVAL RATE CONSTANTS

The removal rate constant is one of the governing factors of pollutant treatment through constructed wetlands. This section listed the equations of both average removal rate constant (k) and average volumetric removal rate constant (k_v) for the steady stage pollutants in subsurface flow wetlands treatment system. The first order plug flow kinetics can be written as (Akratos et al., 2008; Chen et al., 2009):

$$\ln\left(\frac{C_o}{C_i}\right) = -\frac{k}{q} \tag{4.16}$$

$$\ln\left(\frac{C_o}{C_i}\right) = -k_v \times T_r \tag{4.17}$$

$$\ln\left(\frac{C_o - C^*}{C_i - C^*}\right) = -\frac{k}{q} \tag{4.18}$$

$$\ln\left(\frac{C_o - C^*}{C_i - C^*}\right) = -k_v \times T_r \tag{4.19}$$

The mixed flow models can be written with various possible formulae as (Liu, 2002):

$$\frac{C_o}{C_i} = \left(\frac{N}{N + k/q}\right)^N \tag{4.20}$$

$$\frac{C_o}{C_i} = \left(\frac{N}{N + k_v \times T_r}\right)^N \tag{4.21}$$

$$\frac{C_o - C^*}{C_i - C^*} = \left(\frac{N}{N + k/q}\right)^N \tag{4.22}$$

Theoretical Approach

$$\frac{C_o - C^*}{C_i - C^*} = \left(\frac{N}{N + k_v \times T_r}\right)^N \quad (4.23)$$

where: C_i = influent concentration; C_o = effluent concentration; C^* = background concentration; k = removal rate constant, m/d; k_v = volumetric removal rate constant, d⁻¹; q = hydraulic loading rate, m/d; T_r = hydraulic retention time, d; N = number of tanks.

The background concentration for BOD, COD, and TSS pollutants may be given as (Zurita et al., 2009):

$$C^*_{BOD} = C^*_{COD} = 3.5 + 0.053 C_i \quad (4.24)$$

$$C^*_{TSS} = 7.8 + 0.063 C_i \quad (4.25)$$

where: C^*_{BOD} = background concentration for BOD pollutant; C^*_{COD} = background concentration for COD pollutant; C^*_{TSS} = background concentration for TSS pollutant.

In cases that C^* is higher than the effluent concentration then its value is equal to zero. Two methods are used to determine the removal rate constants which represent the experimental data in steady stage as given in the following subsections.

4.3.1 GRAPHICAL METHOD

To estimate the average removal rate constants for these pollutants, relationship between $ln(C_i/C_o)$ or $ln(C_i-C^*/C_o-C^*)$ and $1/q$ is represented by drawings, getting three straight lines for plastic, gravel, and rubber media. These straight lines are forced to pass through the origin, so their gradients are the k values. In the same method, relationship between $ln(C_i/C_o)$ or $ln(C_i-C^*/C_o-C^*)$ and the retention time (T_r) is represented to get the volumetric removal rate constant (k_v) value.

4.3.2 CALCULATION METHOD

The average removal and volumetric removal rate constants for BOD, COD, and TSS pollutants are estimated by applying mixed flow models with C^* equal to zero, by substituting the inlet and outlet concentrations and the corresponding loading rate or retention time in these models (Eq. 4.20 or 4.21) getting k and k_v for each set of data then taking the mean value for these calculations. In the same manner, the removal rate constants are determined using the modified mixed flow model with C^* values (Eqs. 4.22 and 4.23). The number of tanks N that used in these equations is 4 for mixed and modified mixed flow models for all pollutants. More than N values were tested and the number N equal to 4 gave the best representation for the experimental data.

The background concentration values for ammonia, phosphate, and fecal coliforms pollutants are taken as zero and the removal rate constants for plug and mixed flow models are determined using the calculation method. The C^* values for heavy metals under study are taken as zero and the removal rate constants for plug flow models are determined using graphical method while the calculation method is used for mixed flow models.

4.4 ARTIFICIAL NEURAL NETWORKS MODELING

Artificial neural networks are parallel distributed processing systems. They are inspired from the networks of nerve cells (neurons) of the biological central nervous system. Similar to the brain, a neural network is composed of artificial neurons and interconnections. They are complex as computation devices that can accept a large number of inputs and learn solely from training samples (Du and Swamy, 2006; Samarasinghe, 2007).

The individual computational elements that make up most neural system models are called neuron or processing elements. Neurons are connected through a connection called connecting weight (Graupe, 2007). The actions performed by a biological neuron can be mapped in mathematical terms, Figure 4.2 (Demuth et al., 2009).

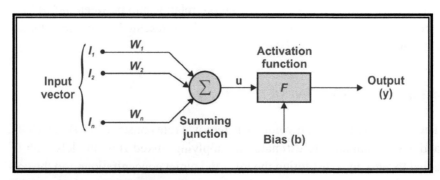

FIGURE 4.2 Simplified neuron model.

4.4.1 TYPES OF NEURAL NETWORKS

Artificial neural networks method is a recommended modeling system in case of many complicated phenomena associated with each other such as a system of horizontal subsurface flow constructed wetlands. The neural networks are composed of many simple neurons like processing units. These networks have the ability to recognize patterns of input and output information and making a group of complex relationships linking these input variables with each other and with the output values. Also, these ANNs have the potential ability to predict the value of a new output if they are fed with the corresponding input variables (Hu and Hwang, 2002).

A properly designed and trained neural network has the ability to provide reasonable results for a set of any possible data. The ANNs are classified into non-recurrent (feed-forward) and recurrent (involving feed back) networks depending on the neuron connections and the flow of data through different layers of the network (Khare and Nagendra, 2007).

The feed-forward network is used in this study in which, the information moves only in the forward direction. Feed-forward neural networks (FFNNs) are the most widely used networks because of the simplicity of their design. The FFNNs could be broadly classified into three categories, single/multilayer perceptron, radial basis networks, and self-organizing maps (Freeman and Skapura, 1991). The multilayer perceptron is the most well-known and most popular neural network among all the existing neural networks. This type of neural networks consist of three types of layers as follows (Hu and Hwang, 2002):

- Input layer is the layer which receives the problem inputs (independent variables) and its number of neurons equals to the number of inputs.
- Output layer is the layer that provides the outputs of the network. The number of neurons in the output layer equals to the number of network output (dependent variables). In this experimental work, the output layer represents the effluent concentration for each studied pollutant.
- Hidden layers are the intermediate layers between the input and output layers. The number of hidden layers and the number of neurons per each hidden layer is selected according to the complexity and the type of data.

In this study, different neural networks layers will be tested for each pollutant in set up and steady stages with input layer and one or more hidden layers to find out the output layer. These layers could be arranged as a general neural network as illustrated in Figure 4.3.

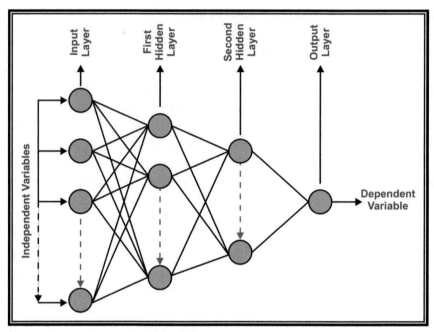

FIGURE 4.3 Schematic diagram for a general ANN.

The best neural networks structure in the study stages required two hidden layers and takes the form of N_i-N_1-N_2-N_o as:

N_i = number of input variables;
N_1 = number of neurons in the first hidden layers;
N_2 = number of neurons in the second hidden layers;
N_o = number of output variables.

Figure 4.4 shows a flow chart for the design of a multilayer feed forward neural network (MFFNN). For a good design of the neural networks, the following arrangements must be taken into consideration (Jain and Fanelli, 2000):

- Preparation of suitable training data set that represent cases of study, which the ANN needs to learn.
- Selection of a suitable artificial neural network structure for a given application purposes.
- Training of the selected artificial neural network.
- Evaluation of the trained ANN using test patterns until its performance is satisfactory.

Theoretical Approach

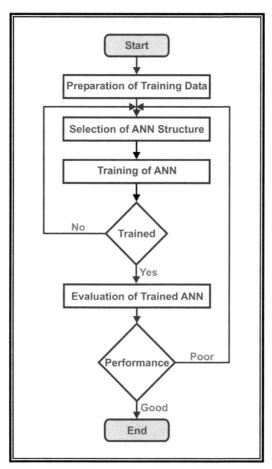

FIGURE 4.4 Flow chart of MFFNN.

4.4.2 TRAINING ALGORITHMS FOR ANNs

It is used in systems where a target output is available. A block diagram of a learning algorithms system is shown in Figure 4.5. The ANN learns during training by adapting to a dataset of inputs and the desired output corresponding to them. The network parameters such as the weights and biases are adjusted according to the error between the desired and obtained outputs in a closed-loop feed back type system (Dreyfus, 2005).

There are many learning algorithms of the artificial neural networks such as back propagation (BP) and Marquardt-Levenberg (ML) algorithms.

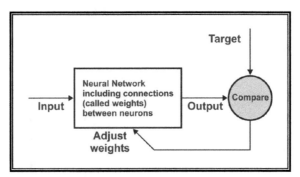

FIGURE 4.5 Block diagram of learning algorithms.

The back propagation is one of the most popular learning algorithms but the convergence of BP learning algorithm progresses slowly in general (Khare and Nagendra, 2007).

If enough memory is available, the Marquardt-Levenberg algorithm can result in dramatically reduced training times in comparison with the required training times for either of the learning algorithms. However, for networks with a few hundred weights the algorithm is very efficient (Khare and Nagendra, 2007). Figure 4.6 illustrates a flow chart for the learning algorithm.

The Marquardt-Levenberg algorithm is used for training the networks in set up and steady stages. The application randomly divides input vectors and target vectors into three sets as follows:

- 60% are used for training.
- 20% are used to validate that the network is generalizing and to stop training before over fitting.
- 20% are used as a completely independent test of network generalization.

4.5 SET-UP STAGE DATA

In ANNs the model accuracy (expect the output with small mean square error) depends on the number of inputs and corresponding outputs data. The experimental data (influent and effluent concentrations) for set up stage are 12 runs. For increasing the number of data used for estimating the weights in the hidden layers, each wetland cell is divided into four basins with the same width and depth of media (2 m and 0.5 m, respectively), and lengths from inlet weir as 2, 5, 8, and 10 m (Figure 4.7).

Theoretical Approach

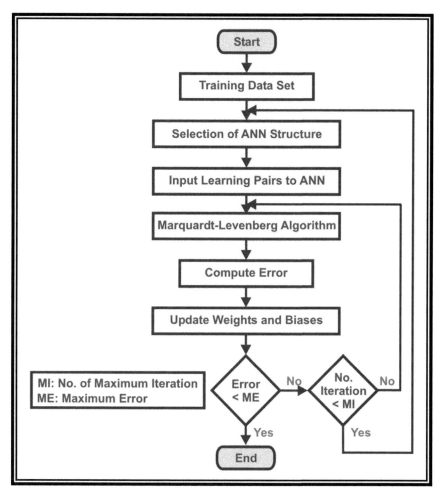

FIGURE 4.6 Learning algorithm of ANNs.

In the set up stage the average discharge Q of each wetland cell is approximately constant, but every cell has four hydraulic loading rates ($Q/4$, $Q/10$, $Q/16$, and $Q/20$). The three cells are filled with rubber, gravel, and plastic media, each one of these bed media has a different surface areas and actual water velocity. For the neural networks used in this stage, the input parameters for all pollutants are hydraulic loading rate (q – m/d), media surface area (A_s – m²), actual velocity (v – m/d), time from start of operation (T_o – d), and the influent concentration (C_i – mg/l); and the output parameter is the effluent concentration (C_o – mg/l). Table 4.1 exhibits the method which is used for increasing the input data for the set up stage.

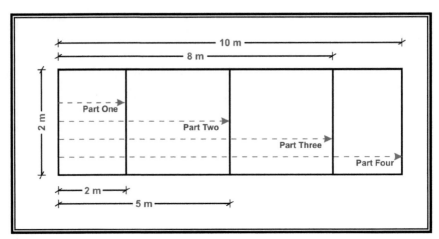

FIGURE 4.7 Schematic diagram for divided wetland cell.

TABLE 4.1 Input and Output Data Used for ANNs and SPSS Modeling

C_i	C_o	q	A_s	v	T_o
	C_{or1}	q_{1r}	A_{sr1}	v_{1r}	
	C_{or2}	q_{2r}	A_{sr2}	v_{2r}	
	C_{or3}	q_{3r}	A_{sr3}	v_{3r}	
	C_{or4}	q_{4r}	A_{sr4}	v_{4r}	
	C_{og1}	q_{1g}	A_{sg1}	v_{1g}	
C_{i1}	C_{og2}	q_{2g}	A_{sg2}	v_{2g}	T_{o1}
	C_{og3}	q_{3g}	A_{sg3}	v_{3g}	
	C_{og4}	q_{4g}	A_{sg4}	v_{4g}	
	C_{op1}	q_{1p}	A_{sp1}	v_{1p}	
	C_{op2}	q_{2p}	A_{sp2}	v_{2p}	
	C_{op3}	q_{3p}	A_{sp3}	v_{3p}	
	C_{op4}	q_{4p}	A_{sp4}	v_{4p}	

Every input concentration at time from start of operation gives the following:

- Four values of outlet concentrations at the four studied lengths (2, 5, 8, and 10 m from cell inlet) for each used media (rubber, gravel, and plastic) which means 12 outlet concentrations data patterns or points.
- Four values of loading rates for each media (12 data patterns).
- Four values of surface area for each media (12 data patterns).
- Four values of actual velocity for each media (12 data patterns).

For computing the surface area at any distance in one wetland cell, first calculating the total volume of media at this length (volume of coarse gravel plus volume of media) then each volume multiplied by the corresponding specific surface area. For computing the actual velocity at any distance in one wetland cell, first the average media porosity for coarse gravel and media type at the specific time from start of operation are calculated (Eq. 4.26) then the actual velocity can be estimated from Eq. (4.27).

$$n_{avg} = \frac{n_{cg} \times V_{cg} + n_m \times V_m}{V_{cg} + V_m} \quad (4.26)$$

where: n_{avg} = average porosity; n_{cg} = porosity of coarse gravel; n_m = porosity of media; V_{cg} = volume of coarse gravel, m³; V_m = volume of media, m³.

$$v = \frac{Q}{n_{avg} \times h \times W} \quad (4.27)$$

where: v = actual water velocity, m/d; Q = discharge to wetland cell, m³/d; h = average depth of water, m; W = width of wetland cell, m.

In the set up stage modeling, 144 data patterns (12 runs × 4 lengths × 3 media) were available from the experimental measurements.

4.6 STEADY STAGE DATA

The experimental data (influent and effluent concentrations) for steady stage are 25 runs. Five average flow rates were used to investigate the effect of loading rate and retention time on treatment performance in steady stage. The input variables for the models are influent concentration (C_i – mg/l), the hydraulic loading rate (q – m/d), media surface area (A_s – m²), and the actual velocity (v – m/d). For the dissolved oxygen pollutant the influent concentration was not taken as its values approximately equal to zero. Five cycles was performed through the steady stage as each cycle represents a specific discharge and water samples were repeated five times during each cycle, so the total number of input sample was 25.

To increase the input data, every influent concentration gave four outlet concentrations, loading rate, surface area, and actual velocity at lengths 2, 5, 8, and 10 m from cell inlet for each media which means 100 data patterns/media for BOD, COD, and TSS pollutants. For the remaining parameters the water samples were taken only at influent and effluent points, so each inlet concentration gave 25 data patterns/media. For BOD, COD, and TSS pollutants, 300 data patterns (25 run × 4 lengths × 3 media) were available from the experimental work. For ammonia, phosphate, dissolved Oxygen, fecal coliforms and some heavy metals, 75 data patterns (25 run × 3 media) were available.

4.7 STATISTICAL MODELING APPLYING SPSS

The SPSS program is among the most widely used program for statistical analysis in social science. This program is used to find out the relationships between the input variables (independent) and the output variable (dependent). There will be a lot of trials to get the most likely equations that represent the whole data for each parameter.

The SPSS contains the systems of multiple linear (more than one independent variable) and nonlinear regression equations. Regression analysis is a statistical tool for the investigation of relationships between variables. The same data and input variables used in neural networks with the same excluded sets for calibration and validation processes were underwent to the SPSS analysis to make to comparison between ANNs and SPSS results. Multiple linear models can be used for prediction or to evaluate whether there is a linear relationship between variables. Such models can be represented by the general linear regression model:

$$Y = \beta_0 + \beta_1 \times X_1 + \beta_2 \times X_2 + \beta_3 \times X_3 + \ldots\ldots\ldots\ldots + \beta_i \times X_i \qquad (4.28)$$

The system of nonlinear equations will be applied in this study to minimize the error between experimental results and the output obtained from these equations. Nonlinear regression analysis can be represented by many formulae and the following equation is the most used in the analysis:

$$Y = \beta_0 + \beta_1 \times X_1^{Z1} + \beta_2 \times X_2^{Z2} + \beta_3 \times X_3^{Z3} + \ldots\ldots\ldots + \beta_i \times X_i^{Zi} \qquad (4.29)$$

where: Y = dependent variable; X = independent variable; Z, β = constants 0, 1, 2, ... i.

The coefficients of Eqs. (4.28) and (4.29) are estimated using many techniques including the least square method in order to minimize the errors between the measured data and predicted ones. Also, SPSS program is used to compare between different outlet concentrations. One way analysis of variance (ANOVA) is used to test the hypothesis that there is no significant difference between the numbers of treatment for the normal distribution of observations, otherwise for the case of abnormal distribution, the chi-square test is applied to determine the significant difference.

4.8 CALIBRATION AND VALIDATION PROCESSES FOR MODELS

In set up stage models, 24 random samples were excluded from the data and used to check the accuracy of the obtained models (validation process). These random samples represent 16.7% of the total data, the remaining 120 data patterns were used for construction and calibration purposes.

In steady stage models (BOD, COD, and TSS pollutants), 60 random samples were chosen from the total data to validate the obtained models, these random samples represent 20% of the total data and the remaining part of data (240 patterns) were used for modeling construction and calibration process. For ammonia, phosphate, dissolved oxygen, fecal coliforms and some heavy metals, 60 data patterns were used for model construction and calibration; and 15 for validation purpose (i.e., 80% for modeling and 20% for test).

4.9 COMPARISON BETWEEN FIELD, ANN, AND SPSS RESULTS

After the models have been constructed, they were graphically analyzed for goodness of fit by plotting the actual against the predicted results. The training performance (the relationship between the measured concentrations and corresponding model ones); the error values and the percentage errors; and the equality diagrams (the experimental versus the model outputs) for these outputs will be discussed in Chapter 8.

The points that give percentage error (difference between measured concentration and model outputs) less than 5% will be considered as a good output result for the model, between 5 and 10% acceptable, and more than 10% not good representation. The error and the percentage error between experimental and model outputs are computed using the following formulae for ANNs and SPSS models:

$$Error = C_{Exp} - C_{ANN} \qquad (4.30)$$

$$E_n = \left(\frac{C_{Exp} - C_{ANN}}{C_{Exp}} \right) \times 100 \qquad (4.31)$$

$$Error = C_{Exp} - C_{SPSS} \qquad (4.32)$$

$$E_s = \left(\frac{C_{Exp} - C_{SPSS}}{C_{Exp}} \right) \times 100 \qquad (4.33)$$

where: E_n = artificial neural network percentage error, %; C_{Exp} = experimental measured output concentration; C_{ANN} = artificial neural network output concentration; E_s = SPSS percentage error, %; C_{spss} = regression equation output concentration.

KEYWORDS

- artificial neural networks modeling
- hydraulic retention time
- loading rate
- removal efficiency
- removal rate constants
- setup stage
- steady stage

CHAPTER 5

FIELD AND EXPERIMENTAL RESULTS

CONTENTS

5.1   Introduction 129
5.2   Porosity Measurements 130
5.3   Hydraulic Parameters for Set-Up Stage 133
5.4   Pollutants Treatment Through Set-Up Stage 136
5.5   Hydraulic Calculations for Steady Stage 143
5.6   Treatment Concentrations of BOD, COD, and TSS 146
5.7   Treatment Concentrations of NH3, PO4, DO, and FC 158
5.8   Treatment Concentrations of Heavy Metals 169
Keywords 180

5.1   INTRODUCTION

This chapter presents the field and experimental data of the three horizontal subsurface flow constructed wetland cells planted with common reeds (*Phragmites Australis*); and having different media which include gravel, pieces of plastic pipes, and shredded tire rubber chips. The experimental work could be divided in two stages, the set up one and steady stage. These two stages were performed during two sampling periods in years 2009 and 2010.

The aim of the set up stage, which continued five months, was to determine the initial pollutant removal performance for the system at the early clean media conditions. Specific objectives of this first study were:

- Obtaining water samples from wetland cells and analyze these samples for BOD, COD, and TSS pollutants.
- Determining the porosity of bed media and the variation of these porosities with time progression.

The aim of the steady stage, which continued over six months, was the comparison between the wastewater treatment performance of the plastic and rubber reed beds and the gravel reed bed. Specific objectives of this second study were:

- Obtaining weekly water samples from inlet, intermediate of the reed bed at three distances along its length, and outlet for each wetland cell.
- Analyzing collected water samples for BOD, COD, TSS, ammonia, phosphate, DO, FC, and some heavy metals.
- Studying the effect of wastewater residence time in reed beds on the treatment performance.

During the whole operation periods (the set up and steady stages), monitoring of water temperature and hydrogen ion were performed (Appendix II). The data obtained from field and experimental works were analyzed in forms of tables and graphs. The selected parameters are covering the most important wastewater properties as: physical (TSS and water temperature); chemical (BOD, COD, DO, and pH); biological (fecal coliforms); and nutrients (ammonia, phosphate, and some heavy metals).

5.2 POROSITY MEASUREMENTS

After the preparation of treatment cells, eight tests were performed at different periods from 5/9/2009 to 10/4/2010. These tests are to determine the porosity of each media, in order to calculate the actual retention time for different used discharges.

The values of measured porosity for the three media are tabulated in detail in Appendix (I). The porosity was measured vertically at 16.67, 33.33, and 50 cm depths; and at 2.5, 5.5, and 7.5 m distances in the longitudinal direction from inlet weir. Table 5.1 gives the average values of the measured porosity.

Notice: $T_o = 0$ means that the wastewater does not flow in the wetland systems and the corresponding values in previous table represent the initial

Field and Experimental Results

TABLE 5.1 Average Values of Porosity for the Coarse Gravel and Used Media

Date	T_o	Average Value of Porosity			
		Plastic Media	Rubber Media	Gravel Media	Coarse Gravel
05/09/2009	0	0.866	0.576	0.431	0.453
19/09/2009	15	0.842	0.558	0.404	0.431
05/10/2009	31	0.827	0.544	0.393	0.411
14/10/2009	40	0.819	0.533	0.381	0.398
20/11/2009	77	0.812	0.527	0.374	0.387
20/12/2009	107	0.799	0.516	0.365	0.376
06/02/2010	155	0.795	0.512	0.362	0.371
10/04/2010	218	0.788	0.505	0.358	0.365

T_o = time from start of operation, d.

porosity. Coarse gravel (in inlet and outlet zones) porosity is the average of porosities at three measured distances in the bottom layer of gravel cell (16.67 cm thickness).

After six months from the start of operation, it was noticed that the difference between porosity values obtained from tests was very small and it could be neglected; this means that the end of set up stage and beginning of steady stage. The differences between the calculated porosities at 06/02/2010 and 10/04/2010 of the used media are given as follows:

Coarse gravel layer = *100 × (0.371–0.365)/0.371 = 1.62%*
Gravel media = *100 × (0.362–0.358)/0.362 = 1.11%*
Rubber media = *100 × (0.512–0.505)/0.512 = 1.37%*
Plastic media = *100 × (0.795–0.788)/0.795 = 0.88%*

Referring to Table 5.1, the final values of porosity were taken as 0.365 for coarse gravel (decreased from initial value, 0.453 by 19.43%), 0.358 for gravel media (decreased from initial value, 0.431 by 16.94%), 0.505 for rubber media (decreased from initial value, 0.576 by 12.33%), and 0.788 for plastic media (decreased from initial value, 0.866 by 9.01%), these percentages illustrate that, the plastic and rubber media have a clogging potential smaller than the gravel one. Figure 5.1 represents the porosity values for the three media versus time progression from the start of operation. The end of set up stage and the start of steady stage are illustrated in this figure.

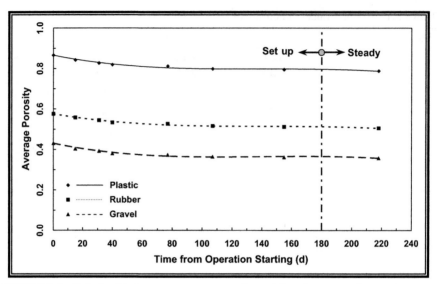

FIGURE 5.1 Relationship between average porosity and time from operation starting.

From Figure 5.1, it is found that the porosity value varies with time in a polynomial function of the third degree for the three media as well as for coarse gravel. The following equation represents the general form of this function. Table 5.2 shows the values of constants in this equation in addition to the determination coefficients (R^2) for the media under study.

$$n_i = b_1 + b_2 \times T_o + b_3 \times T_o^2 + b_4 \times T_o^3 \tag{4.1}$$

where: n_i = porosity value; b_i = constants i = 1, 2, … 4; T_o = time from start of operation, d.

TABLE 5.2 Values of Constants in Eq. (5.1) and R^2 for the Media Under Study

Porosity	B_1	b_2	b_3	b_4	R_2
n_{cg}	0.453	−0.0016	1E-05	3E-08	0.990
n_g	0.431	−0.0016	1E-05	3E-08	0.972
n_r	0.576	−0.0013	9E-06	2E-08	0.984
n_p	0.866	−0.0015	1E-05	2E-08	0.976

n_{cg} = porosity for coarse gravel; n_g = porosity for gravel media; n_r = porosity for rubber media; n_p = porosity for plastic media; R^2 = determination coefficient.

Field and Experimental Results

5.3 HYDRAULIC PARAMETERS FOR SET-UP STAGE

The study started in 31/10/2009 and ended in 3/4/2010 having nearly constant flow rate, by adjusting the inlet weirs for the three treatment cells with an approximate constant retention time for each wetland cell. The set up stage period was related to time that the porosity had reached to steadiness region. At the end of this stage, plants were grown and spread over the treatment cells surfaces and all system was in steady state.

Small-scale wastewater treatment system often does not have a continuous base flow. The wetland system in Samaha is fed intermittently with a flow rate given in two sequences of about 20 to 25 min feeding separated by 35 to 40 min rest period (22.5 min feeding/hour). So, the average daily time of operation varies from 8 to 10 hr. All hydraulic calculations are carried out on an average time of operation equals 9 hr.

Caselles-Osorio (2006) studied the effect of using continuous and intermittent feeding strategy on the performance of horizontal subsurface flow constructed wetlands, from this study it was concluded that intermittent feeding provided more oxidized treatment environment in comparison to continuous feeding.

A constant head on inlet weirs of the treatment cells (H_d = 4.7 cm) was used for the set up stage. Applying this head value in Eqs. (3.4)–(3.6), discharge values of gravel, plastic, and rubber cells can be obtained. Table 5.3 presents the inlet values of discharge and the corresponding values of hydraulic loading rate (q) for the wetland cells (surface area = 20 m²).

$$\text{Discharge for gravel weir} = 4.30 \times (4.7)^{2.5} = 205.93 \text{ cm}^3/\text{s} \quad (3.4)$$

$$\text{Discharge for plastic weir} = 4.11 \times (4.7)^{2.5} = 196.83 \text{ cm}^3/\text{s} \quad (3.5)$$

$$\text{Discharge for rubber weir} = 4.37 \times (4.7)^{2.5} = 209.28 \text{ cm}^3/\text{s} \quad (3.6)$$

The entering load during the nine pump operation hours will be treated during 24 h (one day). To convert from cm³/s to m³/d, the multiplayer is

TABLE 5.3 Values of Discharge and Loading Rate for the Set Up Stage

Media Type	Discharge (m³/d)	q (cm/d)	Surface Area (m²)
Gravel	6.672	33.36	20
Plastic	6.377	31.89	
Rubber	6.781	33.90	

0.0324 (9 × 60 × 60/10⁶ = 0.0324). Figure 5.2 illustrates the longitudinal section of treatment cell for calculating the water volumes inside it. The hydraulic loading rate values (q, cm/d) of cells (based on the input discharge to the wetland cells, assuming no loss or gain of water) are exhibited in Table 5.3, using Eq. (4.1):

$$q = \frac{Q_i \times 100}{A} \quad (4.2)$$

where: Q_i = input discharge, m³/d; A = surface area of treatment cell, m².

Referring to Figure 5.2 and noticing that the submerged volume V_{m1} (calculated under the water surface), it is found that: V_{m1} = 1.29 m³ and V_{m3} = 7.42 m³. The actual retention time of each wetland cell is given by the following Eq. (4.2). The water volumes could be calculated as follows:

$$V_{m1} = 2.0 \times 0.5 \times (1.08 + 1.5)/2 = 1.29 \ m^3$$
$$V_{m3} = 2.0 \times 0.5 \times (7.84 + 7.0)/2 = 7.42 \ m^3$$
$$V_{w10} = 2.0 \times V_{m1} \times n_{cg} + V_{m3} \times n_m$$

$$T_r = \frac{V_{wx}}{Q_i} \quad (4.3)$$

where: V_{wx} = volume of water inside cell at distance x, m³; n_{cg} = porosity of coarse gravel; n_m = porosity of used media; T_r = hydraulic retention time, d.

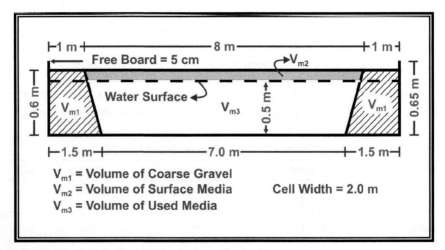

FIGURE 5.2 Longitudinal section in the wetland cell.

Table 5.4 illustrates the actual hydraulic retention time (10 m from inlet) for the set up stage. The porosity values in this table were calculated by interpolation, Figure 5.3 and Eq. (4.4), and related to the time from start of sampling, at 31/10/2009 (the first time that water samples were taken after 56 days from the start of operation). After obtaining the porosity of media, the water volume was calculated at the end of wetland cells and in sequence the actual retention time was determined.

As an example for these calculations of the retention time given in Table 5.4, at time $T_s = 154$ days which corresponds to the start time of operation $T_o = 210$ days, for rubber media the following procedure is used, Figure 5.3:

$$n_{cg} = 0.371 - (0.371 - 0.365) \times (210 - 155)/(218 - 155) = 0.366$$

$$n_m = 0.512 - (0.512 - 0.505) \times (210 - 155)/(218 - 155) = 0.506$$

$$V_{w10} = 2.58 \times 0.366 + 7.42 \times 0.506 = 4.699 \text{ m}^3 \quad \text{(Table 5.4)}$$

$$T_r = 4.699/6.781 = 0.693 \text{ day} \quad (4.4)$$

The average value of T_r may be taken as 0.552 day for gravel, 0.707 day for rubber, and 1.083 day for plastic. The detailed values of T_r at different distances from inlet for set up stage are listed in Appendix (II).

TABLE 5.4 Hydraulic Calculations of the Actual Retention Time for the Set Up Stage

T_s	n_{cg}	Gravel Cell			Rubber Cell			Plastic Cell		
		n_g	V_{w10}	T_r	n_r	V_{w10}	T_r	n_p	V_{w10}	T_r
0	0.393	0.378	3.819	0.572	0.530	4.947	0.729	0.816	7.069	1.108
14	0.389	0.376	3.794	0.569	0.528	4.921	0.726	0.813	7.036	1.103
28	0.384	0.372	3.751	0.562	0.524	4.879	0.719	0.809	6.994	1.097
42	0.379	0.367	3.701	0.555	0.519	4.829	0.712	0.803	6.936	1.088
56	0.375	0.365	3.676	0.551	0.515	4.789	0.706	0.799	6.896	1.081
70	0.374	0.364	3.666	0.549	0.514	4.779	0.705	0.798	6.886	1.080
84	0.372	0.363	3.653	0.548	0.513	4.766	0.703	0.797	6.874	1.078
98	0.371	0.362	3.643	0.546	0.512	4.756	0.701	0.796	6.864	1.076
112	0.370	0.361	3.633	0.545	0.510	4.739	0.699	0.794	6.846	1.074
126	0.368	0.360	3.621	0.543	0.509	4.726	0.697	0.792	6.826	1.070
140	0.367	0.359	3.611	0.541	0.507	4.709	0.694	0.791	6.816	1.069
154	0.366	0.358	3.601	0.540	0.506	4.699	0.693	0.789	6.799	1.066
Average T_r				0.552			0.707			1.083

T_s = time from start of sampling, day.

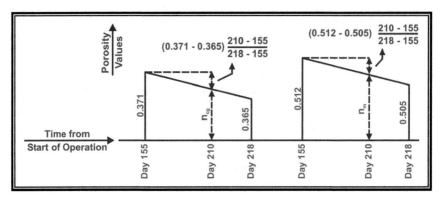

FIGURE 5.3 Sketch for interpolation process for calculating the porosity.

From Table 5.4 it is noticed that the decrease in media porosity leads to a decrease in hydraulic retention time through 5 months (set up period) by 5.59% (0.768 hr) for gravel cell, 4.94% (0.864 hr) for rubber cell, and 3.79% (1.008 hr) for plastic cell. The reduction in porosity for wetland beds is related to the development of reed roots and the growth of attached biofilm on the bed media surfaces in addition to the accumulation of suspended matter on media pores.

Plates 5.1–5.3 show the attached biofilm growth on surfaces of gravel, plastic, and rubber media after three months from the start of operation. The biofilm layer had the biggest surface area in case of plastic media (inside and outside surfaces) compared with gravel and rubber media. Therefore the corrugated plastic media are a good environment for bacterial biofilm growth.

5.4 POLLUTANTS TREATMENT THROUGH SET-UP STAGE

Monitoring water treatment through HSSF wetlands started from October 2009 till April 2010. Twelve runs were carried out during that period. In each run, water samples were collected from the feeding channel which represents the influent concentration. Other water samples were collected from cells and the outlet of each cell. Water samples were tested to determine the pollutant concentrations (BOD, COD, and TSS) at these positions for each cell in order to investigate the pollutant removal behavior of the wetland system.

Table 5.5 exhibits the influent concentrations of set up stage for the tested parameters. The detailed results of water samples analysis for set up stage are presented in tabulated form in Appendix (II). Values of mean, range, and standard deviation (*S.D.*) for influent and effluent concentrations are given

Field and Experimental Results 137

PLATE 5.1 Attached biofilm on gravel media.

PLATE 5.2 Attached biofilm on plastic media.

in Appendix (VI). Pollutant removal efficiency (*RE*) was calculated for all tested pollutants using the following equation:

$$RE = \frac{(C_i - C_o)}{C_i} \times 100 \tag{4.5}$$

where: *RE* = removal efficiency, %; C_i = influent load, mg/l; C_o = effluent load, mg/l.

PLATE 5.3 Attached biofilm on rubber media.

TABLE 5.5 Influent Concentrations of Treatment Parameters for the Set Up Stage

Date	Run No.	T_o (day)	T_s (day)	\multicolumn{3}{c}{Influent Concentration (mg/l)}
				BOD	COD	TSS
31/10/2009	1	56	0	220	344	150
14/11/2009	2	70	14	211	325	148
28/11/2009	3	84	28	232	368	162
12/12/2009	4	98	42	198	300	180
26/12/2009	5	112	56	205	331	146
09/01/2010	6	126	70	191	285	143
23/01/2010	7	140	84	172	257	156
06/02/2010	8	154	98	180	273	166
20/02/2010	9	168	112	194	308	172
06/03/2010	10	182	126	184	297	138
20/03/2010	11	196	140	174	264	144
03/04/2010	12	210	154	168	251	155

5.4.1 BIOCHEMICAL OXYGEN DEMAND

Results of influent and effluent BOD concentration and removal efficiency at distances of 2, 5, 8, 10 m from the cell inlet for set up runs are presented in Tables 5.6–5.8 for rubber, gravel, and plastic media, respectively.

TABLE 5.6 BOD Concentration and *RE* for Rubber Cell at Different Distances

T_s	Influent (mg/l)	BOD Concentration (mg/l)				BOD Removal Efficiency (%)			
		2 m	5 m	8 m	10 m	2 m	5 m	8 m	10 m
0	220	207	190	183	177	5.91	13.64	16.82	19.55
14	211	198	181	175	169	6.16	14.22	17.06	19.91
28	232	217	198	191	184	6.47	14.66	17.67	20.69
42	198	185	167	161	154	6.57	15.66	18.69	22.22
56	205	190	170	163	156	7.32	17.07	20.49	23.90
70	191	176	156	149	142	7.85	18.32	21.99	25.65
84	172	158	138	132	125	8.14	19.77	23.26	27.33
98	180	164	142	134	126	8.89	21.11	25.56	30.00
112	194	174	146	136	127	10.31	24.74	29.90	34.54
126	184	163	135	126	116	11.41	26.63	31.52	36.96
140	174	154	127	118	108	11.49	27.01	32.18	37.93
154	168	148	122	113	103	11.90	27.38	32.74	38.69

TABLE 5.7 BOD Concentration and *RE* for Gravel Cell at Different Distances

T_s	Influent (mg/l)	BOD Concentration (mg/l)				BOD Removal Efficiency (%)			
		2 m	5 m	8 m	10 m	2 m	5 m	8 m	10 m
0	220	205	186	178	173	6.82	15.45	19.09	21.36
14	211	196	176	168	163	7.11	16.59	20.38	22.75
28	232	215	192	183	178	7.33	17.24	21.12	23.28
42	198	182	161	153	148	8.08	18.69	22.73	25.25
56	205	187	164	155	149	8.78	20.00	24.39	27.32
70	191	173	150	141	135	9.42	21.47	26.18	29.32
84	172	155	132	124	117	9.88	23.26	27.91	31.98
98	180	160	135	125	118	11.11	25.00	30.56	34.44
112	194	170	138	126	119	12.37	28.87	35.05	38.66
126	184	159	127	115	106	13.59	30.98	37.50	42.39
140	174	150	119	107	99	13.79	31.61	38.51	43.10
154	168	144	114	103	95	14.29	32.14	38.69	43.45

5.4.2 CHEMICAL OXYGEN DEMAND

Tables 5.9–5.11 show the results of influent and effluent COD values and the calculated removal efficiency for set up runs at different distances from the inlet for rubber, gravel, and plastic media, respectively.

TABLE 5.8 BOD Concentration and *RE* for Plastic Cell at Different Distances

T_s	Influent (mg/l)	BOD Concentration (mg/l)				BOD Removal Efficiency (%)			
		2 m	5 m	8 m	10 m	2 m	5 m	8 m	10 m
0	220	202	183	175	167	8.18	16.82	20.45	24.09
14	211	193	174	166	157	8.53	17.54	21.33	25.59
28	232	211	187	178	168	9.05	19.40	23.28	27.59
42	198	179	156	148	139	9.60	21.21	25.25	29.80
56	205	184	159	149	139	10.24	22.44	27.32	32.20
70	191	170	144	135	125	10.99	24.61	29.32	34.55
84	172	152	128	118	110	11.63	25.58	31.40	36.05
98	180	157	130	119	109	12.78	27.78	33.89	39.44
112	194	165	131	117	106	14.95	32.47	39.69	45.36
126	184	154	119	105	95	16.30	35.33	42.93	48.37
140	174	145	112	99	89	16.67	35.63	43.10	48.85
154	168	140	108	95	85	16.67	35.71	43.45	49.40

TABLE 5.9 COD Concentration and *RE* for Rubber Cell at Different Distances

T_s	Influent (mg/l)	COD Concentration (mg/l)				COD Removal Efficiency (%)			
		2 m	5 m	8 m	10 m	2 m	5 m	8 m	10 m
0	344	325	300	285	278	5.52	12.79	17.15	19.19
14	325	306	281	268	262	5.85	13.54	17.54	19.38
28	368	346	317	302	294	5.98	13.86	17.93	20.11
42	300	282	255	243	235	6.00	15.00	19.00	21.67
56	331	308	277	262	253	6.95	16.31	20.85	23.56
70	285	264	235	221	213	7.37	17.54	22.46	25.26
84	257	237	208	196	188	7.78	19.07	23.74	26.85
98	273	250	217	202	192	8.42	20.51	26.01	29.67
112	308	278	234	215	203	9.74	24.03	30.19	34.09
126	297	264	220	202	188	11.11	25.93	31.99	36.70
140	264	234	194	178	165	11.36	26.52	32.58	37.50
154	251	222	184	168	155	11.55	26.69	33.07	38.25

Field and Experimental Results

TABLE 5.10 COD Concentration and *RE* for Gravel Cell at Different Distances

T_s	Influent (mg/l)	COD Concentration (mg/l)				COD Removal Efficiency (%)			
		2 m	5 m	8 m	10 m	2 m	5 m	8 m	10 m
0	344	323	295	276	272	6.10	14.24	19.77	20.93
14	325	304	275	256	253	6.46	15.38	21.23	22.15
28	368	344	309	288	285	6.52	16.03	21.74	22.55
42	300	278	247	230	226	7.33	17.67	23.33	24.67
56	331	304	268	248	242	8.16	19.03	25.08	26.89
70	285	260	227	209	203	8.77	20.35	26.67	28.77
84	257	233	200	183	176	9.34	22.18	28.79	31.52
98	273	244	207	188	180	10.62	24.18	31.14	34.07
112	308	272	222	198	190	11.69	27.92	35.71	38.31
126	297	259	208	184	172	12.79	29.97	38.05	42.09
140	264	229	183	161	151	13.26	30.68	39.02	42.80
154	251	217	172	152	143	13.55	31.47	39.44	43.03

TABLE 5.11 COD Concentration and *RE* for Plastic Cell at Different Distances

T_s	Influent (mg/l)	COD Concentration (mg/l)				COD Removal Efficiency (%)			
		2 m	5 m	8 m	10 m	2 m	5 m	8 m	10 m
0	344	314	289	270	263	8.72	15.99	21.51	23.55
14	325	296	271	252	244	8.92	16.62	22.46	24.92
28	368	333	300	279	269	9.51	18.48	24.18	26.90
42	300	270	239	222	213	10.00	20.33	26.00	29.00
56	331	295	259	237	226	10.88	21.75	28.40	31.72
70	285	253	217	199	188	11.23	23.86	30.18	34.04
84	257	226	193	174	166	12.06	24.90	32.30	35.41
98	273	237	199	178	167	13.19	27.11	34.80	38.83
112	308	261	210	183	170	15.26	31.82	40.58	44.81
126	297	247	194	167	155	16.84	34.68	43.77	47.81
140	264	219	172	148	136	17.05	34.85	43.94	48.48
154	251	208	163	140	128	17.13	35.06	44.22	49.00

5.4.3 TOTAL SUSPENDED SOLIDS

The inlet and outlet TSS concentration and the removal efficiency for set up stage at different distances from inlet weirs are presented in Tables 5.12–5.14 for rubber, gravel, and plastic media, respectively.

TABLE 5.12 TSS Concentration and *RE* for Rubber Cell at Different Distances

T_s	Influent (mg/l)	TSS Concentration (mg/l)				TSS Removal Efficiency (%)			
		2 m	5 m	8 m	10 m	2 m	5 m	8 m	10 m
0	150	130	106	95	91	13.33	29.33	36.67	39.33
14	148	128	104	93	88	13.51	29.73	37.16	40.54
28	162	140	113	101	96	13.58	30.25	37.65	40.74
42	180	155	125	111	105	13.89	30.56	38.33	41.67
56	146	126	101	89	85	13.70	30.82	39.04	41.78
70	143	123	98	87	82	13.99	31.47	39.16	42.66
84	156	134	106	94	89	14.10	32.05	39.74	42.95
98	166	142	113	99	93	14.46	31.93	40.36	43.98
112	172	147	116	102	96	14.53	32.56	40.70	44.19
126	138	118	93	82	77	14.49	32.61	40.58	44.20
140	144	123	97	85	80	14.58	32.64	40.97	44.44
154	155	132	104	91	85	14.84	32.90	41.29	45.16

TABLE 5.13 TSS Concentration and *RE* for Gravel Cell at Different Distances

T_s	Influent (mg/l)	TSS Concentration (mg/l)				TSS Removal Efficiency (%)			
		2 m	5 m	8 m	10 m	2 m	5 m	8 m	10 m
0	150	130	103	92	86	13.33	31.33	38.67	42.67
14	148	127	101	90	84	14.19	31.76	39.19	43.24
28	162	139	110	97	91	14.20	32.1	40.12	43.83
42	180	154	121	107	100	14.44	32.78	40.56	44.44
56	146	125	98	86	80	14.38	32.88	41.10	45.21
70	143	122	95	83	77	14.69	33.57	41.96	46.15
84	156	133	103	90	83	14.74	33.97	42.31	46.79
98	166	141	109	95	88	15.06	34.34	42.77	46.99
112	172	146	112	98	91	15.12	34.88	43.02	47.09
126	138	117	90	78	72	15.22	34.78	43.48	47.83
140	144	122	93	81	75	15.28	35.42	43.75	47.92
154	155	131	100	87	80	15.48	35.48	43.87	48.39

Field and Experimental Results 143

TABLE 5.14 TSS Concentration and *RE* for Plastic Cell at Different Distances

T_s	Influent (mg/l)	TSS Concentration (mg/l)				TSS Removal Efficiency (%)			
		2 m	5 m	8 m	10 m	2 m	5 m	8 m	10 m
0	150	121	84	70	66	19.33	44.00	53.33	56.00
14	148	119	82	69	64	19.59	44.59	53.38	56.76
28	162	130	89	74	70	19.75	45.06	54.32	56.79
42	180	144	98	82	76	20.00	45.56	54.44	57.78
56	146	116	79	65	61	20.55	45.89	55.48	58.22
70	143	114	77	63	59	20.28	46.15	55.94	58.74
84	156	124	83	68	64	20.51	46.79	56.41	58.97
98	166	131	88	72	67	21.08	46.99	56.63	59.64
112	172	136	91	74	69	20.93	47.09	56.98	59.88
126	138	109	72	59	55	21.01	47.83	57.25	60.14
140	144	114	75	62	57	20.83	47.92	56.94	60.42
154	155	122	81	66	61	21.29	47.74	57.42	60.65

5.5 HYDRAULIC CALCULATIONS FOR STEADY STAGE

The steady stage started after the full growing of reeds and their roots penetrated the cell media. Plant cutting (harvesting) was practiced three times during this study. The steady stage deals with more tested parameters than the set up stage as the constructed wetland system had reached to the stabilization conditions. Twelve parameters were analyzed containing biochemical oxygen demand, chemical oxygen demand, total suspended solids, ammonia, phosphate, dissolved oxygen, fecal coliforms, and some selected elements of heavy metals (Zinc, Iron, Manganese, Lead, and Cadmium).

The values of mean, range, and standard deviation for influent and effluent concentrations for steady stage parameters are given in Appendix (VI). Table 5.15 presents the final porosity for gravel, plastic, rubber, and coarse gravel media. The porosity values were considered as the final values after 218 days from the start of operation. Table 5.16 presents the calculated water volumes at different distances for gravel, rubber, and plastic cells, which were used to calculate the hydraulic retention time (volume of water/Q), T_r.

$$V_{w2} = 1.29 \times n_{cg} + 0.71 \times n_m$$
$$V_{w5} = 1.29 \times n_{cg} + 3.71 \times n_m$$
$$V_{w8} = 1.29 \times n_{cg} + 6.71 \times n_m$$

TABLE 5.15 Final Porosity for the Used Media

Type of Media	Porosity
Gravel	0.358
Plastic	0.788
Rubber	0.505
Coarse Gravel	0.365

TABLE 5.16 Water Volumes at Different Distances for the Three Wetland Cells

Media Type	\multicolumn{4}{c}{Water Volumes Inside Wetland Cells at Different Distances (m^3)}
	V_{w2}	V_{w5}	V_{w8}	V_{w10}
Gravel	0.725	1.799	2.873	3.598
Rubber	0.829	2.344	3.859	4.689
Plastic	1.030	3.394	5.758	6.789

$V_{w10} = 2.58 \times n_{cg} + 7.42 \times n_m$
V_{w2} = water volume at the first 2 m, m
V_{w5} = water volume at the first 5 m, m
V_{w8} = water volume at the first 8 m, m
V_{w10} = water volume in the wetland cell, m
n_{cg} = porosity of coarse grave
n_m = porosity of used media

The water volume can be obtained by applying the above-mentioned formulae. For example V_{w8} for rubber cell equals 3.859 m³ (1.29 × 0.365 + 6.71 × 0.505).

Five discharges were used through the steady stage. Every discharge continued 5 consecutive weeks (five sampling cycles). The hydraulic parameters for this stage (discharge, retention time, and loading rates) at the end of wetland cells are shown in Table 5.17 for the three media.

In the previous table, the hydraulic retention time at 10 m from plastic cell inlet (Cycle No. 5) is calculated as in the following steps:

$$\text{Discharge of plastic weir} = 4.11 \times (2.4)^{2.5} = 36.68 \text{ cm}^3/\text{s} \quad (3.5)$$

$$Q = 36.68 \times 0.0324 = 1.188 \text{ m}^3/\text{d} \quad \text{(Units)}$$

$$q = 1.188/20 = 0.0594 \text{ m/d} = 5.94 \text{ cm/d} \quad (4.1)$$

TABLE 5.17 Hydraulic Calculations for Steady Stage for the Three Media

Cycle No.	H_d (cm)	Gravel Cell Q (m³/d)	q (cm/d)	T_r (day)	Rubber Cell Q (m³/d)	q (cm/d)	T_r (day)	Plastic Cell Q (m³/d)	q (cm/d)	T_r (day)
1	4.2	5.037	25.19	0.714	5.119	25.60	0.916	4.814	24.07	1.410
2	3.6	3.426	17.13	1.050	3.482	17.41	1.347	3.275	16.38	2.073
3	3.1	2.357	11.79	1.527	2.396	11.98	1.957	2.253	11.27	3.013
4	2.7	1.669	8.35	2.156	1.696	8.48	2.765	1.595	7.98	4.256
5	2.4	1.243	6.22	2.895	1.263	6.32	3.713	1.188	5.94	5.715

$$V_{w10} = 6.789 \text{ m}^3 \qquad \text{(Table 5.16)}$$

$$T_r = 6.789/1.188 = 5.715 \text{ day} \qquad (4.2)$$

5.6 TREATMENT CONCENTRATIONS OF BOD, COD, AND TSS

Tables 5.18–5.22 present the sampling date and influent concentrations for BOD, COD, and TSS pollutants in steady stage. These pollutants are the governing ones for evaluating the wetland systems performance. So many samples were analyzed to study the variation effect of discharges and distances on the treatment process.

5.6.1 BIOCHEMICAL OXYGEN DEMAND

Tables 5.23–5.37 exhibit the results of influent and effluents BOD concentrations and the calculated corresponding removal efficiency at distances

TABLE 5.18 Influent Concentrations for BOD, COD, and TSS (Cycle No. 1)

Average Discharge	Run No.	T_o	Date	Influent Concentration (mg/l)		
				BOD	COD	TSS
Cycle No. 1	1	224	17/04/2010	190	297	147
$Q_{av} = 4.99 \text{ m}^3/\text{d}$	2	231	24/04/2010	185	285	140
	3	238	01/05/2010	182	289	135
	4	245	08/05/2010	193	292	153
	5	252	15/05/2010	197	318	149
Average value of influent concentration (mg/l)				189.4	296.2	144.8

TABLE 5.19 Influent Concentrations for BOD, COD, and TSS (Cycle No. 2)

Average Discharge	Run No.	T_o	Date	Influent Concentration (mg/l)		
				BOD	COD	TSS
Cycle No. 2	1	259	22/05/2010	200	303	131
$Q_{av} = 3.39 \text{ m}^3/\text{d}$	2	266	29/05/2010	203	322	139
	3	273	05/06/2010	209	337	170
	4	280	12/06/2010	188	285	151
	5	287	19/06/2010	194	290	141
Average value of influent concentration (mg/l)				198.8	307.4	146.4

Field and Experimental Results

TABLE 5.20 Influent Concentrations for BOD, COD, and TSS (Cycle No. 3)

Average Discharge	Run No.	T_o	Date	Influent Concentration (mg/l) BOD	COD	TSS
Cycle No. 3 $Q_{av} = 2.34$ m³/d	1	294	26/06/2010	180	286	162
	2	301	03/07/2010	192	291	148
	3	308	10/07/2010	198	319	165
	4	315	17/07/2010	187	279	160
	5	322	24/07/2010	182	272	144
Average value of influent concentration (mg/l)				187.8	289.4	155.8

TABLE 5.21 Influent Concentrations for BOD, COD, and TSS (Cycle No. 4)

Average Discharge	Run No.	T_o	Date	Influent Concentration (mg/l) BOD	COD	TSS
Cycle No. 4 $Q_{av} = 1.65$ m³/d	1	329	31/07/2010	178	266	177
	2	336	07/08/2010	162	242	169
	3	343	14/08/2010	176	267	164
	4	350	21/08/2010	199	316	159
	5	357	28/08/2010	189	305	173
Average value of influent concentration (mg/l)				180.8	279.2	168.4

TABLE 5.22 Influent Concentrations for BOD, COD, and TSS (Cycle No. 5)

Average Discharge	Run No.	T_o	Date	Influent Concentration (mg/l) BOD	COD	TSS
Cycle No. 5 $Q_{av} = 1.23$ m³/d	1	378	18/09/2010	183	273	171
	2	385	25/09/2010	177	268	176
	3	392	02/10/2010	193	306	180
	4	399	09/10/2010	187	302	165
	5	406	16/10/2010	179	271	168
Average value of influent concentration (mg/l)				183.8	284.0	172.0

TABLE 5.23 BOD Concentration and *RE* for Rubber Cell (Cycle 1, $Q = 5.119$ m^3/d)

T_o	Influent (mg/l)	BOD Concentration (mg/l)				BOD Removal Efficiency (%)			
		2 m	5 m	8 m	10 m	2 m	5 m	8 m	10 m
224	190	164	131	114	105	13.68	31.05	40.00	44.74
231	185	159	126	110	101	14.05	31.89	40.54	45.41
238	182	156	123	107	98	14.29	32.42	41.21	46.15
245	193	166	132	116	105	13.99	31.61	39.90	45.60
252	197	170	136	119	109	13.71	30.96	39.59	44.67
Mean	189.4	163.0	129.6	113.2	103.6	13.94	31.59	40.25	45.31

TABLE 5.24 BOD Concentration and *RE* for Rubber Cell (Cycle 2, $Q = 3.482$ m^3/d)

T_o	Influent (mg/l)	BOD Concentration (mg/l)				BOD Removal Efficiency (%)			
		2 m	5 m	8 m	10 m	2 m	5 m	8 m	10 m
259	200	165	124	104	92	17.50	38.00	48.00	54.00
266	203	168	126	106	94	17.24	37.93	47.78	53.69
273	209	173	131	110	98	17.22	37.32	47.37	53.11
280	188	155	116	96	85	17.55	38.30	48.94	54.79
287	194	160	120	100	88	17.53	38.14	48.45	54.64
Mean	198.8	164.2	123.4	103.2	91.4	17.41	37.94	48.11	54.05

TABLE 5.25 BOD Concentration and *RE* for Rubber Cell (Cycle 3, $Q = 2.396$ m^3/d)

T_o	Influent (mg/l)	BOD Concentration (mg/l)				BOD Removal Efficiency (%)			
		2 m	5 m	8 m	10 m	2 m	5 m	8 m	10 m
294	180	143	99	76	63	20.56	45.00	57.78	65.00
301	192	153	106	82	68	20.31	44.79	57.29	64.58
308	198	158	110	86	72	20.20	44.44	56.57	63.64
315	187	149	104	81	67	20.32	44.39	56.68	64.17
322	182	145	100	77	64	20.33	45.05	57.69	64.84
Mean	187.8	149.6	103.8	80.4	66.8	20.34	44.74	57.20	64.45

Field and Experimental Results

TABLE 5.26 BOD Concentration and *RE* for Rubber Cell (Cycle 4, $Q = 1.696$ m³/d)

T_o	Influent (mg/l)	BOD Concentration (mg/l)				BOD Removal Efficiency (%)			
		2 m	5 m	8 m	10 m	2 m	5 m	8 m	10 m
329	178	139	92	69	54	21.91	48.31	61.24	69.66
336	162	126	84	62	49	22.22	48.15	61.73	69.75
343	176	138	91	68	54	21.59	48.30	61.36	69.32
350	199	156	104	78	62	21.61	47.74	60.80	68.84
357	189	148	98	73	58	21.69	48.15	61.38	69.31
Mean	180.8	141.4	93.8	70.0	55.4	21.80	48.13	61.30	69.38

TABLE 5.27 BOD Concentration and *RE* for Rubber Cell (Cycle 5, $Q = 1.263$ m³/d)

T_o	Influent (mg/l)	BOD Concentration (mg/l)				BOD Removal Efficiency (%)			
		2 m	5 m	8 m	10 m	2 m	5 m	8 m	10 m
378	183	142	91	65	50	22.40	50.27	64.48	72.68
385	177	138	87	63	48	22.03	50.85	64.41	72.88
392	193	150	96	70	54	22.28	50.26	63.73	72.02
399	187	146	93	67	52	21.93	50.27	64.17	72.19
406	179	139	88	63	48	22.35	50.84	64.80	73.18
Mean	183.8	143.0	91.0	65.6	50.4	22.20	50.50	64.32	72.59

TABLE 5.28 BOD Concentration and *RE* for Gravel Cell (Cycle 1, $Q = 5.037$ m³/d)

T_o	Influent (mg/l)	BOD Concentration (mg/l)				BOD Removal Efficiency (%)			
		2 m	5 m	8 m	10 m	2 m	5 m	8 m	10 m
224	190	162	122	106	97	14.74	35.79	44.21	48.95
231	185	157	119	102	93	15.14	35.68	44.86	49.73
238	182	154	116	100	91	15.38	36.26	45.05	50.00
245	193	164	124	107	97	15.03	35.75	44.56	49.74
252	197	168	127	110	101	14.72	35.53	44.16	48.73
Mean	189.4	161.0	121.6	105.0	95.8	15.00	35.80	44.57	49.43

TABLE 5.29 BOD Concentration and *RE* for Gravel Cell (Cycle 2, $Q = 3.426$ m^3/d)

T_o	Influent (mg/l)	BOD Concentration (mg/l)				BOD Removal Efficiency (%)			
		2 m	5 m	8 m	10 m	2 m	5 m	8 m	10 m
259	200	163	115	94	84	18.50	42.50	53.00	58.00
266	203	166	117	96	86	18.23	42.36	52.71	57.64
273	209	172	122	100	90	17.70	41.63	52.15	56.94
280	188	153	107	87	78	18.62	43.09	53.72	58.51
287	194	158	111	90	80	18.56	42.78	53.61	58.76
Mean	198.8	162.4	114.4	93.4	83.6	18.32	42.47	53.04	57.97

TABLE 5.30 BOD Concentration and *RE* for Gravel Cell (Cycle 3, $Q = 2.357$ m^3/d)

T_o	Influent (mg/l)	BOD Concentration (mg/l)				BOD Removal Efficiency (%)			
		2 m	5 m	8 m	10 m	2 m	5 m	8 m	10 m
294	180	141	89	66	55	21.67	50.56	63.33	69.44
301	192	151	96	72	60	21.35	50.00	62.50	68.75
308	198	156	100	76	64	21.21	49.49	61.62	67.68
315	187	148	94	71	60	20.86	49.73	62.03	67.91
322	182	143	90	68	56	21.43	50.55	62.64	69.23
Mean	187.8	147.8	93.8	70.6	59.0	21.30	50.07	62.42	68.60

TABLE 5.31 BOD Concentration and *RE* for Gravel Cell (Cycle 4, $Q = 1.669$ m^3/d)

T_o	Influent (mg/l)	BOD Concentration (mg/l)				BOD Removal Efficiency (%)			
		2 m	5 m	8 m	10 m	2 m	5 m	8 m	10 m
329	178	137	82	58	46	23.03	53.93	67.42	74.16
336	162	125	74	52	41	22.84	54.32	67.90	74.69
343	176	136	81	58	46	22.73	53.98	67.05	73.86
350	199	154	93	67	54	22.61	53.27	66.33	72.86
357	189	146	88	63	50	22.75	53.44	66.67	73.54
Mean	180.8	139.6	83.6	59.6	47.4	22.79	53.79	67.07	73.82

Field and Experimental Results 151

TABLE 5.32 BOD Concentration and *RE* for Gravel Cell (Cycle 5, Q = 1.243 m³/d)

T_o	Influent (mg/l)	BOD Concentration (mg/l)				BOD Removal Efficiency (%)			
		2 m	5 m	8 m	10 m	2 m	5 m	8 m	10 m
378	183	140	81	56	43	23.50	55.74	69.40	76.50
385	177	136	78	54	41	23.16	55.93	69.49	76.84
392	193	148	87	60	46	23.32	54.92	68.91	76.17
399	187	144	84	58	45	22.99	55.08	68.98	75.94
406	179	137	78	53	40	23.46	56.42	70.39	77.65
Mean	183.8	141.0	81.6	56.2	43.0	23.29	55.62	69.44	76.62

TABLE 5.33 BOD Concentration and *RE* for Plastic Cell (Cycle 1, Q = 4.814 m³/d)

T_o	Influent (mg/l)	BOD Concentration (mg/l)				BOD Removal Efficiency (%)			
		2 m	5 m	8 m	10 m	2 m	5 m	8 m	10 m
224	190	156	111	94	83	17.89	41.58	50.53	56.32
231	185	152	107	90	79	17.84	42.16	51.35	57.30
238	182	149	105	88	77	18.13	42.31	51.65	57.69
245	193	159	113	95	84	17.62	41.45	50.78	56.48
252	197	162	116	98	86	17.77	41.12	50.25	56.35
Mean	189.4	155.6	110.4	93.0	81.8	17.85	41.72	50.91	56.83

TABLE 5.34 BOD Concentration and *RE* for Plastic Cell (Cycle 2, Q = 3.275 m³/d)

T_o	Influent (mg/l)	BOD Concentration (mg/l)				BOD Removal Efficiency (%)			
		2 m	5 m	8 m	10 m	2 m	5 m	8 m	10 m
259	200	158	103	83	70	21.00	48.50	58.50	65.00
266	203	161	106	85	72	20.69	47.78	58.13	64.53
273	209	165	109	88	75	21.05	47.85	57.89	64.11
280	188	148	96	76	64	21.28	48.94	59.57	65.96
287	194	153	99	79	67	21.13	48.97	59.28	65.46
Mean	198.8	157.0	102.6	82.2	69.6	21.03	48.41	58.68	65.01

TABLE 5.35 BOD Concentration and *RE* for Plastic Cell (Cycle 3, Q = 2.253 m³/d)

T_o	Influent (mg/l)	BOD Concentration (mg/l)				BOD Removal Efficiency (%)			
		2 m	5 m	8 m	10 m	2 m	5 m	8 m	10 m
294	180	137	80	58	44	23.89	55.56	67.78	75.56
301	192	146	86	63	48	23.96	55.21	67.19	75.00
308	198	151	89	66	51	23.74	55.05	66.67	74.24
315	187	142	84	61	47	24.06	55.08	67.38	74.87
322	182	138	81	59	45	24.18	55.49	67.58	75.27
Mean	187.8	142.8	84.0	61.4	47.0	23.96	55.28	67.32	74.99

TABLE 5.36 BOD Concentration and *RE* for Plastic Cell (Cycle 4, Q = 1.595 m³/d)

T_o	Influent (mg/l)	BOD Concentration (mg/l)				BOD Removal Efficiency (%)			
		2 m	5 m	8 m	10 m	2 m	5 m	8 m	10 m
329	178	133	73	50	36	25.28	58.99	71.91	79.78
336	162	121	66	45	32	25.31	59.26	72.22	80.25
343	176	131	73	50	36	25.57	58.52	71.59	79.55
350	199	149	83	58	42	25.13	58.29	70.85	78.89
357	189	141	78	54	39	25.40	58.73	71.43	79.37
Mean	180.8	135.0	74.6	51.4	37.0	25.34	58.76	71.60	79.57

TABLE 5.37 BOD Concentration and *RE* for Plastic Cell (Cycle 5, Q = 1.188 m³/d)

T_o	Influent (mg/l)	BOD Concentration (mg/l)				BOD Removal Efficiency (%)			
		2 m	5 m	8 m	10 m	2 m	5 m	8 m	10 m
378	183	135	70	46	30	26.23	61.75	74.86	83.61
385	177	130	68	44	28	26.55	61.58	75.14	84.18
392	193	142	75	49	32	26.42	61.14	74.61	83.42
399	187	138	73	48	31	26.20	60.96	74.33	83.42
406	179	131	68	44	28	26.82	62.01	75.42	84.36
Mean	183.8	135.2	70.8	46.2	29.8	26.45	61.49	74.87	83.80

of 2, 5, 8, and 10 m measured from cell inlet for rubber, gravel, and plastic media. The mean values of the previous parameters for each cycle are given in these tables.

Field and Experimental Results 153

5.6.2 CHEMICAL OXYGEN DEMAND

Chemical oxygen demand does not differentiate between biologically available and inert organic matter, and it is a measure of the total quantity of oxygen required to oxidize all organic material into carbon dioxide and water. The COD values are always greater than BOD values, but COD measurements can be made in a few hours while BOD measurements take five days. The results of the five cycles of steady stage for chemical oxygen demand pollutant are given in Tables 5.38–5.52. These tables present the influent and effluents COD concentrations and the corresponding calculated removal efficiency at longitudinal distances of 2, 5, 8, and 10 m measured from inlet weir for rubber, gravel, and plastic cells. The mean values of the forgoing parameters for each cycle are presented in these tables.

TABLE 5.38 COD Concentration and *RE* for Rubber Cell (Cycle 1, $Q = 5.119$ m³/d)

T_o	Influent (mg/l)	COD Concentration (mg/l)				COD Removal Efficiency (%)			
		2 m	5 m	8 m	10 m	2 m	5 m	8 m	10 m
224	297	257	207	177	165	13.43	30.27	40.38	44.42
231	285	246	196	168	156	13.57	31.14	40.97	45.19
238	289	249	197	169	157	13.81	31.81	41.50	45.65
245	292	253	202	175	160	13.48	30.92	40.16	45.28
252	318	276	221	191	177	13.14	30.45	39.89	44.29
Mean	296.2	256.2	204.6	176.0	163.0	13.49	30.92	40.58	44.97

TABLE 5.39 COD Concentration and *RE* for Rubber Cell (Cycle 2, $Q = 3.482$ m³/d)

T_o	Influent (mg/l)	COD Concentration (mg/l)				COD Removal Efficiency (%)			
		2 m	5 m	8 m	10 m	2 m	5 m	8 m	10 m
259	303	251	190	157	140	17.17	37.30	48.19	53.80
266	322	268	202	167	150	16.83	37.31	48.17	53.45
273	337	280	213	177	159	16.94	36.81	47.49	52.83
280	285	236	177	145	130	17.15	37.86	49.10	54.36
287	290	240	181	149	132	17.11	37.49	48.54	54.41
Mean	307.4	255.0	192.6	159.0	142.2	17.04	37.35	48.30	53.77

TABLE 5.40 COD Concentration and *RE* for Rubber Cell (Cycle 3, Q = 2.396 m³/d)

T_o	Influent (mg/l)	COD Concentration (mg/l)				COD Removal Efficiency (%)			
		2 m	5 m	8 m	10 m	2 m	5 m	8 m	10 m
294	286	228	159	120	101	20.20	44.35	58.00	64.65
301	291	233	162	124	104	19.91	44.31	57.38	64.25
308	319	256	179	138	117	19.84	43.95	56.79	63.36
315	279	223	157	120	101	20.10	43.75	57.01	63.81
322	272	217	151	114	96	20.12	44.41	58.03	64.66
Mean	289.4	231.4	161.6	123.2	103.8	20.03	44.15	57.44	64.15

TABLE 5.41 COD Concentration and *RE* for Rubber Cell (Cycle 4, Q = 1.696 m³/d)

T_o	Influent (mg/l)	COD Concentration (mg/l)				COD Removal Efficiency (%)			
		2 m	5 m	8 m	10 m	2 m	5 m	8 m	10 m
329	266	209	139	103	81	21.33	47.68	61.23	69.51
336	242	189	127	92	74	21.83	47.48	61.95	69.40
343	267	210	139	103	82	21.25	47.88	61.38	69.25
350	316	249	167	123	99	21.17	47.13	61.06	68.66
357	305	240	160	117	94	21.27	47.51	61.62	69.16
Mean	279.2	219.4	146.4	107.6	86.0	21.37	47.54	61.45	69.20

TABLE 5.42 COD Concentration and *RE* for Rubber Cell (Cycle 5, Q = 1.263 m³/d)

T_o	Influent (mg/l)	COD Concentration (mg/l)				COD Removal Efficiency (%)			
		2 m	5 m	8 m	10 m	2 m	5 m	8 m	10 m
378	273	213	137	97	75	22.02	49.84	64.49	72.54
385	268	210	133	95	73	21.69	50.41	64.58	72.78
392	306	239	154	111	86	21.98	49.73	63.77	71.93
399	302	237	151	108	84	21.42	49.94	64.19	72.15
406	271	212	135	95	73	21.83	50.22	64.97	73.08
Mean	284.0	222.2	142.0	101.2	78.2	21.79	50.03	64.40	72.50

Field and Experimental Results

TABLE 5.43 COD Concentration and *RE* for Gravel Cell (Cycle 1, Q = 5.037 m³/d)

T_o	Influent (mg/l)	COD Concentration (mg/l)				COD Removal Efficiency (%)			
		2 m	5 m	8 m	10 m	2 m	5 m	8 m	10 m
224	297	255	193	164	153	14.11	34.99	44.76	48.46
231	285	243	186	155	144	14.62	34.65	45.54	49.41
238	289	246	187	157	146	14.85	35.27	45.65	49.46
245	292	250	190	161	148	14.51	35.03	44.94	49.39
252	318	273	208	176	164	14.08	34.54	44.61	48.39
Mean	296.2	253.4	192.8	162.6	151.0	14.43	34.90	45.10	49.02

TABLE 5.44 COD Concentration and *RE* for Gravel Cell (Cycle 2, Q = 3.426 m³/d)

T_o	Influent (mg/l)	COD Concentration (mg/l)				COD Removal Efficiency (%)			
		2 m	5 m	8 m	10 m	2 m	5 m	8 m	10 m
259	303	249	177	141	128	17.83	41.59	53.47	57.76
266	322	266	188	151	138	17.45	41.66	53.14	57.17
273	337	280	200	160	146	16.94	40.67	52.54	56.69
280	285	234	164	131	119	17.85	42.43	54.01	58.22
287	290	238	168	133	120	17.80	41.98	54.07	58.56
Mean	307.4	253.4	179.4	143.2	130.2	17.57	41.67	53.45	57.68

TABLE 5.45 COD Concentration and *RE* for Gravel Cell (Cycle 3, Q = 2.357 m³/d)

T_o	Influent (mg/l)	COD Concentration (mg/l)				COD Removal Efficiency (%)			
		2 m	5 m	8 m	10 m	2 m	5 m	8 m	10 m
294	286	226	143	104	88	20.90	49.95	63.60	69.20
301	291	231	147	108	92	20.59	49.47	62.88	68.38
308	319	254	164	121	104	20.46	48.65	62.11	67.43
315	279	223	142	105	90	20.10	49.12	62.38	67.75
322	272	215	136	101	84	20.85	49.93	62.82	69.08
Mean	289.4	229.8	146.4	107.8	91.6	20.58	49.42	62.76	68.37

TABLE 5.46 COD Concentration and *RE* for Gravel Cell (Cycle 4, Q = 1.669 m³/d)

T_o	Influent (mg/l)	COD Concentration (mg/l)				COD Removal Efficiency (%)			
		2 m	5 m	8 m	10 m	2 m	5 m	8 m	10 m
329	266	206	124	86	69	22.46	53.33	67.63	74.03
336	242	188	112	77	62	22.25	53.68	68.15	74.36
343	267	208	124	87	70	22.00	53.50	67.38	73.75
350	316	246	150	105	86	22.12	52.51	66.76	72.77
357	305	237	144	101	81	22.25	52.76	66.87	73.43
Mean	279.2	217.0	130.8	91.2	73.6	22.22	53.16	67.36	73.67

TABLE 5.47 COD Concentration and *RE* for Gravel Cell (Cycle 5, Q = 1.243 m³/d)

T_o	Influent (mg/l)	COD Concentration (mg/l)				COD Removal Efficiency (%)			
		2 m	5 m	8 m	10 m	2 m	5 m	8 m	10 m
378	273	211	123	83	65	22.75	54.97	69.61	76.20
385	268	208	120	81	63	22.44	55.25	69.80	76.51
392	306	237	140	94	74	22.64	54.30	69.32	75.84
399	302	234	137	93	73	22.42	54.58	69.17	75.80
406	271	209	120	80	61	22.94	55.75	70.50	77.51
Mean	284.0	219.8	128.0	86.2	67.2	22.64	54.97	69.68	76.37

TABLE 5.48 COD Concentration and *RE* for Plastic Cell (Cycle 1, Q = 4.814 m³/d)

T_o	Influent (mg/l)	COD Concentration (mg/l)				COD Removal Efficiency (%)			
		2 m	5 m	8 m	10 m	2 m	5 m	8 m	10 m
224	297	243	175	145	131	18.15	41.05	51.16	55.87
231	285	233	166	137	123	18.14	41.68	51.86	56.78
238	289	235	169	138	123	18.65	41.50	52.23	57.42
245	292	240	173	142	128	17.93	40.84	51.44	56.23
252	318	260	189	156	140	18.17	40.52	50.90	55.94
Mean	296.2	242.2	174.4	143.6	129.0	18.21	41.12	51.52	56.45

Field and Experimental Results 157

TABLE 5.49 COD Concentration and *RE* for Plastic Cell (Cycle 2, Q = 3.275 m³/d)

T_o	Influent (mg/l)	COD Concentration (mg/l)				COD Removal Efficiency (%)			
		2 m	5 m	8 m	10 m	2 m	5 m	8 m	10 m
259	303	238	158	124	107	21.46	47.86	59.08	64.69
266	322	254	170	133	115	21.17	47.24	58.72	64.31
273	337	265	178	140	122	21.39	47.20	58.47	63.81
280	285	223	147	114	98	21.71	48.39	59.98	65.60
287	290	227	149	117	101	21.60	48.54	59.59	65.12
Mean	307.4	241.4	160.4	125.6	108.6	21.47	47.85	59.17	64.71

TABLE 5.50 COD Concentration and *RE* for Plastic Cell (Cycle 3, Q = 2.253 m³/d)

T_o	Influent (mg/l)	COD Concentration (mg/l)				COD Removal Efficiency (%)			
		2 m	5 m	8 m	10 m	2 m	5 m	8 m	10 m
294	286	216	128	91	71	24.40	55.20	68.15	75.15
301	291	220	132	94	73	24.38	54.63	67.69	74.91
308	319	242	145	105	83	24.22	54.60	67.12	74.01
315	279	211	127	90	71	24.40	54.50	67.75	74.56
322	272	205	122	87	68	24.53	55.09	67.97	74.97
Mean	289.4	218.8	130.8	93.4	73.2	24.39	54.80	67.74	74.72

TABLE 5.51 COD Concentration and *RE* for Plastic Cell (Cycle 4, Q = 1.595 m³/d)

T_o	Influent (mg/l)	COD Concentration (mg/l)				COD Removal Efficiency (%)			
		2 m	5 m	8 m	10 m	2 m	5 m	8 m	10 m
329	266	198	110	74	54	25.47	58.60	72.15	79.67
336	242	180	100	66	48	25.56	58.64	72.70	80.15
343	267	198	112	75	55	25.75	58.00	71.88	79.38
350	316	235	133	91	67	25.60	57.89	71.19	78.79
357	305	226	127	86	64	25.86	58.34	71.79	79.01
Mean	279.2	207.4	116.4	78.4	57.6	25.65	58.29	71.94	79.40

TABLE 5.52 COD Concentration and *RE* for Plastic Cell (Cycle 5, Q = 1.188 m³/d)

T_o	Influent (mg/l)	COD Concentration (mg/l)				COD Removal Efficiency (%)			
		2 m	5 m	8 m	10 m	2 m	5 m	8 m	10 m
378	273	201	106	68	45	26.41	61.19	75.10	83.52
385	268	196	104	66	43	26.92	61.22	75.39	83.97
392	306	224	120	77	51	26.88	60.83	74.87	83.35
399	302	222	119	76	50	26.40	60.55	74.80	83.42
406	271	198	104	66	43	26.99	61.65	75.66	84.15
Mean	284.0	208.2	110.6	70.6	46.4	26.72	61.09	75.16	83.68

5.6.3 TOTAL SUSPENDED SOLIDS

The results of the five cycles of steady stage for total suspended solids pollutant are given in Tables 5.53–5.67. These tables present the influent and effluents concentrations and the corresponding calculated removal efficiency at distances of 2, 5, 8, and 10 m measured from inlet weir for rubber, gravel, and plastic cell. These tables present the mean values of the above mentioned parameters for each cycle.

5.7 TREATMENT CONCENTRATIONS OF NH3, PO4, DO, AND FC

The treatment evaluation of ammonia (NH_3), phosphate (PO_4), dissolved oxygen (DO), and fecal coliforms (FC) parameters are presented in this section. Influent and effluents concentrations at the end of wetland cells were measured.

5.7.1 AMMONIA

Ammonia concentration in wastewater is an indicator of water pollution especially for domestic sewage. Tables 5.68–5.72 demonstrate the results of ammonia concentrations of influent and effluents concentrations for rubber, gravel, and plastic cells and the corresponding calculated removal efficiencies, in addition to the mean values of these parameters for steady stage cycles.

Field and Experimental Results

TABLE 5.53 TSS Concentration and *RE* for Rubber Cell (Cycle 1, $Q = 5.119$ m^3/d)

T_o	Influent (mg/l)	TSS Concentration (mg/l)				TSS Removal Efficiency (%)			
		2 m	*5 m*	*8 m*	*10 m*	*2 m*	*5 m*	*8 m*	*10 m*
224	147	123	93	81	73	16.33	36.73	44.90	50.34
231	140	117	88	76	68	16.43	37.14	45.71	51.43
238	135	113	85	74	66	16.30	37.04	45.19	51.11
245	153	128	98	86	77	16.34	35.95	43.79	49.67
252	149	125	96	84	76	16.11	35.57	43.62	48.99
Mean	144.8	121.2	92.0	80.2	72.0	16.30	36.49	44.64	50.31

TABLE 5.54 TSS Concentration and *RE* for Rubber Cell (Cycle 2, $Q = 3.482$ m^3/d)

T_o	Influent (mg/l)	TSS Concentration (mg/l)				TSS Removal Efficiency (%)			
		2 m	*5 m*	*8 m*	*10 m*	*2 m*	*5 m*	*8 m*	*10 m*
259	131	105	73	58	50	19.85	44.27	55.73	61.83
266	139	112	79	63	55	19.42	43.17	54.68	60.43
273	170	138	98	79	70	18.82	42.35	53.53	58.82
280	151	122	86	69	61	19.21	43.05	54.30	59.60
287	141	114	79	63	55	19.15	43.97	55.32	60.99
Mean	146.4	118.2	83.0	66.4	58.2	19.29	43.36	54.71	60.33

TABLE 5.55 TSS Concentration and *RE* for Rubber Cell (Cycle 3, $Q = 2.396$ m^3/d)

T_o	Influent (mg/l)	TSS Concentration (mg/l)				TSS Removal Efficiency (%)			
		2 m	*5 m*	*8 m*	*10 m*	*2 m*	*5 m*	*8 m*	*10 m*
294	162	124	79	59	50	23.46	51.23	63.58	69.14
301	148	113	72	53	45	23.65	51.35	64.19	69.59
308	165	127	82	62	53	23.03	50.30	62.42	67.88
315	160	123	79	59	50	23.13	50.63	63.13	68.75
322	144	110	69	51	43	23.61	52.08	64.58	70.14
Mean	155.8	119.4	76.2	56.8	48.2	23.38	51.12	63.58	69.10

TABLE 5.56 TSS Concentration and *RE* for Rubber Cell (Cycle 4, Q = 1.696 m^3/d)

T_o	Influent (mg/l)	TSS Concentration (mg/l)				TSS Removal Efficiency (%)			
		2 m	5 m	8 m	10 m	2 m	5 m	8 m	10 m
329	177	131	73	46	33	25.99	58.76	74.01	81.36
336	169	125	69	43	31	26.04	59.17	74.56	81.66
343	164	121	67	41	29	26.22	59.15	75.00	82.32
350	159	117	65	40	28	26.42	59.12	74.84	82.39
357	173	128	72	45	33	26.01	58.38	73.99	80.92
Mean	168.4	124.4	69.2	43.0	30.8	26.14	58.92	74.48	81.73

TABLE 5.57 TSS Concentration and *RE* for Rubber Cell (Cycle 5, Q = 1.263 m^3/d)

T_o	Influent (mg/l)	TSS Concentration (mg/l)				TSS Removal Efficiency (%)			
		2 m	5 m	8 m	10 m	2 m	5 m	8 m	10 m
378	171	122	62	37	24	28.65	63.74	78.36	85.96
385	176	127	65	40	26	27.84	63.07	77.27	85.23
392	180	129	66	40	26	28.33	63.33	77.78	85.56
399	165	118	59	35	22	28.48	64.24	78.79	86.67
406	168	120	60	35	22	28.57	64.29	79.17	86.90
Mean	172.0	123.2	62.4	37.4	24.0	28.37	63.73	78.27	86.06

TABLE 5.58 TSS Concentration and *RE* for Gravel Cell (Cycle 1, Q = 5.037 m^3/d)

T_o	Influent (mg/l)	TSS Concentration (mg/l)				TSS Removal Efficiency (%)			
		2 m	5 m	8 m	10 m	2 m	5 m	8 m	10 m
224	147	119	85	72	66	19.05	42.18	51.02	55.10
231	140	113	79	67	61	19.29	43.57	52.14	56.43
238	135	109	77	65	60	19.26	42.96	51.85	55.56
245	153	125	89	76	70	18.30	41.83	50.33	54.25
252	149	122	87	75	69	18.12	41.61	49.66	53.69
Mean	144.8	117.6	83.4	71.0	65.2	18.80	42.43	51.00	55.01

Field and Experimental Results

TABLE 5.59 TSS Concentration and *RE* for Gravel Cell (Cycle 2, Q = 3.426 m³/d)

T_o	Influent (mg/l)	TSS Concentration (mg/l)				TSS Removal Efficiency (%)			
		2 m	5 m	8 m	10 m	2 m	5 m	8 m	10 m
259	131	104	69	54	48	20.61	47.33	58.78	63.36
266	139	110	74	59	53	20.86	46.76	57.55	61.87
273	170	136	92	73	66	20.00	45.88	57.06	61.18
280	151	120	81	65	58	20.53	46.36	56.95	61.59
287	141	112	75	59	53	20.57	46.81	58.16	62.41
Mean	146.4	116.4	78.2	62.0	55.6	20.51	46.63	57.70	62.08

TABLE 5.60 TSS Concentration and *RE* for Gravel Cell (Cycle 3, Q = 2.357 m³/d)

T_o	Influent (mg/l)	TSS Concentration (mg/l)				TSS Removal Efficiency (%)			
		2 m	5 m	8 m	10 m	2 m	5 m	8 m	10 m
294	162	122	71	52	44	24.69	56.17	67.90	72.84
301	148	111	64	47	39	25.00	56.76	68.24	73.65
308	165	125	73	54	46	24.24	55.76	67.27	72.12
315	160	121	71	52	44	24.38	55.63	67.50	72.50
322	144	108	62	45	37	25.00	56.94	68.75	74.31
Mean	155.8	117.4	68.2	50.0	42.0	24.66	56.25	67.93	73.08

TABLE 5.61 TSS Concentration and *RE* for Gravel Cell (Cycle 4, Q = 1.669 m³/d)

T_o	Influent (mg/l)	TSS Concentration (mg/l)				TSS Removal Efficiency (%)			
		2 m	5 m	8 m	10 m	2 m	5 m	8 m	10 m
329	177	128	66	41	29	27.68	62.71	76.84	83.62
336	169	122	63	38	27	27.81	62.72	77.51	84.02
343	164	118	60	36	25	28.05	63.41	78.05	84.76
350	159	115	59	36	25	27.67	62.89	77.36	84.28
357	173	126	65	41	29	27.17	62.43	76.30	83.24
Mean	168.4	121.8	62.6	38.4	27.0	27.68	62.83	77.21	83.98

TABLE 5.62 TSS Concentration and *RE* for Gravel Cell (Cycle 5, Q = 1.243 m^3/d)

T_o	Influent (mg/l)	*TSS Concentration (mg/l)*				*TSS Removal Efficiency (%)*			
		2 m	*5 m*	*8 m*	*10 m*	*2 m*	*5 m*	*8 m*	*10 m*
378	171	121	57	30	19	29.24	66.67	82.46	88.89
385	176	125	59	32	21	28.98	66.48	81.82	88.07
392	180	127	60	32	20	29.44	66.67	82.22	88.89
399	165	116	54	28	18	29.70	67.27	83.03	89.09
406	168	118	55	28	17	29.76	67.26	83.33	89.88
Mean	172.0	121.4	57.0	30.0	19.0	29.42	66.87	82.57	88.96

TABLE 5.63 TSS Concentration and *RE* for Plastic Cell (Cycle 1, Q = 4.814 m^3/d)

T_o	Influent (mg/l)	*TSS Concentration (mg/l)*				*TSS Removal Efficiency (%)*			
		2 m	*5 m*	*8 m*	*10 m*	*2 m*	*5 m*	*8 m*	*10 m*
224	147	112	71	56	50	23.81	51.70	61.90	65.99
231	140	106	66	52	45	24.29	52.86	62.86	67.86
238	135	103	65	51	45	23.70	51.85	62.22	66.67
245	153	117	75	59	53	23.53	50.98	61.44	65.36
252	149	114	74	59	53	23.49	50.34	60.40	64.43
Mean	144.8	110.4	70.2	55.4	49.2	23.76	51.55	61.76	66.06

TABLE 5.64 TSS Concentration and *RE* for Plastic Cell (Cycle 2, Q = 3.275 m^3/d)

T_o	Influent (mg/l)	*TSS Concentration (mg/l)*				*TSS Removal Efficiency (%)*			
		2 m	*5 m*	*8 m*	*10 m*	*2 m*	*5 m*	*8 m*	*10 m*
259	131	94	48	34	25	28.24	63.36	74.05	80.92
266	139	100	52	37	28	28.06	62.59	73.38	79.86
273	170	124	65	48	37	27.06	61.76	71.76	78.24
280	151	109	57	42	32	27.81	62.25	72.19	78.81
287	141	101	52	37	28	28.37	63.12	73.76	80.14
Mean	146.4	105.6	54.8	39.6	30.0	27.91	62.62	73.03	79.59

Field and Experimental Results

TABLE 5.65 TSS Concentration and *RE* for Plastic Cell (Cycle 3, Q = 2.253 m^3/d)

T_o	Influent (mg/l)	TSS Concentration (mg/l)				TSS Removal Efficiency (%)			
		2 m	5 m	8 m	10 m	2 m	5 m	8 m	10 m
294	162	111	52	30	21	31.48	67.90	81.48	87.04
301	148	101	47	26	19	31.76	68.24	82.43	87.16
308	165	114	54	32	23	30.91	67.27	80.61	86.06
315	160	110	52	30	22	31.25	67.50	81.25	86.25
322	144	98	45	25	17	31.94	68.75	82.64	88.19
Mean	155.8	106.8	50.0	28.6	20.4	31.47	67.93	81.68	86.94

TABLE 5.66 TSS Concentration and *RE* for Plastic Cell (Cycle 4, Q = 1.595 m^3/d)

T_o	Influent (mg/l)	TSS Concentration (mg/l)				TSS Removal Efficiency (%)			
		2 m	5 m	8 m	10 m	2 m	5 m	8 m	10 m
329	177	121	51	30	17	31.64	71.19	83.05	90.40
336	169	115	48	28	16	31.95	71.60	83.43	90.53
343	164	112	46	26	14	31.71	71.95	84.15	91.46
350	159	108	45	26	14	32.08	71.70	83.65	91.19
357	173	118	50	30	17	31.79	71.10	82.66	90.17
Mean	168.4	114.8	48.0	28.0	15.6	31.83	71.51	83.39	90.75

TABLE 5.67 TSS Concentration and *RE* for Plastic Cell (Cycle 5, Q = 1.188 m^3/d)

T_o	Influent (mg/l)	TSS Concentration (mg/l)				TSS Removal Efficiency (%)			
		2 m	5 m	8 m	10 m	2 m	5 m	8 m	10 m
378	171	115	44	20	10	32.75	74.27	88.30	94.15
385	176	119	46	22	12	32.39	73.86	87.50	93.18
392	180	121	47	21	11	32.78	73.89	88.33	93.89
399	165	111	42	19	10	32.73	74.55	88.48	93.94
406	168	112	43	19	9	33.33	74.40	88.69	94.64
Mean	172.0	115.6	44.4	20.2	10.4	32.80	74.19	88.26	93.96

TABLE 5.68 Ammonia Concentration and *RE* (Cycle No. 1, Q_{av} = 4.99 m^3/d)

Date	Influent (mg/l)	Effluent Concentration (mg/l)			NH_3 Removal Efficiency (%)		
		Rubber	*Gravel*	*Plastic*	*Rubber*	*Gravel*	*Plastic*
17/04/2010	19.43	10.29	9.86	8.19	47.04	49.25	57.85
24/04/2010	16.76	9.38	8.99	7.72	44.03	46.36	53.94
01/05/2010	26.16	13.82	13.25	11.09	47.17	49.35	57.61
08/05/2010	22.44	12.50	12.23	9.36	44.30	45.50	58.29
15/05/2010	15.28	8.22	7.73	6.52	46.20	49.41	57.33
Mean	20.01	10.84	10.41	8.58	45.75	47.97	57.00

TABLE 5.69 Ammonia Concentration and *RE* (Cycle No. 2, Q_{av} = 3.39 m^3/d)

Date	Influent (mg/l)	Effluent Concentration (mg/l)			NH_3 Removal Efficiency (%)		
		Rubber	*Gravel*	*Plastic*	*Rubber*	*Gravel*	*Plastic*
22/05/2010	17.79	8.75	8.21	7.10	50.82	53.85	60.09
29/05/2010	18.48	9.32	8.36	7.66	49.57	54.76	58.55
05/06/2010	20.87	10.25	9.23	8.64	50.89	55.77	58.60
12/06/2010	24.86	12.46	11.92	10.45	49.88	52.05	57.96
19/06/2010	23.55	11.14	10.63	10.00	52.70	54.86	57.54
Mean	21.11	10.38	9.67	8.77	50.77	54.26	58.55

TABLE 5.70 Ammonia Concentration and *RE* (Cycle No. 3, Q_{av} = 2.34 m^3/d)

Date	Influent (mg/l)	Effluent Concentration (mg/l)			NH_3 Removal Efficiency (%)		
		Rubber	*Gravel*	*Plastic*	*Rubber*	*Gravel*	*Plastic*
26/06/2010	16.35	7.16	6.85	5.82	56.21	58.10	64.40
03/07/2010	23.74	11.06	9.85	8.33	53.41	58.51	64.91
10/07/2010	18.64	8.51	8.03	6.44	54.35	56.92	65.45
17/07/2010	15.92	6.78	6.39	5.75	57.41	59.86	63.88
24/07/2010	21.31	9.25	8.76	7.55	56.59	58.89	64.57
Mean	19.19	8.55	7.98	6.78	55.59	58.46	64.64

Field and Experimental Results

TABLE 5.71 Ammonia Concentration and *RE* (Cycle No. 4, Q_{av} = 1.65 m³/d)

Date	Influent (mg/l)	Effluent Concentration (mg/l)			NH₃ Removal Efficiency (%)		
		Rubber	*Gravel*	*Plastic*	*Rubber*	*Gravel*	*Plastic*
31/07/2010	15.73	6.68	6.30	4.25	57.53	59.95	72.98
07/08/2010	19.92	8.32	8.00	5.56	58.23	59.84	72.09
14/08/2010	25.38	11.31	9.98	7.18	55.44	60.68	71.71
21/08/2010	16.11	6.75	6.50	4.45	58.10	59.65	72.38
28/08/2010	20.37	8.46	7.99	5.29	58.47	60.78	74.03
Mean	19.50	8.30	7.75	5.35	57.55	60.18	72.64

TABLE 5.72 Ammonia Concentration and *RE* (Cycle No. 5, Q_{av} = 1.23 m³/d)

Date	Influent (mg/l)	Effluent Concentration (mg/l)			NH₃ Removal Efficiency (%)		
		Rubber	*Gravel*	*Plastic*	*Rubber*	*Gravel*	*Plastic*
18/09/2010	20.22	7.44	6.05	4.44	63.20	70.08	78.04
25/09/2010	18.57	6.70	6.29	3.52	63.92	66.13	81.04
02/10/2010	15.96	6.38	5.19	3.65	60.03	67.48	77.13
09/10/2010	21.83	8.07	7.32	4.99	63.03	66.47	77.14
16/10/2010	19.45	7.35	6.17	4.03	62.21	68.28	79.28
Mean	19.21	7.19	6.20	4.13	62.48	67.69	78.53

5.7.2 PHOSPHATE

The most reactive forms of phosphorus compounds are phosphates. Tables 5.73–5.77 present phosphate concentration values of steady stage cycles and the removal efficiencies for rubber, gravel, and plastic cells.

5.7.3 DISSOLVED OXYGEN

Dissolved oxygen analysis measures the amount of gaseous oxygen, dissolved in an aqueous solution. The influent DO for wetland cells approaches zero. The effluents DO concentration results for rubber, gravel, plastic cells are shown in Tables 5.78–5.82.

TABLE 5.73 Phosphate Concentration and *RE* (Cycle No. 1, Q_{av} = 4.99 m^3/d)

Date	Influent (mg/l)	Effluent Concentration (mg/l)			PO$_4$ Removal Efficiency (%)		
		Rubber	*Gravel*	*Plastic*	*Rubber*	*Gravel*	*Plastic*
17/04/2010	3.24	1.72	1.42	0.99	46.88	56.25	69.38
24/04/2010	2.96	1.54	1.27	0.93	47.96	57.16	68.53
01/05/2010	2.77	1.42	1.18	0.86	48.75	57.55	68.81
08/05/2010	3.54	1.88	1.55	1.13	46.95	56.11	68.11
15/05/2010	3.32	1.75	1.44	0.99	47.25	56.77	70.05
Mean	3.17	1.66	1.37	0.98	47.56	56.77	68.98

TABLE 5.74 Phosphate Concentration and *RE* (Cycle No. 2, Q_{av} = 3.39 m^3/d)

Date	Influent (mg/l)	Effluent Concentration (mg/l)			PO$_4$ Removal Efficiency (%)		
		Rubber	*Gravel*	*Plastic*	*Rubber*	*Gravel*	*Plastic*
22/05/2010	2.85	1.17	0.98	0.78	58.89	65.56	72.78
29/05/2010	2.71	1.12	0.92	0.73	58.50	66.01	73.11
05/06/2010	2.53	1.03	0.86	0.70	59.13	65.99	72.28
12/06/2010	3.11	1.26	1.09	0.85	59.48	64.85	72.51
19/06/2010	3.26	1.36	1.09	0.87	58.32	66.51	73.44
Mean	2.89	1.19	0.99	0.79	58.86	65.78	72.82

TABLE 5.75 Phosphate Concentration and *RE* (Cycle No. 3, Q_{av} = 2.34 m^3/d)

Date	Influent (mg/l)	Effluent Concentration (mg/l)			PO$_4$ Removal Efficiency (%)		
		Rubber	*Gravel*	*Plastic*	*Rubber*	*Gravel*	*Plastic*
26/06/2010	1.89	0.72	0.60	0.44	62.07	68.28	76.55
03/07/2010	2.05	0.76	0.64	0.49	63.05	68.68	76.15
10/07/2010	2.33	0.85	0.72	0.54	63.68	69.03	76.97
17/07/2010	2.14	0.77	0.69	0.48	63.98	67.56	77.34
24/07/2010	2.57	0.99	0.79	0.58	61.54	69.22	77.39
Mean	2.20	0.82	0.69	0.51	62.86	68.55	76.88

TABLE 5.76 Phosphate Concentration and *RE* (Cycle No. 4, Q_{av} = 1.65 m³/d)

Date	Influent (mg/l)	Effluent Concentration (mg/l)			PO₄ Removal Efficiency (%)		
		Rubber	*Gravel*	*Plastic*	*Rubber*	*Gravel*	*Plastic*
31/07/2010	2.70	0.90	0.79	0.48	66.79	70.71	82.14
07/08/2010	2.94	0.97	0.86	0.51	67.06	70.79	82.59
14/08/2010	3.05	0.98	0.87	0.52	67.79	71.59	82.94
21/08/2010	3.22	1.10	0.99	0.59	65.90	69.35	81.65
28/08/2010	3.16	1.10	0.95	0.58	65.15	69.78	81.79
Mean	3.01	1.01	0.89	0.54	66.54	70.44	82.22

TABLE 5.77 Phosphate Concentration and *RE* (Cycle No. 5, Q_{av} = 1.23 m³/d)

Date	Influent (mg/l)	Effluent Concentration (mg/l)			PO₄ Removal Efficiency (%)		
		Rubber	*Gravel*	*Plastic*	*Rubber*	*Gravel*	*Plastic*
18/09/2010	3.15	0.84	0.77	0.32	73.41	75.48	89.81
25/09/2010	2.80	0.76	0.67	0.33	73.01	75.99	88.30
02/10/2010	2.66	0.72	0.63	0.31	72.80	76.50	88.33
09/10/2010	2.89	0.79	0.72	0.34	72.52	75.14	88.14
16/10/2010	2.97	0.83	0.72	0.34	72.13	75.90	88.54
Mean	2.89	0.79	0.70	0.33	72.77	75.80	88.62

TABLE 5.78 Outlet Dissolved Oxygen Concentration (Cycle No. 1, Q_{av} = 4.99 m³/d)

Cycle No.	Run No.	Date	T_o (day)	T_s (day)	Outlet DO concentration (mg/l)		
					Rubber	*Gravel*	*Plastic*
1	1	17/04/2010	224	168	1.7	2.8	3.0
	2	24/04/2010	231	175	1.7	2.5	3.1
	3	01/05/2010	238	182	1.9	2.4	3.3
	4	08/05/2010	245	189	2.1	2.9	2.9
	5	15/05/2010	252	196	1.8	2.5	3.4
The average value of dissolved oxygen (mg/l)					1.84	2.62	3.14

TABLE 5.79 Outlet Dissolved Oxygen Concentration (Cycle No. 2, Q_{av} = 3.39 m³/d)

Cycle No.	Run No.	Date	T_o (day)	T_s (day)	Outlet DO concentration (mg/l)		
					Rubber	Gravel	Plastic
2	1	22/05/2010	259	203	2.9	3.4	3.8
	2	29/05/2010	266	210	2.8	3.3	3.9
	3	05/06/2010	273	217	3.0	3.5	3.7
	4	12/06/2010	280	224	3.1	3.4	3.8
	5	19/06/2010	287	231	3.0	3.3	3.7
The average value of dissolved oxygen (mg/l)					2.96	3.38	3.78

TABLE 5.80 Outlet Dissolved Oxygen Concentration (Cycle No. 3, Q_{av} = 2.34 m³/d)

Cycle No.	Run No.	Date	T_o (day)	T_s (day)	Outlet DO concentration (mg/l)		
					Rubber	Gravel	Plastic
3	1	26/06/2010	294	238	3.3	4.0	4.2
	2	03/07/2010	301	245	3.5	4.1	4.3
	3	10/07/2010	308	252	3.4	4.1	4.4
	4	17/07/2010	315	259	3.4	4.2	4.3
	5	24/07/2010	322	266	3.4	4.0	4.3
The average value of dissolved oxygen (mg/l)					3.40	4.08	4.30

TABLE 5.81 Outlet Dissolved Oxygen Concentration (Cycle No. 4, Q_{av} = 1.65 m³/d)

Cycle No.	Run No.	Date	T_o (day)	T_s (day)	Outlet DO concentration (mg/l)		
					Rubber	Gravel	Plastic
4	1	31/07/2010	329	273	3.8	4.2	4.6
	2	07/08/2010	336	280	3.8	4.3	4.7
	3	14/08/2010	343	287	3.9	4.2	4.6
	4	21/08/2010	350	294	3.7	4.4	4.5
	5	28/08/2010	357	301	3.9	4.3	4.4
The average value of dissolved oxygen (mg/l)					3.82	4.28	4.56

Field and Experimental Results 169

TABLE 5.82 Outlet Dissolved Oxygen Concentration (Cycle No. 5, $Q_{av} = 1.23$ m³/d)

Cycle No.	Run No.	Date	T_o (day)	T_s (day)	Outlet DO concentration (mg/l)		
					Rubber	Gravel	Plastic
5	1	18/09/2010	378	322	4.1	4.5	5.2
	2	25/09/2010	385	329	4.2	4.6	5.1
	3	02/10/2010	392	336	4.1	4.5	4.9
	4	09/10/2010	399	343	4.2	4.7	4.9
	5	16/10/2010	406	350	4.2	4.5	5.0
The average value of dissolved oxygen (mg/l)					4.16	4.56	5.02

5.7.4 FECAL COLIFORMS

The influent and effluents fecal coliforms counts (MPN = most probable number) results and the corresponding removal efficiencies for the rubber, gravel, and plastic cells are shown in Tables 5.83–5.87 for each steady stage cycle.

5.8 TREATMENT CONCENTRATIONS OF HEAVY METALS

Five heavy metals were tested by Atomic Absorption Spectrometer. These metals are Zinc, Iron, Manganese, Lead, and Cadmium elements. The following items present the results of the heavy metals in tabular forms. Influent and effluents concentrations at the end of treatment wetland cells were measured for steady stage cycles.

5.8.1 ZINC ELEMENT

Domestic and industrial discharges are probably the two most important sources for zinc in the water environment. The influent and effluents concentrations of Zinc element for the rubber, gravel, and plastic cells are shown in Tables 5.88–5.92 with the corresponding calculated removal efficiencies at the end of cells for the five cycles of steady stage.

TABLE 5.83 FC (MPN/100 ml) and *RE* for the Three Media (Cycle 1, Q_{av} = 4.99 m³/d)

Date	Inlet (MPN)	Outlet Fecal Coliforms (MPN)			FC Removal Efficiency (%)		
		Rubber	*Gravel*	*Plastic*	*Rubber*	*Gravel*	*Plastic*
17/04/2010	210000	6100	5600	5400	97.10	97.33	97.43
24/04/2010	240000	6300	5800	5500	97.38	97.58	97.71
01/05/2010	210000	6000	5800	5500	97.14	97.24	97.38
08/05/2010	250000	6300	6000	5600	97.48	97.60	97.76
15/05/2010	240000	6100	6000	5400	97.46	97.50	97.75
Mean	230000	6160	5840	5480	97.31	97.45	97.61

TABLE 5.84 FC (MPN/100 ml) and *RE* for the Three Media (Cycle 2, Q_{av} = 3.39 m³/d)

Date	Inlet (MPN)	Outlet Fecal Coliforms (MPN)			FC Removal Efficiency (%)		
		Rubber	*Gravel*	*Plastic*	*Rubber*	*Gravel*	*Plastic*
22/05/2010	240000	4800	4500	4300	98.00	98.13	98.21
29/05/2010	260000	4900	4700	4500	98.12	98.19	98.27
05/06/2010	220000	4700	4500	4100	97.86	97.95	98.14
12/06/2010	250000	4800	4600	4300	98.08	98.16	98.28
19/06/2010	270000	4900	4700	4500	98.19	98.26	98.33
Mean	248000	4820	4600	4340	98.05	98.14	98.25

TABLE 5.85 FC (MPN/100 ml) and *RE* for the Three Media (Cycle 3, Q_{av} = 2.34 m³/d)

Date	Inlet (MPN)	Outlet Fecal Coliforms (MPN)			FC Removal Efficiency (%)		
		Rubber	*Gravel*	*Plastic*	*Rubber*	*Gravel*	*Plastic*
26/06/2010	230000	3900	3700	3500	98.30	98.39	98.48
03/07/2010	210000	3900	3600	3400	98.14	98.29	98.38
10/07/2010	200000	3800	3600	3400	98.10	98.20	98.30
17/07/2010	210000	3800	3700	3500	98.19	98.24	98.33
24/07/2010	260000	4000	3800	3600	98.46	98.54	98.62
Mean	222000	3880	3680	3480	98.24	98.33	98.42

Field and Experimental Results

TABLE 5.86 FC (MPN/100 ml) and *RE* for the Three Media (Cycle 4, Q_{av} = 1.65 m³/d)

Date	Inlet (MPN)	Outlet Fecal Coliforms (MPN)			FC Removal Efficiency (%)		
		Rubber	*Gravel*	*Plastic*	*Rubber*	*Gravel*	*Plastic*
31/07/2010	220000	3200	2800	2600	98.55	98.73	98.82
07/08/2010	250000	3200	3100	2700	98.72	98.76	98.92
14/08/2010	260000	3300	3100	2700	98.73	98.81	98.96
21/08/2010	220000	3100	2700	2600	98.59	98.77	98.82
28/08/2010	210000	3100	2700	2500	98.52	98.71	98.81
Mean	232000	3180	2880	2620	98.62	98.76	98.87

TABLE 5.87 FC (MPN/100 ml) and *RE* for the Three Media (Cycle 5, Q_{av} = 1.23 m³/d)

Date	Inlet (MPN)	Outlet Fecal Coliforms (MPN)			FC Removal Efficiency (%)		
		Rubber	*Gravel*	*Plastic*	*Rubber*	*Gravel*	*Plastic*
18/09/2010	270000	2400	2200	2000	99.11	99.19	99.26
25/09/2010	210000	2300	2100	1700	98.90	99.00	99.19
02/10/2010	240000	2400	2300	2000	99.00	99.04	99.17
09/10/2010	200000	2200	2100	1700	98.90	98.95	99.15
16/10/2010	220000	2300	2100	1800	98.95	99.05	99.18
Mean	228000	2320	2160	1840	98.97	99.04	99.19

TABLE 5.88 Zinc Concentration and Removal Efficiency (Cycle 1, Q_{av} = 4.99 m³/d)

Date	Influent (mg/l)	Effluent Concentration (mg/l)			Zn Removal Efficiency (%)		
		Rubber	*Gravel*	*Plastic*	*Rubber*	*Gravel*	*Plastic*
17/04/2010	1.55	0.96	0.88	0.76	38.32	43.25	51.00
24/04/2010	1.85	1.13	1.03	0.90	39.15	44.25	51.12
01/05/2010	1.66	1.00	0.94	0.82	39.69	43.10	50.88
08/05/2010	2.01	1.20	1.15	0.98	40.14	42.69	51.36
15/05/2010	2.16	1.31	1.20	1.03	39.45	44.43	52.23
Mean	1.85	1.12	1.04	0.90	39.35	43.54	51.32

TABLE 5.89 Zinc Concentration and Removal Efficiency (Cycle 2, Q_{av} = 3.39 m³/d)

Date	Influent (mg/l)	Effluent Concentration (mg/l)			Zn Removal Efficiency (%)		
		Rubber	*Gravel*	*Plastic*	*Rubber*	*Gravel*	*Plastic*
22/05/2010	1.42	0.80	0.70	0.53	43.84	50.94	62.99
29/05/2010	1.41	0.79	0.70	0.51	44.20	50.05	63.95
05/06/2010	1.73	0.96	0.84	0.62	44.65	51.56	64.18
12/06/2010	1.91	1.05	0.91	0.71	44.98	52.17	62.78
19/06/2010	1.74	0.95	0.83	0.66	45.25	52.35	62.15
Mean	1.64	0.91	0.80	0.60	44.58	51.41	63.21

TABLE 5.90 Zinc Concentration and Removal Efficiency (Cycle 3, Q_{av} = 2.34 m³/d)

Date	Influent (mg/l)	Effluent Concentration (mg/l)			Zn Removal Efficiency (%)		
		Rubber	*Gravel*	*Plastic*	*Rubber*	*Gravel*	*Plastic*
26/06/2010	1.28	0.66	0.56	0.41	48.09	56.18	67.74
03/07/2010	1.38	0.70	0.58	0.44	49.55	58.14	68.26
10/07/2010	1.57	0.79	0.65	0.49	49.59	58.69	68.65
17/07/2010	1.66	0.83	0.70	0.50	49.88	57.88	69.74
24/07/2010	1.79	0.89	0.78	0.55	50.44	56.48	69.43
Mean	1.54	0.77	0.65	0.48	49.51	57.47	68.76

TABLE 5.91 Zinc Concentration and Removal Efficiency (Cycle 4, Q_{av} = 1.65 m³/d)

Date	Influent (mg/l)	Effluent Concentration (mg/l)			Zn Removal Efficiency (%)		
		Rubber	*Gravel*	*Plastic*	*Rubber*	*Gravel*	*Plastic*
31/07/2010	1.43	0.53	0.47	0.41	62.64	66.83	71.54
07/08/2010	1.85	0.72	0.65	0.52	61.12	64.89	72.15
14/08/2010	1.79	0.69	0.63	0.49	61.45	64.85	72.35
21/08/2010	1.92	0.72	0.66	0.52	62.33	65.63	72.79
28/08/2010	1.64	0.64	0.56	0.46	61.13	65.75	72.00
Mean	1.73	0.66	0.59	0.48	61.73	65.59	72.17

Field and Experimental Results   173

TABLE 5.92 Zinc Concentration and Removal Efficiency (Cycle 5, Q_{av} = 1.23 m³/d)

Date	Influent (mg/l)	Effluent Concentration (mg/l)			Zn Removal Efficiency (%)		
		Rubber	Gravel	Plastic	Rubber	Gravel	Plastic
18/09/2010	1.50	0.45	0.36	0.28	69.94	75.69	81.44
25/09/2010	1.69	0.54	0.42	0.32	68.15	75.12	81.33
02/10/2010	1.44	0.46	0.36	0.27	68.22	75.16	81.10
09/10/2010	1.38	0.43	0.36	0.24	68.48	74.00	82.46
16/10/2010	1.87	0.59	0.48	0.32	68.47	74.59	82.81
Mean	1.58	0.49	0.40	0.29	68.65	74.91	81.83

5.8.2 IRON ELEMENT

The influent and effluents concentrations of Iron element for the rubber, gravel, and plastic cells are shown in Tables 5.93–5.97 and the corresponding removal efficiencies for the five cycles of steady stage.

5.8.3 MANGANESE ELEMENT

The influent and effluent results of Manganese element and the corresponding removal efficiencies for the rubber, gravel, and plastic cells are shown respectively in Tables 5.98–5.102 for steady stage cycles.

TABLE 5.93 Iron Concentration and Removal Efficiency (Cycle 1, Q_{av} = 4.99 m³/d)

Date	Influent (mg/l)	Effluent Concentration (mg/l)			Fe Removal Efficiency (%)		
		Rubber	Gravel	Plastic	Rubber	Gravel	Plastic
17/04/2010	0.72	0.58	0.50	0.47	19.72	30.56	34.72
24/04/2010	0.82	0.65	0.57	0.54	20.72	30.00	33.75
01/05/2010	0.63	0.50	0.43	0.41	20.60	31.16	34.13
08/05/2010	0.65	0.52	0.46	0.42	20.35	29.63	35.88
15/05/2010	0.95	0.76	0.67	0.61	19.97	29.12	35.44
Mean	0.75	0.60	0.53	0.49	20.27	30.09	34.78

TABLE 5.94 Iron Concentration and Removal Efficiency (Cycle 2, Q_{av} = 3.39 m^3/d)

Date	Influent (mg/l)	Effluent Concentration (mg/l)			Fe Removal Efficiency (%)		
		Rubber	Gravel	Plastic	Rubber	Gravel	Plastic
22/05/2010	1.12	0.74	0.72	0.51	33.78	36.10	54.39
29/05/2010	1.05	0.69	0.66	0.47	34.23	37.50	54.99
05/06/2010	0.96	0.63	0.60	0.43	34.55	37.96	55.26
12/06/2010	0.84	0.56	0.53	0.39	33.77	37.23	54.01
19/06/2010	0.79	0.53	0.50	0.36	33.06	36.48	54.05
Mean	0.95	0.63	0.60	0.43	33.88	37.05	54.54

TABLE 5.95 Iron Concentration and Removal Efficiency (Cycle 3, Q_{av} = 2.34 m^3/d)

Date	Influent (mg/l)	Effluent Concentration (mg/l)			Fe Removal Efficiency (%)		
		Rubber	Gravel	Plastic	Rubber	Gravel	Plastic
26/06/2010	0.85	0.46	0.37	0.26	45.56	56.43	69.37
03/07/2010	0.77	0.42	0.34	0.23	45.50	55.54	70.44
10/07/2010	0.64	0.34	0.27	0.19	47.12	57.44	70.74
17/07/2010	0.66	0.35	0.28	0.21	46.95	57.86	68.63
24/07/2010	0.87	0.47	0.38	0.28	46.42	56.78	67.91
Mean	0.76	0.41	0.33	0.23	46.31	56.81	69.42

TABLE 5.96 Iron Concentration and Removal Efficiency (Cycle 4, Q_{av} = 1.65 m^3/d)

Date	Influent (mg/l)	Effluent Concentration (mg/l)			Fe Removal Efficiency (%)		
		Rubber	Gravel	Plastic	Rubber	Gravel	Plastic
31/07/2010	1.19	0.58	0.49	0.33	51.15	58.85	72.15
07/08/2010	0.89	0.43	0.38	0.24	51.14	57.10	72.52
14/08/2010	0.64	0.31	0.27	0.17	51.58	57.90	73.94
21/08/2010	0.97	0.46	0.41	0.28	52.95	58.22	71.46
28/08/2010	0.83	0.40	0.33	0.22	52.23	60.43	73.64
Mean	0.90	0.44	0.37	0.25	51.81	58.50	72.74

Field and Experimental Results

TABLE 5.97 Iron Concentration and Removal Efficiency (Cycle 5, Q_{av} = 1.23 m³/d)

Date	Influent (mg/l)	Effluent Concentration (mg/l)			Fe Removal Efficiency (%)		
		Rubber	Gravel	Plastic	Rubber	Gravel	Plastic
18/09/2010	0.98	0.36	0.22	0.19	63.17	77.24	80.12
25/09/2010	0.76	0.28	0.18	0.15	63.77	76.89	80.86
02/10/2010	0.74	0.27	0.18	0.15	62.95	76.26	79.68
09/10/2010	0.89	0.31	0.22	0.19	64.74	75.47	79.12
16/10/2010	0.88	0.32	0.22	0.19	63.85	74.49	78.36
Mean	0.85	0.31	0.20	0.17	63.70	76.07	79.63

TABLE 5.98 Manganese Concentration and RE (Cycle No. 1, Q_{av} = 4.99 m³/d)

Date	Influent (mg/l)	Effluent Concentration (mg/l)			Mn Removal Efficiency (%)		
		Rubber	Gravel	Plastic	Rubber	Gravel	Plastic
17/04/2010	0.41	0.34	0.32	0.28	17.07	21.95	31.71
24/04/2010	0.27	0.22	0.21	0.18	18.18	22.42	32.37
01/05/2010	0.22	0.18	0.17	0.15	18.95	22.63	32.33
08/05/2010	0.30	0.25	0.24	0.21	17.85	21.25	30.48
15/05/2010	0.28	0.23	0.22	0.19	16.55	20.98	31.44
Mean	0.30	0.24	0.23	0.20	17.72	21.85	31.67

TABLE 5.99 Manganese Concentration and RE (Cycle No. 2, Q_{av} = 3.39 m³/d)

Date	Influent (mg/l)	Effluent Concentration (mg/l)			Mn Removal Efficiency (%)		
		Rubber	Gravel	Plastic	Rubber	Gravel	Plastic
22/05/2010	0.15	0.11	0.11	0.07	25.93	29.63	51.85
29/05/2010	0.48	0.36	0.33	0.24	25.05	31.22	50.50
05/06/2010	0.26	0.19	0.18	0.13	25.91	30.60	51.22
12/06/2010	0.42	0.32	0.30	0.20	24.33	28.28	53.36
19/06/2010	0.50	0.38	0.35	0.23	24.13	29.55	53.14
Mean	0.36	0.27	0.25	0.17	25.07	29.86	52.01

TABLE 5.100 Manganese Concentration and *RE* (Cycle No. 3, Q_{av} = 2.34 m^3/d)

Date	Influent (mg/l)	Effluent Concentration (mg/l)			Mn Removal Efficiency (%)		
		Rubber	*Gravel*	*Plastic*	*Rubber*	*Gravel*	*Plastic*
26/06/2010	0.36	0.21	0.16	0.13	40.91	54.55	63.64
03/07/2010	0.45	0.27	0.21	0.17	40.40	53.98	62.28
10/07/2010	0.29	0.17	0.13	0.10	41.16	53.62	63.91
17/07/2010	0.38	0.22	0.17	0.13	41.89	54.51	64.55
24/07/2010	0.17	0.10	0.08	0.06	42.85	55.81	64.77
Mean	0.33	0.19	0.15	0.12	41.44	54.49	63.83

TABLE 5.101 Manganese Concentration and *RE* (Cycle No. 4, Q_{av} = 1.65 m^3/d)

Date	Influent (mg/l)	Effluent Concentration (mg/l)			Mn Removal Efficiency (%)		
		Rubber	*Gravel*	*Plastic*	*Rubber*	*Gravel*	*Plastic*
31/07/2010	0.28	0.14	0.12	0.09	50.00	56.67	66.67
07/08/2010	0.23	0.11	0.10	0.07	51.50	57.42	67.42
14/08/2010	0.24	0.12	0.10	0.08	51.20	57.59	67.23
21/08/2010	0.34	0.17	0.15	0.11	49.87	56.33	67.25
28/08/2010	0.37	0.18	0.16	0.12	52.11	56.48	66.94
Mean	0.29	0.14	0.13	0.10	50.94	56.90	67.10

TABLE 5.102 Manganese Concentration and *RE* (Cycle No. 5, Q_{av} = 1.23 m^3/d)

Date	Influent (mg/l)	Effluent Concentration (mg/l)			Mn Removal Efficiency (%)		
		Rubber	*Gravel*	*Plastic*	*Rubber*	*Gravel*	*Plastic*
18/09/2010	0.26	0.10	0.07	0.06	60.71	71.43	78.57
25/09/2010	0.25	0.10	0.07	0.05	59.86	72.72	79.22
02/10/2010	0.31	0.12	0.09	0.06	59.74	72.45	79.54
09/10/2010	0.41	0.16	0.12	0.09	61.46	71.85	78.94
16/10/2010	0.44	0.17	0.12	0.10	61.73	71.64	77.16
Mean	0.33	0.13	0.09	0.07	60.70	72.02	78.69

5.8.4 LEAD ELEMENT

Lead is often found in wastewater from printed circuit board factories, electronics assembly plants, and battery recycling plants. The influent and effluents concentrations of Lead with the removal efficiencies for wetland cells are shown in Tables 5.103–5.107.

5.8.5 CADMIUM ELEMENT

The influent and effluent results of Cadmium element for the rubber, gravel, and plastic cells and the corresponding removal efficiencies are shown in Tables 5.108–5.112 for steady stage cycles.

TABLE 5.103 Lead Concentration and Removal Efficiency (Cycle 1, Q_{av} = 4.99 m³/d)

Date	Influent (mg/l)	Effluent Concentration (mg/l)			Pb Removal Efficiency (%)		
		Rubber	Gravel	Plastic	Rubber	Gravel	Plastic
17/04/2010	0.040	0.032	0.029	0.024	20.00	27.50	40.00
24/04/2010	0.043	0.035	0.031	0.026	19.17	28.11	39.64
01/05/2010	0.061	0.049	0.044	0.037	19.50	27.15	39.82
08/05/2010	0.031	0.024	0.023	0.018	21.33	26.98	41.15
15/05/2010	0.021	0.017	0.015	0.012	20.79	26.84	40.87
Mean	0.039	0.031	0.028	0.023	20.16	27.32	40.30

TABLE 5.104 Lead Concentration and Removal Efficiency (Cycle 2, Q_{av} = 3.39 m³/d)

Date	Influent (mg/l)	Effluent Concentration (mg/l)			Pb Removal Efficiency (%)		
		Rubber	Gravel	Plastic	Rubber	Gravel	Plastic
22/05/2010	0.033	0.024	0.022	0.017	27.91	32.56	48.84
29/05/2010	0.039	0.028	0.026	0.020	28.28	33.15	49.73
05/06/2010	0.025	0.018	0.017	0.013	28.51	33.57	48.11
12/06/2010	0.027	0.019	0.018	0.014	29.61	32.60	48.02
19/06/2010	0.048	0.035	0.032	0.024	27.25	32.99	49.88
Mean	0.034	0.025	0.023	0.018	28.31	32.97	48.92

TABLE 5.105 Lead Concentration and Removal Efficiency (Cycle 3, Q_{av} = 2.34 m^3/d)

Date	Influent (mg/l)	Effluent Concentration (mg/l)			Pb Removal Efficiency (%)		
		Rubber	*Gravel*	*Plastic*	*Rubber*	*Gravel*	*Plastic*
26/06/2010	0.036	0.018	0.017	0.015	49.18	54.10	57.38
03/07/2010	0.044	0.022	0.020	0.018	50.50	53.58	58.22
10/07/2010	0.027	0.014	0.012	0.011	49.73	53.99	58.88
17/07/2010	0.064	0.032	0.029	0.026	49.78	55.16	59.76
24/07/2010	0.047	0.023	0.021	0.019	51.73	55.53	59.48
Mean	0.044	0.022	0.020	0.018	50.18	54.47	58.74

TABLE 5.106 Lead Concentration and Removal Efficiency (Cycle 4, Q_{av} = 1.65 m^3/d)

Date	Influent (mg/l)	Effluent Concentration (mg/l)			Pb Removal Efficiency (%)		
		Rubber	*Gravel*	*Plastic*	*Rubber*	*Gravel*	*Plastic*
31/07/2010	0.027	0.012	0.011	0.010	54.84	58.06	61.29
07/08/2010	0.071	0.031	0.030	0.027	55.64	58.11	61.83
14/08/2010	0.061	0.029	0.026	0.023	53.12	57.12	61.98
21/08/2010	0.055	0.025	0.022	0.022	53.74	59.22	60.32
28/08/2010	0.043	0.020	0.017	0.016	54.50	59.77	62.74
Mean	0.051	0.023	0.021	0.020	54.37	58.46	61.63

TABLE 5.107 Lead Concentration and Removal Efficiency (Cycle 5, Q_{av} = 1.23 m^3/d)

Date	Influent (mg/l)	Effluent Concentration (mg/l)			Pb Removal Efficiency (%)		
		Rubber	*Gravel*	*Plastic*	*Rubber*	*Gravel*	*Plastic*
18/09/2010	0.072	0.031	0.027	0.024	57.14	61.90	66.67
25/09/2010	0.045	0.019	0.017	0.015	57.22	62.85	66.31
02/10/2010	0.061	0.026	0.023	0.020	57.36	62.94	66.78
09/10/2010	0.032	0.014	0.012	0.011	57.23	62.73	66.91
16/10/2010	0.066	0.028	0.025	0.021	58.12	62.49	67.89
Mean	0.055	0.023	0.021	0.018	57.41	62.58	66.91

TABLE 5.108 Cadmium Concentration and *RE* (Cycle No. 1, Q_{av} = 4.99 m³/d)

Date	Influent (mg/l)	Effluent Concentration (mg/l)			Cd Removal Efficiency (%)		
		Rubber	*Gravel*	*Plastic*	*Rubber*	*Gravel*	*Plastic*
17/04/2010	0.00173	0.00142	0.00132	0.00122	17.65	23.53	29.41
24/04/2010	0.00252	0.00206	0.00191	0.00175	18.20	24.15	30.61
01/05/2010	0.00217	0.00178	0.00164	0.00154	18.13	24.23	29.11
08/05/2010	0.00306	0.00251	0.00237	0.00219	17.95	22.68	28.33
15/05/2010	0.00211	0.00175	0.00164	0.00150	16.89	22.14	28.79
Mean	0.00232	0.00191	0.00178	0.00164	17.76	23.35	29.25

TABLE 5.109 Cadmium Concentration and *RE* (Cycle No. 2, Q_{av} = 3.39 m³/d)

Date	Influent (mg/l)	Effluent Concentration (mg/l)			Cd Removal Efficiency (%)		
		Rubber	*Gravel*	*Plastic*	*Rubber*	*Gravel*	*Plastic*
22/05/2010	0.00191	0.00153	0.00138	0.00130	20.00	28.00	32.10
29/05/2010	0.00223	0.00178	0.00160	0.00149	20.24	28.16	33.15
05/06/2010	0.00298	0.00237	0.00212	0.00198	20.56	28.95	33.46
12/06/2010	0.00212	0.00167	0.00152	0.00146	21.36	28.31	31.26
19/06/2010	0.00233	0.00182	0.00167	0.00160	21.87	28.47	31.48
Mean	0.00231	0.00183	0.00166	0.00156	20.81	28.38	32.29

TABLE 5.110 Cadmium Concentration and *RE* (Cycle No. 3, Q_{av} = 2.34 m³/d)

Date	Influent (mg/l)	Effluent Concentration (mg/l)			Cd Removal Efficiency (%)		
		Rubber	*Gravel*	*Plastic*	*Rubber*	*Gravel*	*Plastic*
26/06/2010	0.00196	0.00149	0.00140	0.00131	23.81	28.57	33.33
03/07/2010	0.00185	0.00139	0.00130	0.00122	24.64	29.64	33.89
10/07/2010	0.00267	0.00200	0.00187	0.00174	24.98	29.87	34.76
17/07/2010	0.00284	0.00215	0.00201	0.00185	24.36	29.31	34.78
24/07/2010	0.00290	0.00224	0.00202	0.00191	22.87	30.48	34.12
Mean	0.00244	0.00186	0.00172	0.00161	24.13	29.57	34.18

TABLE 5.111 Cadmium Concentration and *RE* (Cycle No. 4, Q_{av} = 1.65 m³/d)

Date	Influent (mg/l)	Effluent Concentration (mg/l)			Cd Removal Efficiency (%)		
		Rubber	*Gravel*	*Plastic*	*Rubber*	*Gravel*	*Plastic*
31/07/2010	0.00255	0.00170	0.00161	0.00145	33.33	36.67	43.33
07/08/2010	0.00273	0.00183	0.00170	0.00153	33.00	37.65	44.13
14/08/2010	0.00261	0.00172	0.00162	0.00146	33.98	37.94	44.15
21/08/2010	0.00235	0.00154	0.00152	0.00130	34.61	35.46	44.58
28/08/2010	0.00247	0.00167	0.00159	0.00141	32.32	35.78	42.94
Mean	0.00254	0.00169	0.00161	0.00143	33.45	36.70	43.83

TABLE 5.112 Cadmium Concentration and *RE* (Cycle No. 5, Q_{av} = 1.23 m³/d)

Date	Influent (mg/l)	Effluent Concentration (mg/l)			Cd Removal Efficiency (%)		
		Rubber	*Gravel*	*Plastic*	*Rubber*	*Gravel*	*Plastic*
18/09/2010	0.00175	0.00112	0.00087	0.00078	35.78	50.12	55.26
25/09/2010	0.00130	0.00082	0.00064	0.00057	36.64	50.55	55.99
02/10/2010	0.00126	0.00080	0.00062	0.00055	36.48	50.63	56.48
09/10/2010	0.00182	0.00116	0.00089	0.00079	36.18	50.89	56.78
16/10/2010	0.00241	0.00152	0.00116	0.00111	36.94	51.84	54.12
Mean	0.00171	0.00109	0.00084	0.00076	36.40	50.81	55.73

KEYWORDS

- **BOD**
- **COD**
- **coliform**
- **heavy metals**
- **porosity**
- **setup stage**
- **steady stage**
- **TSS**

CHAPTER 6

SET UP STAGE ANALYSIS

CONTENTS

6.1 Introduction 181
6.2 Biochemical Oxygen Demand 182
6.3 Chemical Oxygen Demand 200
6.4 Total Suspended Solids 213
Keywords 228

6.1 INTRODUCTION

Set up stage represents the interval from system start of operation till the stability in media porosity and completion of the attached biofilm and plant growth. Through this stage BOD, COD, and TSS were the main pollutants. The treatment performance of the horizontal subsurface flow constructed wetland systems can be evaluated. A comparison study is presented between the three deferent media. Two main aspects are provided on this part, the first one is the effect of wetland cell length and the second aspect is the effect of plant and attached biofilm growth on treatment efficiency. Figure 6.1 illustrates the logical expected variation of pollutants concentration along a cell length. The water sampling started at 31/10/2009 till 3/4/2010 (end of this stage). Measured discharges of 6.377, 6.672, and 6.781 m^3/d were passed through plastic, gravel, and rubber cells, respectively.

Parts of this chapter have been reprinted from Zidan, A. A., Rashed, A. A., El-Gamal, M. A, and Abdel Hadi, M., "BOD Treatment in HSSF Constructed Wetlands Using Different Media (Set Up Stage)," Mansoura Engineering Journal, Vol.38, No 3,pp C36-C46, Sept. 2013. Reprinted with permission.

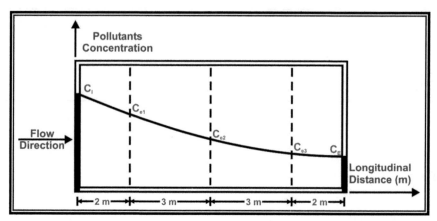

FIGURE 6.1 Definition sketch for the logical expected variation of pollutants along cell length.

6.2 BIOCHEMICAL OXYGEN DEMAND

The effluent BOD concentration with distance, hydraulic loading rate, and influent concentration is studied. The variation of BOD removal efficiency with distance, hydraulic retention time, and time from the start of sampling is also determined.

6.2.1 EFFECT OF DISTANCE ON BOD TREATMENT

Tables 6.1 shows the BOD concentrations during the start, average, and end of the set up stage collected from cell entrance and at 2, 5, 8, and 10 m distances measured from inlet.

Figures 6.2 and 6.3 give the variation of BOD concentration with longitudinal distance from inlet for start and end dates for this stage. Figure 6.4 illustrates this variation for average values of the whole stage.

From Table 6.1 and Figures 6.2–6.4, it is noticed that:

- In plastic cell, BOD concentration reduces with distance better than other media. Gravel cell takes the second order in reduction.
- At the start of the stage, the inlet BOD concentration is 220 mg/l which reduces to 202, 205, and 207 mg/l for plastic, gravel, and rubber media, respectively at 2 m distance from inlet. These values are 183, 186, and 190 mg/l at 5 m and 175, 178, and 183 mg/l at 8 m. At outlets, the BOD concentration values are 167, 173, and 177 mg/l.

Set Up Stage Analysis

TABLE 6.1 Summary of BOD Concentration Sampled at Different Distances

Media	Inlet (mg/l)	Outlet at 2 m			Outlet at 5 m			Outlet at 8 m			Outlet at 10 m		
		Start	Avg.	End	Start	Avg.	End	Start	Avg.	End	Start	Avg.	End
Rubber	220	207	178	148	190	156	122	183	148	113	177	141	103
Gravel	194	205	175	144	186	150	114	178	140	103	173	133	95
Plastic	168	202	171	140	183	144	108	175	134	95	167	124	85

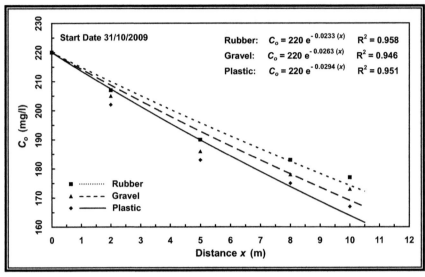

FIGURE 6.2 Relationship between C_o and distance for BOD (start values of set-up stage).

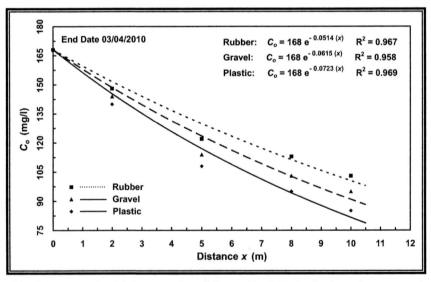

FIGURE 6.3 Relationship between C_o and distance for BOD (end values of set-up stage).

- At the end of the stage, the inlet BOD concentration is 168 mg/l which gives concentration of 140, 144, and 148 mg/l for plastic, gravel, and rubber cells at 2 m; and at 5 m 108, 114, and 122 mg/l, in the same order. These values are 95, 103, and 113 mg/l at 8 m and 85, 95, and 103 mg/l at outlets.

Set Up Stage Analysis

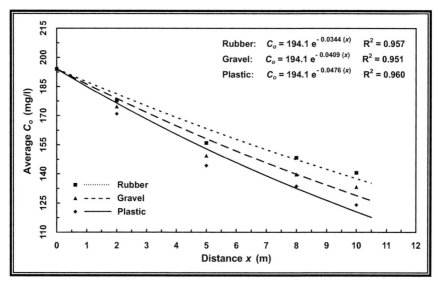

FIGURE 6.4 Relationship between C_o and distance for BOD (average values of set-up stage).

- During the whole stage, the average inlet BOD concentration is 194 mg/l which gives average concentration for plastic, gravel, and rubber cells, respectively of 171, 175, and 178 mg/l at 2 m; 144, 150, and 156 mg/l at 5 m; 134, 140, and 148 mg/l at 8 m; and 124, 133, and 141 mg/l at outlets.

Concentration of BOD, which is gradually reduced with distance for the three media, may be represented by an exponential function that gives the best determination coefficient. The exponential equations for the average values are written as:

$$\text{Plastic: } C_o = 194.1 e^{-0.0476 x} \quad R^2 = 0.960 \tag{6.1a}$$

$$\text{Gravel: } C_o = 194.1 e^{-0.0409 x} \quad R^2 = 0.951 \tag{6.1b}$$

$$\text{Rubber: } C_o = 194.1 e^{-0.0344 x} \quad R^2 = 0.957 \tag{6.1c}$$

where: x = distance from cell inlet, m; C_o = BOD outlet concentration, mg/l.

The BOD treatment efficiency at a certain distance depending on the concentration of the influent flow and the measured effluent concentration at this specified distance. The removal efficiency could be calculated as follows:

$$Removal\ Efficiency = \left(1 - \frac{C_o}{C_i}\right) \times 100 \qquad (4.5)$$

where: C_i = inlet concentration, mg/l; C_o = outlet concentration, mg/l.

Figure 6.5 illustrates the variation of BOD removal efficiency with treatment length for the whole stage.

Tables 5.6–5.8 and Figure 6.5 exhibit that:

- At 2 m from inlet, BOD removal efficiency enhances from 8.18 to 16.67% in plastic cell; from 6.82 to 14.29% in gravel; and from 5.91 to 11.90% in rubber from the start to the end of the stage.
- At 5 m, BOD removal efficiency increases from 16.82 to 35.71% in plastic cell; from 15.45 to 32.14% in gravel; and from 13.64 to 27.38% in rubber from the start to the end of the stage.
- At 8 m, BOD removal efficiency improves from 20.45 to 43.45% in plastic cell; from 19.09 to 38.69% in gravel; and from 16.82 to 32.74% in rubber from the start to the end of the stage.
- At outlets, BOD removal efficiency raises from 24.09 to 49.40% in plastic; from 21.36 to 43.45% in gravel; and from 19.55 to 38.69% in rubber cell at start and end of the stage, respectively.
- At the end of set up stage the treatment of BOD is improved for the three media and at all distances more than the start of the stage. This

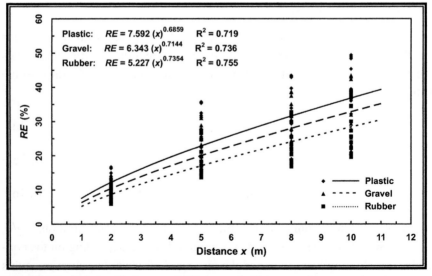

FIGURE 6.5 Relationship between BOD removal efficiency and treatment length.

may be due to the plants and attached biofilm growth during the stage period.
- The better removal efficiency happens at the cells middle part (5 to 8 m) comparing with the other parts. This means that BOD treatment mainly exists at the cell middle zone.

The BOD removal efficiency is gradually increased with treatment length for the three media according to a power function. The power equations are written as following:

Plastic: $RE = 7.592x^{0.6859}$ $R^2 = 0.719$ (6.2a)

Gravel: $RE = 6.343x^{0.7144}$ $R^2 = 0.736$ (6.2b)

Rubber: $RE = 5.227x^{0.7354}$ $R^2 = 0.755$ (6.2c)

where: RE = BOD removal efficiency, %.

6.2.2 IMPACT OF Q ON BOD TREATMENT

The wetland cell is considered as four basins 2 m width and surface areas of 4, 10, 16, and 20 m² giving four corresponding values for loading rate (Eq. 4.1) at distances of 2, 5, 8, and 10 m from cell entrance as shown in Table 6.2. This table gives the BOD concentration and the discharge values for the three media at different distances.

$$q = \frac{Q \times 100}{A} \quad (4.1)$$

where: q = loading rate, cm/d; Q = discharge, m³/d; A = surface area, m².

Figure 6.6 illustrates the relationship between the outlet BOD concentration and the q values.

From Table 6.2 and Figure 6.6, it is found that:
- High value of q occurs near the cell entrance and it decreases with increasing the surface area of cells. After 2 m from inlet, the q values are 159.4, 166.8, and 169.5 cm/d for plastic, gravel, and rubber cells with corresponding effluent BOD concentration of 171, 174.7, and 177.8 mg/l, respectively. At outlets, the q values are 31.9, 33.4, and 33.9 cm/d and the corresponding effluent BOD concentration values are 124.1, 133.3, and 140.6 mg/l.

TABLE 6.2 BOD Concentration and Hydraulic Loading Rate at Different Distances

Distance From Inlet	Plastic Cell $Q = 6.377$ m³/d		Gravel Cell $Q = 6.672$ m³/d		Rubber Cell $Q = 6.781$ m³/d	
	C_o (mg/l)	q (cm/d)	C_o (mg/l)	q (cm/d)	C_o (mg/l)	q (cm/d)
2 m	171.0	159.4	174.7	166.8	177.8	169.5
5 m	144.3	63.8	149.5	66.7	156.0	67.8
8 m	133.7	39.9	139.8	41.7	148.4	42.4
10 m	124.1	31.9	133.3	33.4	140.6	33.9

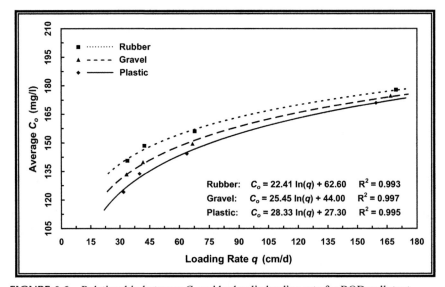

FIGURE 6.6 Relationship between C_o and hydraulic loading rate for BOD pollutant.

- The differences between BOD concentrations for the three media are small at high BOD loads, while they are relatively big at small loads.
- Plastic media is the best in treating performance of BOD followed by gravel media.

The BOD outlet concentration is directly proportional to hydraulic loading rate according to a logarithmic function. The logarithmic equations are given as:

Plastic: $C_o = 27.30 + 28.33 \ln q$ $R^2 = 0.995$ (6.3a)

Gravel: $C_o = 44.00 + 25.45 \ln q$ $R^2 = 0.997$ (6.3b)

Rubber: $C_o = 62.60 + 22.41 \ln q$ $R^2 = 0.993$ (6.3c) where: C_o = BOD outlet concentration, mg/l; q = hydraulic loading rate, cm/d.

These equations can be used to estimate C_o concentration based on q in the range between 31.9 and 169.5 cm/d.

6.2.3 INLET AND OUTLET BOD CONCENTRATIONS

Figures 6.7–6.10 give the relationship between effluent and influent concentrations (C_o and C_i) of BOD for the three media at 2, 5, 8, and 10 m distances from inlet. The C_i and C_o values in these figures for cells media are listed in Tables 5.6–5.8, Chapter 5.

Effluent and influent BOD relationship for the three media is varying according to an exponential function as the best fit. The exponential equations at distance 10 m from inlet are as follows:

From Figures 6.7–6.10, it is noticed that:

- Through set up stage the outlet BOD concentration of plastic media is smaller than the other media (better BOD treatment) at all distances, followed by the gravel media.
- The difference between BOD concentrations for the three media is small at the first 2 m, then this difference increases with moving toward cell outlet, due to the little volume of media in this zone. In the first 2 m of cell length, coarse gravel occupies 64.5% of the zone volume

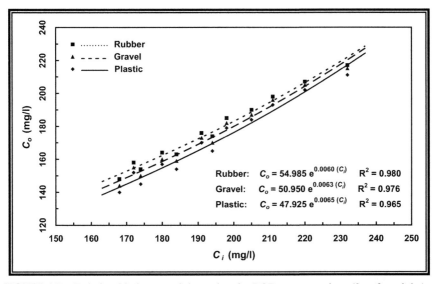

FIGURE 6.7 Relationship between inlet and outlet BOD concentrations (2 m from inlet).

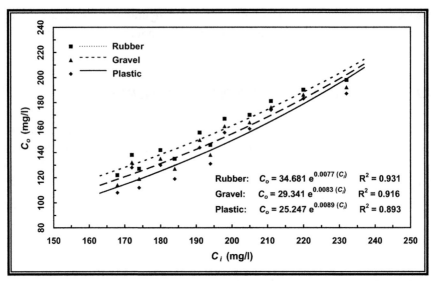

FIGURE 6.8 Relationship between inlet and outlet BOD concentrations (5 m from inlet).

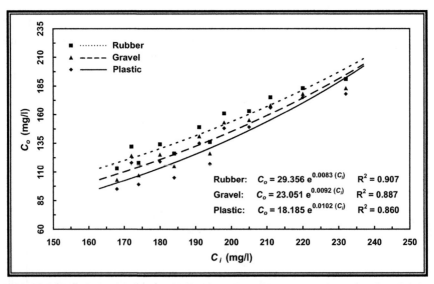

FIGURE 6.9 Relationship between inlet and outlet BOD concentrations (8 m from inlet).

(1.29 m^3) while the treatment media occupies 35.5% of the zone volume (0.71 m^3), Figure 5.2 (Chapter 5). At 5, 8, and 10 m distances from inlet, the media occupies about 74.2, 83.9, and 74.2% of the zone volume, respectively.

Set Up Stage Analysis

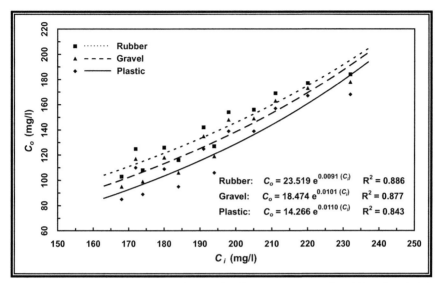

FIGURE 6.10 Relationship between inlet and outlet BOD concentrations (10 m from inlet).

$$\text{Plastic:} \quad C_o = 14.266 e^{0.0110 C_i} \quad R^2 = 0.843 \tag{6.4a}$$

$$\text{Gravel:} \quad C_o = 18.474 e^{0.0101 C_i} \quad R^2 = 0.877 \tag{6.4b}$$

$$\text{Rubber:} \quad C_o = 23.519 e^{0.0091 C_i} \quad R^2 = 0.886 \tag{6.4c}$$

where: C_o = BOD outlet concentration, mg/l; C_i = BOD inlet concentration, mg/l.

These equations are valid at the inlet BOD concentration ranges from 168 to 232 mg/l.

6.2.4 IMPACT OF T_R ON BOD REMOVAL EFFICIENCY

Table 6.3 presents the volume of coarse gravel existed at start and end of each wetland cell in addition to the volume of media at different distances. From these volumes and by knowing the porosities, the water volume inside media pores through set up stage could be calculated at each distance (Eq.(4.3) and Figure 5.2). The retention time at any distance could be calculated by dividing the water volume at that distance by the discharge (Eq. 4.2).

$$V_{wx} = (V_{cg} \times n_{cg}) + (V_m \times n_m) \tag{4.3}$$

TABLE 6.3 Volume of Treatment Media Calculated for Different Distances

Distance from Inlet	Coarse Gravel Volume (V_{cg})	Media Volume (V_m)	Total Volume
2 m	1.29 m³	0.71 m³	2.0 m³
5 m	1.29 m³	3.71 m³	5.0 m³
8 m	1.29 m³	6.71 m³	8.0 m³
10 m	2.58 m³	7.42 m³	10 m³

where: V_{wx} = volume of water inside cell at distance x, m³; V_{cg} = volume of coarse gravel, m³; V_m = volume of media in wetland cell, m³; n_{cg} = porosity of coarse gravel; n_m = porosity of used media.

$$T_r = \frac{24 \times V_{wx}}{Q} \qquad (4.2)$$

where: Q = discharge to wetland cell, m³/d; T_r = hydraulic retention time, hr.

Tables 6.4–6.6 exhibit the BOD removal efficiency and the actual T_r at different distances. Each row in these tables represents the results of one run of set up stage at time T_s (time from start of sampling). The detailed calculations for retention time (based on the porosities of the media and the calculated water volumes) of the three media are listed in Appendix II for set up stage.

Figures 6.11–6.14 illustrate the relationship between BOD removal efficiency and retention time values for wetland cells at distances of 2, 5, 8, and 10 m from inlet.

From Tables 6.6–6.8 and Figures 6.11–6.14, it is found that:

- The gravel cell has the lowest T_r followed by the rubber and then the plastic cell. This is compatible with the porosity of the three media types. The gravel media has the smallest porosity followed by rubber media, while plastic media has the greatest one.
- The BOD removal efficiency increases with the decrease of T_r value. For example, for gravel cell (at the start of the stage, $T_s = 0$) the removal efficiency at 2 m length from inlet equals 6.82% ($T_r = 2.784$ hr) while at the end of the stage ($T_s = 154$ day) the removal efficiency equals 14.29% ($T_r = 2.616$ hr). These values are 5.91% ($T_r = 3.120$ hr) and 11.90% ($T_r = 2.952$ hr) for rubber cell also, 8.18% ($T_r = 4.080$ hr) and 16.67% ($T_r = 3.888$ hr) for plastic cell. The positive effect on treatment

Set Up Stage Analysis

TABLE 6.4 BOD Removal Efficiency and T_r for Plastic Cell at Different Distances

T_s (day)	2 m Dis. RE (%)	T_r (hr)	5 m Dis. RE (%)	T_r (hr)	8 m Dis. RE (%)	T_r (hr)	10 m Dis. RE (%)	T_r (hr)
0	8.18	4.080	16.82	13.296	20.45	22.512	24.09	26.592
14	8.53	4.056	17.54	13.248	21.33	22.416	25.59	26.472
28	9.05	4.032	19.40	13.152	23.28	22.296	27.59	26.304
42	9.60	3.984	21.21	13.056	25.25	22.104	29.80	26.088
56	10.24	3.960	22.44	12.984	27.32	21.984	32.20	25.944
70	10.99	3.936	24.61	12.960	29.32	21.960	34.55	25.920
84	11.63	3.936	25.58	12.936	31.40	21.936	36.05	25.872
98	12.78	3.936	27.78	12.912	33.89	21.888	39.44	25.824
112	14.95	3.912	32.47	12.888	39.69	21.840	45.36	25.752
126	16.30	3.888	35.33	12.840	42.93	21.792	48.37	25.680
140	16.67	3.888	35.63	12.816	43.10	21.744	48.85	25.656
154	16.67	3.888	35.71	12.792	43.45	21.696	49.40	25.584

TABLE 6.5 BOD Removal Efficiency and T_r for Gravel Cell at Different Distances

T_s (day)	2 m Dis. RE (%)	T_r (hr)	5 m Dis. RE (%)	T_r (hr)	8 m Dis. RE (%)	T_r (hr)	10 m Dis. RE (%)	T_r (hr)
0	6.82	2.784	15.45	6.864	19.09	10.944	21.36	13.728
14	7.11	2.760	16.59	6.816	20.38	10.872	22.75	13.656
28	7.33	2.736	17.24	6.744	21.12	10.752	23.28	13.488
42	8.08	2.688	18.69	6.648	22.73	10.608	25.25	13.320
56	8.78	2.664	20.00	6.600	24.39	10.560	27.32	13.224
70	9.42	2.664	21.47	6.600	26.18	10.512	29.32	13.176
84	9.88	2.664	23.26	6.576	27.91	10.488	31.98	13.152
98	11.11	2.640	25.00	6.552	30.56	10.464	34.44	13.104
112	12.37	2.640	28.87	6.528	35.05	10.440	38.66	13.080
126	13.59	2.616	30.98	6.504	37.50	10.392	42.39	13.032
140	13.79	2.616	31.61	6.504	38.51	10.368	43.10	12.984
154	14.29	2.616	32.14	6.480	38.69	10.344	43.45	12.960

due to the plants and attached biofilm growth are higher than the negative effect of decrease T_r (the decrease of T_r in set up stage is due to the decrease of media porosities).

TABLE 6.6 BOD Removal Efficiency and T_r for Rubber Cell at Different Distances

T_s (day)	2 m Dis. RE (%)	T_r (hr)	5 m Dis. RE (%)	T_r (hr)	8 m Dis. RE (%)	T_r (hr)	10 m Dis. RE (%)	T_r (hr)
0	5.91	3.120	13.64	8.760	16.82	14.376	19.55	17.496
14	6.16	3.096	14.22	8.712	17.06	14.304	19.91	17.424
28	6.47	3.072	14.66	8.640	17.67	14.208	20.69	17.256
42	6.57	3.024	15.66	8.544	18.69	14.064	22.22	17.088
56	7.32	3.000	17.07	8.472	20.49	13.944	23.90	16.944
70	7.85	3.000	18.32	8.448	21.99	13.920	25.65	16.920
84	8.14	2.976	19.77	8.424	23.26	13.872	27.33	16.872
98	8.89	2.976	21.11	8.424	25.56	13.848	30.00	16.824
112	10.31	2.976	24.74	8.376	29.90	13.800	34.54	16.776
126	11.41	2.952	26.63	8.352	31.52	13.776	36.96	16.728
140	11.49	2.952	27.01	8.328	32.18	13.704	37.93	16.656
154	11.90	2.952	27.38	8.304	32.74	13.680	38.69	16.632

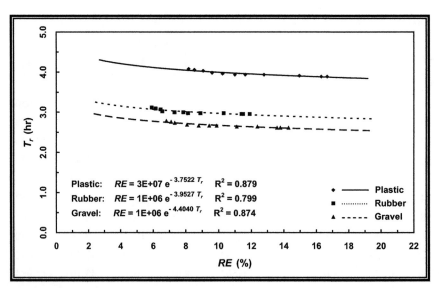

FIGURE 6.11 Relationship between BOD removal efficiency and retention time (2 m from inlet).

- Generally the retention time decreases with time from the start of the stage to the end. At the outlets, the T_r decreases from 26.592 to 25.584 hr for plastic, from 13.728 to 12.960 hr for gravel, and from 17.496

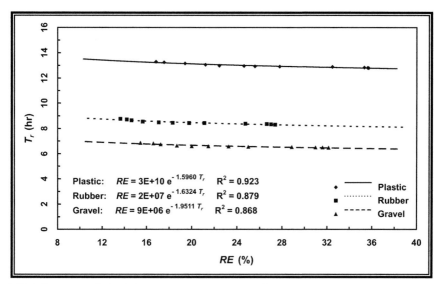

FIGURE 6.12 Relationship between BOD removal efficiency and retention time (5 m from inlet).

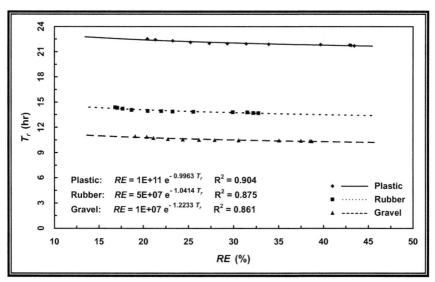

FIGURE 6.13 Relationship between BOD removal efficiency and retention time (8 m from inlet).

to 16.632 hr for rubber media. The decrease of T_r is due to the reduction in pore spaces volume of media with accumulating of suspended

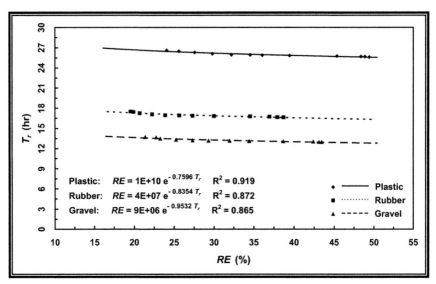

FIGURE 6.14 Relationship between BOD removal efficiency and retention time (10 m from inlet).

matter and growth of biofilm thickness around particles.
- For different media types, the removal efficiency is found the best in plastic followed by gravel media. The treatment status of cells outlets show that the plastic cell reaches 49.40% at (T_s = 154 day), while at the same time from start of sampling, the rubber and gravel cells reach 38.69 and 43.45%.

The BOD removal efficiency for the three media is inversely proportion to retention time and follows an exponential function. The exponential equations at distance 10 m from inlet are written as:

$$\text{Plastic:} \quad RE = 1E+10 \; e^{-0.7596 \, T_r} \quad R^2 = 0.919 \quad (6.5a)$$

$$\text{Gravel:} \quad RE = 9E+06 \; e^{-0.9532 \, T_r} \quad R^2 = 0.872 \quad (6.5b)$$

$$\text{Rubber:} \quad RE = 4E+07 \; e^{-0.8354 \, T_r} \quad R^2 = 0.865 \quad (6.5c)$$

where: RE = BOD removal efficiency, %; T_r = hydraulic retention time, hr.

All relationships between removal efficiency and T_r at cells outlets are valid at retention time ranges from 25.584 to 26.592 hr for plastic, from 12.960 to 13.728 hr for gravel, and from 16.632 to 17.496 hr for rubber.

Set Up Stage Analysis

TABLE 6.7 Summary of COD Concentration Sampled at Different Distances

Media	Inlet (mg/l)	Outlet at 2 m			Outlet at 5 m			Outlet at 8 m			Outlet at 10 m		
		Start	Avg.	End	Start	Avg.	End	Start	Avg.	End	Start	Avg.	End
Rubber	344	325	276	222	300	244	184	285	229	168	278	219	155
Gravel	300	323	272	217	295	234	172	276	214	152	272	208	143
Plastic	251	314	263	208	289	226	163	270	204	140	263	194	128

TABLE 6.8 COD Concentration and Hydraulic Loading Rate at Different Distances

Distance from Inlet	Plastic Cell $Q = 6.377$ m^3/d		Gravel Cell $Q = 6.672$ m^3/d		Rubber Cell $Q = 6.781$ m^3/d	
	C_o (mg/l)	q (cm/d)	C_o (mg/l)	q (cm/d)	C_o (mg/l)	q (cm/d)
2 m	263.3	159.4	272.3	166.8	276.3	169.5
5 m	225.5	63.8	234.4	66.7	243.5	67.8
8 m	204.1	39.9	214.4	41.7	228.5	42.4
10 m	193.8	31.9	207.8	33.4	218.8	33.9

6.2.5 OPERATING TIME EFFECT ON BOD TREATMENT EFFICIENCY

The development of BOD removal efficiency with time from start of sampling, T_s at distances of 2 and 8 m from inlet is given in Figure 6.15, for 5 and 10 m in Figure 6.16. These binaries are chosen to prevent the overlapping of lines which represent the best fit function (third order polynomial). The values drawn in these figures for rubber, gravel, and plastic are given in Chapter 5.

The average difference between pollutant removal efficiency through stage runs for plastic cell and both gravel and rubber cells is given by Eqs. (4.7) and (4.8). This average difference between gravel and rubber cells is represented by Eq. (4.9). The values of average difference for BOD pollutant between the three media are illustrated in Appendix II.

$$\text{Average Difference (Plastic \& Gravel)} = \frac{\sum(RE_{xp} - RE_{xg})}{\text{No. of Runs}} \quad (4.7)$$

$$\text{Average Difference (Plastic \& Rubber)} = \frac{\sum(RE_{xp} - RE_{xr})}{\text{No. of Runs}} \quad (4.8)$$

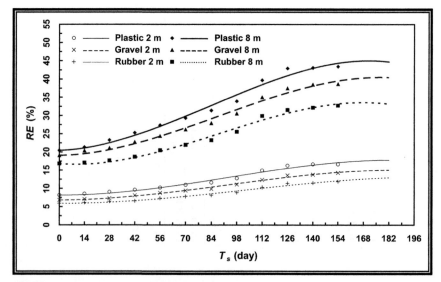

FIGURE 6.15 Relationship between BOD removal efficiency and T_s at 2 and 8 m from inlet.

Set Up Stage Analysis

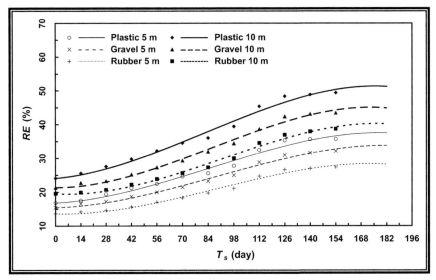

FIGURE 6.16 Relationship between BOD removal efficiency and T_s at 5 and 10 m from inlet.

$$\text{Average Difference (Gravel \& Rubber)} = \frac{\sum (RE_{xg} - RE_{xr})}{\text{No. of Runs}} \quad (4.9)$$

where: RE_{xp} = removal efficiency of plastic cell at distance x, %; RE_{xg} = removal efficiency of gravel cell at distance x, %; RE_{xr} = removal efficiency of rubber cell at distance x, %.

From Figures 6.15 and 6.16, it is appeared that:

- At 2 m from inlet, plastic cell gives average removal efficiency higher than gravel and rubber cells by about 1.92 and 3.60%, respectively. These values at 5 m are about 2.77 and 6.19%, at 8 m are about 3.28 and 7.79%, and at cells outlets are about 4.83 and 8.66%.
- Gravel cell gives average removal efficiency higher than rubber cell by about 1.68, 3.42, 4.52, and 3.83% at distances of 2, 5, 8, and 10 m from cell inlet.
- The average difference between plastic removal efficiency and both gravel and rubber removal efficiencies increases with increasing treatment distance.

The set up stage lasted for 154 days (after which the porosity had approximately steady value), through which the growth of reeds roots through the cells media and the increase of bacterial biofilm attached to media and plant parts, which enhances BOD biodegradation and the treatment efficiency (Zidan et al., 2013).

6.3 CHEMICAL OXYGEN DEMAND

The outlet COD concentration is studied with distance, loading rate, and inlet concentration. The variation of COD treatment efficiency with distance, retention time, and time from start of sampling is also discussed.

6.3.1 EFFECT OF DISTANCE ON COD TREATMENT

Table 6.7 shows the start, average, and end concentrations of COD pollutant for rubber, gravel, and plastic media at cell entrance and at distances of 2, 5, 8, and 10 m measured from the cell inlet.

Figures 6.17 and 6.18 give the variation of COD concentration with distance from inlet for start and end of set up stage. Figure 6.19 illustrates this variation for the average value of the whole stage. The values in these figures were listed in Chapter 5 for the used media.

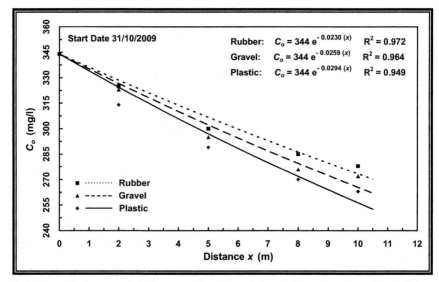

FIGURE 6.17 Relationship between C_o and distance for COD (start values of set-up stage).

Set Up Stage Analysis

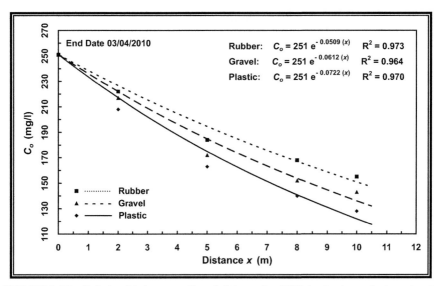

FIGURE 6.18 Relationship between C_o and distance for COD (end values of set-up stage).

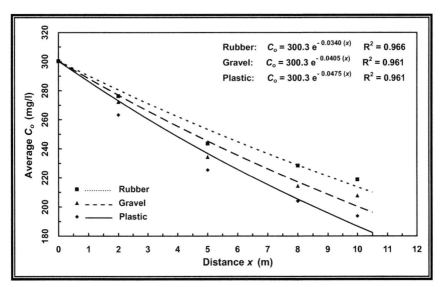

FIGURE 6.19 Relationship between C_o and distance for COD (average values of set-up stage).

From Tables 5.9–5.11 and Figures 6.17–6.19, it is found that:

- In plastic media, COD concentration value reduces with distance better than other media. Gravel takes the second order in COD reduction.

- At the start of the stage, the influent COD concentration is 344 mg/l which is reduced to 314, 323, and 325 mg/l at 2 m for plastic, gravel, and rubber cells, respectively. These values are 289, 295, and 300 mg/l at 5 m; and 270, 276, and 285 mg/l at 8 m. At outlets, the COD concentration values are 263, 272, and 278 mg/l.
- At the end of the stage, the influent COD concentration is 251 mg/l which gives effluent concentration of 208, 217, and 222 mg/l at 2 m for plastic, gravel, and rubber cells, respectively; and 163, 172, and 184 mg/l at 5 m. These values are 140, 152, and 168 mg/l at 8 m; and 128, 143, and 155 mg/l at 10 m (outlets).
- During the whole stage, the average influent COD concentration is 300 mg/l which gives average effluent concentration for plastic, gravel, and rubber cells of 263, 272, and 276 mg/l at 2 m; 226, 234, and 244 mg/l at 5 m; 204, 214, and 229 mg/l at 8 m; and 194, 208, and 219 mg/l at outlets, respectively.

Concentration of COD, which is gradually reduced with distance for the three cells, may be represented by an exponential function. The exponential equations for the average values of the stage are:

Plastic: $\quad C_o = 300.3 \ e^{-0.0405\,x} \quad R^2 = 0.961 \quad$ (6.6b)

Plastic: $\quad C_o = 300.3 \ e^{-0.0405\,x} \quad R^2 = 0.961 \quad$ (6.6b)

Plastic: $\quad C_o = 300.3 \ e^{-0.0340\,x} \quad R^2 = 0.961 \quad$ (6.6c)

Figure 6.20 exhibits the variation of COD removal efficiency with distance from the cell inlet.

Tables 5.9–5.11 and Figure 6.20 show that:

- At 2 m from inlet, COD removal efficiency increases from 8.72 to 17.13% in plastic, from 6.10 to 13.55% in gravel, and from 5.52 to 11.55% in rubber cell; from the start to the end of the stage.
- At middle distance, COD removal efficiency raises from 15.99 to 35.06% in plastic, from 14.24 to 31.47% in gravel, and from 12.79 to 26.69% in rubber; from the start to the end of the stage.
- At 8 m from inlet, COD removal efficiency enhances from 21.51 to 44.22% in plastic, from 19.77 to 39.44% in gravel, and from 17.15 to 33.07% in rubber; from the start to the end of the stage.
- At outlet, COD removal efficiency improves from 23.55 to 49.0% in plastic, from 20.93 to 43.03% in gravel, and from 19.19 to 38.25% in rubber cell; from the start to the end of the stage.

Set Up Stage Analysis

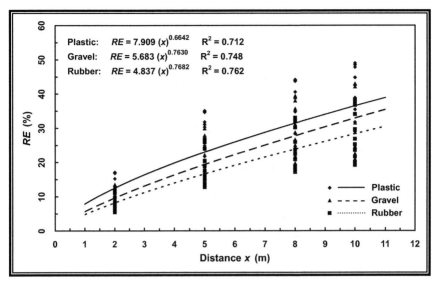

FIGURE 6.20 Relationship between COD removal efficiency and treatment length.

- The treatment of COD enhances at the end of the stage more than at the start for different distances. This may be due to the increase of plant and attached biofilm growth.
- The maximum rate of treatment occurs at the middle of the basin. After that this rate decreases towards the end of the cell.

The COD removal efficiency, which is gradually increased with distance for the three media, may be represented by a power function, which gives the best determination coefficient. The power equations are written as:

Plastic: $\quad RE = 7.909 \, x^{0.6642} \quad R^2 = 0.712$ (6.7a)

Gravel: $\quad RE = 5.683 \, x^{0.7630} \quad R^2 = 0.748$ (6.7b)

Rubber: $\quad RE = 4.837 \, x^{0.7682} \quad R^2 = 0.762$ (6.7c)

where: RE = COD removal efficiency, %; x = distance from cell inlet, m.

6.3.2 IMPACT OF Q ON COD TREATMENT

The wetland cell is considered as four basins with surface areas of 4, 10, 16, and 20 m² giving four values for q at distances 2, 5, 8, 10 m from inlet, respectively as shown in Table 6.8.

Figure 6.21 presents the relationship between outlet COD concentration (average C_o) and loading rate (q).

From Table 6.8 and Figure 6.21, it is appeared that:

- The average effluent COD value increases with increasing the loading rate, which means a good treatment of COD.
- High value of q occurs near the cell inlet and it decreases with increasing the treatment distance. At 2 m from inlet, the q values are 159.4, 166.8, and 169.5 cm/d for plastic, gravel, and rubber cells with corresponding COD concentration of 263.3, 272.3, and 276.3 mg/l, respectively. At outlets, the q values are 31.9, 33.4, and 33.9 cm/d and the COD concentration values are 193.8, 207.8, and 218.8 mg/l.
- The differences between average COD concentration for the three media are small at high q, while they are relatively big at small q.
- Plastic media acts better in COD treating performance followed by gravel and then rubber media.

The average effluent COD concentration is directly proportional to loading rate according to a logarithmic function. The logarithmic equations are given as:

$$\textit{Plastic:} \quad C_o = 45.22 + 43.10 \ln q \quad R^2 = 0.999 \quad (6.8a)$$

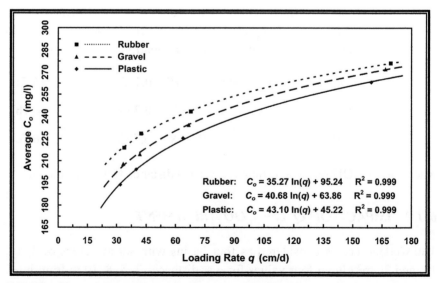

FIGURE 6.21 Relationship between C_o and hydraulic loading rate for COD pollutant.

Plastic: $C_o = 63.86 + 40.68 \ln q$ $R^2 = 0.999$ (6.8a)

Plastic: $C_o = 95.24 + 35.27 \ln q$ $R^2 = 0.999$ (6.8a)

where: C_o = COD outlet concentration, mg/l; q = hydraulic loading rate, cm/d.

The above-mentioned equations are valid for loading rate ranges from 31.9 to 169.5 cm/d.

6.3.3 INLET AND OUTLET COD CONCENTRATIONS

Figures 6.22–6.25 give the relationship between inlet and outlet concentrations (C_i and C_o) of COD for the three cells at 2, 5, 8, and 10 m, from inlet. The C_i and C_o in these figures for rubber, gravel, and plastic media were listed in Tables 5.9–5.11, Chapter 5.

Tables 5.9–5.11 and Figures 6.22–6.25 show that:

- Through set up stage the effluent COD concentration of plastic media is smaller than the other media at distances of 2, 5, 8, and 10 m (better COD treatment), followed by the gravel media.
- The difference between COD concentration for gravel and rubber media is very small at the first 2 m then this difference increases towards outlets of cells, due to the little volume of media in the inlet zone.

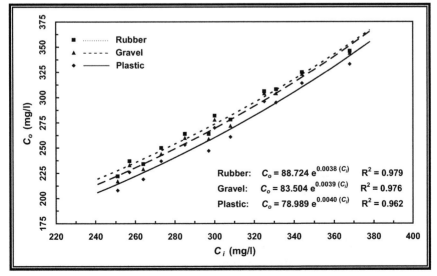

FIGURE 6.22 Relationship between inlet and outlet COD concentrations (2 m from inlet).

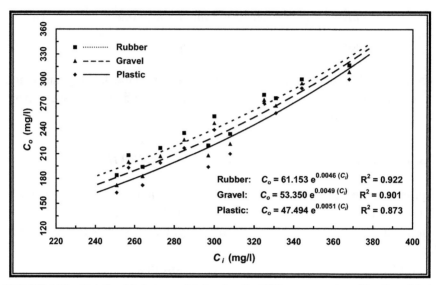

FIGURE 6.23 Relationship between inlet and outlet COD concentrations (5 m from inlet).

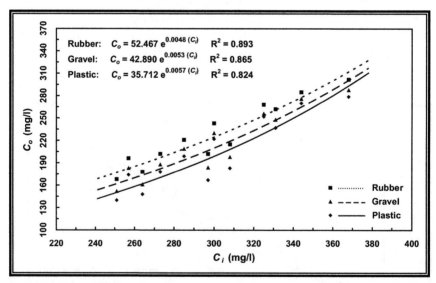

FIGURE 6.24 Relationship between inlet and outlet COD concentrations (8 m from inlet).

Effluent and influent COD concentrations relationship for the three media follows an exponential function. The exponential equations at outlets are written as:

Set Up Stage Analysis

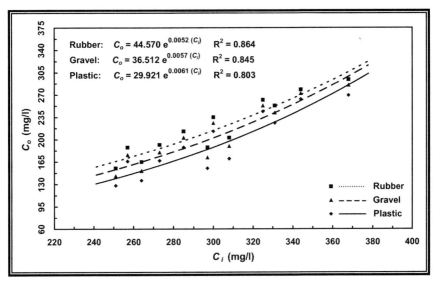

FIGURE 6.25 Relationship between inlet and outlet COD concentrations (10 m from inlet).

$$\text{Plastic:} \quad C_o = 29.921\, e^{0.0061 C_i} \quad R^2 = 0.803 \tag{6.7a}$$

$$\text{Gravel:} \quad C_o = 36.512\, e^{0.0057 C_i} \quad R^2 = 0.845 \tag{6.7b}$$

$$\text{Rubber:} \quad C_o = 44.570\, e^{0.0052 C_i} \quad R^2 = 0.864 \tag{6.7c}$$

where: C_o = COD outlet concentration, mg/l; C_i = COD inlet concentration, mg/l.

These equations are valid for the influent COD concentration ranges from 251 to 368 mg/l.

6.3.4 IMPACT OF TR ON COD REMOVAL EFFICIENCY

Through the set up stage the three-wetland cells were operated with relatively constant discharge. The difference in T_r value was due to the variation in media porosity. Tables 6.9–6.11 present the COD removal efficiencies and the T_r values at different distances from inlet.

Figures 6.26–6.29 show the relationships between COD removal efficiency and hydraulic retention time at distances of 2, 5, 8, and 10 m from inlet.

Tables 6.9–6.11 and Figures 6.26–6.29 exhibit that:

- The plastic cell has the higher T_r value than the values of the two other cells followed by the rubber and then the gravel. This is in line with the

TABLE 6.9 COD Removal Efficiency and T_r for Plastic Cell at Different Distances

T_s (day)	2 m Dis. RE (%)	T_r (hr)	5 m Dis. RE (%)	T_r (hr)	8 m Dis. RE (%)	T_r (hr)	10 m Dis. RE (%)	T_r (hr)
0	8.72	4.080	15.99	13.296	21.51	22.512	23.55	26.592
14	8.92	4.056	16.62	13.248	22.46	22.416	24.92	26.472
28	9.51	4.032	18.48	13.152	24.18	22.296	26.90	26.304
42	10.00	3.984	20.33	13.056	26.00	22.104	29.00	26.088
56	10.88	3.960	21.75	12.984	28.40	21.984	31.72	25.944
70	11.23	3.936	23.86	12.960	30.18	21.960	34.04	25.920
84	12.06	3.936	24.90	12.936	32.30	21.936	35.41	25.872
98	13.19	3.936	27.11	12.912	34.80	21.888	38.83	25.824
112	15.26	3.912	31.82	12.888	40.58	21.840	44.81	25.752
126	16.84	3.888	34.68	12.840	43.77	21.792	47.81	25.680
140	17.05	3.888	34.85	12.816	43.94	21.744	48.48	25.656
154	17.13	3.888	35.06	12.792	44.22	21.696	49.00	25.584

TABLE 6.10 COD Removal Efficiency and T_r for Gravel Cell at Different Distance

T_s (day)	2 m Dis. RE (%)	T_r (hr)	5 m Dis. RE (%)	T_r (hr)	8 m Dis. RE (%)	T_r (hr)	10 m Dis. RE (%)	T_r (hr)
0	6.10	2.784	14.24	6.864	19.77	10.944	20.93	13.728
14	6.46	2.760	15.38	6.816	21.23	10.872	22.15	13.656
28	6.52	2.736	16.03	6.744	21.74	10.752	22.55	13.488
42	7.33	2.688	17.67	6.648	23.33	10.608	24.67	13.320
56	8.16	2.664	19.03	6.600	25.08	10.560	26.89	13.224
70	8.77	2.664	20.35	6.600	26.67	10.512	28.77	13.176
84	9.34	2.664	22.18	6.576	28.79	10.488	31.52	13.152
98	10.62	2.640	24.18	6.552	31.14	10.464	34.07	13.104
112	11.69	2.640	27.92	6.528	35.71	10.440	38.31	13.080
126	12.79	2.616	29.97	6.504	38.05	10.392	42.09	13.032
140	13.26	2.616	30.68	6.504	39.02	10.368	42.80	12.984
154	13.55	2.616	31.47	6.480	39.44	10.344	43.03	12.960

porosity of the three media types. The plastic media has the highest value of porosity followed by rubber, while gravel has the smallest one.
- The COD removal efficiency decreases with increasing T_r through this stage. For example, for plastic cell (at the start of the stage $T_s = 0$) the

Set Up Stage Analysis

TABLE 6.11 COD Removal Efficiency and T_r for Rubber Cell at Different Distance

T_s (day)	2 m Dis. RE (%)	2 m Dis. T_r (hr)	5 m Dis. RE (%)	5 m Dis. T_r (hr)	8 m Dis. RE (%)	8 m Dis. T_r (hr)	10 m Dis. RE (%)	10 m Dis. T_r (hr)
0	5.52	3.120	12.79	8.760	17.15	14.376	19.19	17.496
14	5.85	3.096	13.54	8.712	17.54	14.304	19.38	17.424
28	5.98	3.072	13.86	8.640	17.93	14.208	20.11	17.256
42	6.00	3.024	15.00	8.544	19.00	14.064	21.67	17.088
56	6.95	3.000	16.31	8.472	20.85	13.944	23.56	16.944
70	7.37	3.000	17.54	8.448	22.46	13.920	25.26	16.920
84	7.78	2.976	19.07	8.424	23.74	13.872	26.85	16.872
98	8.42	2.976	20.51	8.424	26.01	13.848	29.67	16.824
112	9.74	2.976	24.03	8.376	30.19	13.800	34.09	16.776
126	11.11	2.952	25.93	8.352	31.99	13.776	36.70	16.728
140	11.36	2.952	26.52	8.328	32.58	13.704	37.50	16.656
154	11.55	2.952	26.69	8.304	33.07	13.680	38.25	16.632

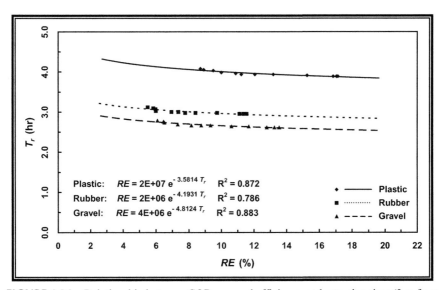

FIGURE 6.26 Relationship between COD removal efficiency and retention time (2 m from inlet).

removal efficiency at 2 m from inlet equals 8.72% (T_r = 4.080 hr), while at the end of the stage (T_s = 154 day) the removal efficiency equals 17.13% (T_r = 3.888 hr). For rubber cell the values are 5.52% (T_r = 3.120 hr) and 11.55% (T_r = 2.952 hr). For gravel cell the values

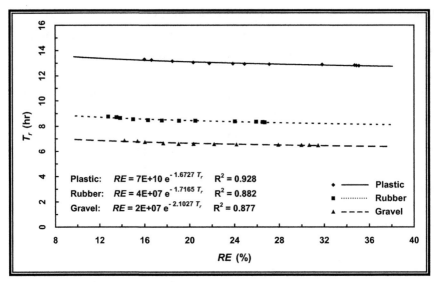

FIGURE 6.27 Relationship between COD removal efficiency and retention time (5 m from inlet).

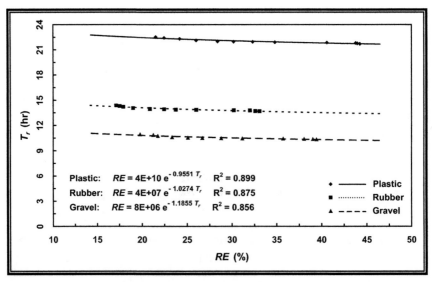

FIGURE 6.28 Relationship between COD removal efficiency and retention time (8 m from inlet).

are 6.10% ($T_r = 2.784$ hr) and 13.55% ($T_r = 2.616$ hr). This means that the treatment performance decreases with increasing T_r values. This could be concluded in the set up stage, the plant and biofilm growth is more effective than retention time.

Set Up Stage Analysis

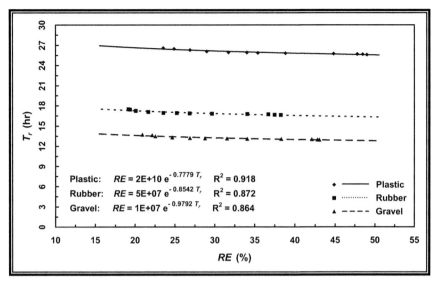

FIGURE 6.29 Relationship between COD removal efficiency and retention time (10 m from inlet).

- The removal efficiency is found better in plastic cell than the two other cells followed by gravel. Values of removal efficiency at T_s equal to 154 days are 49.0, 43.03, and 38.25% for plastic, gravel, and rubber cells at outlets, respectively.

The COD removal efficiency for the three media reversely is proportional to T_r according to an exponential function. The exponential equations at outlets are as follows:

$$\text{Plastic:} \quad RE = 2E + 10e^{-0.7779 T_r} \quad R^2 = 0.918 \quad (6.10a)$$

$$\text{Gravel:} \quad RE = 1E + 07e^{-0.9792 T_r} \quad R^2 = 0.872 \quad (6.10b)$$

$$\text{Rubber:} \quad RE = 5E + 07e^{-0.8542 T_r} \quad R^2 = 0.864 \quad (6.10c)$$

where: RE = COD removal efficiency, %; T_r = hydraulic retention time, hr.

6.3.5 OPERATING TIME EFFECT ON COD TREATMENT EFFICIENCY

The variation of COD removal efficiency with time progression from start of sampling at distances of 2 and 8 m from inlet is illustrated in Figure 6.30, and at distances of 5 and 10 m in Figure 6.31. These binaries are chosen to prevent the overlapping of lines which represent the best fit function (third order polynomial). The values of average difference for COD pollutant between the three media are illustrated in Appendix II.

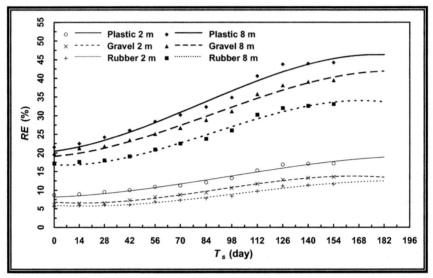

FIGURE 6.30 Relationship between COD removal efficiency and T_s at 2 and 8 m from inlet.

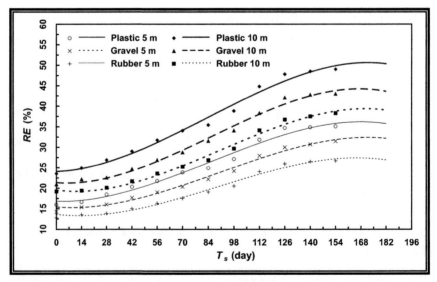

FIGURE 6.31 Relationship between COD removal efficiency and T_s at 5 and 10 m from inlet.

As mentioned before, the set up stage lasted for 154 days, through which the growth of reeds roots through the media and the increase of attached biofilm. These enhance pollutant removal efficiency.

From Tables 5.9–5.11 and Figures 6.30 and 6.31, it is found that:

- At distance of 2 m from inlet, rubber cell gives lower average removal efficiency than gravel and plastic cells by 1.41 and 4.43%, respectively. These values at 5 m are 3.11 and 6.14%, 4.79 and 8.32% at 8 m, and at outlets are 3.80 and 8.52%.
- Gravel cell gives average removal efficiency lower than plastic cell by about 3.02, 3.03, 3.53, and 4.72% at distances of 2, 5, 8, and 10 m from cell inlet, respectively.
- The average differences between plastic removal efficiency and both gravel and rubber removal efficiencies increase with the increasing treatment distance.

6.4 TOTAL SUSPENDED SOLIDS

The effluent TSS concentration was studied with distance, loading rate, and input concentration. The variation of TSS removal efficiency with distance, retention time, and time from start of sampling was also determined.

6.4.1 EFFECT OF DISTANCE ON TSS TREATMENT

Table 6.12 shows the start, average, and end concentrations of TSS pollutant for rubber, gravel, and plastic media at cell entrance and at distances of 2, 5, 8, and 10 m measured from inlet.

Figures 6.32 and 6.33 give the variation of TSS concentration with longitudinal distance for start and end of the set up stage for the three media. Figure 6.34 illustrates this variation for average value of the whole stage. Values in these figures for rubber, gravel, and plastic media were listed in Chapter 5.

From Tables 5.12–5.14 and Figures 6.32–6.34, it is noticed that:

- In plastic cell, TSS concentration reduces with distance better than other media. Gravel takes the second order in TSS reduction. The difference between gravel and rubber media is small in comparison with other pollutants (BOD and COD).
- At the start of the stage, the inlet TSS concentration is 150 mg/l which

TABLE 6.12 Summary of TSS Concentration Sampled at Different Distances

Media	Inlet (mg/l)	Outlet at 2 m Start	Avg.	End	Outlet at 5 m Start	Avg.	End	Outlet at 8 m Start	Avg.	End	Outlet at 10 m Start	Avg.	End
Rubber	150	130	133	132	106	106	104	95	94	91	91	89	85
Gravel	155	130	132	131	103	103	100	92	90	87	86	84	80
Plastic	155	121	123	122	84	83	81	70	69	66	66	64	61

Set Up Stage Analysis

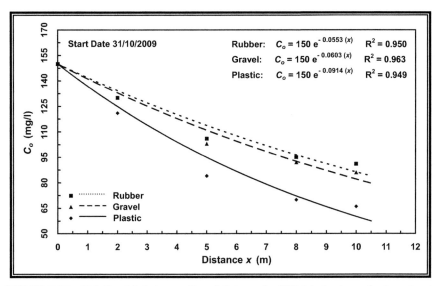

FIGURE 6.32 Relationship between C_o and distance for TSS (start values of set-up stage).

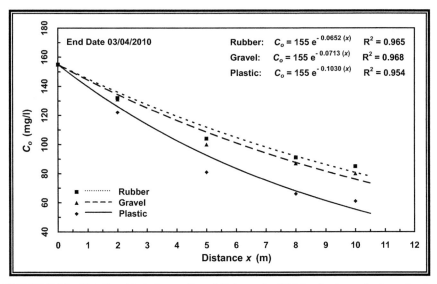

FIGURE 6.33 Relationship between C_o and distance for TSS (end values of set-up stage).

reduces to 121, 130, and 130 mg/l at 2 m for plastic, gravel, and rubber cells, respectively. These values are 84, 103, and 106 mg/l at 5 m, and 70, 92, and 95 mg/l at 8 m. At the end of cells the TSS concentrations are 66, 86, and 91 mg/l.

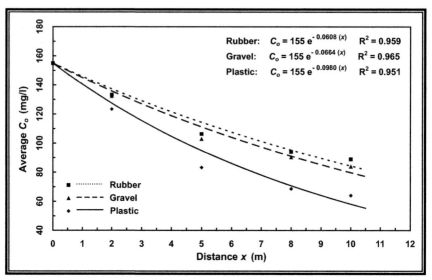

FIGURE 6.34 Relationship between C_o and distance for TSS (average values of set-up stage).

- At the end of the stage, the inlet TSS concentration is 155 mg/l which gives outlet concentration of 122, 131, and 132 mg/l at 2 m for plastic, gravel, and rubber cells, respectively, and 81, 100, and 104 mg/l at 5 m. These values are 66, 87, and 91 mg/l at 8 m, and 61, 80, and 85 mg/l at 10 m at outlets.
- During the whole stage, the average inlet TSS concentration is 155 mg/l which gives average concentration for plastic, gravel, and rubber cells, respectively of 123, 132, and 133 mg/l at 2 m; 83, 103, and 106 mg/l at 5 m; 69, 90, and 94 mg/l at 8 m; and 64, 84, and 89 mg/l at the end of cells.

Concentration of TSS, which is gradually reduced with distance for the three used media, could be represented by an exponential function which gives the best determination coefficient. The exponential equations for the average values are as follows:

$$\text{Plastic:} \quad C_o = 155\, e^{-0.0908x} \quad R^2 = 0.951 \tag{6.11a}$$

$$\text{Gravel:} \quad C_o = 155\, e^{-0.0664x} \quad R^2 = 0.965 \tag{6.11b}$$

$$\text{Rubber:} \quad C_o = 155\, e^{-0.0608x} \quad R^2 = 0.959 \tag{6.11c}$$

where: C_o = TSS outlet concentration, mg/l; x = distance from cell inlet, m.

Figure 6.35 presents the variation of TSS removal efficiency with treatment length for the whole stage.

From Tables 5.12–5.14, and Figure 6.35, it is obtained that:

- At 2 m from plastic cell inlet, TSS removal efficiency increases from 19.33 to 21.29%, from 13.33 to 15.48% in gravel, and from 13.33 to 14.84% in rubber; from the start to the end of the stage.
- At 5 m from plastic cell inlet, TSS removal efficiency raises from 44.0 to 47.74%, from 31.33 to 35.48% in gravel, and from 29.33 to 32.90% in rubber; from the start to the end of the set up stage.
- At 8 m from plastic cell inlet, TSS removal efficiency improves from 53.33 to 57.42%, from 38.67 to 43.87% in gravel, and from 36.67 to 41.29% in rubber; from the start to the end of the stage.
- At outlets, TSS removal efficiency enhances from 56.0 to 60.65% in plastic, from 42.67 to 48.39% in gravel, and from 39.33 to 45.16% in rubber; from the start to the end of the set up stage.
- The rate of treatment increases till the middle of the cells followed by a decrease in this rate towards the end of the cells.

The TSS removal efficiency, which is gradually increased with treatment length for the three used media, may be represented according to power function. The power equations are as following:

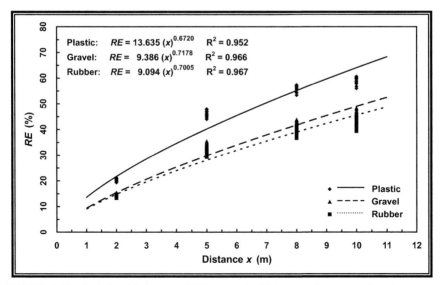

FIGURE 6.35 Relationship between TSS removal efficiency and treatment length.

Plastic: $RE = 13.635\, x^{0.6720}$ $R^2 = 0.952$ (6.12a)

Gravel: $RE = 9.386\, x^{0.7178}$ $R^2 = 0.966$ (6.12b)

Rubber: $RE = 9.094\, x^{0.7005}$ $R^2 = 0.967$ (6.12c)

where: RE = TSS removal efficiency, %; x = distance from cell inlet, m.

6.4.2 IMPACT OF Q ON TSS TREATMENT

The wetland cell is considered as four basins with surface areas of 4, 10, 16, and 20 m² giving four values for q at distances 2, 5, 8, 10 m from inlet, respectively as shown in Table 6.13.

Figure 6.36 gives the relationship between the average effluent TSS concentration and the q values.

From Table 6.13 and Figure 6.36, it is concluded that:

- The outlet TSS concentration increases with increasing q which means the treatment performance decreases. Plastic media treats TSS better than the other media followed by gravel media.
- High value of q occurs near the cell entrance and it decreases with the increase of surface area of cells. After 2 m from inlet, the q values

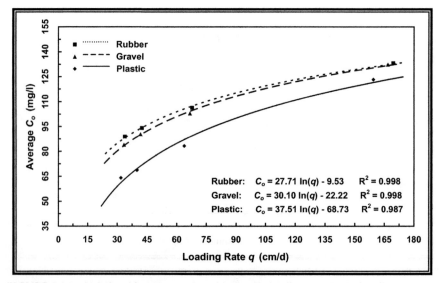

FIGURE 6.36   Relationship between C_o and hydraulic loading rate for TSS pollutant.

Set Up Stage Analysis

TABLE 6.13 Average TSS Concentration and Loading Rate at Different Distances

Distance From Inlet	Plastic Cell $Q = 6.377$ m³/d		Gravel Cell $Q = 6.672$ m³/d		Rubber Cell $Q = 6.781$ m³/d	
	C_o (mg/l)	q (cm/d)	C_o (mg/l)	q (cm/d)	C_o (mg/l)	q (cm/d)
2 m	123.3	159.4	132.3	166.8	133.2	169.5
5 m	83.3	63.8	102.9	66.7	106.3	67.8
8 m	68.7	39.9	90.3	41.7	94.1	42.4
10 m	64.1	31.9	83.9	33.4	88.9	33.9

are 159.4, 166.8, and 169.5 cm/d for plastic, gravel, and rubber cells with corresponding effluent TSS concentration of 123.3, 132.3, and 133.2 mg/l, respectively. At outlets, the q values are 31.9, 33.4, and 33.9 cm/d and the effluent TSS concentrations are 64.1, 83.9, and 88.9 mg/l.

- The C_o difference between the three media is relatively small after short treatment distance (cell inlet region), while it is bigger at the cells exit. This difference is very small between gravel and rubber media along the wetland cell; this may be due to the combined influence of big surface area (high value of biofilm) and small porosity (low value of actual velocity) of gravel media and contrary in rubber media.

The TSS outlet concentration is directly proportional to the loading rate according to a logarithmic function giving the best determination coefficient. The logarithmic equations are written as:

$$\text{Plastic: } \quad C_o = -68.73 + 37.51 \ln q \quad R^2 = 0.987 \quad (6.13a)$$

$$\text{Gravel: } \quad C_o = -22.22 + 30.10 \ln q \quad R^2 = 0.998 \quad (6.13b)$$

$$\text{Rubber: } \quad C_o = -9.53 + 27.71 \ln q \quad R^2 = 0.998 \quad (6.13c)$$

where: C_o = TSS outlet concentration, mg/l; q = hydraulic loading rate, cm/d.

These equations are valid for hydraulic loading rate varies between 31.9 and 169.5 cm/d.

6.4.3 INLET–OUTLET TSS CONCENTRATIONS

Figures 6.37–6.40 give the relationship between effluent and influent concentrations (C_o and C_i) of TSS for the three media at 2, 5, 8, and 10 m, from

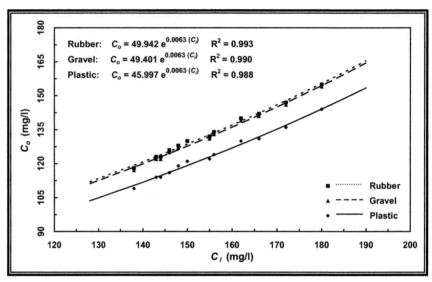

FIGURE 6.37 Relationship between inlet and outlet TSS concentrations (2 m from inlet).

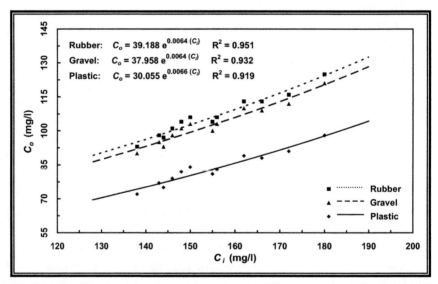

FIGURE 6.38 Relationship between inlet and outlet TSS concentrations (5 m from inlet).

inlet. The C_o and C_i values in these figures for rubber, gravel, and plastic media were listed in Tables 5.12–5.14.

Tables 5.12–5.14 and Figures 6.37–6.40 declare that:

Set Up Stage Analysis

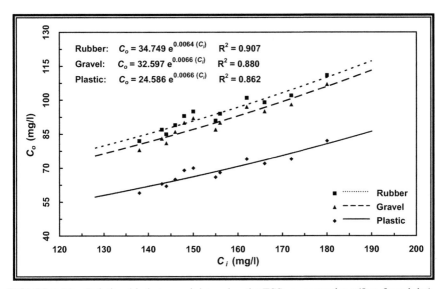

FIGURE 6.39 Relationship between inlet and outlet TSS concentrations (8 m from inlet).

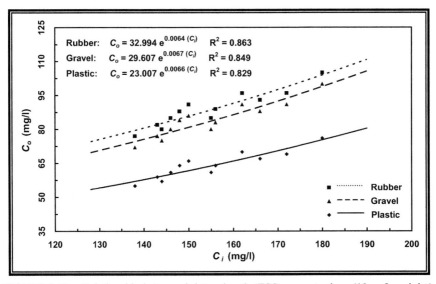

FIGURE 6.40 Relationship between inlet and outlet TSS concentrations (10 m from inlet).

- Through set up stage the outlet TSS values of plastic media are lower than the other media at all distances (better TSS treatment performance), followed by the gravel and then rubber.
- The difference between TSS outlet for gravel and rubber is small at

all distances (gravel have higher surface area than rubber while rubber media have higher porosity than gravel) the mixed effect of these two parameters may be the reason for this small difference.

The relationship between effluent and influent TSS concentrations for the three media varies according to an exponential function. The exponential equations at outlets are as follows:

$$Plastic: \quad C_o = -23.007 \, e^{0.0066 \, C_i} \quad R^2 = 0.829 \qquad (6.14a)$$

$$Gravel: \quad C_o = -29.607 \, e^{0.0067 \, C_i} \quad R^2 = 0.849 \qquad (6.14b)$$

$$Rubber: \quad C_o = -32.994 \, e^{0.0064 \, C_i} \quad R^2 = 0.863 \qquad (6.14c)$$

These equations are valid at the inlet TSS concentration ranges from 138 to 180 mg/l.

6.4.4 IMPACT OF T_R ON TSS REMOVAL EFFICIENCY

Tables 6.14–6.16 present the TSS removal efficiency and retention time at different distances.

TABLE 6.14 TSS Removal Efficiency and T_r for Plastic Cell at Different Distances

T_s (day)	2 m Dis. RE (%)	2 m Dis. T_r (hr)	5 m Dis. RE (%)	5 m Dis. T_r (hr)	8 m Dis. RE (%)	8 m Dis. T_r (hr)	10 m Dis. RE (%)	10 m Dis. T_r (hr)
0	19.33	4.080	44.00	13.296	53.33	22.512	56.00	26.592
14	19.59	4.056	44.59	13.248	53.38	22.416	56.76	26.472
28	19.75	4.032	45.06	13.152	54.32	22.296	56.79	26.304
42	20.00	3.984	45.56	13.056	54.44	22.104	57.78	26.088
56	20.55	3.960	45.89	12.984	55.48	21.984	58.22	25.944
70	20.28	3.936	46.15	12.960	55.94	21.960	58.74	25.920
84	20.51	3.936	46.79	12.936	56.41	21.936	58.97	25.872
98	21.08	3.936	46.99	12.912	56.63	21.888	59.64	25.824
112	20.93	3.912	47.09	12.888	56.98	21.840	59.88	25.752
126	21.01	3.888	47.83	12.840	57.25	21.792	60.14	25.680
140	20.83	3.888	47.92	12.816	56.94	21.744	60.42	25.656
154	21.29	3.888	47.74	12.792	57.42	21.696	60.65	25.584

Set Up Stage Analysis

TABLE 6.15 TSS Removal Efficiency and T_r for Gravel Cell at Different Distances

T_s (day)	2 m Dis. RE (%)	T_r (hr)	5 m Dis. RE (%)	T_r (hr)	8 m Dis. RE (%)	T_r (hr)	10 m Dis. RE (%)	T_r (hr)
0	13.33	2.784	31.33	6.864	38.67	10.944	42.67	13.728
14	14.19	2.760	31.76	6.816	39.19	10.872	43.24	13.656
28	14.20	2.736	32.1	6.744	40.12	10.752	43.83	13.488
42	14.44	2.688	32.78	6.648	40.56	10.608	44.44	13.320
56	14.38	2.664	32.88	6.600	41.10	10.560	45.21	13.224
70	14.69	2.664	33.57	6.600	41.96	10.512	46.15	13.176
84	14.74	2.664	33.97	6.576	42.31	10.488	46.79	13.152
98	15.06	2.640	34.34	6.552	42.77	10.464	46.99	13.104
112	15.12	2.640	34.88	6.528	43.02	10.440	47.09	13.080
126	15.22	2.616	34.78	6.504	43.48	10.392	47.83	13.032
140	15.28	2.616	35.42	6.504	43.75	10.368	47.92	12.984
154	15.48	2.616	35.48	6.480	43.87	10.344	48.39	12.960

TABLE 6.16 TSS Removal Efficiency and T_r for Rubber Cell at Different Distances

T_s (day)	2 m Dis. RE (%)	T_r (hr)	5 m Dis. RE (%)	T_r (hr)	8 m Dis. RE (%)	T_r (hr)	10 m Dis. RE (%)	T_r (hr)
0	13.33	3.120	29.33	8.760	36.67	14.376	39.33	17.496
14	13.51	3.096	29.73	8.712	37.16	14.304	40.54	17.424
28	13.58	3.072	30.25	8.640	37.65	14.208	40.74	17.256
42	13.89	3.024	30.56	8.544	38.33	14.064	41.67	17.088
56	13.70	3.000	30.82	8.472	39.04	13.944	41.78	16.944
70	13.99	3.000	31.47	8.448	39.16	13.920	42.66	16.920
84	14.10	2.976	32.05	8.424	39.74	13.872	42.95	16.872
98	14.46	2.976	31.93	8.424	40.36	13.848	43.98	16.824
112	14.53	2.976	32.56	8.376	40.70	13.800	44.19	16.776
126	14.49	2.952	32.61	8.352	40.58	13.776	44.20	16.728
140	14.58	2.952	32.64	8.328	40.97	13.704	44.44	16.656
154	14.84	2.952	32.90	8.304	41.29	13.680	45.16	16.632

Figures 6.41–6.44 illustrate the relationship between TSS removal efficiency and the hydraulic retention time for wetland cells at different distances.

Tables 6.14–6.16 and Figures 6.41–6.44 show that:

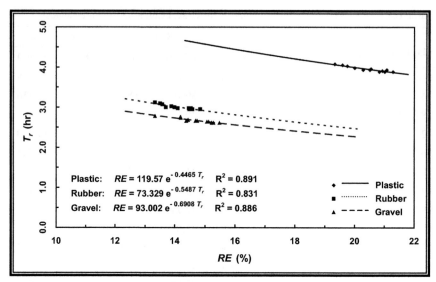

FIGURE 6.41 Relationship between TSS removal efficiency and retention time (2 m from inlet).

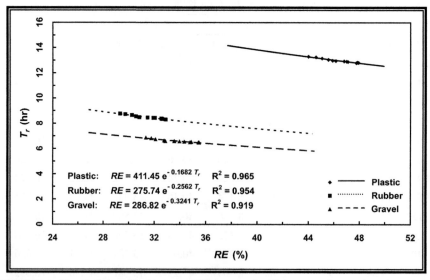

FIGURE 6.42 Relationship between TSS removal efficiency and retention time (5 m from inlet).

- The gravel cell has small porosity, gives small water volumes which in turn leads to low T_r followed by rubber, while plastic cell has the biggest value of T_r.
- The TSS removal efficiency increases with decreasing T_r through set

Set Up Stage Analysis 225

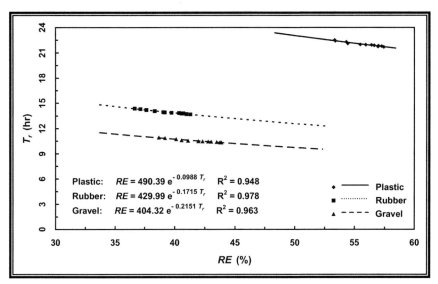

FIGURE 6.43 Relationship between TSS removal efficiency and retention time (8 m from inlet).

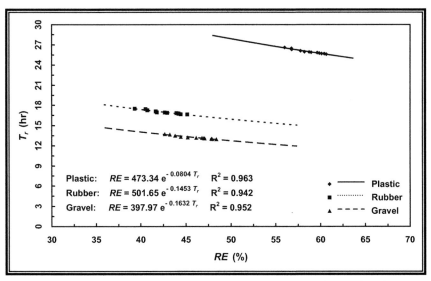

FIGURE 6.44 Relationship between TSS removal efficiency and retention time (10 m from inlet).

up stage for example, for gravel (at the start of the stage) the removal efficiency at 2 m from inlet equals 13.33% (T_r = 2.784 hr) while at the end of the stage, the removal efficiency equal to 15.48% (T_r = 2.616 hr).

These values are 13.33% ($T_r = 3.120$ hr) and 14.84% ($T_r = 2.952$ hr) for rubber, 19.33% ($T_r = 4.080$ hr) and 21.29% ($T_r = 3.888$ hr) for plastic.
- For different media types the removal efficiency is higher in plastic than the other media followed by gravel and then rubber. The treatment status of cells outlets shows that the plastic cell reaches 60.65% at ($T_s = 154$ day) while at the same time from the start of sampling, the rubber and gravel cells reach 45.16 and 48.39%, respectively.
- Through set up stage, there are many factors that improve the treatment performance such as the decrease of actual water velocity through the bed media, plant roots growth, and attached biofilm increase on media surface.

The TSS removal efficiency for the three media is inversely proportional to the retention time according to an exponential function as the best-fit relationship. The exponential equations at outlets are:

$$\text{Plastic:} \quad RE = 473.34\ e^{-0.0804\ T_r} \quad R^2 = 0.963 \quad (6.15a)$$

$$\text{Gravel:} \quad RE = 397.97\ e^{-0.1632\ T_r} \quad R^2 = 0.952 \quad (6.15b)$$

$$\text{Rubber:} \quad RE = 501.65\ e^{-0.1453\ T_r} \quad R^2 = 0.942 \quad (6.15c)$$

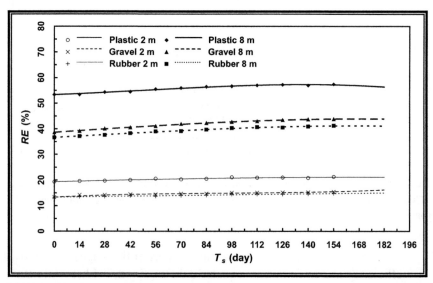

FIGURE 6.45 Relationship between TSS removal efficiency and T_s at 2 and 8 m from inlet.

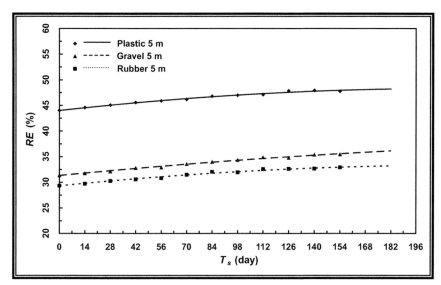

FIGURE 6.46 Relationship between TSS removal efficiency and T_s at 5 from inlet.

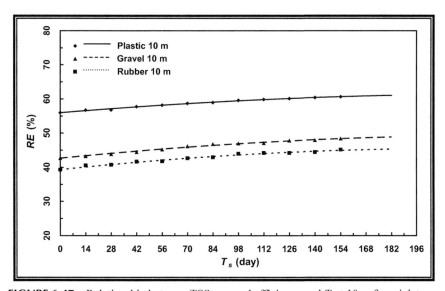

FIGURE 6.47 Relationship between TSS removal efficiency and T_s at 10 m from inlet.

6.4.5 OPERATING TIME EFFECT ON TSS TREATMENT EFFICIENCY

The development of TSS removal efficiency with time progression from start of sampling at distances of 2 and 8 m from inlet is illustrated in Figure 6.45, while at distances of 5 and 10 m in Figures 6.46 and 6.47, respectively. These choices are chosen to prevent the overlapping of lines, which represent the best-fit function (second order polynomial). The values of average difference for TSS pollutant between the three media at different distances are illustrated in Appendix II.

From Tables 5.12–5.14 and Figures 6.45–6.47, it is noticed that:

- At distance of 2 m from inlet, gravel gives higher average removal efficiency than rubber by 0.59% and lowers than plastic by 5.75%. These values at 5 m distance from inlet are 2.20 and 12.69%, at 8 m are 2.43 and 13.98%, and at outlets are about 3.24 and 12.79%.
- Plastic gives average removal efficiency higher than rubber by about 6.35, 14.9, 16.41, and 16.03% at distances of 2, 5, 8, and 10 m from inlet, respectively.
- The gravel and rubber media have approximately the same removal efficiency of TSS but the plastic media gives the highest one.

It can be concluded that the set up stage lasted for 154 days (about 5 months, 12 runs bi-weekly) through which the growth of reeds roots through the cells media and the increase of bacterial biofilm attached to media and root parts. These roots and biofilm improve the TSS removal efficiency.

KEYWORDS

- **plastic media**
- **porosity**
- **rubber media**
- **setup stage**
- **treatment efficiency**

CHAPTER 7

STEADY STAGE ANALYSIS

CONTENTS

7.1 Introduction 229
7.2 Biochemical Oxygen Demand 230
7.3 Chemical Oxygen Demand 248
7.4 Total Suspended Solids 264
7.5 Dissolved Oxygen 282
7.6 Ammonia Treatment 287
7.7 Phosphate Treatment 294
7.8 Fecal Coliforms 301
7.9 Zinc Treatment 308
7.10 Iron Treatment 314
7.11 Manganese Treatment 321
7.12 Lead Treatment 327
7.13 Cadmium Treatment 334
7.14 Removal Rate Constants for Experimental Data 340
Keywords 355

7.1 INTRODUCTION

Steady stage started after all wetland system components reached the steady conditions including plants, biofilm bacteria, and media porosity. This chapter mainly deals with the analysis and discussion of experimental results of parameters measured on the steady stage. The stage parameters consists of biochemical oxygen demand (BOD), chemical oxygen demand (COD), total

---

Parts of this chapter have been reprinted from Zidan, A. A., Rashed, A. A., El-Gamal, M. A., and Abdel Hadi, M. A., "The Effect of Using Different Media n BOD treatment in HSSF Constructed Wetlands (Steady State)," International Water Technology Journal, Vol. 4, No. 1, pp,53-67, March 2014. Reprinted with permission.

suspended solids (TSS), ammonia (NH_3), phosphate (PO_4), dissolved oxygen (DO), fecal coliforms (FC), and some of heavy metals as Zinc (Zn), Iron (Fe), Manganese (Mn), Lead (Pb), and Cadmium (Cd).

A comparative study is presented between the three used media (plastic, gravel, and rubber) in treating the influent primary treated wastewaters. Two main factors affecting the treatment performance; the first one is the effect of constructed wetland cell length (treatment length), and the second factor is the effect of discharge variation. In the steady stage, different sewage loads were applied to the wetland cells in order to study the effects of changing retention time and loads on treatment. Five average sewage loads (discharges) of 4.99, 3.39, 2.34, 1.65, and 1.23 m^3/d (five sampling cycles) were applied during this stage for 6 months started from 17/4/2010 and ended on 16/10/2010.

7.2 BIOCHEMICAL OXYGEN DEMAND

The effluent BOD concentration was studied with distance from cell inlet, loading rate, and influent concentration. The variation of BOD removal efficiency with distance, retention time, and discharge was also determined.

7.2.1 EFFECT OF DISTANCE ON BOD TREATMENT

Tables 7.1–7.3 exhibit the discharge values, the average inlet concentration, the corresponding average outlet concentration, and the removal efficiency for BOD pollutant at different distances from inlet. The average values in these tables were given in Chapter 5.

TABLE 7.1 Average BOD Concentration and Removal Efficiency for Plastic Cell

$Q(m^3/d)$	Influent (mg/l)	Average Concentration (mg/l)				Average Removal Efficiency (%)			
		2 m	5 m	8 m	10 m	2 m	5 m	8 m	10 m
4.814	189.4	155.6	110.4	93.0	81.8	17.85	41.72	50.91	56.83
3.275	198.8	157.0	102.6	82.2	69.6	21.03	48.41	58.68	65.01
2.253	187.8	142.8	84.0	61.4	47.0	23.96	55.28	67.32	74.99
1.595	180.8	135.0	74.6	51.4	37.0	25.34	58.76	71.60	79.57
1.188	183.8	135.2	70.8	46.2	29.8	26.45	61.49	74.87	83.80
Mean	188.1	145.1	88.5	66.8	53.0	22.93	53.13	64.68	72.04

TABLE 7.2 Average BOD Concentration and Removal Efficiency for Gravel Cell

Q(m³/d)	Influent (mg/l)	Average Concentration (mg/l)				Average Removal Efficiency (%)			
		2 m	5 m	8 m	10 m	2 m	5 m	8 m	10 m
5.037	189.4	161.0	121.6	105.0	95.8	15.00	35.80	44.57	49.43
3.426	198.8	162.4	114.4	93.4	83.6	18.32	42.47	53.04	57.97
2.357	187.8	147.8	93.8	70.6	59.0	21.30	50.07	62.42	68.60
1.669	180.8	139.6	83.6	59.6	47.4	22.79	53.79	67.07	73.82
1.243	183.8	141.0	81.6	56.2	43.0	23.29	55.62	69.44	76.62
Mean	188.1	150.4	99.0	77.0	65.8	20.14	47.55	59.31	65.29

TABLE 7.3 Average BOD Concentration and Removal Efficiency for Rubber Cell

Q(m³/d)	Influent (mg/l)	Average Concentration (mg/l)				Average Removal Efficiency (%)			
		2 m	5 m	8 m	10 m	2 m	5 m	8 m	10 m
5.119	189.4	163.0	129.6	113.2	103.6	13.94	31.59	40.25	45.31
3.482	198.8	164.2	123.4	103.2	91.4	17.41	37.94	48.11	54.05
2.396	187.8	149.6	103.8	80.4	66.8	20.34	44.74	57.20	64.45
1.696	180.8	141.4	93.8	70.0	55.4	21.80	48.13	61.30	69.38
1.263	183.8	143.0	91.0	65.6	50.4	22.20	50.50	64.32	72.59
Mean	188.1	152.2	108.3	86.5	73.5	19.14	42.58	54.24	61.16

Figures 7.1 and 7.2 show the variation of BOD concentration with treatment length from inlet at maximum and minimum discharges for plastic, gravel, and rubber media. Figure 6.3 illustrates this variation in case of the average discharge.

From Tables 7.1–7.3 and Figures 7.1–7.3, it is noticed that:

- In plastic cell, BOD concentration is reduced with distance more than the other cells. Gravel takes the second order in BOD reduction.
- For the maximum discharge, the influent BOD concentration is 189.4 mg/l which is reduced at 2 m from inlet to 155.6, 161, and 163 mg/l for plastic, gravel, and rubber cells, respectively. These values at 5 m become 110.4, 121.6, and 129.6 mg/l. The corresponding values at 8 m from inlet are 93, 105, and 113.2 mg/l. At outlets, the BOD concentrations are 81.8, 95.8, and 103.6 mg/l.
- At the minimum discharge, the inlet BOD concentration of 183.8 mg/l is reduced to 135.2, 141, and 143 mg/l for plastic, gravel, and rubber

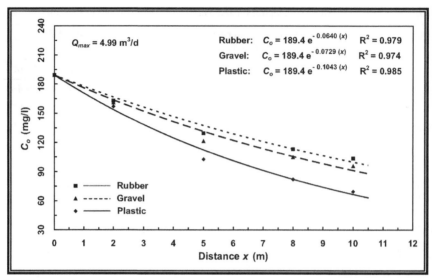

FIGURE 7.1 Relationship between C_o and distance for BOD pollutant (Q_{max} – steady stage).

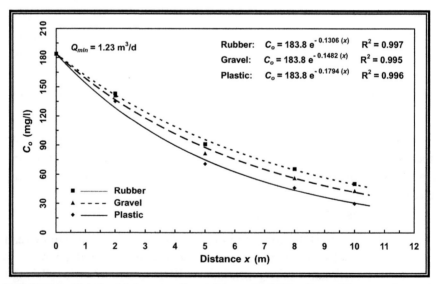

FIGURE 7.2 Relationship between C_o and distance for BOD pollutant (Q_{min} – steady stage).

cells, in the same order at 2 m then to 70.8, 81.6, and 91 mg/l at 5 m distance from inlet. These values are improved to 46.2, 56.2, and 65.6 mg/l at 8 m and 29.8, 43, and 50.4 mg/l at outlets.
- At the average discharge, the average inlet BOD concentration is 188.1 mg/l which is treated to 145.1, 150.4, and 152.2 mg/l at 2 m;

Steady Stage Analysis

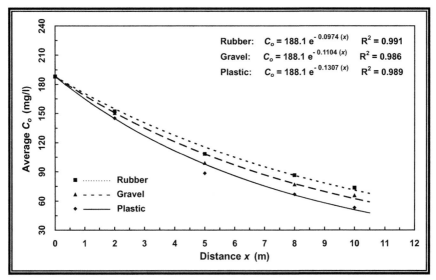

FIGURE 7.3 Relationship between C_o and distance for BOD pollutant (Q_{avg} – steady stage).

88.5, 99, and 108.3 mg/l at 5 m; 66.8, 77, and 86.5 mg/l at 8 m; and 53.0, 65.8, and 73.5 mg/l at outlets for plastic, gravel, and rubber cells, in the same sequence.

Concentration of BOD, which is gradually reduced with distance for the used media, may be represented by an exponential function. This function gives the best determination coefficient (R^2). The exponential equations for the average value of discharge are given as follows:

$$\text{Plastic:} \quad C_o = 188.1 e^{-0.1307x} \quad R^2 = 0.989 \quad (7.1a)$$

$$\text{Gravel:} \quad C_o = 188.1 e^{-0.1104x} \quad R^2 = 0.986 \quad (7.1b)$$

$$\text{Rubber:} \quad C_o = 188.1 e^{-0.0974x} \quad R^2 = 0.991 \quad (7.1c)$$

where:
x = distance from cell inlet, m;
C_o = average BOD concentration, mg/l.

Using the relationships between C_o and distance, the average pollutant concentration may be determined in-between distances at each wetland cell. Another important application of these relationships lies in estimating the required cell length for any of the three media types to reach a certain pollutant effluent concentration. Figure 7.4 presents the variation of BOD removal efficiency with treatment length in case of the used discharges.

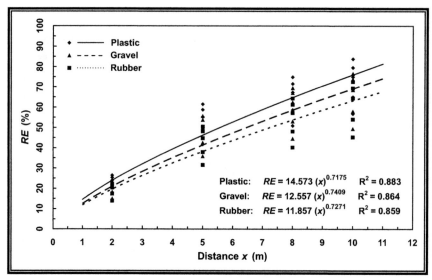

FIGURE 7.4 Relationship between BOD removal efficiency and treatment length.

From Tables 7.1–7.3 and Figure 7.4, it is remarked that:

- At 2 m from plastic cell inlet, BOD removal efficiency for Q_{max} is 17.85% and increases to 26.45% at Q_{min}. This efficiency changes from 15 to 23.29% for gravel and from 13.94 to 22.2% for rubber.
- At 5 m from plastic cell inlet, BOD removal efficiency raises from 41.72 to 61.49% at Q_{max} and Q_{min}, respectively, from 35.8 to 55.62% for gravel, and from 31.59 to 50.5% for rubber.
- At 8 m from plastic cell inlet, BOD removal efficiency for Q_{max} is 50.91% and increases to 74.87% at Q_{min}. This efficiency changes from 44.57 to 69.44% for gravel and from 40.25 to 64.32% for rubber, in the same sequence.
- At outlet, BOD removal efficiency raises from 56.83 to 83.8% in plastic cell at Q_{max} and Q_{min}, in the same order, from 49.43 to 76.62% for gravel, and from 45.31 to 72.59% for rubber.
- Generally, in all media cells, the treatment efficiency improves in case of smaller discharges (loads) compared with higher ones.
- The maximum rate of removal exists at the middle of basin (5 m from inlet), this rate decreases gradually towards the end of cells, which means that, the first five meters are the most effective length in BOD treatment performance through the steady stage.

The BOD removal efficiency is gradually increased with treatment length for the three used media according to a power function. The power equations are written as follows:

$$Plastic: \quad RE = 14.573 x^{0.7175} \quad R^2 = 0.883 \quad (7.2a)$$

$$Gravel: \quad RE = 12.557 x^{0.7409} \quad R^2 = 0.864 \quad (7.2b)$$

$$Rubber: \quad RE = 11.857 x^{0.7271} \quad R^2 = 0.859 \quad (7.2c)$$

where:
x = distance from inlet, m;
RE = BOD removal efficiency, %.

7.2.2 INLET AND OUTLET BOD CONCENTRATIONS RELATIONSHIPS

Figures 7.5–7.8 give the relationship between effluent and influent concentrations (C_o and C_i) of BOD for the three media at 2, 5, 8, and 10 m from inlet. The C_i and C_o in these figures for plastic, gravel, and rubber media are listed in Tables 7.1–7.3.

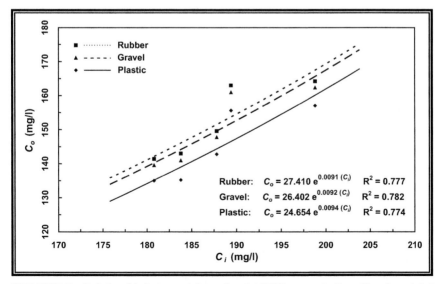

FIGURE 7.5 Relationship between inlet and outlet BOD concentrations (2 m from inlet).

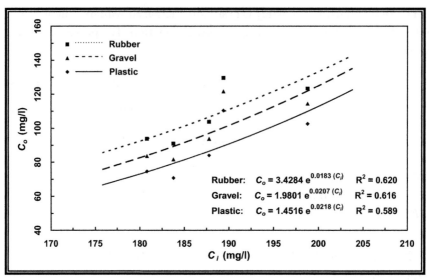

FIGURE 7.6 Relationship between inlet and outlet BOD concentrations (5 m from inlet).

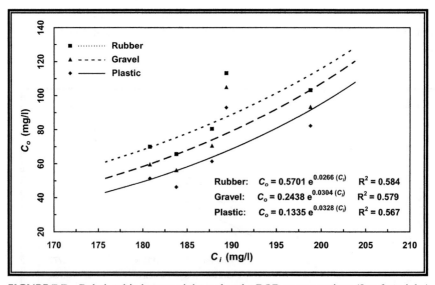

FIGURE 7.7 Relationship between inlet and outlet BOD concentrations (8 m from inlet).

From Tables 7.1–7.3 and Figures 7.5–7.8, it is found that:

- Through steady stage the outlet BOD concentration of plastic media is smaller than both other media (better BOD treatment performance) at all distances, followed by gravel.

Steady Stage Analysis

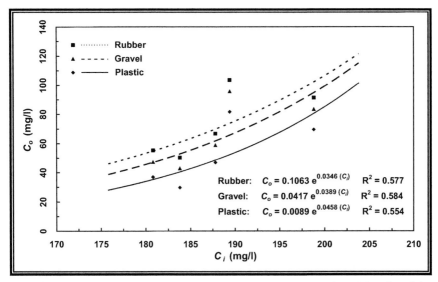

FIGURE 7.8 Relationship between inlet and outlet BOD concentrations (10 m from inlet).

- For plastic media, the BOD concentration at outlet is in the allowable limit (60 mg/l) of Law No. 48 of 1982 (NAWQAM, 2002) at discharge lower than 2.253 m^3/d that corresponds to q value of 11.27 cm/d.
- For gravel media the BOD concentration at the end of cell is in the permitted limit at discharge lower than 2.357 m^3/d that corresponds to q value of 11.79 cm/d.
- For rubber media the BOD concentration at 10 m distance from inlet weir is in the admissible limit at discharge lower than 1.696 m^3/d which corresponds to q value of 8.48 cm/d.

The relationship between the effluent and influent concentrations of BOD for the three media is varying according to an exponential function. The exponential equations at outlets are:

$$\textit{Plastic:} \quad C_o = 0.0089 e^{0.0458 C_i} \quad R^2 = 0.554 \tag{7.3a}$$

$$\textit{Gravel:} \quad C_o = 0.0417 e^{0.0389 C_i} \quad R^2 = 0.584 \tag{7.3b}$$

$$\textit{Rubber:} \quad C_o = 0.1063 e^{0.0346 C_i} \quad R^2 = 0.577 \tag{7.3c}$$

These equations are valid for the inlet BOD concentration ranges from 180.8 to 198.8 mg/l (q ranges from 5.94 to 25.6 cm/d).

7.2.3 IMPACT OF LOADING RATE ON BOD TREATMENT

To study the impact of q on BOD treatment, each wetland cell is considered as four basins with 2, 5, 8, and 10 m lengths and 2 m width (surface areas of 4, 10, 16, and 20 m²). So each discharge (Q, m³/d) gives four loading rates correspond to each of the four basins area (Eq. 4.1):

$$q = \frac{Q \times 100}{A} \qquad (4.1)$$

Tables 7.4–7.8 show the average outlet BOD concentration and the loading rate for plastic, gravel, and rubber cells at different distances. The average values of influent BOD concentration are given in these tables.

Figure 7.9 illustrates the relationship between the effluent BOD concentration and the loading rate. The data used in this figure are listed in Tables 7.4–7.8 and the five cycles are represented together. Five discharges, at four different distances through the cell (four surface areas), give 20 loading rates.

TABLE 7.4 Effluent BOD and q at Different Distances (Cycle No.1 – C_i = 189.4 mg/l)

Distance From Inlet	Plastic Cell Q = 4.814 m³/d		Gravel Cell Q = 5.037 m³/d		Rubber Cell Q = 5.119 m³/d	
	C_o(mg/l)	q(cm/d)	C_o(mg/l)	q(cm/d)	C_o(mg/l)	q(cm/d)
2 m	155.6	120.35	161.0	125.93	163.0	127.98
5 m	110.4	48.14	121.6	50.37	129.6	51.19
8 m	93.0	30.09	105.0	31.48	113.2	31.99
10 m	81.8	24.07	95.8	25.19	103.6	25.60

TABLE 7.5 Effluent BOD and q at Different Distances (Cycle No.2 – C_i = 198.8 mg/l)

Distance From Inlet	Plastic Cell Q = 3.275 m³/d		Gravel Cell Q = 3.426 m³/d		Rubber Cell Q = 3.482 m³/d	
	C_o(mg/l)	q(cm/d)	C_o(mg/l)	q(cm/d)	C_o(mg/l)	q(cm/d)
2 m	157.0	81.88	162.4	85.65	164.2	87.05
5 m	102.6	32.75	114.4	34.26	123.4	34.82
8 m	82.2	20.47	93.4	21.41	103.2	21.76
10 m	69.6	16.38	83.6	17.13	91.4	17.41

TABLE 7.6 Effluent BOD and q at Different Distances (Cycle No.3 – C_i = 187.8 mg/l)

Distance From Inlet	Plastic Cell Q = 2.253 m³/d		Gravel Cell Q = 2.357 m³/d		Rubber Cell Q = 2.396 m³/d	
	C_o(mg/l)	q(cm/d)	C_o(mg/l)	q(cm/d)	C_o(mg/l)	q(cm/d)
2 m	142.8	56.33	147.8	58.93	149.6	59.90
5 m	84.0	22.53	93.8	23.57	103.8	23.96
8 m	61.4	14.08	70.6	14.73	80.4	14.98
10 m	47.0	11.27	59.0	11.79	66.8	11.98

TABLE 7.7 Effluent BOD and q at Different Distances (Cycle No.4 – C_i = 180.8 mg/l)

Distance From Inlet	Plastic Cell Q = 1.595 m³/d		Gravel Cell Q = 1.669 m³/d		Rubber Cell Q = 1.696 m³/d	
	Co(mg/l)	q(cm/d)	Co(mg/l)	q(cm/d)	Co(mg/l)	q(cm/d)
2 m	135.0	39.88	139.6	41.73	141.4	42.40
5 m	74.6	15.95	83.6	16.69	93.8	16.96
8 m	51.4	9.97	59.6	10.43	70.0	10.60
10 m	37.0	7.98	47.4	8.35	55.4	8.48

TABLE 7.8 Effluent BOD and q at Different Distances (Cycle No.5 – C_i = 183.8 mg/l)

Distance From Inlet	Plastic Cell Q = 1.188 m³/d		Gravel Cell Q = 1.243 m³/d		Rubber Cell Q = 1.263 m³/d	
	Co(mg/l)	q(cm/d)	Co(mg/l)	q(cm/d)	Co(mg/l)	q(cm/d)
2 m	135.2	29.70	141.0	31.08	143.0	31.58
5 m	70.8	11.88	81.6	12.43	91.0	12.63
8 m	46.2	7.43	56.2	7.77	65.6	7.89
10 m	29.8	5.94	43.0	6.22	50.4	6.32

From Tables 7.4–7.8 and Figure 7.9, it is concluded that:

- The effluent BOD concentration decreases with the decrease of loading rate indicating analogous reduction in BOD treatment performance for the three used media.
- For cycle number one (Q_{max} = 4.814, 5.037, and 5.119 m³/d), the influent BOD concentration is 189.4 mg/l. At 2 m from inlet, the loading rates are 120.4, 125.9, and 128 cm/d for plastic, gravel, and rubber

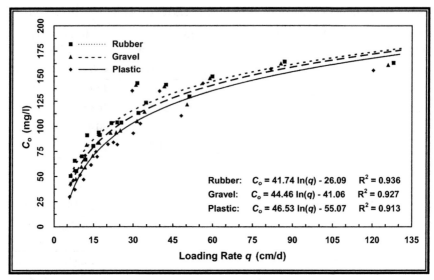

FIGURE 7.9 Relationship between C_o and hydraulic loading rate for BOD pollutant.

cells with corresponding effluent BOD concentrations of 155.6, 161, and 163 mg/l, respectively. After 10 m from inlet weir, the q values are 24.1, 25.2, and 25.6 cm/d and the outlet BOD concentrations are 81.8, 95.8, and 103.6 mg/l.

- For cycle number five (Q_{min} = 1.188, 1.243, and 1.263 m³/d and C_i = 183.8 mg/l) at 2 m from inlet, the q values are 29.7, 31.1, and 31.6 cm/d for plastic, gravel, and rubber cells with corresponding effluent BOD concentrations of 135.2, 141, and 143 mg/l, respectively. At outlet, the q values are 5.9, 6.2, and 6.3 cm/d and BOD concentrations are 29.8, 43.0, and 50.4 mg/l.
- At low q values, the difference between the outlet BOD concentrations for the three media is bigger than at high q values. Plastic cell treats BOD better than the other two cells followed by gravel.

The BOD effluent concentration directly proportions with the loading rate for the three cells according to a logarithmic function, as a best fit relationship. The logarithmic equations are given as follows:

$$\textit{Plastic}: \quad C_o = -55.07 + 46.53 \ln q \quad R^2 = 0.913 \quad (7.4a)$$

$$\textit{Gravel}: \quad C_o = -41.06 + 44.46 \ln q \quad R^2 = 0.927 \quad (7.4b)$$

$$\textit{Rubber}: \quad C_o = -26.09 + 41.74 \ln q \quad R^2 = 0.936 \quad (7.4c)$$

Steady Stage Analysis

The relationships between outlet pollutant concentration and loading rate are valid for q values range from 5.94 to 127.98 cm/d.

7.2.4 IMPACT OF TR ON BOD REMOVAL EFFICIENCY

At the end of set up stage the porosity values for the used media reached to the steady state where porosity and biofilm layer thickness became nearly fixed values. The gravel media has the smallest porosity of 0.358 followed by the rubber media porosity which equals 0.505, while plastic media has the greatest porosity of 0.788. The coarse gravel media in the inlet and outlet parts of each wetland cell has a porosity of 0.365, Chapter 5.

Tables 7.9–7.11 give the values of T_r which correspond to the applied discharges for plastic, gravel, and rubber media, respectively.

Figures 7.10–7.12 illustrate the relationship between BOD removal efficiency and T_r at different distances from inlet of cells.

TABLE 7.9 BOD Removal Efficiency and T_r for Plastic Media at Different Distances

Cycle No.	Q(m³/d)	Hydraulic Retention Time (hr)				Average Removal Efficiency (%)			
		2 m	5 m	8 m	10 m	2 m	5 m	8 m	10 m
1	4.814	5.14	16.92	28.70	33.84	17.85	41.72	50.91	56.83
2	3.275	7.54	24.86	42.19	49.73	21.03	48.41	58.68	65.01
3	2.253	10.97	36.14	61.32	72.31	23.96	55.28	67.32	74.99
4	1.595	15.48	51.07	86.64	102.12	25.34	58.76	71.60	79.57
5	1.188	20.81	68.54	116.30	137.11	26.45	61.49	74.87	83.80

TABLE 7.10 BOD Removal Efficiency and T_r for Gravel Media at Different Distances

Cycle No.	Q(m³/d)	Hydraulic Retention Time (hr)				Average Removal Efficiency (%)			
		2 m	5 m	8 m	10 m	2 m	5 m	8 m	10 m
1	5.037	3.46	8.57	13.68	17.16	15.00	35.80	44.57	49.43
2	3.426	5.09	12.60	20.14	25.20	18.32	42.47	53.04	57.97
3	2.357	7.39	18.31	29.26	36.65	21.30	50.07	62.42	68.60
4	1.669	10.42	25.87	41.30	51.74	22.79	53.79	67.07	73.82
5	1.243	13.99	34.73	55.46	69.48	23.29	55.62	69.44	76.62

TABLE 7.11 BOD Removal Efficiency and T_r for Rubber Media at Different Distances

Cycle No.	Q(m³/d)	Hydraulic Retention Time (hr) 2 m	5 m	8 m	10 m	Average Removal Efficiency (%) 2 m	5 m	8 m	10 m
1	5.119	3.89	10.99	18.10	21.98	13.94	31.59	40.25	45.31
2	3.482	5.71	16.15	26.59	32.30	17.41	37.94	48.11	54.05
3	2.396	8.30	23.47	38.66	46.97	20.34	44.74	57.20	64.45
4	1.696	11.74	33.17	54.60	66.34	21.80	48.13	61.30	69.38
5	1.263	15.74	44.54	73.34	89.09	22.20	50.50	64.32	72.59

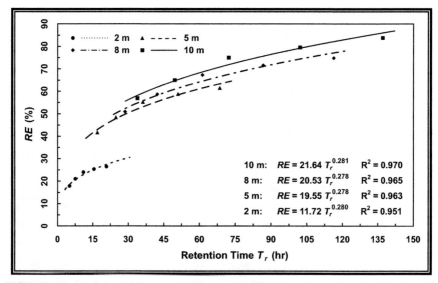

FIGURE 7.10 Relationship between BOD removal efficiency and retention time (plastic cell).

From Tables 7.9–7.11 and Figures 7.10–7.12, it is noticed that:

- The gravel cell has the lowest retention time followed by the rubber and then the plastic cell. This is compatible with the three media porosities, where gravel has the smallest porosity followed by rubber, while plastic has the greatest porosity.
- The BOD removal efficiency increases with the increase of T_r through steady stage. For example for gravel cell, at T_r equals 8.57 hr for $Q = 5.037$ m³/d (5 m from inlet), the removal efficiency equals 35.8%, while at T_r equals 34.73 hr for $Q = 1.243$ m³/d, the removal efficiency equals 55.62% at the same distance.

Steady Stage Analysis

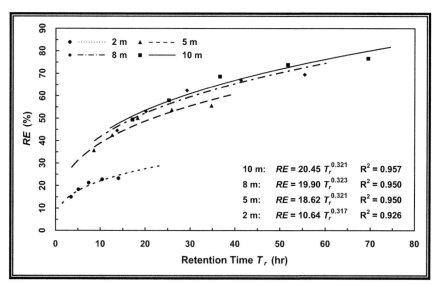

FIGURE 7.11 Relationship between BOD removal efficiency and retention time (gravel cell).

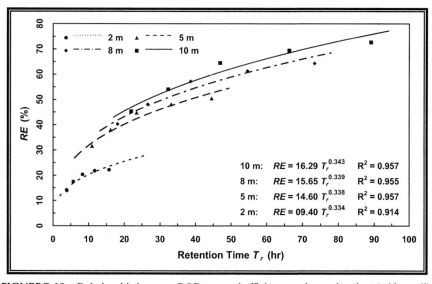

FIGURE 7.12 Relationship between BOD removal efficiency and retention time (rubber cell).

- Generally the T_r increases with the decreasing value of discharge. At 2 m from inlet weir, the T_r increases from 3.46 to 13.99 hr for gravel cell and from 3.89 to 15.74 hr for rubber and from 5.14 to 20.81 hr for plastic as the average discharge decreases from 4.99 to 1.23 m³/d.

- At any point inside the wetland cell, the longer T_r produces a better RE of BOD.
- For different media types the BOD removal efficiency is found better in plastic cell followed by gravel and then rubber cell. For example when applying the smallest discharge at cells inlet, the BOD removal efficiency values are 83.8, 76.62, and 72.59% at the outlet for plastic, gravel, and rubber cells, respectively.

The BOD removal efficiency is directly proportional to the retention time according to a power function. The power equations at outlets are:

$$\text{Plastic:} \quad RE = 21.64 \, T_r^{0.281} \quad R^2 = 0.970 \tag{7.5a}$$

$$\text{Gravel:} \quad RE = 20.45 \, T_r^{0.321} \quad R^2 = 0.957 \tag{7.5b}$$

$$\text{Rubber:} \quad RE = 16.29 \, T_r^{0.343} \quad R^2 = 0.957 \tag{7.5c}$$

All deduced relationships between pollutant removal efficiency and retention time are valid for Q_{avg} varies between 4.99 and 1.23 m³/d which gives T_r ranges from 33.84 to 137.11 hr for plastic cell, from 17.16 to 69.48 hr for gravel cell, and from 21.98 to 89.09 hr for rubber cell.

7.2.5 IMPACT OF DISCHARGE ON BOD TREATMENT EFFICIENCY

This study focuses on the effect of changing discharge on treatment. Figures 7.13–7.16 show the relationship between BOD removal efficiency and the corresponding Q at 2, 5, 8, and 10 m from inlet. Values in these figures are listed in Tables 7.9–7.11.

Tables 7.9–7.11 and Figures 7.13–7.16 highlight that:

- As the discharge decreases, the removal efficiency of BOD increases at 2, 5, 8, and 10 m distance from inlet and for plastic, gravel, and rubber media.
- The plastic cell has highest removal efficiency followed by the gravel and then the rubber. This may be due to the higher surface area (high amount of attached biofilm) of the plastic media comparing with the other used media.
- As for plastic cell, at the biggest Q (4.814 m³/d), the BOD removal efficiency values are 17.85, 41.72, 50.91, and 56.83% at distances of 2, 5, 8, and 10 m from inlet, respectively. While these removal efficiencies are 26.45, 61.49, 74.87, and 83.8% at the smallest Q (1.188 m³/d).

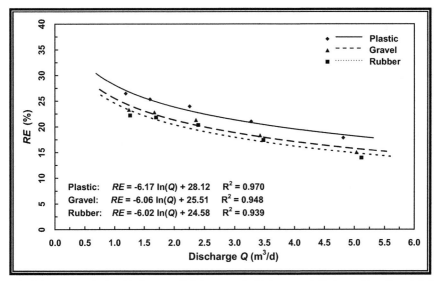

FIGURE 7.13 Relationship between BOD removal efficiency and discharge (2 m from inlet).

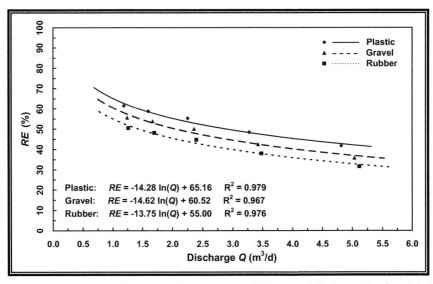

FIGURE 7.14 Relationship between BOD removal efficiency and discharge (5 m from inlet).

- For gravel cell, at the biggest Q (5.037 m³/d), the BOD removal efficiency values are 15.0, 35.8, 44.57, and 49.43% at distances of 2, 5, 8, and 10 m from inlet, in the same sequence. These removal efficiencies are 23.29, 55.62, 69.44, and 76.62% at the smallest Q (1.243 m³/d).

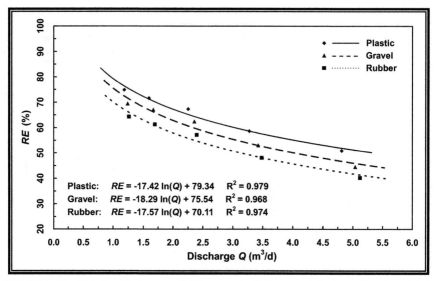

FIGURE 7.15 Relationship between BOD removal efficiency and discharge (8 m from inlet).

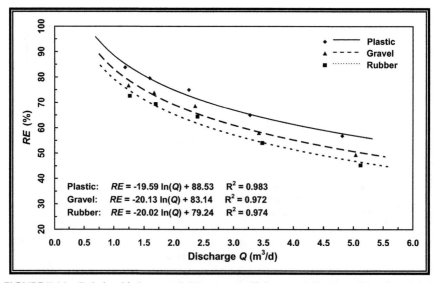

FIGURE 7.16 Relationship between BOD removal efficiency and discharge (10 m from inlet).

- For rubber cell, at the biggest Q (5.119 m³/d), the BOD removal efficiency values are 13.94, 31.59, 40.25, and 45.31% at distances of 2, 5, 8, and 10 m from inlet, in the same order. While these removal efficiencies are 22.2, 50.5, 64.32, and 72.59% at the smallest Q (1.263 m³/d).

- The difference between rubber and gravel removal efficiencies is relatively small at the first 2 m from inlet. For example at this distance, the average BOD removal efficiency is 19.14 and 20.14% for rubber and gravel media, respectively. This may be due to the small difference between media surface area at this distance.

The average difference for pollutant removal efficiency through 5 discharge cycles between plastic cell and both gravel and rubber cells is given in Eqs. (4.10) and (4.11). The average difference between gravel and rubber cells is presented in Eq. (4.12). The values of average difference for BOD pollutant between the three media at different distances are illustrated in Appendix II.

$$Average\ Difference\ (Plastic\ \&\ Gravel) = \frac{\sum(RE_{xp} - RE_{xg})}{No.\ of\ Cycles} \quad (4.10)$$

$$Average\ Difference\ (Plastic\ \&\ Rubber) = \frac{\sum(RE_{xp} - RE_{xr})}{No.\ of\ Cycles} \quad (4.11)$$

$$Average\ Difference\ (Gravel\ \&\ Rubber) = \frac{\sum(RE_{xg} - RE_{xr})}{No.\ of\ Cycles} \quad (4.12)$$

where: RE_{xp} = removal efficiency of plastic at distance x, %; RE_{xg} = removal efficiency of gravel at distance x, %; RE_{xr} = removal efficiency of rubber at distance x, %.

At distance of 2 m from inlet, plastic cell gives average difference higher than both gravel and rubber cells by about 2.79 and 3.79%, respectively. These values at 5 m are about 5.58 and 10.55%, at 8 m are about 5.37 and 10.44%, and at outlets are about 6.75 and 10.88%. Whereas gravel cell gives an average difference higher than rubber by about 1.0, 4.97, 5.07, and 4.13% at distances of 2, 5, 8, and 10 m from inlet, in the same sequence. The BOD removal efficiency (RE, %) is gradually reduced with the increase of discharge at different distances according to a logarithmic function. The logarithmic equations at outlets are as follows:

Plastic: $RE = 88.53 - 19.59 \ln Q$ $R^2 = 0.983$ (7.6a)

Gravel: $RE = 83.14 - 20.13 \ln Q$ $R^2 = 0.972$ (7.6b)

Rubber: $RE = 79.24 - 20.02 \ln Q$ $R^2 = 0.974$ (7.6c)

The relationships between BOD removal efficiency and discharge are valid for Q values range from 1.188 to 5.119 m^3/d.

7.3 CHEMICAL OXYGEN DEMAND

The study highlights the effect of distance, loading rate, and influent concentration on the effluent COD, in addition to the variation of COD treatment efficiency with distance, retention time, and discharge.

7.3.1 EFFECT OF DISTANCE ON COD TREATMENT

Tables 7.12–7.14 present the discharge values, the average inlet concentration, the average outlet concentration, and the corresponding removal efficiency

TABLE 7.12 Average COD Concentration and Removal Efficiency for Plastic Cell

Q(m^3/d)	Influent (mg/l)	Average Concentration (mg/l)				Average Removal Efficiency (%)			
		2 m	5 m	8 m	10 m	2 m	5 m	8 m	10 m
4.814	296.2	242.2	174.4	143.6	129.0	18.21	41.12	51.52	56.45
3.275	307.4	241.4	160.4	125.6	108.6	21.47	47.85	59.17	64.71
2.253	289.4	218.8	130.8	93.4	73.2	24.39	54.80	67.74	74.72
1.595	279.2	207.4	116.4	78.4	57.6	25.65	58.29	71.94	79.40
1.188	284.0	208.2	110.6	70.6	46.4	26.72	61.09	75.16	83.68
Mean	291.2	223.6	138.5	102.3	83.0	23.29	52.63	65.11	71.79

TABLE 7.13 Average COD Concentration and Removal Efficiency for Gravel Cell

Q(m^3/d)	Influent (mg/l)	Average Concentration (mg/l)				Average Removal Efficiency (%)			
		2 m	5 m	8 m	10 m	2 m	5 m	8 m	10 m
5.037	296.2	253.4	192.8	162.6	151.0	14.43	34.90	45.10	49.02
3.426	307.4	253.4	179.4	143.2	130.2	17.57	41.67	53.45	57.68
2.357	289.4	229.8	146.4	107.8	91.6	20.58	49.42	62.76	68.37
1.669	279.2	217.0	130.8	91.2	73.6	22.22	53.16	67.36	73.67
1.243	284.0	219.8	128.0	86.2	67.2	22.64	54.97	69.68	76.37
Mean	291.2	234.7	155.5	118.2	102.7	19.49	46.82	59.67	65.02

Steady Stage Analysis

TABLE 7.14 Average COD Concentration and Removal Efficiency for Rubber Cell

Q(m³/d)	Influent (mg/l)	Average Concentration (mg/l)				Average Removal Efficiency (%)			
		2 m	5 m	8 m	10 m	2 m	5 m	8 m	10 m
5.119	296.2	256.2	204.6	176.0	163.0	13.49	30.92	40.58	44.97
3.482	307.4	255.0	192.6	159.0	142.2	17.04	37.35	48.30	53.77
2.396	289.4	231.4	161.6	123.2	103.8	20.03	44.15	57.44	64.15
1.696	279.2	219.4	146.4	107.6	86.0	21.37	47.54	61.45	69.20
1.263	284.0	222.2	142.0	101.2	78.2	21.79	50.03	64.40	72.50
Mean	291.2	236.8	169.4	133.4	114.6	18.74	42.00	54.43	60.92

for COD pollutant at distances of 2, 5, 8, and 10 m from inlet. These tables present the five cycles mean values. The average concentration and removal efficiency were given in Chapter 5.

Figures 7.17 and 7.18 give the variation of COD concentration with treatment length from inlet at maximum and minimum discharges for the three used media. Figure 7.19 illustrates this variation at the average discharge. Figure 7.20 presents the variation of COD removal efficiency with treatment length in case of the used discharges.

From Tables 7.12–7.14 and Figures 7.17–7.19, it is noticed that:

- In plastic cell, COD concentration is reduced with distance better than other cells. Gravel takes the second order in reduction.
- At the maximum discharge, the inlet COD concentration is 296.2 mg/l which decreases to 242.2, 253.4, and 256.2 mg/l at 2 m from inlet for plastic, gravel, and rubber cells, respectively. These values improve to 174.4, 192.8, and 204.6 mg/l at 5 m; and then to 143.6, 162.6, and 176 mg/l at 8 m. At outlets the COD concentrations are 129, 151, and 163 mg/l, in the same sequence.
- At the minimum discharge, the inlet COD concentration is 284 mg/l reduces to 208.2, 219.8, and 222.2 mg/l for plastic, gravel, and rubber cells, in the same order at 2 m from inlet then to 110.6, 128, and 142 mg/l at 5 m. These values improve to 70.6, 86.2, and 101.2 mg/l at 8 m and 46.4, 67.2, and 78.2 mg/l at outlets.
- During the stage, the average inlet COD is 291.2 mg/l which is treated to 223.6, 234.7, and 236.8 mg/l at 2 m; 138.6, 155.5, and 169.4 mg/l at 5 m; 102.3, 118.2, and 133.4 mg/l at 8 m; and 83, 102.7, and 114.6 mg/l at outlets for plastic, gravel, and rubber.

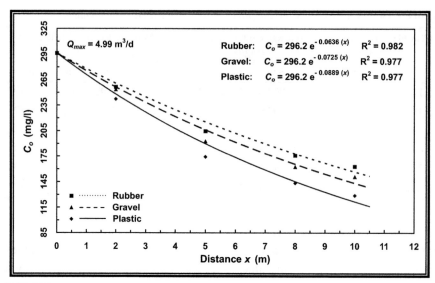

FIGURE 7.17 Relationship between C_o and distance for COD pollutant (Q_{max} – steady stage).

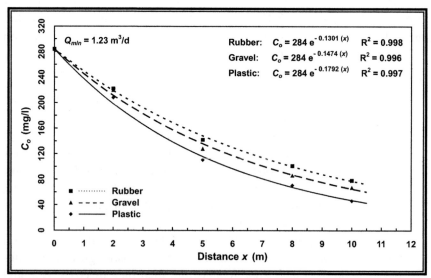

FIGURE 7.18 Relationship between C_o and distance for COD pollutant (Q_{min} – steady stage).

The outlet COD concentration is gradually reduced with distance for the three used media according to an exponential function. The exponential equations for the average values are written as:

Steady Stage Analysis

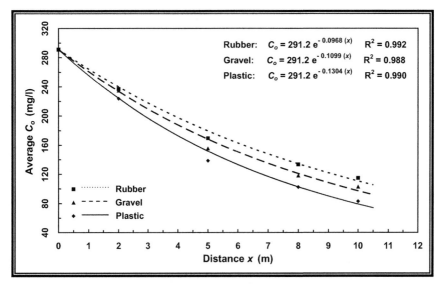

FIGURE 7.19 Relationship between C_o and distance for COD pollutant (Q_{avg} – steady stage).

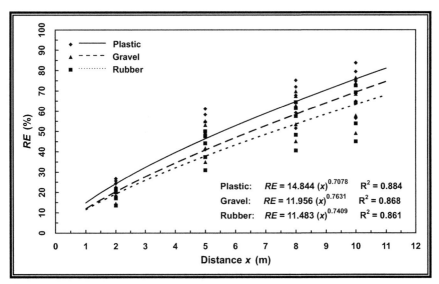

FIGURE 7.20 Relationship between COD removal efficiency and treatment length.

$$\text{Plastic:} \quad C_o = 291.2e^{-0.1304x} \quad R^2 = 0.990 \quad (7.7a)$$

$$\text{Gravel:} \quad C_o = 291.2e^{-0.1099x} \quad R^2 = 0.988 \quad (7.7b)$$

$$\text{Rubber:} \quad C_o = 291.2e^{-0.0968x} \quad R^2 = 0.992 \quad (7.7c)$$

From Tables 7.12–7.14 and Figure 7.20, it is observed that:

- At 2 m from plastic cell inlet, COD removal efficiency raises from 18.21 to 26.72% at Q_{max} and Q_{min}, respectively, from 14.43 to 22.64% in gravel, and from 13.49 to 21.79% in rubber.
- At 5 m from plastic cell inlet, COD removal efficiency increases from 41.12 to 61.09% at Q_{max} and Q_{min}, in the same order, from 34.9 to 54.97% in gravel, and from 30.92 to 50.03% in rubber.
- At 8 m from plastic cell inlet, COD removal efficiency raises from 51.52 to 75.16% at Q_{max} and Q_{min}, respectively, from 45.1 to 69.68% in gravel, and from 40.58 to 64.4% in rubber.
- At outlet, COD removal efficiency improves from 56.45 to 83.68% in plastic cell at Q_{max} and Q_{min}, respectively, from 49.02 to 76.37% in gravel, and from 44.97 to 72.5% in rubber.

The COD removal efficiency is gradually increased with distance for the three media according to a power function which gives the best determination coefficient. The power equations are as follows:

$$\text{Plastic:} \quad RE = 14.844 x^{0.7078} \quad R^2 = 0.884 \quad (7.8a)$$

$$\text{Gravel:} \quad RE = 11.956 x^{0.7631} \quad R^2 = 0.868 \quad (7.8b)$$

$$\text{Rubber:} \quad RE = 11.483 x^{0.7409} \quad R^2 = 0.861 \quad (7.8c)$$

7.3.2 INLET AND OUTLET COD CONCENTRATIONS RELATIONSHIPS

Figures 7.21–7.24 give the relationship between effluent and influent COD concentrations (C_o and C_i) at 2, 5, 8, and 10 m distances from inlet. The C_i and C_o values in these figures are listed in Tables 7.12–7.14.

Tables 7.12–7.14 and Figures 7.21–7.24 exhibit that:

- Through steady stage the outlet COD values of plastic media are smaller than the other used media (better COD treatment) at all distances, followed by the gravel and then the rubber.
- For plastic media, the effluent COD concentration at outlet is in the allowable limit (80 mg/l) of Law No. 48 of 1982 (NAWQAM, 2002) at discharge lower than 2.253 m³/d that corresponds to loading rate equals 11.27 cm/d and T_r equal to 72.31 hr.

Steady Stage Analysis

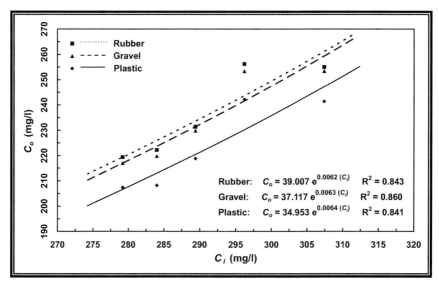

FIGURE 7.21 Relationship between inlet and outlet COD concentrations (2 m from inlet).

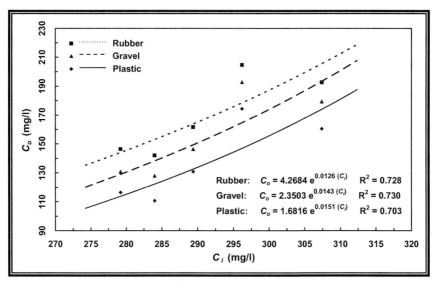

FIGURE 7.22 Relationship between inlet and outlet COD concentrations (5 m from inlet).

- For gravel media the effluent COD concentration at the end of cell is in the permitted limit at discharge value lower than 1.669 m^3/d that corresponds to q value of 8.35 cm/d and retention time equal to 51.74 hr.

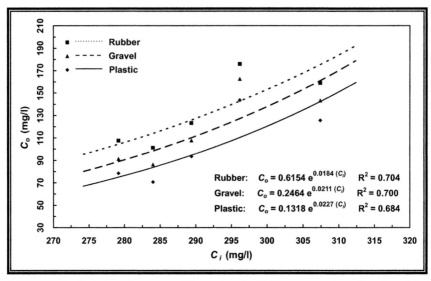

FIGURE 7.23 Relationship between inlet and outlet COD concentrations (8 m from inlet).

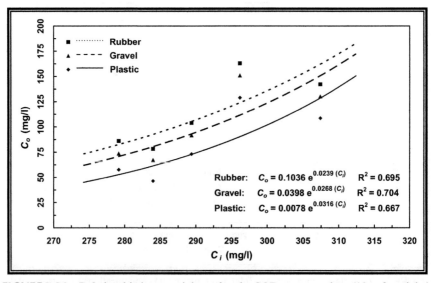

FIGURE 7.24 Relationship between inlet and outlet COD concentrations (10 m from inlet).

- For rubber media the effluent COD concentration at 10 m from inlet is in the admissible limit, at discharge value lower than 1.263 m³/d which corresponds to q value of 6.32 cm/d and retention time equal 89.09 hr.

Steady Stage Analysis

The best fit of effluent and influent COD relationship for the three media is an exponential function. The exponential equations at outlets are given as follows:

$$Plastic:\quad C_o = 0.0078 e^{0.0316 C_i} \quad R^2 = 0.667 \tag{7.9a}$$

$$Gravel:\quad C_o = 0.0398 e^{0.0268 C_i} \quad R^2 = 0.704 \tag{7.9b}$$

$$Rubber:\quad C_o = 0.1036 e^{0.0239 C_i} \quad R^2 = 0.695 \tag{7.9c}$$

where:
C_o = COD outlet concentration, mg/l;
C_i = COD inlet concentration, mg/l.

These equations are valid for the inlet COD concentration ranges from 242 to 337 mg/l (q ranges from 5.94 to 25.6 cm/d).

7.3.3 IMPACT OF LOADING RATE ON COD TREATMENT

Tables 7.15–7.19 show the average outlet COD concentration and the loading rate at different distances for steady stage cycles.

TABLE 7.15 Outlet COD and q at Different Distances (Cycle No.1 – C_i = 296.2 mg/l)

Distance From Inlet	Plastic Cell $Q = 4.814$ m³/d		Gravel Cell $Q = 5.037$ m³/d		Rubber Cell $Q = 5.119$ m³/d	
	C_o(mg/l)	q(cm/d)	C_o(mg/l)	q(cm/d)	C_o(mg/l)	q(cm/d)
2 m	242.2	120.35	253.4	125.93	256.2	127.98
5 m	174.4	48.14	192.8	50.37	204.6	51.19
8 m	143.6	30.09	162.6	31.48	176.0	31.99
10 m	129.0	24.07	151.0	25.19	163.0	25.60

TABLE 7.16 Outlet COD and q at Different Distances (Cycle No.2 – C_i = 307.4 mg/l)

Distance From Inlet	Plastic Cell $Q = 3.275$ m³/d		Gravel Cell $Q = 3.426$ m³/d		Rubber Cell $Q = 3.482$ m³/d	
	C_o(mg/l)	q(cm/d)	C_o(mg/l)	q(cm/d)	C_o(mg/l)	q(cm/d)
2 m	241.4	81.88	253.4	85.65	255.0	87.05
5 m	160.4	32.75	179.4	34.26	192.6	34.82
8 m	125.6	20.47	143.2	21.41	159.0	21.76
10 m	108.6	16.38	130.2	17.13	142.2	17.41

TABLE 7.17 Outlet COD and q at Different Distances (Cycle No.3 – C_i = 289.4 mg/l)

Distance From Inlet	Plastic Cell Q = 2.253 m³/d		Gravel Cell Q = 2.357 m³/d		Rubber Cell Q = 2.396 m³/d	
	C_o(mg/l)	q(cm/d)	C_o(mg/l)	q(cm/d)	C_o(mg/l)	q(cm/d)
2 m	218.8	56.33	229.8	58.93	231.4	59.90
5 m	130.8	22.53	146.4	23.57	161.6	23.96
8 m	93.4	14.08	107.8	14.73	123.2	14.98
10 m	73.2	11.27	91.6	11.79	103.8	11.98

TABLE 7.18 Outlet COD and q at Different Distances (Cycle No.4 – C_i = 279.2 mg/l)

Distance From Inlet	Plastic Cell Q = 1.595 m³/d		Gravel Cell Q = 1.669 m³/d		Rubber Cell Q = 1.696 m³/d	
	C_o(mg/l)	q(cm/d)	C_o(mg/l)	q(cm/d)	C_o(mg/l)	q(cm/d)
2 m	207.4	39.88	217.0	41.73	219.4	42.40
5 m	116.4	15.95	130.8	16.69	146.4	16.96
8 m	78.4	9.97	91.2	10.43	107.6	10.60
10 m	57.6	7.98	73.6	8.35	86.0	8.48

TABLE 7.19 Outlet COD and q at Different Distances (Cycle No.5 – C_i = 284 mg/l)

Distance From Inlet	Plastic Cell Q = 1.188 m³/d		Gravel Cell Q = 1.243 m³/d		Rubber Cell Q = 1.263 m³/d	
	C_o(mg/l)	q(cm/d)	C_o(mg/l)	q(cm/d)	C_o(mg/l)	q(cm/d)
2 m	208.2	29.70	219.8	31.08	222.2	31.58
5 m	110.6	11.88	128.0	12.43	142.0	12.63
8 m	70.6	7.43	86.2	7.77	101.2	7.89
10 m	46.4	5.94	67.2	6.22	78.2	6.32

Figure 7.25 exhibits the relationship between outlet COD concentration and loading rate for the three wetland cells. The data used in this figure are listed in Tables 7.15–7.19 and the five cycles are represented together.

From Tables 7.15–7.19 and Figure 7.25, it is noticed that:

- The effluent COD concentration decreases with the decrease of loading rate which means the enhancement of the treatment performance.
- The q value is maximum near inlet and it gradually decreases in the direction of cell outlet due to the increase of the applied treatment area. For cycle number one (Q_{max} = 4.814, 5.037, and 5.119 m³/d and

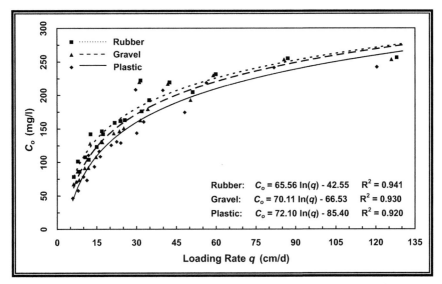

FIGURE 7.25 Relationship between C_o and hydraulic loading rate of COD pollutant.

$C_i = 296.2$ mg/l) at 2 m from inlet, the q values are 120.35, 125.93, and 127.98 cm/d for plastic, gravel, and rubber cells with corresponding effluent COD concentration of 242.2, 253.4, and 256.2 mg/l. At outlet, the q values are 24.07, 25.19, and 25.6 cm/d and the effluent COD concentrations are 129, 151, and 163 mg/l, in the same sequence.

- As for cycle number five (Q_{min} = 1.188, 1.243, and 1.263 m³/d and C_i = 284 mg/l) at 2 m from inlet, the q values are 29.7, 31.08, and 31.58 cm/d for plastic, gravel, and rubber cells which corresponds to outlet COD concentration of 208.2, 219.8, and 222.2 mg/l. After 10 m, q values are 5.94, 6.22, and 6.32 cm/d and outlet COD concentrations are 46.4, 67.2, and 78.2 mg/l.
- At low q values, the difference between outlet COD concentrations for the three media is big, whereas at high q values, the difference between these media is small especially between rubber and gravel. Plastic cell treats COD better followed by gravel and then rubber cell.

The relationship between COD outlet concentration and loading rate follows a logarithmic function. The logarithmic equations are:

Plastic: $C_o = -85.40 + 72.10 \ln q$ $R^2 = 0.920$ (7.10a)

Gravel: $C_o = -66.53 + 70.11 \ln q$ $R^2 = 0.930$ (7.10b)

Rubber: $C_o = -42.55 + 65.56 \ln q$ $R^2 = 0.941$ (7.10c) where:

q = hydraulic loading rate, cm/d;
C_o = COD outlet concentration, mg/l.

The relationships between outlet concentration and loading rate are valid for q values range from 5.94 to 127.98 cm/d.

TABLE 7.20 COD Removal Efficiency and T_r for Plastic Media at Different Distances

Cycle No.	Q(m³/d)	Hydraulic Retention Time (hr)				Average Removal Efficiency (%)			
		2 m	5 m	8 m	10 m	2 m	5 m	8 m	10 m
1	4.814	5.14	16.92	28.70	33.84	18.21	41.12	51.52	56.45
2	3.275	7.54	24.86	42.19	49.73	21.47	47.85	59.17	64.71
3	2.253	10.97	36.14	61.32	72.31	24.39	54.80	67.74	74.72
4	1.595	15.48	51.07	86.64	102.12	25.65	58.29	71.94	79.40
5	1.188	20.81	68.54	116.30	137.11	26.72	61.09	75.16	83.68

TABLE 7.21 COD Removal Efficiency and T_r for Gravel Media at Different Distances

Cycle No.	Q(m³/d)	Hydraulic Retention Time (hr)				Average Removal Efficiency (%)			
		2 m	5 m	8 m	10 m	2 m	5 m	8 m	10 m
1	5.037	3.46	8.57	13.68	17.16	14.43	34.90	45.10	49.02
2	3.426	5.09	12.60	20.14	25.20	17.57	41.67	53.45	57.68
3	2.357	7.39	18.31	29.26	36.65	20.58	49.42	62.76	68.37
4	1.669	10.42	25.87	41.30	51.74	22.22	53.16	67.36	73.67
5	1.243	13.99	34.73	55.46	69.48	22.64	54.97	69.68	76.37

TABLE 7.22 COD Removal Efficiency and T_r for Rubber Media at Different Distances

Cycle No.	Q(m³/d)	Hydraulic Retention Time (hr)				Average Removal Efficiency (%)			
		2 m	5 m	8 m	10 m	2 m	5 m	8 m	10 m
1	5.119	3.89	10.99	18.10	21.98	13.49	30.92	40.58	44.97
2	3.482	5.71	16.15	26.59	32.30	17.04	37.35	48.30	53.77
3	2.396	8.30	23.47	38.66	46.97	20.03	44.15	57.44	64.15
4	1.696	11.74	33.17	54.60	66.34	21.37	47.54	61.45	69.20
5	1.263	15.74	44.54	73.34	89.09	21.79	50.03	64.40	72.50

7.3.4 IMPACT OF T_r ON COD REMOVAL EFFICIENCY

Tables 7.20–7.22 give the values of calculated hydraulic retention time which correspond to the applied discharges at distances of 2, 5, 8, and 10 m from inlet.

Figures 7.26–7.28 show the relationship between COD removal efficiency and retention time at different distances from inlet.

From Tables 7.20–7.22 and Figures 7.26–7.28, it is noticed that:

- The plastic cell has the highest retention time followed by the rubber and then the gravel. This is in line with the three media porosities, where the plastic media has the biggest porosity followed by rubber, while gravel media has the smallest one.
- The hydraulic retention time decreases as the discharge increases. For example at outlets with Q_{avg} increases from 1.23 to 4.99 m³/d, T_r value decreases from 69.48 to 17.16 hr for gravel and from 89.09 to 21.98 hr for rubber and from 137.11 to 33.84 hr for plastic, respectively.
- The COD removal efficiency increases with the increasing value of T_r through steady stage. For example in case of the rubber cell, at T_r equals 18.1 hr (8 m from inlet), the removal efficiency equals 40.58%, while at T_r equals 73.34 hr at the same distance, the removal efficiency equals 64.4%, which means improvement of the treatment with the increase of T_r.

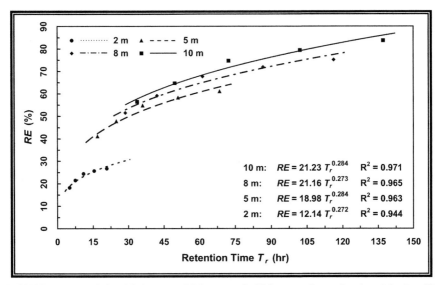

FIGURE 7.26 Relationship between COD removal efficiency and retention time (plastic cell).

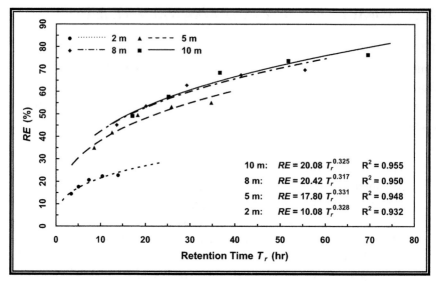

FIGURE 7.27 Relationship between COD removal efficiency and retention time (gravel cell).

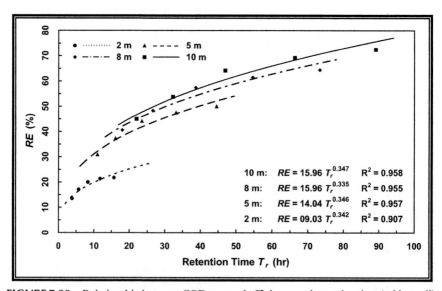

FIGURE 7.28 Relationship between COD removal efficiency and retention time (rubber cell).

- It is found that the COD removal efficiency is better in plastic cell than in both gravel and rubber cells. For example at outlets, COD treatment at the greatest discharge has a removal efficiency of 56.45, 49.02, and 44.97 for plastic, gravel, and rubber cells, in the same sequence.

- The COD removal efficiency decreases with the decrease of retention time for the three media and at all distances.
- The difference between the COD removal efficiencies from inlet till 5 m length is big for the used media which means that, these five meters are the most effective parts in the three wetland cells in COD treatment performance.

The COD removal efficiency for the three media is directly proportional to T_r according to a power function as the best-fit relationship. The power equations at outlets are written as:

$$Plastic: \quad RE = 21.23\, T_r^{0.284} \quad R^2 = 0.971 \quad (7.11a)$$

$$Gravel: \quad RE = 20.08\, T_r^{0.325} \quad R^2 = 0.955 \quad (7.11b)$$

$$Rubber: \quad RE = 15.96\, T_r^{0.347} \quad R^2 = 0.958 \quad (7.11c)$$

7.3.5 IMPACT OF DISCHARGE ON COD TREATMENT EFFICIENCY

Figures 7.29–7.32 show the relationship between RE of COD and Q at distances of 2, 5, 8, and 10 m from inlet. The used values in these figures are listed in Tables 7.20–7.22.

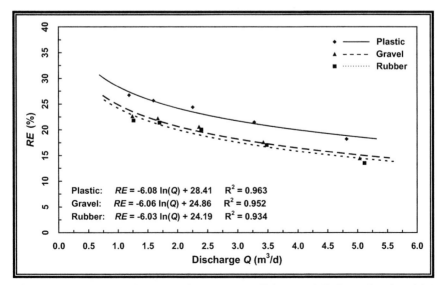

FIGURE 7.29 Relationship between COD removal efficiency and discharge (2 m from inlet).

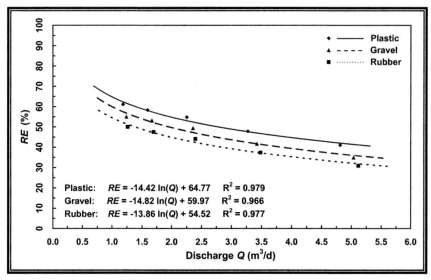

FIGURE 7.30 Relationship between COD removal efficiency and discharge (5 m from inlet).

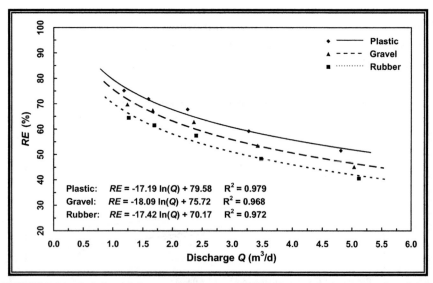

FIGURE 7.31 Relationship between COD removal efficiency and discharge (8 m from inlet).

Tables 7.20–7.22 and Figures 7.29–7.32 show that:

- A reversing relationship controls the effect of discharge on COD removal efficiency. As the discharge increases, the *RE* decreases at all distances.

Steady Stage Analysis

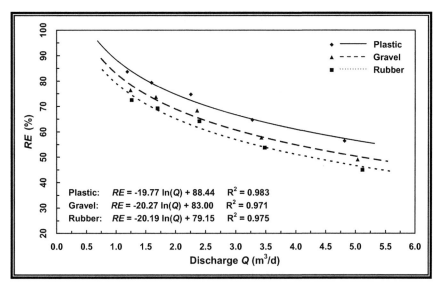

FIGURE 7.32 Relationship between COD removal efficiency and discharge (10 m from inlet).

- The rubber cell has a lower removal efficiency followed by the gravel and then the plastic. This may be due to the lower surface area (low amount of attached biofilm) of rubber media comparing with the other used media.
- For plastic cell, at the biggest Q (4.814 m^3/d), the COD removal efficiency values are about 18.21, 41.12, 51.52, and 56.45% at 2, 5, 8, and 10 m, respectively. While *REs* are 26.72, 61.09, 75.16, and 83.68% at the smallest Q (1.188 m^3/d).
- For gravel cell, at the biggest Q (5.037 m^3/d), the COD removal efficiency is 14.43, 34.9, 45.1, and 49.02% at distances of 2, 5, 8, and 10 m from inlet, in the same sequence. While removal efficiencies are 22.64, 54.97, 69.68, and 76.37% at the smallest Q (1.243 m^3/d).
- For rubber cell, at the biggest Q (5.119 m^3/d), the COD removal efficiency is 13.49, 30.92, 40.58, and 44.97% at distances of 2, 5, 8, and 10 m from inlet, in the same order. While *REs* are 21.79, 50.03, 64.4, and 72.5% at the smallest Q (1.263 m^3/d).

After 2 m from inlet, rubber cell gives average difference lower than both gravel and plastic cells by about 0.74 and 4.54%, respectively. At 5 m from inlet these values are about 4.83 and 10.63%, at 8 m 5.24 and 10.67%, and at outlets 4.10 and 10.87%. Gravel cell gives average difference lower than plastic cell by about 3.80, 5.81, 5.44, and 6.77% at distances of 2, 5, 8, and 10 m from inlet, in the same sequence.

The difference between rubber and gravel removal efficiencies is small at the first 2 m from inlet. This slight difference in removal efficiency may be due to the small difference in surface area values at this length. The first two meters of the cells has 1.29 m³ of coarse gravel and the residual volume represents the media is only 0.71 m³. The values of average difference for COD pollutant between the three media at different distances are illustrated in Appendix II.

The best relationship between the COD removal efficiency and discharge at different distances is a logarithmic function. The logarithmic equations at outlets are as follows:

$$\textit{Plastic}: \quad RE = 88.44 - 19.77 \ln Q \quad R^2 = 0.983 \quad (7.12a)$$

$$\textit{Gravel}: \quad RE = 83.0 - 20.27 \ln Q \quad R^2 = 0.971 \quad (7.12b)$$

$$\textit{Rubber}: \quad RE = 79.15 - 20.19 \ln Q \quad R^2 = 0.975 \quad (7.12c)$$

These relationships are valid for Q ranges from 1.188 to 5.119 m³/d.

7.4 TOTAL SUSPENDED SOLIDS

The effluent TSS concentration was studied with distance, loading rate, and influent concentration. The variation of TSS removal efficiency with distance, retention time, and discharge was also determined.

7.4.1 EFFECT OF DISTANCE ON TSS TREATMENT

The effect of longitudinal distance on average outlet concentration of TSS for the three used media was studied. Tables 7.23–7.25 present the discharge, the average influent and effluent concentrations, and the corresponding average removal efficiency for TSS pollutant at different distances. The average concentration and removal efficiency were mentioned in Chapter 5.

Figures 7.33 and 7.34 give the variation of TSS concentration with treatment length at maximum and minimum discharges. Figure 7.35 illustrates this variation for the average discharge.

From Tables 7.23–7.25 and Figures 7.33–7.35, it is noticed that:
- In plastic cell, TSS concentration is reduced with distance more than other used media cells. Gravel comes in the second order.

TABLE 7.23 Average TSS Concentration and Removal Efficiency for Plastic Cell

Q(m³/d)	Influent (mg/l)	Average Concentration (mg/l)				Average Removal Efficiency (%)			
		2 m	5 m	8 m	10 m	2 m	5 m	8 m	10 m
4.814	144.8	110.4	70.2	55.4	49.2	23.76	51.55	61.76	66.06
3.275	146.4	105.6	54.8	39.6	30.0	27.91	62.62	73.03	79.59
2.253	155.8	106.8	50.0	28.6	20.4	31.47	67.93	81.68	86.94
1.595	168.4	114.8	48.0	28.0	15.6	31.83	71.51	83.39	90.75
1.188	172.0	115.6	44.4	20.2	10.4	32.80	74.19	88.26	93.96
Mean	157.5	110.6	53.5	34.4	25.1	29.55	65.56	77.62	83.46

TABLE 7.24 Average TSS Concentration and Removal Efficiency for Gravel Cell

Q(m³/d)	Influent (mg/l)	Average Concentration (mg/l)				Average Removal Efficiency (%)			
		2 m	5 m	8 m	10 m	2 m	5 m	8 m	10 m
5.037	144.8	117.6	83.4	71.0	65.2	18.80	42.43	51.00	55.01
3.426	146.4	116.4	78.2	62.0	55.6	20.51	46.63	57.70	62.08
2.357	155.8	117.4	68.2	50.0	42.0	24.66	56.25	67.93	73.08
1.669	168.4	121.8	62.6	38.4	27.0	27.68	62.83	77.21	83.98
1.243	172.0	121.4	57.0	30.0	19.0	29.42	66.87	82.57	88.96
Mean	157.5	118.9	69.9	50.3	41.8	24.21	55.00	67.28	72.62

TABLE 7.25 Average TSS Concentration and Removal Efficiency for Rubber Cell

Q(m³/d)	Influent (mg/l)	Average Concentration (mg/l)				Average Removal Efficiency (%)			
		2 m	5 m	8 m	10 m	2 m	5 m	8 m	10 m
5.119	144.8	121.2	92.0	80.2	72.0	16.30	36.49	44.64	50.31
3.482	146.4	118.2	83.0	66.4	58.2	19.29	43.36	54.71	60.33
2.396	155.8	119.4	76.2	56.8	48.2	23.38	51.12	63.58	69.10
1.696	168.4	124.4	69.2	43.0	30.8	26.14	58.92	74.48	81.73
1.263	172.0	123.2	62.4	37.4	24.0	28.37	63.73	78.27	86.06
Mean	157.5	121.3	76.6	56.8	46.6	22.70	50.72	63.14	69.51

- At the maximum discharge, the inlet TSS concentration is 144.8 mg/l which decreases to 110.4, 117.6, and 121.2 mg/l for plastic, gravel, and rubber cells, respectively at 2 m from inlet. These values have more

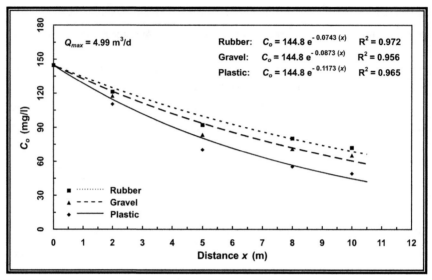

FIGURE 7.33 Relationship between C_o and distance for TSS pollutant (Q_{max} – steady stage).

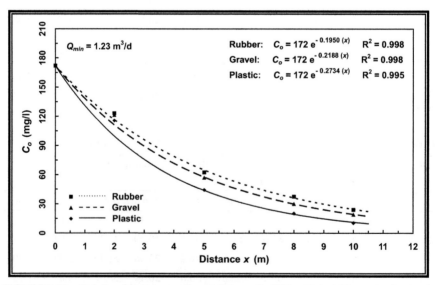

FIGURE 7.34 Relationship between C_o and distance for TSS pollutant (Q_{min} – steady stage).

decrease as going towards the cell outlet, at 5 m the values are 70.2, 83.4, and 92 mg/l. at 8 m the values are 55.4, 71, and 80.2 mg/l. At outlets, TSS concentrations are 49.2, 65.2, and 72 mg/l.

- At the minimum discharge, the inlet TSS concentration is 172 mg/l and reduces to 115.6, 121.4, and 123.2 mg/l at 2 m from inlet for plastic,

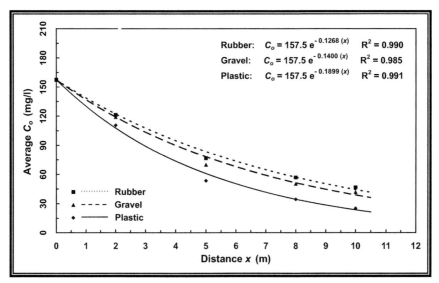

FIGURE 7.35 Relationship between C_o and distance for TSS pollutant (Q_{avg} – steady stage).

gravel, and rubber cells, respectively then to 44.4, 57, and 62.4 mg/l at 5 m. These values improve to 20.2, 30, and 37.4 mg/l at 8 m; and 10.4, 19, and 24 mg/l at outlets.
- For the average inlet TSS value is 157.5 mg/l which treats to 110.6, 118.9, and 121.3 mg/l at 2 m; 53.5, 69.9, and 76.6 mg/l at 5 m; 34.4, 50.3, and 56.8 mg/l at 8 m; and 25.1, 41.8, and 46.6 mg/l at 10 m from inlet for plastic, gravel, and rubber cells, in the same order.

Concentration of BOD, COD, and TSS which is gradually reduced with distance for the three media may be represented by an exponential function as given by the following general formula:

$$C_o = C_i\, e^{\omega x} \quad (7.13)$$

where:
x = distance from wetland cell inlet, m;
ω = constant depends on media type and discharge.

Concentration of TSS, which is gradually reduced with distance, follows an exponential function. The exponential equations for the average discharge are given as:

$$\text{Plastic:} \quad C_o = 157.5 e^{-0.1899x} \quad R^2 = 0.991 \quad (7.14a)$$

$$Gravel: \quad C_o = 157.5e^{-0.140x} \quad R^2 = 0.985 \quad (7.14b)$$

$$Rubber: \quad C_o = 157.5e^{-0.1268x} \quad R^2 = 0.990 \quad (7.14c)$$

Figure 7.36 presents the variation of TSS removal efficiency with treatment distance for the used discharges.

From Tables 7.23–7.25 and Figure 7.36, it may be concluded that:

- At 2 m from plastic cell inlet, TSS removal efficiency raises from 23.76 to 32.8% at Q_{max} and Q_{min}, respectively, from 18.8 to 29.42% for gravel, and from 16.3 to 28.37% for rubber.
- At 5 m from plastic cell inlet, TSS removal efficiency increases from 51.55 to 74.19% at Q_{max} and Q_{min}, respectively, from 42.43 to 66.87% for gravel, and from 36.49 to 63.73% for rubber.
- At 8 m from plastic cell inlet, TSS removal efficiency improves from 61.76 to 88.26% at Q_{max} and Q_{min}, in the same order, from 51 to 82.57% for gravel, and from 44.64 to 78.27% for rubber.
- At outlet, TSS removal efficiency enhances from 66.06 to 93.96 for plastic cell at Q_{max} and Q_{min}, respectively, from 55.01 to 88.96% for gravel, and from 50.31 to 86.06% for rubber.

The TSS removal efficiency is gradually increased with treatment length for the three used media according to a power function which gives the best determination coefficient. The power equations are:

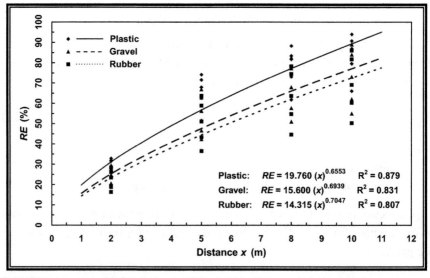

FIGURE 7.36 Relationship between TSS removal efficiency and treatment length.

Steady Stage Analysis

$$\text{Plastic:} \quad RE = 19.76x^{0.6553} \quad R^2 = 0.879 \quad (7.14a)$$

$$\text{Gravel:} \quad RE = 15.60x^{0.6939} \quad R^2 = 0.831 \quad (7.14b)$$

$$\text{Rubber:} \quad RE = 14.315x^{0.7047} \quad R^2 = 0.807 \quad (7.14c)$$

7.4.2 WEIGHTS OF INLET AND OUTLET SUSPENDED SOLIDS

Tables 7.26–7.28 give the inlet and outlet weights of suspended solids (Eqn. (4.13) and 4.14) and the computed rate of sediment storage (Eq. 4.15) for plastic, gravel, and rubber cells.

$$W_i = Q \times C_i \quad (4.13)$$

$$W_o = Q \times C_o \quad (4.14)$$

$$R_s = \frac{W_i - W_o}{x_o - x_i} \quad (4.15)$$

TABLE 7.26 Inlet and Outlet Weights of Suspended Solids and R_s for Plastic Cell

W_i(gm/d)	W_o(gm/d)				R_s(gm/m.d)			
	2 m	5 m	8 m	10 m	0–2 m	2–5 m	5–8 m	8–10 m
697.1	531.5	337.9	266.7	236.8	82.8	64.5	23.7	14.9
479.5	345.8	179.5	129.7	98.3	66.8	55.5	16.6	15.7
351.0	240.6	112.7	64.4	46.0	55.2	42.7	16.1	9.2
268.6	183.1	76.6	44.7	24.9	42.7	35.5	10.6	9.9
204.3	137.3	52.7	24.0	12.4	33.5	28.2	9.6	5.8

TABLE 7.27 Inlet and Outlet Weights of Suspended Solids and R_s for Gravel Cell

W_i(gm/d)	W_o(gm/d)				R_s(gm/m.d)			
	2 m	5 m	8 m	10 m	0–2 m	2–5 m	5–8 m	8–10 m
729.4	592.4	420.1	357.6	328.4	68.5	57.4	20.8	14.6
501.6	398.8	267.9	212.4	190.5	51.4	43.6	18.5	11.0
367.2	276.7	160.7	117.9	99.0	45.3	38.7	14.3	9.4
281.1	203.3	104.5	64.1	45.1	38.9	32.9	13.5	9.5
213.8	150.9	70.9	37.3	23.6	31.4	26.7	11.2	6.8

TABLE 7.28 Inlet and Outlet Weights of Suspended Solids and R_s for Rubber Cell

W_i(gm/d)	W_o(gm/d)				R_s(gm/m.d)			
	2 m	5 m	8 m	10 m	0–2 m	2–5 m	5–8 m	8–10 m
741.2	620.4	470.9	410.5	368.6	60.4	49.8	20.1	21.0
509.8	411.6	289.0	231.2	202.7	49.1	40.9	19.3	14.3
373.3	286.1	182.6	136.1	115.5	43.6	34.5	15.5	10.3
285.6	211.0	117.4	72.9	52.2	37.3	31.2	14.8	10.3
217.2	155.6	78.8	47.2	30.3	30.8	25.6	10.5	8.5

where:

W_i = weight of inlet suspended solids, gm/d;
W_o = weight of outlet suspended solids, gm/d;
R_s = rate of sediment storage, gm/m.d;
C_i, C_o = inlet and outlet TSS concentrations, gm/m^3;
Q = discharge, m^3/d;
x_i, x_o = inlet and outlet distances, m.

Figures 7.37–7.40 give the relationship between inlet and outlet weights of suspended solids (W_i and W_o) for the three media at 2, 5, 8, and 10 m distances from inlet.

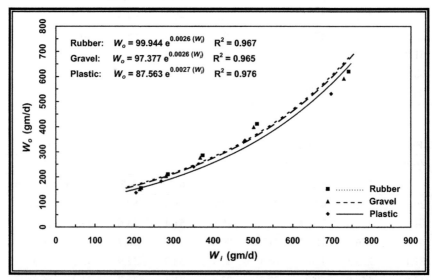

FIGURE 7.37 Relationship between inlet and outlet weights of suspended solids (2 m from inlet).

Steady Stage Analysis

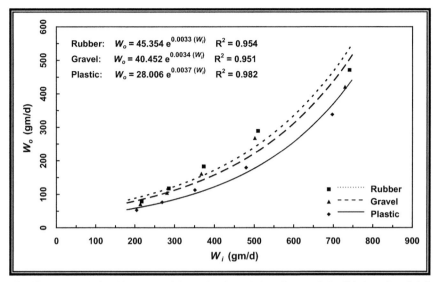

FIGURE 7.38 Relationship between inlet and outlet weights of suspended solids (5 m from inlet).

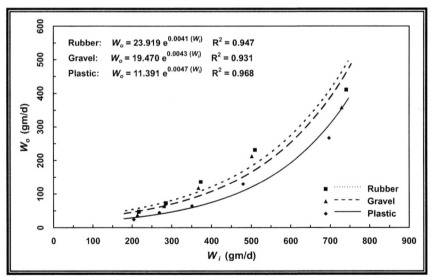

FIGURE 7.39 Relationship between inlet and outlet weights of suspended solids (8 m from inlet).

Tables 7.23–7.28 and Figures 7.37–7.40 show that:

- Through steady stage, the outlet weight of suspended solids for plastic media cell is smaller than other media substances (better TSS treatment) at the studied distances, followed by the gravel media.

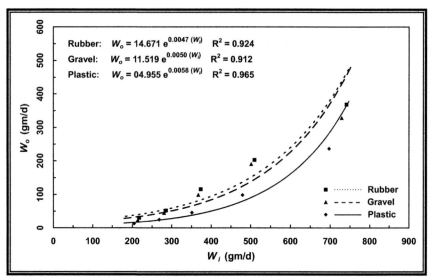

FIGURE 7.40 Relationship between inlet and outlet weights of suspended solids (10 m from inlet).

- At the first 2 m, the rate of sediment storage is 82.8, 68.5, and 60.4 gm/m/d at the maximum inlet weight of suspended solids for plastic, gravel, and rubber cells, respectively. These values are 64.5, 57.4, and 49.8 at distance from 2 to 5 m; 23.7, 20.8, and 20.1 at distance from 5 to 8 m; and 14.9, 14.6, and 21 gm/m/d at distance from 8 to 10 m, in the same sequence.
- At the first 2 m, the rate of sediment storage is 33.5, 31.4, and 30.8 gm/m/d at the minimum inlet weight of suspended solids for plastic, gravel, and rubber cells, respectively. These values are 28.2, 26.7, and 25.6 at distance from 2 to 5 m; 9.6, 11.2, and 10.5 at distance from 5 to 8 m; and 5.8, 6.8, and 8.5 gm/m/d at distance from 8 to 10 m, in the same sequence.
- For plastic media, the TSS concentration at outlet is in the allowable limit (50 mg/l) of Law No. 48 of 1982 (NAWQAM, 2002) at discharge lower than 4.814 m^3/d that corresponds to q value of 24.07 cm/d, and $T_r = 33.84$ hr.
- For gravel media, the TSS concentration at the end of wetland cell is in the permitted limit at discharge value lower than 2.357 m^3/d that corresponds to q value of 11.79 cm/d and retention time of 36.65 hr.
- For rubber media, the TSS concentration at 10 m distance from inlet weir is in the admissible limit at discharge value lower than 2.396 m^3/d which corresponds to q value of 11.98 cm/d and T_r equal to 46.97 hr.

Steady Stage Analysis

Inlet and outlet weights of suspended solids relationship for the three media have an exponential function. The exponential equations at outlets are as follows:

$$\text{Plastic: } W_o = 4.955e^{-0.0058W_i} \quad R^2 = 0.965 \quad (7.16a)$$

$$\text{Gravel: } W_o = 11.519e^{-0.0050W_i} \quad R^2 = 0.912 \quad (7.16b)$$

$$\text{Rubber: } W_o = 14.671e^{-0.0047W_i} \quad R^2 = 0.924 \quad (7.16c)$$

These equations are valid for the inlet weight of suspended solids ranges from 204.3 to 741.2 gm/d, and Q ranges from 1.188 to 5.119 m3/d.

7.4.3 IMPACT OF LOADING RATE ON TSS TREATMENT

Tables 7.29–7.33 show the average outlet TSS concentration and the loading rate for the three wetland cells at different distances from inlet for the five cycles of steady stage.

TABLE 7.29 Effluent TSS and q at Different Distances (Cycle No.1 – C_i = 144.8 mg/l)

Distance From Inlet	Plastic Cell Q = 4.814 m³/d		Gravel Cell Q = 5.037 m³/d		Rubber Cell Q = 5.119 m³/d	
	C_o(mg/l)	q(cm/d)	C_o(mg/l)	q(cm/d)	C_o(mg/l)	q(cm/d)
2 m	110.4	120.35	117.6	125.93	121.2	127.98
5 m	70.2	48.14	83.4	50.37	92.0	51.19
8 m	55.4	30.09	71.0	31.48	80.2	31.99
10 m	49.2	24.07	65.2	25.19	72.0	25.60

TABLE 7.30 Effluent TSS and q at Different Distances (Cycle No.2 – C_i = 146.4 mg/l)

Distance From Inlet	Plastic Cell Q = 3.275 m³/d		Gravel Cell Q = 3.426 m³/d		Rubber Cell Q = 3.482 m³/d	
	C_o(mg/l)	q(cm/d)	C_o(mg/l)	q(cm/d)	C_o(mg/l)	q(cm/d)
2 m	105.6	81.88	116.4	85.65	118.2	87.05
5 m	54.8	32.75	78.2	34.26	83.0	34.82
8 m	39.6	20.47	62.0	21.41	66.4	21.76
10 m	30.0	16.38	55.6	17.13	58.2	17.41

TABLE 7.31 Effluent TSS and q at Different Distances (Cycle No.3 – C_i = 155.8 mg/l)

Distance From Inlet	Plastic Cell Q = 2.253 m³/d		Gravel Cell Q = 2.357 m³/d		Rubber Cell Q = 2.396 m³/d	
	C_o(mg/l)	q(cm/d)	C_o(mg/l)	q(cm/d)	C_o(mg/l)	q(cm/d)
2 m	106.8	56.33	117.4	58.93	119.4	59.90
5 m	50.0	22.53	68.2	23.57	76.2	23.96
8 m	28.6	14.08	50.0	14.73	56.8	14.98
10 m	20.4	11.27	42.0	11.79	48.2	11.98

TABLE 7.32 Effluent TSS and q at Different Distances (Cycle No.4 – C_i = 168.4 mg/l)

Distance From Inlet	Plastic Cell Q = 1.595 m³/d		Gravel Cell Q = 1.669 m³/d		Rubber Cell Q = 1.696 m³/d	
	C_o(mg/l)	q(cm/d)	C_o(mg/l)	q(cm/d)	C_o(mg/l)	q(cm/d)
2 m	114.8	39.88	121.8	41.73	124.4	42.40
5 m	48.0	15.95	62.6	16.69	69.2	16.96
8 m	28.0	9.97	38.4	10.43	43.0	10.60
10 m	15.6	7.98	27.0	8.35	30.8	8.48

TABLE 7.33 Effluent TSS and q at Different Distances (Cycle No.5 – C_i = 172 mg/l)

Distance From Inlet	Plastic Cell Q = 1.188 m³/d		Gravel Cell Q = 1.243 m³/d		Rubber Cell Q = 1.263 m³/d	
	C_o(mg/l)	q(cm/d)	C_o(mg/l)	q(cm/d)	C_o(mg/l)	q(cm/d)
2 m	115.6	29.70	121.4	31.08	123.2	31.58
5 m	44.4	11.88	57.0	12.43	62.4	12.63
8 m	20.2	7.43	30.0	7.77	37.4	7.89
10 m	10.4	5.94	19.0	6.22	24.0	6.32

Figure 7.41 illustrates the relationship between the effluent TSS concentration and loading rate for the three cells. The data used in this figure are listed in Tables 7.29–7.33 and the five cycles are represented together.

From Tables 7.29–7.33 and Figure 7.41, it is appeared that:

- Plastic media treats TSS better at small values of loading rates. This is due to the high amount of attached biofilm and low actual velocity through cell. Gravel media takes the second order in treatment and followed by rubber cell.

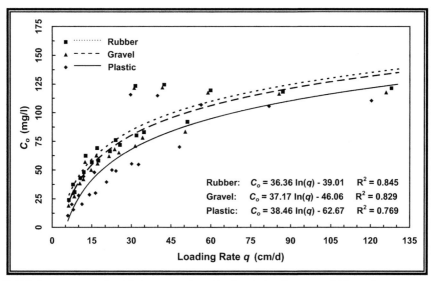

FIGURE 7.41 Relationship between C_o and hydraulic loading rate of TSS pollutant.

- For cycle number one (Q_{max} = 4.814, 5.037, and 5.119 m³/d and C_i = 144.8 mg/l) at 2 m from inlet, the q values are 120.35, 125.93, and 127.98 cm/d for plastic, gravel, and rubber cells, respectively which correspond to TSS concentration of 110.4, 117.6, and 121.2 mg/l. At outlet, the q values are 24.07, 25.19, and 25.6 cm/d and TSS are 49.2, 65.2, and 72 mg/l.
- As for cycle number five (Q_{min} = 1.188, 1.243, and 1.263 m³/d and C_i = 172 mg/l) at 2 m from inlet, the q values are 29.7, 31.08, and 31.58 cm/d for plastic, gravel, and rubber cells, in the same sequence which corresponds to TSS values of 115.6, 121.4, and 123.2 mg/l. At cell outlet, the q values are 5.94, 6.22, and 6.32 cm/d and outlet TSS are 10.4, 19, and 24 mg/l for plastic, gravel, and rubber cells, respectively.
- The difference between effluent TSS concentration for gravel and rubber media is small and both cells have big difference comparing with the plastic one. This is due to the close values of surface areas of both gravel and rubber and the higher surface area in case of plastic. Increase of media surface area is accompanied with increasing the attached biofilm bacteria which enhance the treatment processes.

The TSS outlet concentration is directly proportional to the loading rate according to a logarithmic function. The logarithmic equations at outlets are given as follows (q values range from 5.94 to 127.98 cm/d):

Plastic: $C_o = -62.67 + 38.46 \ln q$ $R^2 = 0.769$ (7.17a)

Gravel: $C_o = -46.06 + 37.17 \ln q$ $R^2 = 0.829$ (7.17b)

Rubber: $C_o = -39.01 + 36.36 \ln q$ $R^2 = 0.845$ (7.17c)

7.4.4 IMPACT OF T_r ON TSS REMOVAL EFFICIENCY

Tables 7.34–7.36 illustrate the calculated T_r corresponding to the applied discharges at distances of 2, 5, 8, and 10 m from inlet. Also these tables include the cycle number and the TSS removal efficiency.

Figures 7.42–7.44 exhibit the relationship between TSS removal efficiency and retention time at different distances.

From Tables 7.34–7.36 and Figures 7.42–7.44, it is noticed that:

- The gravel cell has small porosity that gives small water volumes which in turn leads to low value of T_r followed by rubber, while plastic has the greatest T_r.

TABLE 7.34 TSS Removal Efficiency and T_r for Plastic Media at Different Distances

Cycle No.	Q(m³/d)	Hydraulic Retention Time (hr)				Average Removal Efficiency (%)			
		2 m	5 m	8 m	10 m	2 m	5 m	8 m	10 m
1	4.814	5.14	16.92	28.70	33.84	23.76	51.55	61.76	66.06
2	3.275	7.54	24.86	42.19	49.73	27.91	62.62	73.03	79.59
3	2.253	10.97	36.14	61.32	72.31	31.47	67.93	81.68	86.94
4	1.595	15.48	51.07	86.64	102.12	31.83	71.51	83.39	90.75
5	1.188	20.81	68.54	116.30	137.11	32.80	74.19	88.26	93.96

TABLE 7.35 TSS Removal Efficiency and T_r for Gravel Media at Different Distances

Cycle No.	Q(m³/d)	Hydraulic Retention Time (hr)				Average Removal Efficiency (%)			
		2 m	5 m	8 m	10 m	2 m	5 m	8 m	10 m
1	5.037	3.46	8.57	13.68	17.16	18.80	42.43	51.00	55.01
2	3.426	5.09	12.60	20.14	25.20	20.51	46.63	57.70	62.08
3	2.357	7.39	18.31	29.26	36.65	24.66	56.25	67.93	73.08
4	1.669	10.42	25.87	41.30	51.74	27.68	62.83	77.21	83.98
5	1.243	13.99	34.73	55.46	69.48	29.42	66.87	82.57	88.96

TABLE 7.36 TSS Removal Efficiency and T_r for Rubber Media at Different Distances

Cycle No.	$Q(m^3/d)$	Hydraulic Retention Time (hr)				Average Removal Efficiency (%)			
		2 m	5 m	8 m	10 m	2 m	5 m	8 m	10 m
1	5.119	3.89	10.99	18.10	21.98	16.30	36.49	44.64	50.31
2	3.482	5.71	16.15	26.59	32.30	19.29	43.36	54.71	60.33
3	2.396	8.30	23.47	38.66	46.97	23.38	51.12	63.58	69.10
4	1.696	11.74	33.17	54.60	66.34	26.14	58.92	74.48	81.73
5	1.263	15.74	44.54	73.34	89.09	28.37	63.73	78.27	86.06

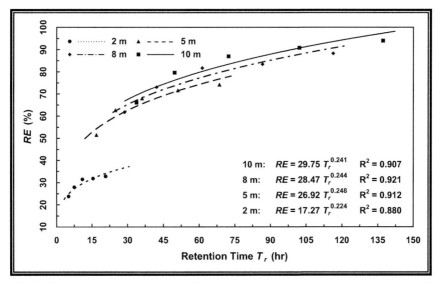

FIGURE 7.42 Relationship between TSS removal efficiency and retention time (plastic cell).

- The TSS removal efficiency increases with the increase of T_r value. For example for plastic cell, at T_r equals 5.14 hr (2 m from inlet), the removal efficiency is 23.76%, while at T_r equals 20.81 hr at the same length, the removal efficiency is 32.8%.
- The TSS treatment enhances with the increase of retention time for the used media and along the wetland cells length.
- The retention time is inversely proportional to inflow discharge. For example at 8 m from inlet, the T_r increases from 13.68 to 55.46 hr for gravel cell and from 18.1 to 73.34 hr for rubber and from 28.7 to 116.3 hr for plastic when Q_{avg} decreases from 4.99 to 1.23 m³/d, respectively.

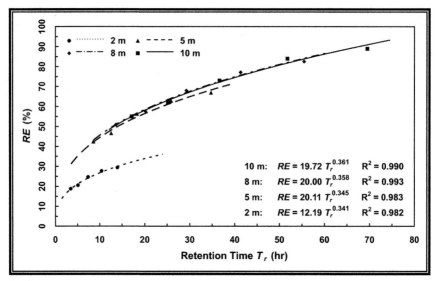

FIGURE 7.43 Relationship between TSS removal efficiency and retention time (gravel cell).

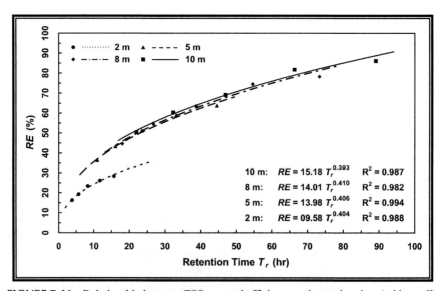

FIGURE 7.44 Relationship between TSS removal efficiency and retention time (rubber cell).

- The difference between the TSS removal efficiency at 5 m till 10 m from inlet weir is small for the gravel and rubber media which mean that, the last half length of these cells have no effective influence on TSS treatment process.

- The difference between the TSS removal efficiency from inlet till 5 m length is big for the three used media which means that, these five meters are the most effective length in the three cells in TSS treatment performance.

The TSS removal efficiency is directly proportionate to T_r according to a power function. The power equations at outlets are as follows:

$$\text{Plastic:} \quad RE = 29.75 \, T_r^{0.241} \quad R^2 = 0.907 \quad (7.18a)$$

$$\text{Gravel:} \quad RE = 19.72 \, T_r^{0.361} \quad R^2 = 0.990 \quad (7.18b)$$

$$\text{Rubber:} \quad RE = 15.18 \, T_r^{0.393} \quad R^2 = 0.987 \quad (7.18c)$$

where:
RE = TSS removal efficiency, %;
T_r = hydraulic retention time, hr.

These equations are valid for the discharge values range from 1.188 to 5.119 m³/d for the three used media.

7.4.5 IMPACT OF DISCHARGE ON TSS TREATMENT EFFICIENCY

Figures 7.45–7.48 show the relationship between TSS removal efficiency and discharge for the three media at different distances. The used values in these figures are listed in Tables 7.34–7.36.

Tables 7.34–7.36 and Figures 7.45–7.48 declare that:

- The discharge values are inversely proportional to the TSS removal efficiency at 2, 5, 8, and 10 m distance from inlet and for the three media types.
- The gravel cell has lower removal efficiency than plastic and higher than rubber. This may be due to the surface area of gravel media that laid in-between rubber and plastic media.
- For plastic cell, at the biggest Q (4.814 m³/d), the values of TSS removal efficiency are 23.76, 51.55, 61.76, and 66.06% at distances of 2, 5, 8, and 10 m from inlet, respectively. While removal efficiencies are 32.8, 74.19, 88.26, and 93.96% at the smallest Q value (1.188 m³/d), in the same order.
- For gravel cell, at the biggest Q (5.037 m³/d), the values of TSS removal efficiency are 18.8, 42.43, 51, and 55.01% at distances of 2, 5, 8, and 10 m from inlet, in the same sequence. While removal efficiencies were 29.42, 66.87, 82.57, and 88.96% at the smallest Q value (1.243 m³/d), in the same order.

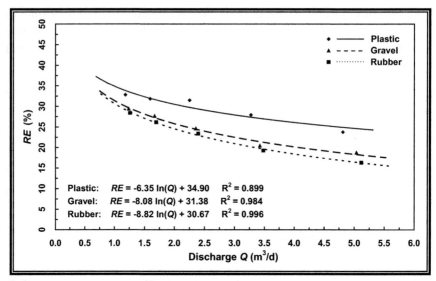

FIGURE 7.45 Relationship between TSS removal efficiency and discharge (2 m from inlet).

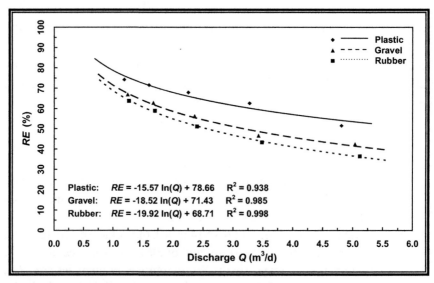

FIGURE 7.46 Relationship between TSS removal efficiency and discharge (5 m from inlet).

- For rubber cell, at the biggest Q (5.119 m³/d), the values of TSS removal efficiency are 16.3, 36.49, 44.64, and 50.31% at distances of 2, 5, 8, and 10 m from inlet weir, in the same order. While removal efficiencies are 28.37, 63.73, 78.27, and 86.06% at the smallest Q value (1.263 m³/d), respectively.

Steady Stage Analysis

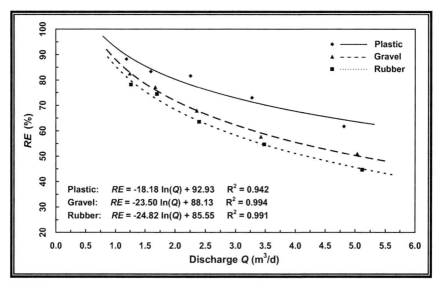

FIGURE 7.47 Relationship between TSS removal efficiency and discharge (8 m from inlet).

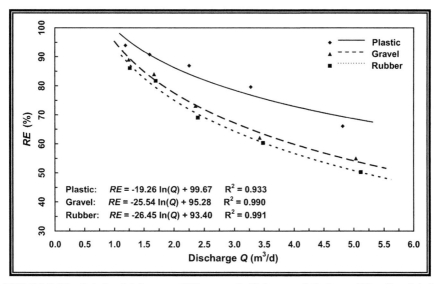

FIGURE 7.48 Relationship between TSS removal efficiency and discharge (10 m from inlet).

- The difference between TSS removal efficiency of both gravel and rubber is small in case of small Q and this difference increases with increasing Q.
- After 2 m from inlet, rubber cell gives removal efficiency lower than gravel and plastic cells by about 1.52 and 6.86%, respectively.

These values at 5 m are about 4.28 and 14.84%, at 8 m 4.15 and 14.49%, and at outlets 3.12 and 13.95%.
- Gravel cell gives removal efficiency lower than plastic by about 5.34, 10.56, 10.34, and 10.84% at distances of 2, 5, 8, and 10 m from inlet.

The values of average difference for TSS pollutant between the three media at different distances are illustrated in Appendix II. The TSS removal efficiency is gradually reduced with discharge increase at different distances according to a logarithmic function which gives the best determination coefficient. The logarithmic equations at outlets for steady stage cycles are as follows:

$$Plastic: \quad RE = 99.67 - 19.26\ln Q \quad R^2 = 0.933 \quad (7.19a)$$

$$Gravel: \quad RE = 95.28 - 25.54\ln Q \quad R^2 = 0.990 \quad (7.19b)$$

$$Rubber: \quad RE = 93.40 - 26.45\ln Q \quad R^2 = 0.991 \quad (7.19c)$$

These equations are valid for Q ranges from 1.188 to 5.119 m^3/d.

7.5 DISSOLVED OXYGEN

The effluent dissolved oxygen (DO) is analyzed with both loading rate and retention time. A comparison between outlets DO of the three media is discussed.

7.5.1 OUTLET DO CONCENTRATION

Table 7.37 shows the average outlet DO concentration with the discharge and loading rate at outlets (surface area = 20 m^2). Values of outlet DO concentration were listed in Chapter 5.

Through the steady stage, the influent DO concentration at inlet is approximately equal zero, while the effluent DO concentration at outlets is presented as clustered columns in Figure 7.49.

Figure 7.49, shows that; for plastic cell, the DO concentration at outlets is in the permitted limit (4.0 mg/l) according to law No. 48 of 1982 (NAWQAM, 2002) at discharge lower than 2.253 m^3/d, which gives q value of 11.27 cm/d and T_r value equals 72.31 hr. For gravel cell this result could be at discharge lower than 2.357 m^3/d (q equals 11.79 cm/d and T_r equals

TABLE 7.37 Effluent DO Concentration, Discharge, and q Values for Wetland Cells

Cycle No.	Plastic Cell			Gravel Cell			Rubber Cell		
	Q(m³/d)	q(cm/d)	C_o(mg/l)	Q(m³/d)	q(cm/d)	C_o(mg/l)	Q(m³/d)	q(cm/d)	C_o(mg/l)
1	4.814	24.07	3.14	5.037	25.19	2.62	5.119	25.60	1.84
2	3.275	16.38	3.78	3.426	17.13	3.38	3.482	17.41	2.96
3	2.253	11.27	4.30	2.357	11.79	4.08	2.396	11.98	3.40
4	1.595	7.98	4.56	1.669	8.35	4.28	1.696	8.48	3.82
5	1.188	5.94	5.02	1.243	6.22	4.56	1.263	6.32	4.16

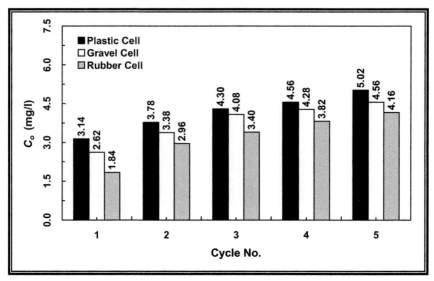

FIGURE 7.49 Effluent dissolved oxygen concentration of various cycles for the three media.

36.65 hr), while for rubber cell at discharge lower than 1.263 m^3/d (q equals 6.32 cm/d and T_r equals 89.09 hr).

7.5.2 IMPACT OF Q ON DO TREATMENT

Figure 7.50 represents the variation of effluent DO concentration with the loading rate at outlets.

From Table 7.37 and Figure 7.50, it could be concluded:

- The outlet DO concentration increases with decreasing loading rate decrease which means the improvement of the treatment performance.
- For cycle number one (Q_{max} = 4.814, 5.037, and 5.119 m^3/d), the loading rate (q) values are 24.07, 25.19, and 25.6 cm/d for plastic, gravel, and rubber cells, respectively which produce outlet DO concentrations of 3.14, 2.62, and 1.84 mg/l.
- For cycle number five (Q_{min} = 1.188, 1.243, and 1.263 m^3/d), the q values are 5.94, 6.22, and 6.32 cm/d with corresponding outlet DO concentrations of 5.02, 4.56, and 4.16 mg/l for plastic, gravel, and rubber cells, in the same sequence.
- At low q values, the differences between outlet DO concentrations for the three media are nearly small whereas at high values of q, these differences are big.

Steady Stage Analysis

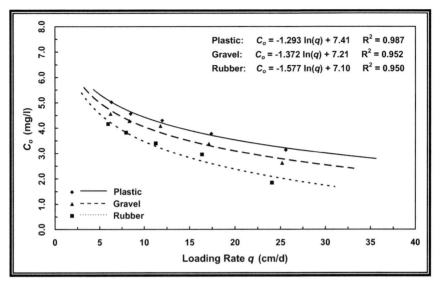

FIGURE 7.50 Relationship between dissolved oxygen concentration and hydraulic loading rate.

- Plastic media gives effluent DO value higher than both the other two media followed by gravel.

The DO effluent concentration is inversely proportional to the loading rate according to a logarithmic function as the best fit relationship. The logarithmic equations at outlets are as follows:

$$\text{Plastic:} \quad C_o = 7.41 - 1.293 \ln q \quad R^2 = 0.987 \quad (7.20a)$$

$$\text{Gravel:} \quad C_o = 7.21 - 1.372 \ln q \quad R^2 = 0.952 \quad (7.20b)$$

$$\text{Rubber:} \quad C_o = 7.10 - 1.577 \ln q \quad R^2 = 0.950 \quad (7.20c)$$

The deduced relationships between outlet parameters concentrations and loading rate are valid for q values range from 5.94 to 25.6 cm/d.

7.5.3 IMPACT OF T_r ON DO CONCENTRATION

Table 7.38 illustrates the calculated retention time corresponding to the discharges at outlets and the values of effluent dissolved oxygen concentration.

Figure 7.51 shows the relationship between effluent DO concentration and T_r at outlets of the three-wetland cells.

TABLE 7.38 Average DO Concentration and T_r Values at Outlets of Wetland Cells

Cycle No.	Effluent Dissolved Oxygen (mg/l)			Hydraulic Retention Time (hr)		
	Rubber	Gravel	Plastic	Rubber	Gravel	Plastic
1	1.84	2.62	3.14	21.98	17.16	33.84
2	2.96	3.38	3.78	32.30	25.20	49.73
3	3.40	4.08	4.30	46.97	36.65	72.31
4	3.82	4.28	4.56	66.34	51.74	102.12
5	4.16	4.56	5.02	89.09	69.48	137.11

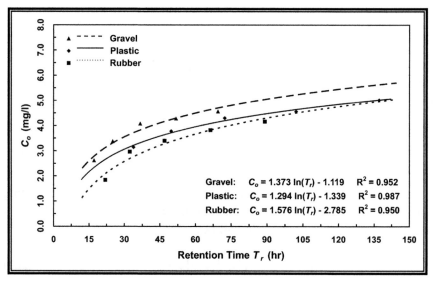

FIGURE 7.51 Relationship between C_o and hydraulic retention time for dissolved oxygen.

From Table 7.38 and Figure 7.51, it is noticed that:

- The effluent concentration of DO increases with the increase of T_r through steady stage at the end of the three cells. The treatment performance enhances with increasing T_r value. For example for gravel cell at T_r equals 17.16 hr, the outlet DO at 10 m equal to 2.62 mg/l. while T_r equal to 69.48 hr, the outlet DO for the same distance equals 4.56 mg/l.
- The difference between effluent concentration of DO for plastic and rubber cells is relatively small at big values of T_r (higher than 105 hr). Gravel cell gives the highest effluent concentration of DO followed by plastic cell.

Steady Stage Analysis

The effluent concentration of DO for the three media directly proportions with T_r value according to a logarithmic function. The logarithmic equations at outlets are as follows:

Plastic: $C_o = -1.339 + 1.294 \ln T_r$ $R^2 = 0.987$ (7.21a)

Gravel: $C_o = -1.119 + 1.373 \ln T_r$ $R^2 = 0.952$ (7.21b)

Rubber: $C_o = -2.785 + 1.576 \ln T_r$ $R^2 = 0.950$ (7.21c)

These equations are valid for Q ranges from 1.188 to 5.119 m³/d.

7.6 AMMONIA TREATMENT

Influent and outlets effluent samples were analyzed in steady stage. The effluent ammonia was studied with both loading rate and influent concentration. The variation of pollutant removal efficiency and both retention time and sewage load are discussed.

7.6.1 INLET AND OUTLET AMMONIA RELATIONSHIPS

Table 7.39 presents the average influent and effluent ammonia concentrations, and the corresponding removal efficiency for plastic, gravel, and rubber media at wetland cells outlets. The average ammonia concentration and removal efficiency were given in Chapter 5.

Figure 7.52 gives the relationship between effluent and influent concentrations (C_o and C_i) of ammonia for the three media at outlets.

TABLE 7.39 Average Ammonia Concentration and Removal Efficiency for Cells

Cycle No.	Influent (mg/l)	Effluent Concentration (mg/l)			NH₃ Removal Efficiency (%)		
		Rubber	Gravel	Plastic	Rubber	Gravel	Plastic
1	20.01	10.84	10.41	8.58	45.75	47.97	57.00
2	21.11	10.38	9.67	8.77	50.77	54.26	58.55
3	19.19	8.55	7.98	6.78	55.59	58.46	64.64
4	19.50	8.30	7.75	5.35	57.55	60.18	72.64
5	19.21	7.19	6.20	4.13	62.48	67.69	78.53

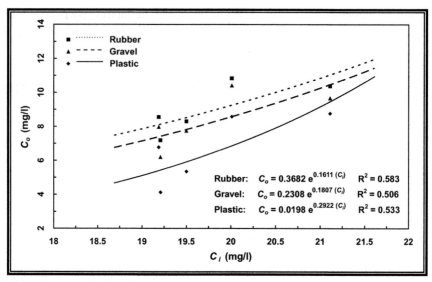

FIGURE 7.52 Relationship between influent and effluent ammonia concentrations.

Table 7.39 and Figure 7.52 exhibit that:

- Through steady stage, the effluent ammonia concentration of plastic media is smaller than the other media (better ammonia treatment) at outlets, followed by the gravel media.
- For plastic, gravel, and rubber media, the outlet concentrations of ammonia at outlets are in the allowable limit (less than 10:12 mg/l) of Law No. 48 of 1982 (NAWQAM, 2002) for discharge lower than 4.814, 5.037, 5.119 m³/d with loading rates equal 24.07, 25.19, and 25.6 cm/d, respectively.

Effluent and influent ammonia relationship for the three media is varying according to an exponential function as the best fit. The exponential equations at outlets are as follows:

$$\text{Plastic:} \quad C_o = 0.0198 e^{0.2922 C_i} \quad R^2 = 0.533 \quad (7.22a)$$

$$\text{Gravel:} \quad C_o = 0.2308 e^{0.1807 C_i} \quad R^2 = 0.506 \quad (7.22b)$$

$$\text{Rubber:} \quad C_o = 0.3682 e^{0.1611 C_i} \quad R^2 = 0.583 \quad (7.22c)$$

where:
C_o = ammonia outlet concentration, mg/l;
C_i = ammonia inlet concentration, mg/l.

Steady Stage Analysis

These equations are valid for the inlet ammonia concentration ranges from 19.19 to 21.11 mg/l, and q varies between 5.94 and 25.6 cm/d.

7.6.2 IMPACT OF Q ON AMMONIA TREATMENT

Table 7.40 illustrates the average inlet and outlet ammonia concentrations with discharge, and q values at the end of cells.

Figure 7.53 represents the variation of average effluent ammonia concentration and the loading rate at distance of 10 m from inlet.

From Table 7.40 and Figure 7.53, it is noticed that:

- The ammonia outlet concentration decreases with the decrease of q which means, the enhancement of the treatment for the three media at outlets with smaller q.
- For cycle number one (Q_{max} = 4.814, 5.037, and 5.119 m³/d), the ammonia influent concentration is 20.01 mg/l and the q values are 24.07, 25.19, and 25.6 cm/d for plastic, gravel, and rubber cells produce corresponding outlet concentrations of 8.58, 10.41, and 10.84 mg/l, respectively.
- For cycle number five (Q_{min} = 1.188, 1.243, and 1.263 m³/d), the ammonia influent concentration is 19.21 mg/l and the q values are 5.94, 6.22, and 6.32 cm/d for plastic, gravel, and rubber cells which produce corresponding outlet concentrations of 4.13, 6.2, and 7.19 mg/l, in the same sequence.
- At low loading rate, the difference between outlet ammonia values for plastic, gravel, and rubber media are big and these differences decrease at high values of loading rate. Plastic cell gives ammonia outlet concentration lower than other cells followed by gravel.

The ammonia effluent concentration is directly proportional to the loading rate according to a logarithmic function. The logarithmic equations at outlets are as follows (q ranges from 5.94 to 25.6 cm/d):

Plastic: $\quad C_o = -1.796 + 3.474 \ln q \quad R^2 = 0.921 \quad$ (7.23a)

Gravel: $\quad C_o = 1.098 + 2.926 \ln q \quad R^2 = 0.959 \quad$ (7.23b)

Rubber: $\quad C_o = 2.356 + 2.666 \ln q \quad R^2 = 0.952 \quad$ (7.23c)

7.6.3 IMPACT OF TR ON AMMONIA REMOVAL EFFICIENCY

Table 7.41 gives the retention time which corresponds to the used discharges at outlets for the three media and ammonia removal efficiency for steady stage cycles.

TABLE 7.40 Influent and Effluent Ammonia Concentrations and q Values for Cells

C_i(mg/l)	Plastic Cell			Gravel Cell			Rubber Cell		
	Q(m³/d)	q(cm/d)	C_o(mg/l)	Q(m³/d)	q(cm/d)	C_o(mg/l)	Q(m³/d)	q(cm/d)	C_o(mg/l)
20.01	4.814	24.07	8.58	5.037	25.19	10.41	5.119	25.60	10.84
21.11	3.275	16.38	8.77	3.426	17.13	9.67	3.482	17.41	10.38
19.19	2.253	11.27	6.78	2.357	11.79	7.98	2.396	11.98	8.55
19.50	1.595	7.98	5.35	1.669	8.35	7.75	1.696	8.48	8.30
19.21	1.188	5.94	4.13	1.243	6.22	6.20	1.263	6.32	7.19

Steady Stage Analysis

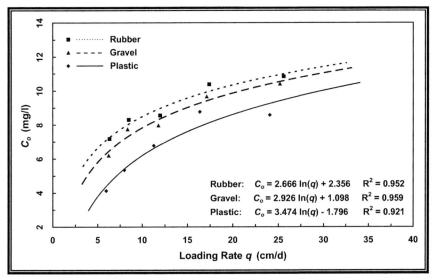

FIGURE 7.53 Relationship between C_o and hydraulic loading rate for ammonia pollutant.

TABLE 7.41 Ammonia Removal Efficiency and Retention Time for the Used Media

Cycle No.	NH$_3$ Removal Efficiency (%)			Hydraulic Retention Time (hr)		
	Rubber	Gravel	Plastic	Rubber	Gravel	Plastic
1	45.75	47.97	57.00	21.98	17.16	33.84
2	50.77	54.26	58.55	32.30	25.20	49.73
3	55.59	58.46	64.64	46.97	36.65	72.31
4	57.55	60.18	72.64	66.34	51.74	102.12
5	62.48	67.69	78.53	89.09	69.48	137.11

Figure 7.54 illustrates the relationship between ammonia removal efficiency and retention time at outlets for plastic, gravel, and rubber cells.

From Table 7.41 and Figure 7.54, it is noticed that:

- The gravel cell gives the highest removal efficiency for ammonia treatment followed by plastic cell.
- The ammonia removal efficiency increases with the increase of retention time through steady stage for the three media at outlets. The treatment improves with the increase of T_r for each media. For example for plastic cell at T_r value equals 33.84 hr, the removal efficiency is 57%, while at T_r value equals 137.11 hr, the removal efficiency is 78.53%.

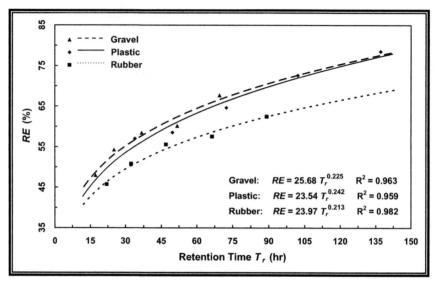

FIGURE 7.54 Relationship between ammonia removal efficiency and hydraulic retention time.

- The difference between removal efficiency of ammonia for gravel and plastic cells is small but it becomes bigger between these two cells and rubber. At low retention time, the difference between rubber and both gravel and plastic cells is small and increases as T_r increases.

The ammonia removal efficiency for the three media is directly proportionate to retention time according to a power function. The power equations at outlets are:

$$\text{Plastic:} \quad RE = 23.54\, T_r^{0.242} \quad R^2 = 0.959 \quad (7.24a)$$

$$\text{Gravel:} \quad RE = 23.54\, T_r^{0.225} \quad R^2 = 0.963 \quad (7.24b)$$

$$\text{Rubber:} \quad RE = 23.97\, T_r^{0.213} \quad R^2 = 0.982 \quad (7.24c)$$

These equations are valid for Q ranges from 1.188 to 5.119 m³/d.

7.6.4 IMPACT OF DISCHARGE ON NH$_3$ TREATMENT EFFICIENCY

Table 7.42 illustrates the used discharges and the corresponding ammonia removal efficiency at the end of wetland cells. Figure 7.55 shows the relationship between the two parameters.

TABLE 7.42 Ammonia Removal Efficiency and Discharge for the Three Media

Cycle No.	NH$_3$ Removal Efficiency (%)			Discharge (m³/d)		
	Rubber	Gravel	Plastic	Rubber	Gravel	Plastic
1	45.75	47.97	57.00	5.119	5.037	4.814
2	50.77	54.26	58.55	3.482	3.426	3.275
3	55.59	58.46	64.64	2.396	2.357	2.253
4	57.55	60.18	72.64	1.696	1.669	1.595
5	62.48	67.69	78.53	1.263	1.243	1.188

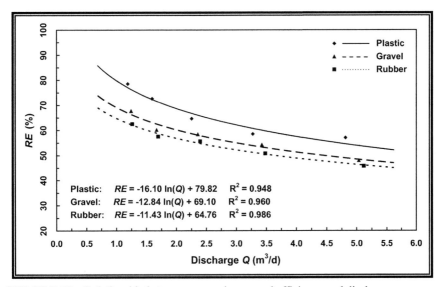

FIGURE 7.55 Relationship between ammonia removal efficiency and discharge.

From Table 7.42 and Figure 7.55, it could be concluded that:

- As the discharge increases, the removal efficiency of ammonia decreases for the three media, indicating deterioration of treatment performance.
- The plastic cell has the highest removal efficiency followed by the gravel. This may be because of the higher surface area (high amount of attached biofilm) of plastic media than the other media.
- At the biggest Q (4.814, 5.037, and 5.119 m³/d), the ammonia removal efficiency values are about 57, 47.97, and 45.75% for plastic, gravel, and rubber cells, respectively. While these values are 78.53, 67.69, and 62.48% for the smallest Q (1.188, 1.243, and 1.263 m³/d).

- The ammonia removal efficiency for plastic media increases by an average value of 8.56 and 11.84% higher than gravel and rubber media. While in gravel cell the ammonia removal efficiency is higher than rubber cell by about 3.28%. These removal efficiency differences are computed in Appendix II.

The ammonia removal efficiency gradually decreases with the increasing discharge value for the three media, according to a logarithmic function. The logarithmic equations at outlets are written as:

$$\text{Plastic}: \quad RE = 79.82 - 16.1 \ln Q \quad R^2 = 0.948 \quad (7.25a)$$

$$\text{Gravel}: \quad RE = 69.1 - 12.84 \ln Q \quad R^2 = 0.960 \quad (7.25b)$$

$$\text{Rubber}: \quad RE = 64.76 - 11.43 \ln Q \quad R^2 = 0.986 \quad (7.25c)$$

where:
Q = discharge, m^3/d;
RE = ammonia removal efficiency, %.

These equations are valid for discharge values range from 1.188 to 5.119 m^3/d for the used media.

7.7 PHOSPHATE TREATMENT

Influent and outlets effluent samples were analyzed for steady stage. The influent phosphate concentration was analyzed with both effluent concentrations and loading rate. The variation of pollutant removal efficiencies and both retention time and sewage loads were studied.

7.7.1 INLET AND OUTLET PHOSPHATE CONCENTRATIONS

Table 7.43 presents the average influent and effluent phosphate concentrations, and the corresponding removal efficiency for plastic, gravel, and rubber at outlets. The average phosphate concentration and removal efficiency were calculated in Chapter 5.

The influent and effluent phosphate concentrations at outlets for steady stage cycles are presented as clustered columns in Figure 7.56 for plastic, gravel, and rubber media.

TABLE 7.43 Average Phosphate Concentration and Removal Efficiency for Cells

Cycle No.	Influent (mg/l)	Effluent Concentration (mg/l)			PO$_4$ Removal Efficiency (%)		
		Rubber	Gravel	Plastic	Rubber	Gravel	Plastic
1	3.17	1.66	1.37	0.98	47.56	56.77	68.98
2	2.89	1.19	0.99	0.79	58.86	65.78	72.82
3	2.20	0.82	0.69	0.51	62.86	68.55	76.88
4	3.01	1.01	0.89	0.54	66.54	70.44	82.22
5	2.89	0.79	0.70	0.33	72.77	75.80	88.62

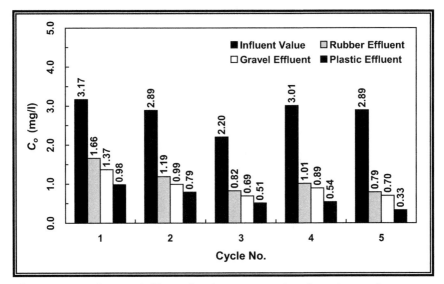

FIGURE 7.56 Influent and effluent phosphate concentrations for various cycles.

Table 7.43 and Figure 7.56 declare that, the effluent concentration of phosphate at outlets are in an acceptable limit (less than 5 mg/l) of Law No. 48 of 1982 (NAWQAM, 2002) at discharge lower than 4.814, 5.037, 5.119 m^3/d with loading rate equal 24.07, 25.19, and 25.6 cm/d for plastic, gravel, and rubber media, in the same sequence.

7.7.2 IMPACT OF Q ON PHOSPHATE TREATMENT

Table 7.44 illustrates the average influent and effluent phosphate concentrations with the discharge and loading rate at outlets for plastic, gravel, and rubber media.

TABLE 7.44 Influent and Effluent Phosphate Concentrations and q Values for Cells

C_i(mg/l)	Plastic Cell			Gravel Cell			Rubber Cell		
	Q(m³/d)	q(cm/d)	C_o(mg/l)	Q(m³/d)	q(cm/d)	C_o(mg/l)	Q(m³/d)	q(cm/d)	C_o(mg/l)
3.17	4.814	24.07	0.98	5.037	25.19	1.37	5.119	25.60	1.66
2.89	3.275	16.38	0.79	3.426	17.13	0.99	3.482	17.41	1.19
2.20	2.253	11.27	0.51	2.357	11.79	0.69	2.396	11.98	0.82
3.01	1.595	7.98	0.54	1.669	8.35	0.89	1.696	8.48	1.01
2.89	1.188	5.94	0.33	1.243	6.22	0.70	1.263	6.32	0.79

Figure 7.57 represents the variation of phosphate outlet concentration and the loading rate at outlets for plastic, gravel, and rubber media.

From Table 7.44 and Figure 7.57, it is noticed that:

- The outlet phosphate concentration decreases with the loading rate decreasing which means, the improvement of the treatment performance for the three media at outlets. Plastic media gives phosphate outlet concentration lower than other media followed by gravel.
- For cycle one (Q_{max} = 4.814, 5.037, and 5.119 m³/d), the phosphate influent concentration is 3.17 mg/l and the q values are 24.07, 25.19, and 25.6 cm/d for plastic, gravel, and rubber cells with corresponding outlet concentrations of 0.98, 1.37, and 1.66 mg/l, in the same order.
- For cycle five (Q_{min} = 1.188, 1.243, and 1.263 m³/d), C_i concentration is 2.89 mg/l and the q values are 5.94, 6.22, and 6.32 cm/d for plastic, gravel, and rubber cells with corresponding outlet concentrations of 0.33, 0.7, and 0.79 mg/l, respectively.
- At low loading rates, the difference between outlet phosphate values for gravel and rubber cells is small and this difference increases at high values of loading rate.

The phosphate outlet concentration is directly proportional to q values according to a logarithmic function as the best fit. The logarithmic equations at outlets are as follows (q ranges from 5.94 to 25.6 cm/d):

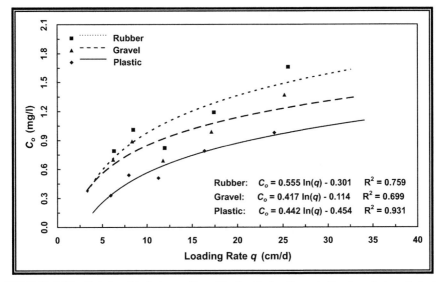

FIGURE 7.57 Relationship between C_o and hydraulic loading rate for phosphate pollutant.

Plastic: $C_o = -0.454 + 0.442 \ln q$ $R^2 = 0.931$ (7.26a)

Gravel: $C_o = -0.114 + 0.417 \ln q$ $R^2 = 0.699$ (7.26b)

Rubber: $C_o = -0.301 + 0.555 \ln q$ $R^2 = 0.759$ (7.26c)

where:
q = hydraulic loading rate, cm/d;
C_o = effluent phosphate concentration, mg/l.

7.7.3 IMPACT OF T_r ON PHOSPHATE REMOVAL EFFICIENCY

Table 7.45 illustrates the retention time, T_r and the phosphate removal efficiency at outlets.

Figure 7.58 shows the relationship between phosphate removal efficiency and T_r at the outlets for plastic, gravel, and rubber media.

From Table 6.45 and Figure 6.58, it is noticed that:

- Plastic cell gives the highest removal efficiency for phosphate treatment followed by gravel.
- The phosphate removal efficiency increases with the increase of retention time for the three media at outlets. The treatment enhances as T_r value increases. For example, for rubber cell at retention time equals 21.98 hr, the removal efficiency becomes 47.56%, and at T_r equals 89.09 hr, the removal efficiency has a value of 72.77%.
- The difference between removal efficiency of phosphate for plastic and gravel cells is small while this difference is big between these two cells and a rubber cell. At big retention time, the difference between rubber and both plastic and gravel cells is small and increases with the decrease of retention time.

TABLE 7.45 Phosphate Removal Efficiency and Retention Time for the Used Media

Cycle No.	PO₄ Removal Efficiency (%)			Hydraulic Retention Time (hr)		
	Rubber	Gravel	Plastic	Rubber	Gravel	Plastic
1	47.56	56.77	68.98	21.98	17.16	33.84
2	58.86	65.78	72.82	32.30	25.20	49.73
3	62.86	68.55	76.88	46.97	36.65	72.31
4	66.54	70.44	82.22	66.34	51.74	102.12
5	72.77	75.80	88.62	89.09	69.48	137.11

Steady Stage Analysis

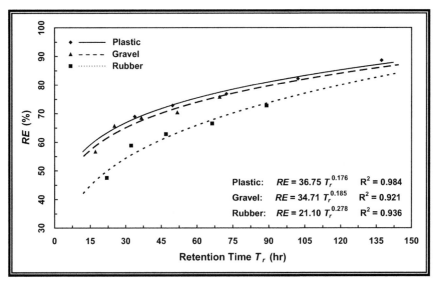

FIGURE 7.58 Relationship between phosphate removal efficiency and hydraulic retention time.

The phosphate removal efficiency directly proportions with the retention time according to a power function. The power equations at outlets are written as follows (Q ranges from 1.188 to 5.119 m^3/d):

$$\text{Plastic:} \quad RE = 36.75 \, T_r^{0.176} \quad R^2 = 0.984 \quad (7.27a)$$

$$\text{Gravel:} \quad RE = 34.71 \, T_r^{0.185} \quad R^2 = 0.921 \quad (7.27b)$$

$$\text{Rubber:} \quad RE = 21.10 \, T_r^{0.278} \quad R^2 = 0.936 \quad (7.27c)$$

7.7.4 IMPACT OF DISCHARGE ON PO$_4$ TREATMENT EFFICIENCY

Table 7.46 gives the used discharges for plastic, gravel, and rubber cells and the corresponding phosphate removal efficiency. Figure 7.59 illustrates the relationship between these two parameters.

From Table 7.46 and Figure 7.59, it is obtained that:

- As the discharge decreases, the removal efficiency of phosphate increases for the three media, which means an improvement of the treatment performance.
- For plastic cell at the biggest Q equal 4.814, 5.037, and 5.119 m^3/d, the phosphate removal efficiency is about 68.98, 56.77, and 47.56% for

TABLE 7.46 Phosphate Removal Efficiency and Discharge for the Three Media

Cycle No.	PO$_4$ Removal Efficiency (%)			Discharge (m³/d)		
	Rubber	Gravel	Plastic	Rubber	Gravel	Plastic
1	47.56	56.77	68.98	5.119	5.037	4.814
2	58.86	65.78	72.82	3.482	3.426	3.275
3	62.86	68.55	76.88	2.396	2.357	2.253
4	66.54	70.44	82.22	1.696	1.669	1.595
5	72.77	75.80	88.62	1.263	1.243	1.188

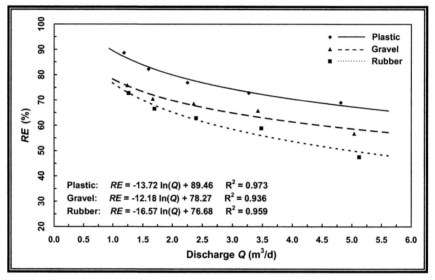

FIGURE 7.59 Relationship between phosphate removal efficiency and discharge.

plastic, gravel, and rubber cells, respectively. These removal efficiencies are 88.62, 75.8, and 72.77% at the smallest Q having a value of 1.188, 1.243, and 1.263 m³/d.

- The phosphate removal efficiency for rubber media decreases by an average value of 5.75% lower than gravel media and 16.19% lower than plastic media. The gravel media gives phosphate removal efficiency lower than plastic media by an average value of 10.44%. These removal efficiency differences are calculated in Appendix II.

The phosphate removal efficiency is gradually decreased with the increase of discharge for the three media according to a logarithmic function. The logarithmic equations at outlets are as follows:

Steady Stage Analysis 301

$$Plastic: \quad RE = 89.46 - 13.72 \ln Q \quad R^2 = 0.973 \quad (7.28a)$$

$$Gravel: \quad RE = 78.27 - 12.18 \ln Q \quad R^2 = 0.936 \quad (7.28b)$$

$$Rubber: \quad RE = 76.68 - 16.57 \ln Q \quad R^2 = 0.959 \quad (7.28c)$$

where: RE = phosphate removal efficiency, %.

These equations are valid for Q ranges from 1.188 to 5.119 m^3/d for the used media.

7.8 FECAL COLIFORMS

Influent and outlets effluent samples were analyzed for fecal coliforms bacteria in steady stage. The influent fecal coliforms were discussed with both loading rate and effluent concentration. The variation of pollutant removal efficiency and both retention time and sewage loads were studied.

7.8.1 INLET AND OUTLET FECAL COLIFORMS CONCENTRATIONS

Table 7.47 presents the influent and effluent fecal coliforms count and the corresponding removal efficiency for plastic, gravel, and rubber media at cells outlets. The fecal coliforms count and removal efficiency were calculated in Chapter 5.

The influent and effluent fecal coliforms count at outlets for steady stage cycles are presented as clustered columns in Figure 7.60.

Table 7.47 and Figure 7.60 show that, the effluent of fecal coliforms counts at outlets are in the allowable limit (5000 MPN/100 ml) according

TABLE 7.47 Average FC Values and Removal Efficiency for Wetland Cells

Cycle No.	Influent (MPN)	Effluent FC (MPN/100 ml)			FC Removal Efficiency (%)		
		Rubber	Gravel	Plastic	Rubber	Gravel	Plastic
1	230,000	6160	5840	5480	97.31	97.45	97.61
2	248,000	4820	4600	4340	98.05	98.14	98.25
3	222,000	3880	3680	3480	98.24	98.33	98.42
4	232,000	3180	2880	2620	98.62	98.76	98.87
5	228,000	2320	2160	1840	98.97	99.04	99.19

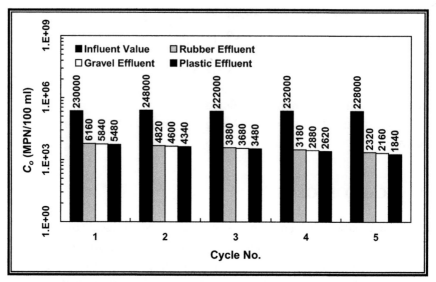

FIGURE 7.60 Influent and effluent fecal coliforms concentrations for various cycles.

to Law No. 48 of 1982 (NAWQAM, 2002) at discharge lower than 3.275, 3.426, and 3.482 m³/d with q equal 16.38, 17.13, and 17.41 cm/d for plastic, gravel, and rubber media, in the same sequence.

7.8.2 IMPACT OF Q ON FECAL COLIFORMS TREATMENT

Table 7.48 demonstrates the average inlet and outlet fecal coliforms counts (MPN/100 ml) with the discharge and loading rate at outlets.

Figure 7.61 represents the variation of effluent fecal coliforms count and loading rate at outlets for plastic, gravel, and rubber cells.

From Table 7.48 and Figure 7.61, it is noticed that:

- The outlet fecal coliforms count decreases with the decrease of loading rate which means, the enhancement of the treatment performance for the three media at outlets.
- For cycle one (Q_{max} = 4.814, 5.037, and 5.119 m³/d), the influent fecal coliforms count is 230000 MPN/100 ml and the q values are 24.07, 25.19, and 25.6 cm/d for plastic, gravel, and rubber cells, respectively which gives corresponding effluent counts of 5480, 5840, and 6160 MPN/100 ml.
- For cycle five (Q_{min} = 1.188, 1.243, and 1.263 m³/d), the influent fecal coliforms count is 228000 MPN/100 ml and the q values are 5.94, 6.22,

TABLE 7.48 Influent and Effluent Fecal Coliforms Values and q for Wetland Cells

Influent (MPN)	Plastic Wetland Cell $Q(m^3/d)$	q(cm/d)	C_o(MPN)	Gravel Cell $Q(m^3/d)$	q(cm/d)	C_o(MPN)	Rubber Cell $Q(m^3/d)$	q(cm/d)	C_o(MPN)
230,000	4.814	24.07	5480	5.037	25.19	5840	5.119	25.60	6160
248,000	3.275	16.38	4340	3.426	17.13	4600	3.482	17.41	4820
222,000	2.253	11.27	3480	2.357	11.79	3680	2.396	11.98	3880
232,000	1.595	7.98	2620	1.669	8.35	2880	1.696	8.48	3180
228,000	1.188	5.94	1840	1.243	6.22	2160	1.263	6.32	2320

MPN = most probable number.

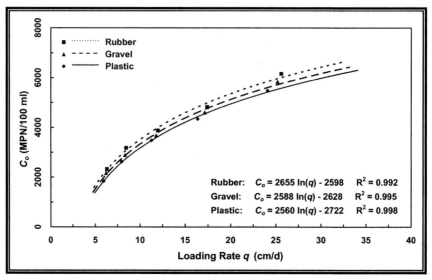

FIGURE 7.61 Relationship between C_o and hydraulic loading rate for fecal coliforms pollutant.

and 6.32 cm/d for plastic, gravel, and rubber cells gives corresponding effluent counts of 1840, 2160, and 2320 MPN/100 ml, in the same sequence.
- At low loading rate, the differences between outlet fecal coliforms counts for the three media are small and these differences increase at high values of loading rate.
- Plastic cell gives outlet fecal coliforms count smaller than other cells followed by gravel then rubber cell.

The fecal coliforms count is directly proportional to the loading rates according to a logarithmic function. The logarithmic equations at outlets are given as follows (q ranges from 5.94 to 25.6 cm/d):

$$\textit{Plastic:} \quad C_o = -2722 + 2560 \ln q \quad R^2 = 0.998 \quad (7.29a)$$

$$\textit{Gravel:} \quad C_o = -2628 + 2588 \ln q \quad R^2 = 0.995 \quad (7.29b)$$

$$\textit{Rubber:} \quad C_o = -2598 + 2655 \ln q \quad R^2 = 0.992 \quad (7.29c)$$

where:
q = hydraulic loading rate, cm/d;
C_o = effluent fecal coliforms count, MPN/100 ml.

7.8.3 IMPACT OF T_r ON FECAL COLIFORMS REMOVAL EFFICIENCY

Table 7.49 illustrates the T_r values which correspond to the used discharges at outlets and fecal coliforms removal efficiency.

Figure 7.62 illustrates the relationship between fecal coliforms removal efficiency and the actual retention time at outlets for plastic, gravel, and rubber media.

From Table 7.49 and Figure 7.62, it is noticed that:

- The fecal coliforms removal efficiency increases with the increase of T_r through steady stage for plastic, gravel, and rubber media at outlets.

TABLE 7.49 Fecal Coliforms Removal Efficiency and T_r Values for the Used Media

Cycle No.	FC Removal Efficiency (%)			Hydraulic Retention Time (hr)		
	Rubber	Gravel	Plastic	Rubber	Gravel	Plastic
1	97.31	97.45	97.61	21.98	17.16	33.84
2	98.05	98.14	98.25	32.30	25.20	49.73
3	98.24	98.33	98.42	46.97	36.65	72.31
4	98.62	98.76	98.87	66.34	51.74	102.12
5	98.97	99.04	99.19	89.09	69.48	137.11

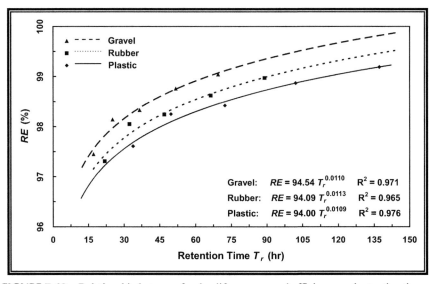

FIGURE 7.62 Relationship between fecal coliforms removal efficiency and retention time.

For example, for gravel cell at retention time equal to 17.16 hr, the removal efficiency is 97.45%, while at T_r equal to 69.48 hr, the removal efficiency is 99.04%.
- The differences between removal efficiency of fecal coliforms for the three cells are relatively small and may be neglected.

The fecal coliforms removal efficiency for the three media is directly proportional to the retention time according to a power function. The power equations at outlets are as follows (Q ranges from 1.188 to 5.119 m³/d):

$$\text{Plastic}: \quad RE = 94.0\ T_r^{0.0109} \quad R^2 = 0.976 \quad (7.30a)$$

$$\text{Gravel}: \quad RE = 94.54\ T_r^{0.0110} \quad R^2 = 0.971 \quad (7.30b)$$

$$\text{Rubber}: \quad RE = 94.09\ T_r^{0.0113} \quad R^2 = 0.965 \quad (7.30c)$$

where:
RE = FC removal efficiency, %;
T_r = hydraulic retention time, hr.

7.8.4 IMPACT OF DISCHARGE ON FC TREATMENT EFFICIENCY

Table 7.50 illustrates the used discharges for the three-wetland cells and the corresponding fecal coliforms removal efficiency.

Figure 7.63 shows the relationship between fecal coliforms removal efficiency and Q for plastic, gravel, and rubber at the end of cells.

From Table 7.50 and Figure 7.63, it is obtained that:

- At the maximum applied Q (4.814, 5.037, and 5.119 m³/d), the fecal coliforms removal efficiency value is about 97.61, 97.45, and 97.31%

TABLE 7.50 Fecal Coliforms Removal Efficiency and Discharge for the Three Media

Cycle No.	FC Removal Efficiency (%)			Discharge (m³/d)		
	Rubber	Gravel	Plastic	Rubber	Gravel	Plastic
1	97.31	97.45	97.61	5.119	5.037	4.814
2	98.05	98.14	98.25	3.482	3.426	3.275
3	98.24	98.33	98.42	2.396	2.357	2.253
4	98.62	98.76	98.87	1.696	1.669	1.595
5	98.97	99.04	99.19	1.263	1.243	1.188

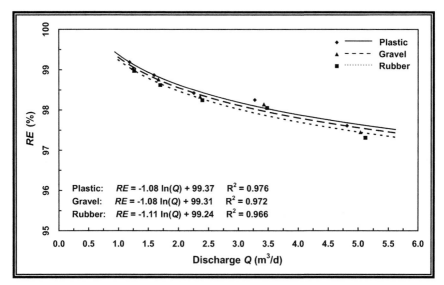

FIGURE 7.63 Relationship between fecal coliforms removal efficiency and discharge.

for plastic, gravel, and rubber cells, in the same sequence. While these values are 99.19, 99.04, and 98.97% at the minimum applied Q (1.188, 1.243, and 1.263 m³/d).

- The fecal coliforms removal efficiency for gravel cell is higher than the rubber one by an average value of 0.11% and lower than plastic cell by an average value of 0.12%. Plastic cell gives fecal coliforms removal efficiency higher than rubber by an average value of 0.23%. These removal efficiency differences are calculated in Appendix II.

The fecal coliforms removal efficiency is gradually reduced by the increase of discharge for the three media according to a logarithmic function which gives the best determination coefficient. The logarithmic equations at outlets are given as follows:

Plastic: $\quad RE = 99.37 - 1.08 \ln Q \quad R^2 = 0.976$ \hfill (7.31a)

Gravel: $\quad RE = 99.31 - 1.08 \ln Q \quad R^2 = 0.972$ \hfill (7.31b)

Rubber: $\quad RE = 99.24 - 1.11 \ln Q \quad R^2 = 0.966$ \hfill (7.31c)

where:
RE = FC removal efficiency, %;
Q = discharge to wetland cells, m³/d.

These equations are valid for discharge values range from 1.188 to 5.119 m³/d for the used media.

7.9 ZINC TREATMENT

The influent Zinc was analyzed with both loading rate and effluent concentration. The variation of pollutant removal efficiency and both retention time and sewage loads were studied.

7.9.1 INLET AND OUTLET ZINC CONCENTRATIONS

Table 7.51 presents the average inlet and outlet Zinc concentrations and the corresponding removal efficiency for plastic, gravel, and rubber media at outlets. This average concentration and removal efficiency were calculated in Chapter 5.

The influent and effluent Zinc concentrations at outlets are presented as clustered columns in Figure 7.64 for the three media.

Table 7.51 and Figure 7.64 demonstrate that, Zinc effluent concentrations are in the allowable limit (less than 5 mg/l) of law No. 48 of 1982 (NAWQAM, 2002) at discharge lower than 4.814, 5.037, 5.119 m³/d (q equal 24.07, 25.19, 25.6 cm/d) for plastic, gravel, and rubber media at outlets, respectively.

7.9.2 IMPACT OF Q ON ZINC TREATMENT

Table 7.52 illustrates the average inlet and outlet Zinc concentrations with discharge and loading rate at outlets.

TABLE 7.51 Zinc Concentration and Removal Efficiency for Wetland Cells

Cycle No.	Influent (mg/l)	Effluent Concentration (mg/l)			Zn Removal Efficiency (%)		
		Rubber	Gravel	Plastic	Rubber	Gravel	Plastic
1	1.85	1.12	1.04	0.90	39.35	43.54	51.32
2	1.64	0.91	0.80	0.60	44.58	51.41	63.21
3	1.54	0.77	0.65	0.48	49.51	57.47	68.76
4	1.73	0.66	0.59	0.48	61.73	65.59	72.17
5	1.58	0.49	0.40	0.29	68.65	74.91	81.83

Steady Stage Analysis

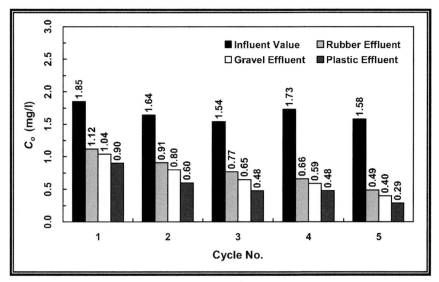

FIGURE 7.64 Influent and effluent Zinc concentrations of used media for various cycles.

Figure 7.65 represents the variation of Zinc outlet concentration and loading rate at outlets for plastic, gravel, and rubber media.

From Table 7.52 and Figure 7.65, it could be concluded that:

- The Zinc effluent concentration decreases with the decrease of loading rate, which means the enhancement of the treatment performance for the three media at outlets.
- For cycle one (Q_{max} = 4.814, 5.037, and 5.119 m³/d), the Zinc influent concentration is 1.85 mg/l and the q values are 24.07, 25.19, and 25.6 cm/d for plastic, gravel, and rubber cells, in the same order with corresponding outlet concentrations of 0.9, 1.04, and 1.12 mg/l.
- For cycle five (Q_{min} = 1.188, 1.243, and 1.263 m³/d), the Zinc influent concentration is 1.58 mg/l and the q values are 5.94, 6.22, and 6.32 cm/d for plastic, gravel, and rubber cells, respectively with corresponding effluent concentrations of 0.29, 0.4, and 0.49 mg/l.
- At low hydraulic loading rate, the differences between outlet Zinc concentrations for the used media are small and these differences increase at high values of q.
- Plastic cell gives the lowest Zinc outlet concentration followed by gravel cell.

The Zinc effluent concentration directly proportions with the loading rate according to a logarithmic function. The logarithmic equations at outlets are (q ranges from 5.94 to 25.6 cm/d):

TABLE 7.52 Inlet and Outlet Zinc Concentrations and q Values for Wetland Cells

C_i(mg/l)	Plastic Cell				Gravel Cell				Rubber Cell		
	Q(m³/d)	q(cm/d)	C_o(mg/l)		Q(m³/d)	q(cm/d)	C_o(mg/l)	Q(m³/d)	q(cm/d)	C_o(mg/l)
1.85	4.814	24.07	0.90		5.037	25.19	1.04	5.119	25.60	1.12
1.64	3.275	16.38	0.60		3.426	17.13	0.80	3.482	17.41	0.91
1.54	2.253	11.27	0.48		2.357	11.79	0.65	2.396	11.98	0.77
1.73	1.595	7.98	0.48		1.669	8.35	0.59	1.696	8.48	0.66
1.58	1.188	5.94	0.29		1.243	6.22	0.40	1.263	6.32	0.49

Steady Stage Analysis

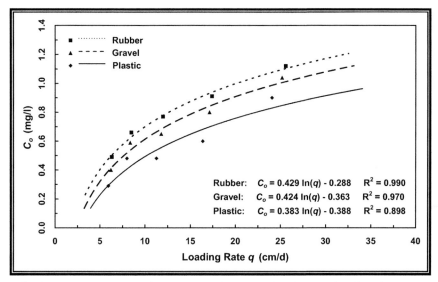

FIGURE 7.65 Relationship between C_o and hydraulic loading rate for Zinc element.

$$\text{Plastic:} \quad C_o = -0.388 + 0.383 \ln q \quad R^2 = 0.898 \quad (7.32a)$$

$$\text{Gravel:} \quad C_o = -0.363 + 0.424 \ln q \quad R^2 = 0.970 \quad (7.32b)$$

$$\text{Rubber:} \quad C_o = -0.288 + 0.429 \ln q \quad R^2 = 0.990 \quad (7.32c)$$

where:
C_o = Zinc effluent concentration, mg/l.

7.9.3 IMPACT OF T_r ON ZINC REMOVAL EFFICIENCY

Table 7.53 exhibits the retention time values and Zinc removal efficiency at outlets.

Figure 7.66 shows the relationship between Zinc removal efficiency and T_r for plastic, gravel, and rubber media at outlets.

From Table 7.53 and Figure 7.66, it is remarked that:

- Gravel cell gives the highest Zinc removal efficiency followed by plastic cell.
- The Zinc removal efficiency increases with the increase of T_r values for the used media at outlets. The treatment improves with the increase of retention time for each media. For example for plastic

TABLE 7.53 Zinc Removal Efficiency and Retention Time for the Used Media

Cycle No.	Zn Removal Efficiency (%)			Hydraulic Retention Time (hr)		
	Rubber	Gravel	Plastic	Rubber	Gravel	Plastic
1	39.35	43.54	51.32	21.98	17.16	33.84
2	44.58	51.41	63.21	32.30	25.20	49.73
3	49.51	57.47	68.76	46.97	36.65	72.31
4	61.73	65.59	72.17	66.34	51.74	102.12
5	68.65	74.91	81.83	89.09	69.48	137.11

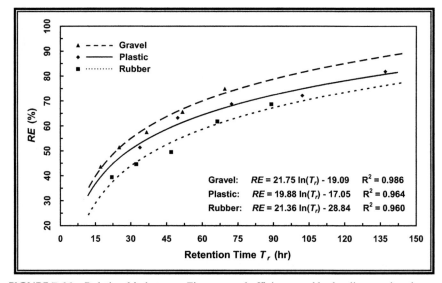

FIGURE 7.66 Relationship between Zinc removal efficiency and hydraulic retention time.

cell at retention time equals 33.84 hr, the removal efficiency equal to 51.32%, while at T_r equals 137.11 hr, the removal efficiency becomes 81.83%.
- The difference between removal efficiency of Zinc for gravel and plastic cells is small at low retention time and increases with the increase of T_r values.

The Zinc removal efficiency for the three media is directly proportional to retention time according to a logarithmic function as the best fit relationship. The logarithmic equations at outlets are written as follows (Q ranges from 1.188 to 5.119 m³/d):

Steady Stage Analysis

$$\textit{Plastic:} \quad RE = -17.05 + 19.88 \ln T_r \quad R^2 = 0.964 \quad (7.33a)$$

$$\textit{Gravel:} \quad RE = -19.09 + 21.75 \ln T_r \quad R^2 = 0.976 \quad (7.33b)$$

$$\textit{Rubber:} \quad RE = -28.84 + 21.36 \ln T_r \quad R^2 = 0.960 \quad (7.33c)$$

7.9.4 IMPACT OF DISCHARGE ON Zn TREATMENT EFFICIENCY

Table 7.54 presents the discharges for the used media and the corresponding Zinc removal efficiency. Figure 7.67 shows the relationship between these two parameters.

Table 7.54 and Figure 7.67 it could be concluded that:

- As the discharge decreases, the removal efficiency of Zinc increases for the three media and the treatment enhances.
- At the biggest Q equals 4.814, 5.037, and 5.119 m³/d, the Zinc removal efficiency is about 51.32, 43.54, and 39.35% for plastic, gravel, and rubber cells. At the smallest Q equal to 1.188, 1.243, and 1.263 m³/d the removal efficiencies are 81.83, 74.91, and 68.65%, respectively.
- The plastic cell has the highest Zinc removal efficiency followed by gravel cell. The Zinc removal efficiency for rubber media decreases by average values of 5.82 and 14.69% lower than gravel and plastic media, in the same order. Plastic cell gives Zinc removal efficiency higher than gravel cell by an average of 8.87%. These average removal efficiency differences are computed in Appendix II.

The Zinc removal efficiency is gradually decreased by the increase of discharge for the used media according to a logarithmic function. The logarithmic equations at outlets are as follows:

$$\textit{Plastic:} \quad RE = 84.19 - 19.88 \ln Q \quad R^2 = 0.964 \quad (7.34a)$$

TABLE 7.54 Zinc Removal Efficiency and Discharge for the Three Media

Cycle No.	Zn Removal Efficiency (%)			Discharge (m³/d)		
	Rubber	Gravel	Plastic	Rubber	Gravel	Plastic
1	39.35	43.54	51.32	5.119	5.037	4.814
2	44.58	51.41	63.21	3.482	3.426	3.275
3	49.51	57.47	68.76	2.396	2.357	2.253
4	61.73	65.59	72.17	1.696	1.669	1.595
5	68.65	74.91	81.83	1.263	1.243	1.188

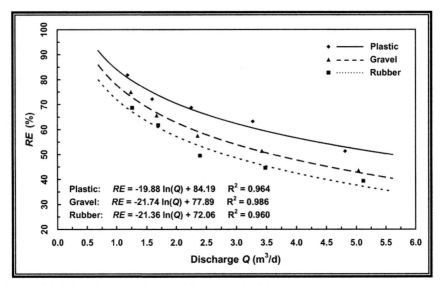

FIGURE 7.67 Relationship between Zinc removal efficiency and discharge for wetland cells.

$$Gravel: \quad RE = 77.89 - 21.74 \ln Q \quad R^2 = 0.986 \quad (7.34b)$$

$$Rubber: \quad RE = 72.06 - 21.36 \ln Q \quad R^2 = 0.960 \quad (7.34c)$$

These equations are valid for Q ranges from 1.188 to 5.119 m³/d for the used media.

7.10 IRON TREATMENT

The inlet Iron concentration was analyzed with both loading rate and outlet concentration. The variation of pollutant removal efficiency and both retention time and sewage loads were studied.

7.10.1 INLET AND OUTLET IRON CONCENTRATIONS

Table 7.55 presents the average inlet and outlet Iron concentrations and the corresponding removal efficiency. This average concentration and removal efficiency were given in Chapter 5.

The influent and effluent Iron concentrations at outlets are presented as clustered columns in Figure 7.68 for plastic, gravel, and rubber media.

Steady Stage Analysis

TABLE 7.55 Iron Concentration and Removal Efficiency for Wetland Cells

Cycle No.	Influent (mg/l)	Effluent Concentration (mg/l)			Fe Removal Efficiency (%)		
		Rubber	Gravel	Plastic	Rubber	Gravel	Plastic
1	0.75	0.60	0.53	0.49	20.27	30.09	34.78
2	0.95	0.63	0.60	0.43	33.88	37.05	54.54
3	0.76	0.41	0.33	0.23	46.31	56.81	69.42
4	0.90	0.44	0.37	0.25	51.81	58.50	72.74
5	0.85	0.31	0.20	0.17	63.70	76.07	79.63

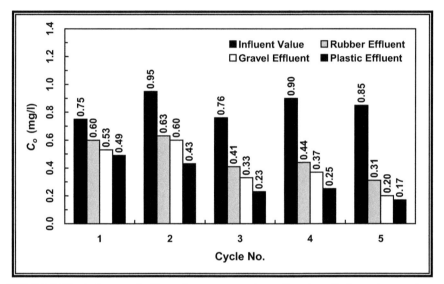

FIGURE 7.68 Influent and effluent Iron concentrations of used media for various cycles.

Table 7.55 and Figure 7.68 declare that, Iron effluents concentration is in the allowable limit (less than 1.5 mg/l) of law No. 48 of 1982 (NAWQAM, 2002) at discharge lower than 4.814, 5.037, and 5.119 m³/d (q = 24.07, 25.19, and 25.6 cm/d) for plastic, gravel, and rubber media at outlets, respectively.

7.10.2 IMPACT OF Q ON IRON TREATMENT

Table 7.56 illustrates the average inlet and outlet Iron concentrations with discharge and loading rate at outlets.

Figure 7.69 represents the variation of the effluent Iron concentration with the loading rate at outlets for plastic, gravel, and rubber media.

TABLE 7.56 Inlet and Outlet Iron Concentrations and q Values for Wetland Cells

C_i(mg/l)	Plastic Cell			Gravel Cell			Rubber Cell		
	Q(m³/d)	q(cm/d)	C_o(mg/l)	Q(m³/d)	q(cm/d)	C_o(mg/l)	Q(m³/d)	q(cm/d)	C_o(mg/l)
0.75	4.814	24.07	0.49	5.037	25.19	0.53	5.119	25.60	0.60
0.95	3.275	16.38	0.43	3.426	17.13	0.60	3.482	17.41	0.63
0.76	2.253	11.27	0.23	2.357	11.79	0.33	2.396	11.98	0.41
0.90	1.595	7.98	0.25	1.669	8.35	0.37	1.696	8.48	0.44
0.85	1.188	5.94	0.17	1.243	6.22	0.20	1.263	6.32	0.31

Steady Stage Analysis

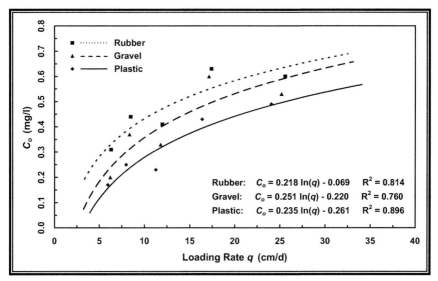

FIGURE 7.69 Relationship between C_o and hydraulic loading rate for Iron element.

From Table 7.56 and Figure 7.69, it is declared that:

- The Iron effluent concentration decreases with decreasing the loading rate which means, the improvement of the treatment performance for the three media at outlets.
- For cycle one (Q_{max} = 4.814, 5.037, and 5.119 m³/d), the Iron influent concentration is 0.75 mg/l and the q values are 24.07, 25.19, and 25.6 cm/d for plastic, gravel, and rubber cells, in the same sequence with corresponding outlet concentrations of 0.49, 0.53, and 0.6 mg/l.
- For cycle five (Q_{min} = 1.188, 1.243, and 1.263 m³/d), the Iron influent concentration is 0.85 mg/l and the q values are 5.94, 6.22, and 6.32 cm/d for plastic, gravel, and rubber cells, in the same order with corresponding effluent concentrations of 0.17, 0.2, and 0.31 mg/l.

The Iron effluent concentration at outlets directly proportions with the loading rate according to a logarithmic function as the best fit relationship. The logarithmic equations at the end of wetland cells are as follows:

$$\textit{Plastic}: \quad C_o = -0.261 + 0.235 \ln q \quad R^2 = 0.896 \quad (7.35a)$$

$$\textit{Gravel}: \quad C_o = -0.220 + 0.251 \ln q \quad R^2 = 0.760 \quad (7.35b)$$

$$\textit{Rubber}: \quad C_o = -0.069 + 0.218 \ln q \quad R^2 = 0.814 \quad (7.35c)$$

These equations are valid for q values range from 5.94 to 25.6 cm/d.

7.10.3 IMPACT OF T_r ON IRON REMOVAL EFFICIENCY

Table 7.57 gives the retention time and Iron removal efficiency at outlets. Figure 7.70 shows the relationship between these two parameters.

From Table 7.57 and Figure 7.70, it is noticed that:

- The gravel cell gives the highest value of Iron removal efficiency followed by plastic.
- The Iron removal efficiency increases with the increase of retention time for the three media at outlets. The treatment develops better with increasing value of the retention time. For example for

TABLE 7.57 Iron Removal Efficiency and Retention Time for the Used Media

Cycle No.	Fe Removal Efficiency (%)			Hydraulic Retention Time (hr)		
	Rubber	Gravel	Plastic	Rubber	Gravel	Plastic
1	20.27	30.09	34.78	21.98	17.16	33.84
2	33.88	37.05	54.54	32.30	25.20	49.73
3	46.31	56.81	69.42	46.97	36.65	72.31
4	51.81	58.50	72.74	66.34	51.74	102.12
5	63.70	76.07	79.63	89.09	69.48	137.11

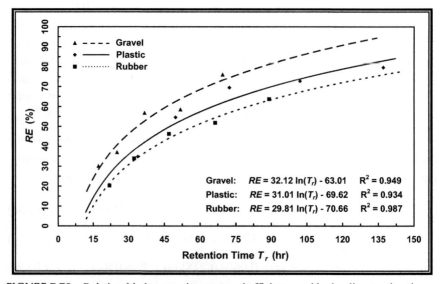

FIGURE 7.70 Relationship between iron removal efficiency and hydraulic retention time.

rubber cell at retention time equals 21.98 hr, the removal efficiency equal to 20.27%, while at T_r equals 89.09 hr, the removal efficiency becomes 63.7%.
- The difference between removal efficiency of Iron for gravel and plastic cells is small at low retention time, and increases with the increase of T_r value.

The Iron removal efficiency for the three media is directly in proportion to retention time according to a logarithmic function. The logarithmic equations at outlets are as follows:

$$\text{Plastic:} \quad RE = -69.62 + 31.01 \ln T_r \quad R^2 = 0.934 \quad (7.36a)$$

$$\text{Gravel:} \quad RE = -63.01 + 32.12 \ln T_r \quad R^2 = 0.949 \quad (7.36b)$$

$$\text{Rubber:} \quad RE = -70.66 + 29.81 \ln T_r \quad R^2 = 0.987 \quad (7.36c)$$

These equations are valid for Q ranges from 1.188 to 5.119 m³/d.

7.10.4 IMPACT OF DISCHARGE ON FE TREATMENT EFFICIENCY

Table 7.58 exhibits the discharges and the corresponding Iron removal efficiencies at the end of wetland cells. Figure 7.71 shows the relationship between these two parameters.

From Table 7.58 and Figure 7.71, it is remarked that:

- As the discharge decreases, the removal efficiency of Iron increases for the used media, which means the improvement of the treatment performance.

TABLE 7.58 Iron Removal Efficiency and Discharge for the Three Media

Cycle No.	Fe Removal Efficiency (%)			Discharge (m³/d)		
	Rubber	Gravel	Plastic	Rubber	Gravel	Plastic
1	20.27	30.09	34.78	5.119	5.037	4.814
2	33.88	37.05	54.54	3.482	3.426	3.275
3	46.31	56.81	69.42	2.396	2.357	2.253
4	51.81	58.50	72.74	1.696	1.669	1.595
5	63.70	76.07	79.63	1.263	1.243	1.188

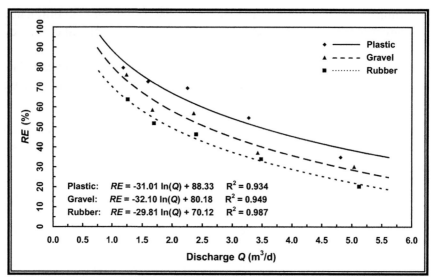

FIGURE 7.71 Relationship between Iron removal efficiency and discharge for wetland cells.

- At the biggest Q equal 4.814, 5.037, and 5.119 m³/d; the Iron removal efficiency is about 34.78, 30.09, and 20.27% for plastic, gravel, and rubber cells. At the smallest average Q equal to 1.188, 1.243, and 1.263 m³/d the removal efficiencies are 79.63, 76.07, and 63.7%, respectively.
- The plastic cell has the highest Iron removal efficiency followed by gravel. The Iron removal efficiency for plastic media increases by an average value of 10.52 and 19.03% higher than efficiencies of gravel and rubber media. Gravel cell gives Iron removal efficiency higher than rubber cell by an average value of 8.51%. These average removal efficiency differences are calculated in Appendix II.

The Iron removal efficiency is gradually reduced by increasing the discharge for the three media according to a logarithmic function. The logarithmic equations at outlets are as follows (Q ranges from 1.188 to 5.119 m³/d):

$$\textit{Plastic:} \quad RE = 88.33 - 31.01 \ln Q \quad R^2 = 0.934 \quad (7.37a)$$

$$\textit{Gravel:} \quad RE = 80.18 - 32.10 \ln Q \quad R^2 = 0.949 \quad (7.37b)$$

$$\textit{Rubber:} \quad RE = 70.12 - 29.81 \ln Q \quad R^2 = 0.987 \quad (7.37c)$$

7.11 MANGANESE TREATMENT

The effluent Manganese concentration was analyzed with both loading rate and influent concentration. The variation of pollutant removal efficiency and both retention time and sewage loads were studied.

7.11.1 INLET AND OUTLET MANGANESE CONCENTRATIONS

Table 7.59 presents the average influent and effluent Manganese concentrations and the corresponding removal efficiency at outlets. The Manganese removal efficiency was calculated in Chapter 5.

The influent and effluent Manganese concentrations at outlets are presented as clustered columns in Figure 7.72.

Table 7.59 and Figure 7.72 show that, Manganese effluents concentration are in the allowable limit (less than 1.0 mg/l) of law No. 48 of 1982 (NAWQAM, 2002) at discharge lower than 4.814, 5.037, and 5.119 m^3/d (loading rate equals 24.07, 25.19, and 25.6 cm/d) for plastic, gravel, and rubber media at outlets, respectively.

7.11.2 IMPACT OF Q ON MANGANESE TREATMENT

Table 7.60 illustrates the average influent and effluent Manganese concentrations with Q and q at outlets for plastic, gravel, and rubber media.

Figure 7.73 represents the variation of Manganese concentration with the loading rate at outlets.

TABLE 7.59 Manganese Concentration and Removal Efficiency for Wetland Cells

Cycle No.	Influent (mg/l)	Effluent Concentration (mg/l)			Mn Removal Efficiency (%)		
		Rubber	Gravel	Plastic	Rubber	Gravel	Plastic
1	0.30	0.24	0.23	0.20	17.72	21.85	31.67
2	0.36	0.27	0.25	0.17	25.07	29.86	52.01
3	0.33	0.19	0.15	0.12	41.44	54.49	63.83
4	0.29	0.14	0.13	0.10	50.94	56.90	67.10
5	0.33	0.13	0.09	0.07	60.70	72.02	78.69

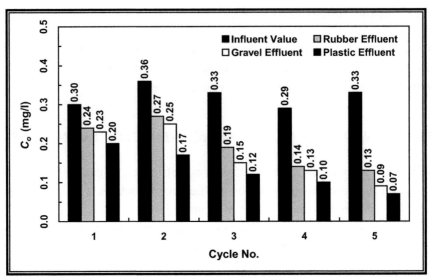

FIGURE 7.72 Inlet and outlet Manganese concentrations of used media for various cycles.

From Table 7.60 and Figure 7.73, it could be concluded that:

- The Manganese effluent concentration decreases with the decrease of loading rate, which means the enhancement of the treatment for the used media at outlets.
- For cycle one (Q_{max} = 4.814, 5.037, and 5.119 m^3/d), the Manganese influent concentration is 0.3 mg/l and the q values are 24.07, 25.19, and 25.6 cm/d for plastic, gravel, and rubber cells with corresponding outlet concentrations of 0.2, 0.23, and 0.24 mg/l, respectively.
- For cycle five (Q_{min} = 1.188, 1.243, and 1.263 m^3/d), the Manganese influent concentration is 0.33 mg/l and the q values are 5.94, 6.22, and 6.32 cm/d for plastic, gravel, and rubber cells with corresponding effluent concentrations of 0.07, 0.09, and 0.13 mg/l, in the same order.
- At low loading rate, the difference between outlet Manganese values for plastic and gravel media is small and this difference increases at high values of q. This status is reversed between gravel and rubber media.
- Plastic media gives Manganese effluent concentration lower than other cells followed by gravel media.

The Manganese effluent concentration is directly proportional to the loading rate according to a logarithmic function. The logarithmic equations at outlets are as follows (q ranges from 5.94 to 25.6 cm/d):

TABLE 7.60 Inlet and Outlet Manganese Concentrations and q Values for Cells

C_i(mg/l)	Plastic Cell				Gravel Cell				Rubber Cell		
	Q(m³/d)	q(cm/d)	C_o(mg/l)	Q(m³/d)	q(cm/d)	C_o(mg/l)	Q(m³/d)	q(cm/d)	C_o(mg/l)
0.30	4.814	24.07	0.20	5.037	25.19	0.23	5.119	25.60	0.24
0.36	3.275	16.38	0.17	3.426	17.13	0.25	3.482	17.41	0.27
0.33	2.253	11.27	0.12	2.357	11.79	0.15	2.396	11.98	0.19
0.29	1.595	7.98	0.10	1.669	8.35	0.13	1.696	8.48	0.14
0.33	1.188	5.94	0.07	1.243	6.22	0.09	1.263	6.32	0.13

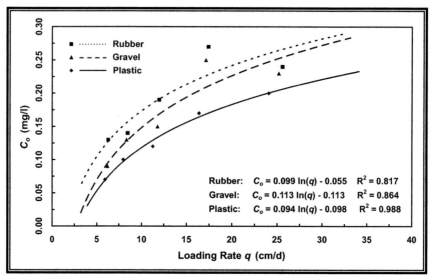

FIGURE 7.73 Relationship between C_o and hydraulic loading rate for Manganese element.

$$\text{Plastic:} \quad C_o = -0.098 + 0.094 \ln q \quad R^2 = 0.988 \quad (7.38a)$$

$$\text{Gravel:} \quad C_o = -0.113 + 0.113 \ln q \quad R^2 = 0.864 \quad (7.38b)$$

$$\text{Rubber:} \quad C_o = -0.055 + 0.099 \ln q \quad R^2 = 0.817 \quad (7.38c)$$

where:
 q = hydraulic loading rate, cm/d;
 C_o = Manganese effluent concentration, mg/l.

7.11.3 IMPACT OF T_r ON MANGANESE TREATMENT EFFICIENCY

Table 7.61 gives the retention time and Manganese removal efficiency at cells outlets.

Figure 7.74 illustrates the relationship between Manganese removal efficiency and actual retention time at outlets for plastic, gravel, and rubber media.

From Table 7.61 and Figure 7.74, it is noticed that:

- The gravel cell gives the highest Manganese removal efficiency followed by the plastic cell.

Steady Stage Analysis

TABLE 7.61 Manganese Removal Efficiency and Retention Time for the Used Media

Cycle No.	Mn Removal Efficiency (%)			Hydraulic Retention Time (hr)		
	Rubber	Gravel	Plastic	Rubber	Gravel	Plastic
1	17.72	21.85	31.67	21.98	17.16	33.84
2	25.07	29.86	52.01	32.30	25.20	49.73
3	41.44	54.49	63.83	46.97	36.65	72.31
4	50.94	56.90	67.10	66.34	51.74	102.12
5	60.70	72.02	78.69	89.09	69.48	137.11

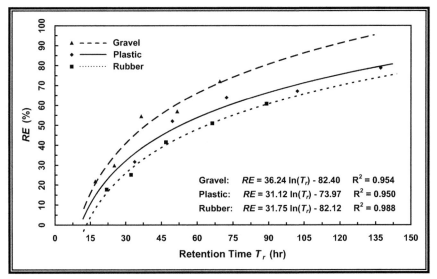

FIGURE 7.74 Relationship between Manganese removal efficiency and hydraulic retention time.

- Manganese removal efficiency increases with the increase of retention time. For example for gravel cell at retention time equals 17.16 hr, the removal efficiency becomes 21.85%, while at T_r equals 69.48 hr, the RE equal to 72.02%.
- The difference between removal efficiency of Manganese for gravel and plastic cells is small at low retention time and increases with the T_r increase.

The Manganese removal efficiency for the three media directly in proportion to the retention time according to a logarithmic function. The logarithmic equations at outlets are (Q ranges from 1.188 to 5.119 m³/d):

Plastic: $RE = -73.97 + 31.12 \ln T_r$ $R^2 = 0.950$ (7.39a)

Gravel: $RE = -82.40 + 36.24 \ln T_r$ $R^2 = 0.954$ (7.39b)

Rubber: $RE = -82.12 + 31.75 \ln T_r$ $R^2 = 0.988$ (7.39c)

where:
T_r = hydraulic retention time, hr;
RE = Manganese removal efficiency, %.

7.11.4 IMPACT OF DISCHARGE ON MN REMOVAL EFFICIENCY

Table 7.62 presents the discharges for the used media and the corresponding Manganese removal efficiency.

Figure 7.75 shows the relationship between Manganese removal efficiency with Q for plastic, gravel, and rubber cells at outlets.

From Table 7.62 and Figure 7.75, it is declared that for steady stage:

- The plastic cell has the highest Manganese removal efficiency followed by gravel. At the biggest $Q = 4.814$, 5.037, and 5.119 m³/d, the Manganese removal efficiency is 31.67, 21.85, and 17.72% for plastic, gravel, and rubber cells. At the smallest average $Q = 1.188$, 1.243, and 1.263 m³/d, these removal efficiencies are 78.69, 72.02, and 60.7%, respectively which means that the treatment enhances by decreasing the discharge.
- The manganese removal efficiency for plastic media increases by an average values of 11.64 and 19.49% higher than gravel and rubber media, in the same order. The gravel cell gives Manganese removal efficiency higher than rubber by an average value of 7.85%. These average RE differences are computed in Appendix II.

TABLE 7.62 Manganese Removal Efficiency and Discharge for the Three Media

Cycle No.	Mn Removal Efficiency (%)			Discharge (m³/d)		
	Rubber	Gravel	Plastic	Rubber	Gravel	Plastic
1	17.72	21.85	31.67	5.119	5.037	4.814
2	25.07	29.86	52.01	3.482	3.426	3.275
3	41.44	54.49	63.83	2.396	2.357	2.253
4	50.94	56.90	67.10	1.696	1.669	1.595
5	60.70	72.02	78.69	1.263	1.243	1.188

Steady Stage Analysis

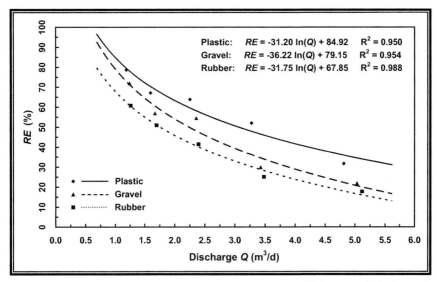

FIGURE 7.75 Relationship between Manganese removal efficiency and discharge for wetland cells.

The Manganese removal efficiency is gradually reduced by increasing the discharge for the used media according to a logarithmic function which gives the best determination coefficient. The logarithmic equations at outlets are given as follows:

Plastic: $\quad RE = 84.92 - 31.20 \ln Q \quad R^2 = 0.950 \quad$ (7.40a)

Gravel: $\quad RE = 79.15 - 36.22 \ln Q \quad R^2 = 0.954 \quad$ (7.40b)

Rubber: $\quad RE = 67.85 - 31.75 \ln Q \quad R^2 = 0.988 \quad$ (7.40c)

These equations are valid for Q values range from 1.188 to 5.119 m³/d for used media.

7.12 LEAD TREATMENT

The influent lead was analyzed with both loading rate and effluent concentration. The variation of pollutant removal efficiency and both retention time and sewage loads were studied.

7.12.1 INLET AND OUTLET LEAD CONCENTRATIONS

Table 7.63 presents the average inlet and outlet Lead concentration and the corresponding removal efficiency for plastic, gravel, and rubber media at outlets. This average concentration and removal efficiency were calculated in Chapter 5.

The influent and effluent Lead concentrations at outlets are presented as clustered columns in Figure 7.76 for the used media.

Table 7.63 and Figure 7.76 exhibit that, Lead effluent concentration is in the permitted limit (less than 0.5 mg/l) of law No. 48 of 1982 (NAWQAM, 2002) at discharge lower than 4.814, 5.037, and 5.119 m^3/d (loading rate

TABLE 7.63 Lead Concentration and Removal Efficiency for Wetland Cells

Cycle No.	Influent (mg/l)	Effluent Concentration (mg/l)			Pb Removal Efficiency (%)		
		Rubber	Gravel	Plastic	Rubber	Gravel	Plastic
1	0.039	0.031	0.028	0.023	20.16	27.32	40.30
2	0.034	0.025	0.023	0.018	28.31	32.97	48.92
3	0.044	0.022	0.020	0.018	50.18	54.47	58.74
4	0.051	0.023	0.021	0.020	54.37	58.46	61.63
5	0.055	0.023	0.021	0.018	57.41	62.58	66.91

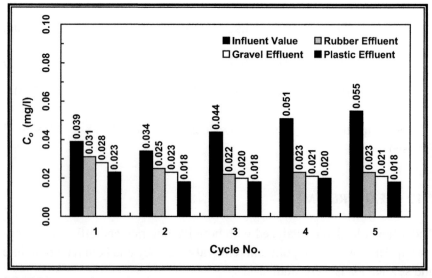

FIGURE 7.76 Influent and effluent Lead concentrations of used media for various cycles.

equals 24.07, 25.19, and 25.6 cm/d) for plastic, gravel, and rubber media at the end of wetland cells, respectively.

7.12.2 IMPACT OF Q ON LEAD TREATMENT

Table 7.64 gives the average influent and effluent Lead concentrations with discharge and loading rate at outlets.

Figure 7.77 represents the variation of the outlet Lead concentration and the values of loading rate at 10 m from inlet for plastic, gravel, and rubber cells.

From Table 7.64 and Figure 7.77, it is noticed that:

- The Lead effluent concentration decreases with the decrease of loading rate which means an improvement of the treatment.
- For cycle one (Q_{max} = 4.814, 5.037, and 5.119 m³/d), the Lead influent concentration is 0.039 mg/l and the q values are 24.07, 25.19, and 25.6 cm/d for plastic, gravel, and rubber cells with corresponding outlet concentrations of 0.023, 0.028, and 0.031 mg/l, respectively.
- For cycle five (Q_{min} = 1.188, 1.243, and 1.263 m³/d), the Lead influent concentration is 0.055 mg/l and the q values are 5.94, 6.22, and 6.32 cm/d for plastic, gravel, and rubber cells with corresponding effluent concentrations of 0.018, 0.021, and 0.023 mg/l, in the same order.
- At low loading rate, the differences between effluent Lead values for the three media are small and these differences increase at high values of q. Plastic cell gives Lead outlet concentration lower than other cells followed by gravel.

The Lead effluent concentration is directly in proportion to the loading rate according to a logarithmic function. The logarithmic equations at outlets are written as (q ranges from 5.94 to 25.6 cm/d):

Plastic: $C_o = 0.0137 + 0.0023 \ln q$ $R^2 = 0.353$ (7.41a)

Gravel: $C_o = 0.0109 + 0.0047 \ln q$ $R^2 = 0.664$ (7.41b)

Rubber: $C_o = 0.0115 + 0.0053 \ln q$ $R^2 = 0.656$ (7.41c)

7.12.3 IMPACT OF T_r ON LEAD REMOVAL EFFICIENCY

Table 7.65 exhibits the retention time and Lead removal efficiency at outlets. Figure 7.78 illustrates the relationship between the two parameters.

TABLE 7.64 Inlet and Outlet Lead Concentrations and q Values for Wetland Cells

C_i(mg/l)	Plastic Cell				Gravel Cell				Rubber Cell		
	Q(m³/d)	q(cm/d)	C_o(mg/l)	Q(m³/d)	q(cm/d)	C_o(mg/l)	Q(m³/d)	q(cm/d)	C_o(mg/l)		
0.039	4.814	24.07	0.023	5.037	25.19	0.028	5.119	25.60	0.031		
0.034	3.275	16.38	0.018	3.426	17.13	0.023	3.482	17.41	0.025		
0.044	2.253	11.27	0.018	2.357	11.79	0.020	2.396	11.98	0.022		
0.051	1.595	7.98	0.020	1.669	8.35	0.021	1.696	8.48	0.023		
0.055	1.188	5.94	0.018	1.243	6.22	0.021	1.263	6.32	0.023		

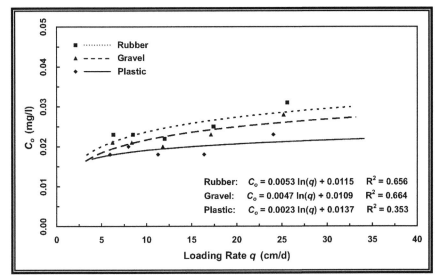

FIGURE 7.77 Relationship between C_o and hydraulic loading rate for Lead element.

TABLE 7.65 Lead Removal Efficiency and Retention Time for the Used Media

Cycle No.	Pb Removal Efficiency (%)			Hydraulic Retention Time (hr)		
	Rubber	Gravel	Plastic	Rubber	Gravel	Plastic
1	20.16	27.32	40.30	21.98	17.16	33.84
2	28.31	32.97	48.92	32.30	25.20	49.73
3	50.18	54.47	58.74	46.97	36.65	72.31
4	54.37	58.46	61.63	66.34	51.74	102.12
5	57.41	62.58	66.91	89.09	69.48	137.11

From Table 7.65 and Figure 7.78, it could be concluded that:

- The treatment improves with the increase of retention time. For example in the case of plastic cell at retention time equals 33.84 hr, the removal efficiency becomes 40.3%, while at T_r equals 137.11 hr, the removal efficiency equal to 66.91%.
- At retention time equal about 75 hr plastic and rubber cells give the same removal efficiency.

The Lead removal efficiency is directly proportional to the retention time according to a logarithmic function as the best-fit relationship. The logarithmic equations at outlets are as follows (discharge ranges from 1.188 to 5.119 m³/d):

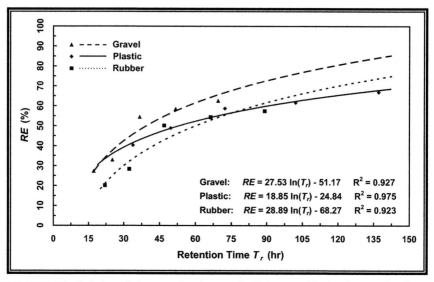

FIGURE 7.78 Relationship between Lead removal efficiency and hydraulic retention time.

$$\text{Plastic:} \quad RE = -24.84 + 18.85 \ln T_r \quad R^2 = 0.975 \quad (7.42a)$$

$$\text{Gravel:} \quad RE = -51.17 + 27.53 \ln T_r \quad R^2 = 0.927 \quad (7.42b)$$

$$\text{Rubber:} \quad RE = -68.27 + 28.89 \ln T_r \quad R^2 = 0.923 \quad (7.42c)$$

7.12.4 IMPACT OF DISCHARGE ON PB TREATMENT EFFICIENCY

Table 7.66 presents the discharges and the corresponding Lead removal efficiency at outlet. Figure 7.79 shows the relationship between these two parameters.

Table 7.66 and Figure 7.79, exhibit that:

- The plastic cell has the highest Lead removal efficiency. Gravel comes after and then rubber cell. At the highest Q equals 4.814, 5.037, and 5.119 m³/d, the Lead removal efficiency is about 40.3, 27.32, and 20.16% for plastic, gravel, and rubber cells. At the lowest Q equals 1.188, 1.243, and 1.263 m³/d these Lead removal efficiency is 66.91, 62.58, and 57.41%, respectively. The treatment performance of wetland system improves with the decrease of discharge.
- The Lead removal efficiency for plastic media increases by an average value of 8.14 and 13.21% higher than gravel and rubber media, in the

Steady Stage Analysis

TABLE 7.66 Lead Removal Efficiency and Discharge for the Three Media

Cycle No.	Pb Removal Efficiency (%)			Discharge (m³/d)		
	Rubber	Gravel	Plastic	Rubber	Gravel	Plastic
1	20.16	27.32	40.30	5.119	5.037	4.814
2	28.31	32.97	48.92	3.482	3.426	3.275
3	50.18	54.47	58.74	2.396	2.357	2.253
4	54.37	58.46	61.63	1.696	1.669	1.595
5	57.41	62.58	66.91	1.263	1.243	1.188

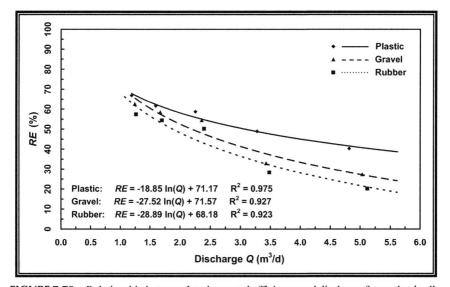

FIGURE 7.79 Relationship between Lead removal efficiency and discharge for wetland cells.

same sequence. Gravel cell gives Lead removal efficiency higher than rubber cell by an average value of 5.07%. These average *RE* differences are calculated in Appendix II.

The Lead removal efficiency is gradually reduced with the increase of discharge for the three media at outlets according to a logarithmic function. The logarithmic equations at outlets are as follows:

Plastic: $RE = 71.17 - 18.85 \ln Q$ $R^2 = 0.975$ (7.43a)

Gravel: $RE = 71.57 - 27.52 \ln Q$ $R^2 = 0.927$ (7.43b)

Rubber: $RE = 68.18 - 28.89 \ln Q$ $R^2 = 0.923$ (7.43c)

where:
 RE = Lead removal efficiency, %;
 Q = discharge to wetland cells, m³/d.

These equations are valid for Q ranges from 1.188 to 5.119 m³/d for the used media.

7.13 CADMIUM TREATMENT

The influent Cadmium concentration was analyzed with both loading rate and effluent concentration. The variation of pollutant removal efficiency and both retention time and discharge were studied.

7.13.1 INLET AND OUTLET CADMIUM CONCENTRATIONS

Table 7.67 presents the average influent and effluent Cadmium concentrations and the corresponding removal efficiency at outlets. These average values and removal efficiency were calculated in Chapter 5.

The influent and effluent Cadmium concentrations at outlets for steady stage cycles are presented as clustered columns in Figure 7.80.

Table 7.67 and Figure 7.80 declare that, Cadmium effluents concentration is in the allowable limit (less than 0.05 mg/l) of law No. 48 of 1982 (NAWQAM, 2002) at discharge lower than 4.814, 5.037, and 5.119 m³/d (loading rate equals 24.07, 25.19, and 25.6 cm/d) for plastic, gravel, and rubber media, respectively at the end of wetland cells.

7.13.2 IMPACT OF Q ON CADMIUM TREATMENT

Table 7.68 illustrates the average influent and effluent Cadmium concentrations, for the five cycles of the steady stage, with discharge and loading rate at outlets.

Figure 7.81 represents the variation of the outlet Cadmium concentration with the value of loading rate at outlets for plastic, gravel, and rubber media.

From Table 7.68 and Figure 7.81, it is noticed that:

- The Cadmium effluent concentration decreases with the decrease of loading rate, which means an improvement of the treatment performance for the three media at outlets.

Steady Stage Analysis

TABLE 7.67 Cadmium Concentration and Removal Efficiency for Wetland Cells

Cycle No.	Influent (µg/l)	Effluent Concentration (µg/l)			Cd Removal Efficiency (%)		
		Rubber	Gravel	Plastic	Rubber	Gravel	Plastic
1	2.32	1.91	1.78	1.64	17.76	23.35	29.25
2	2.31	1.83	1.66	1.56	20.81	28.38	32.29
3	2.44	1.86	1.72	1.61	24.13	29.57	34.18
4	2.54	1.69	1.61	1.43	33.45	36.70	43.83
5	1.71	1.09	0.84	0.76	36.40	50.81	55.73

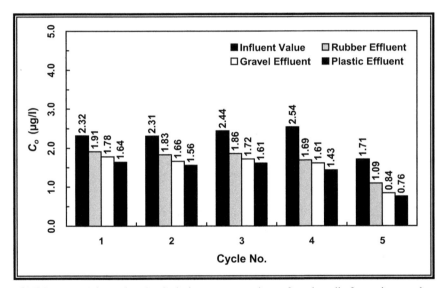

FIGURE 7.80 Inlet and outlet Cadmium concentrations of used media for various cycles.

- For cycle one (Q_{max} = 4.814, 5.037, and 5.119 m³/d), the Cadmium influent concentration is 2.32 µg/l and the loading rates are 24.07, 25.19, and 25.6 cm/d for plastic, gravel, and rubber cells with corresponding outlet concentrations of 1.64, 1.78, and 1.91 µg/l, in the same sequence.
- For cycle five (Q_{min} = 1.188, 1.243, and 1.263 m³/d), the Cadmium influent concentration is 1.71 µg/l and the q values are 5.94, 6.22, and 6.32 cm/d for plastic, gravel, and rubber cells with corresponding effluent concentrations of 0.76, 0.84, and 1.09 µg/l, respectively.
- At low loading rate, the difference between outlet Cadmium concentrations for plastic, and gravel media is small and this difference increases at high values of loading rate.

TABLE 7.68 Inlet and Outlet Cadmium Concentrations and q Values for Cells

$C_i(\mu g/l)$	Plastic Cell			Gravel Cell			Rubber Cell		
	$Q(m^3/d)$	$q(cm/d)$	$C_o(\mu g/l)$	$Q(m^3/d)$	$q(cm/d)$	$C_o(\mu g/l)$	$Q(m^3/d)$	$q(cm/d)$	$C_o(\mu g/l)$
2.32	4.814	24.07	1.64	5.037	25.19	1.78	5.119	25.60	1.91
2.31	3.275	16.38	1.56	3.426	17.13	1.66	3.482	17.41	1.83
2.44	2.253	11.27	1.61	2.357	11.79	1.72	2.396	11.98	1.86
2.54	1.595	7.98	1.43	1.669	8.35	1.61	1.696	8.48	1.69
1.71	1.188	5.94	0.76	1.243	6.22	0.84	1.263	6.32	1.09

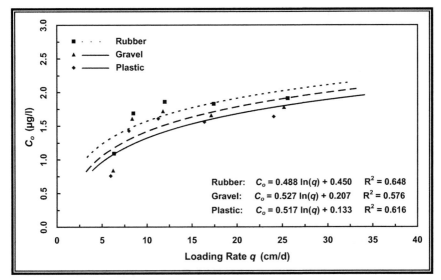

FIGURE 7.81 Relationship between C_o and hydraulic loading rate for Cadmium element.

- Plastic cell gives the lowest Cadmium outlet concentration followed by gravel cell.

The Cadmium effluent concentration is directly in proportion to the loading rate follows a logarithmic function which gives the best determination coefficient. The logarithmic equations at outlets are as (q values range from 5.94 to 25.6 cm/d):

$$\text{Plastic:} \quad C_o = 0.133 + 0.517 \ln q \quad R^2 = 0.616 \quad (7.44a)$$

$$\text{Gravel:} \quad C_o = 0.207 + 0.527 \ln q \quad R^2 = 0.576 \quad (7.44b)$$

$$\text{Rubber:} \quad C_o = 0.450 + 0.488 \ln q \quad R^2 = 0.648 \quad (7.44c)$$

where:
q = hydraulic loading rate, cm/d;
C_o = Cadmium outlet concentration, µg/l.

7.13.3 IMPACT OF T_r ON CADMIUM REMOVAL EFFICIENCY

Table 7.69 gives T_r and Cadmium removal efficiency at outlets.

Figure 7.82 shows the relationship between Cadmium removal efficiency and T_r at the end of cells for plastic, gravel, and rubber media.

TABLE 7.69 Cadmium Removal Efficiency and Retention Time for the Used Media

Cycle No.	Cd Removal Efficiency (%)			Hydraulic Retention Time (hr)		
	Rubber	Gravel	Plastic	Rubber	Gravel	Plastic
1	17.76	23.35	29.25	21.98	17.16	33.84
2	20.81	28.38	32.29	32.30	25.20	49.73
3	24.13	29.57	34.18	46.97	36.65	72.31
4	33.45	36.70	43.83	66.34	51.74	102.12
5	36.40	50.81	55.73	89.09	69.48	137.11

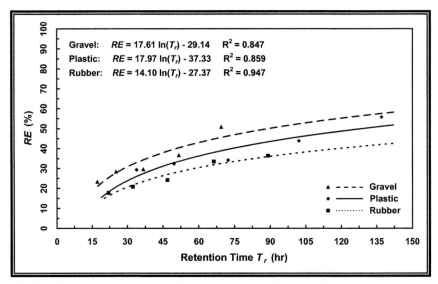

FIGURE 7.82 Relationship between Cadmium removal efficiency and hydraulic retention time.

From Table 7.69 and Figure 7.82, it is noticed that:

- The gravel cell gives the highest Cadmium removal efficiency followed by plastic.
- The Cadmium removal efficiency increases with the increase of retention time for the used media at outlets. The treatment improves with the retention time increase. For example for gravel cell at retention time equals 17.16 hr, the removal efficiency equal to 23.35%, while at T_r equals 69.48 hr, the *RE* becomes 50.81%.
- The difference between removal efficiency of Cadmium for plastic and rubber cells is small at low retention time and increases as T_r increases.

Steady Stage Analysis

The Cadmium RE is directly proportional to the retention time according to a logarithmic function. The logarithmic equations at outlets are (Q ranges from 1.188 to 5.119 m³/d):

$$\text{Plastic:} \quad RE = -37.33 + 17.97 \ln T_r \quad R^2 = 0.859 \quad (7.45a)$$

$$\text{Gravel:} \quad RE = -29.14 + 17.61 \ln T_r \quad R^2 = 0.847 \quad (7.45b)$$

$$\text{Rubber:} \quad RE = -27.37 + 14.10 \ln T_r \quad R^2 = 0.947 \quad (7.45c)$$

7.13.4 IMPACT OF DISCHARGE ON Cd TREATMENT EFFICIENCY

Table 7.70 presents the discharge and Cadmium removal efficiency. Figure 7.83 shows the relationship between these two parameters.

From Table 7.70 and Figure 7.83, it is remarked that:

- The plastic cell has higher Cadmium removal efficiency than the gravel which is in turn higher than the rubber.
- For plastic cell at Q_{max} and Q_{min} equal to 4.814 and 1.188 m³/d the removal efficiency is 29.25 and 55.73%, respectively.
- For gravel cell at Q_{max} equal to 5.037 m³/d the removal efficiency is 23.35% and for Q_{min} equal to 1.243 m³/d the removal efficiency is 50.81%.
- For rubber cell at Q_{max} equal to 5.119 m³/d the removal efficiency is 17.76% and for Q_{min} equal to 1.263 m³/d the removal efficiency is 36.4%. It could be concluded that the treatment performance of wetland system enhances with the decrease of Q.
- The Cadmium removal efficiency for plastic media increases by an average value of 5.29 and 12.55% higher than gravel and rubber media,

TABLE 7.70 Cadmium Removal Efficiency and Discharge for the Three Media

Cycle No.	Cd Removal Efficiency (%)			Discharge (m³/d)		
	Rubber	Gravel	Plastic	Rubber	Gravel	Plastic
1	17.76	23.35	29.25	5.119	5.037	4.814
2	20.81	28.38	32.29	3.482	3.426	3.275
3	24.13	29.57	34.18	2.396	2.357	2.253
4	33.45	36.70	43.83	1.696	1.669	1.595
5	36.40	50.81	55.73	1.263	1.243	1.188

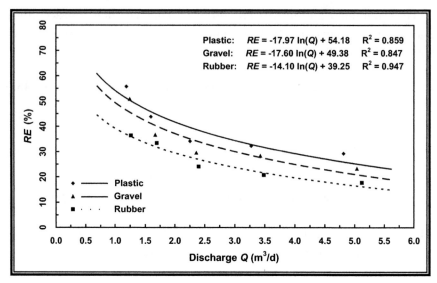

FIGURE 7.83 Relationship between Cadmium removal efficiency and discharge for wetland cells.

respectively. The gravel cell gives Cadmium removal efficiency higher than rubber cell by an average value of 7.25%, Appendix II.

The Cadmium removal efficiency is gradually reduced with the increase of discharge according to a logarithmic function. The logarithmic equations at outlets are as follows (Q ranges from 1.118 to 5.119 m³/d):

$$\textit{Plastic}: \quad RE = 54.18 - 17.97 \ln Q \quad R^2 = 0.859 \quad (7.46a)$$

$$\textit{Gravel}: \quad RE = 49.38 - 17.60 \ln Q \quad R^2 = 0.847 \quad (7.46b)$$

$$\textit{Rubber}: \quad RE = 39.25 - 14.10 \ln Q \quad R^2 = 0.947 \quad (7.46c)$$

7.14 REMOVAL RATE CONSTANTS FOR EXPERIMENTAL DATA

The estimation of both average removal (k) and volumetric removal (k_v) rate constants for the steady stage pollutants in subsurface flow wetlands system in Egypt (moderate conditions) are provided in this section. The first order plug flow kinetics (Eqs. 4.16 and 4.17) and modified plug flow (Eqs. 4.18 and 4.19) were applied to the experimental data. Also, the mixed flow models (Eqs. 4.20 and 4.21) and the modified mixed flow (Eqs. 4.22 and 4.23) were used also to these data.

Steady Stage Analysis 341

7.14.1 ESTIMATING K, K_V BY PLUG FLOW MODEL (BOD – COD – TSS)

Figures 7.84–7.86 illustrate the relationship between $ln(C_i/C_o)$ and $1/q$ for plastic, gravel, and rubber cells for BOD, COD, and TSS pollutants, respectively. Figures 7.87–7.89 show the relationship between $ln(C_i/C_o)$ and the retention time for these pollutants. Whereas Figures 7.90–7.92 give the relationship between $ln(C_i-C^*/C_o-C^*)$ and $1/q$ for BOD, COD, and TSS pollutants, respectively. Figures 7.93–7.95 exhibit the relationship between $ln(C_i-C^*/C_o-C^*)$ and the retention time for these pollutants.

Table 7.71 summarizes the average removal and volumetric removal rate constants for the three media applying plug flow models with zero background concentrations (C^*) for BOD, COD, and TSS pollutants. While Table 7.72 summarizes these average rate constants applying modified plug flow models with C^* values (Eqs. 4.24 and 4.25). The values of removal rate constants in Table 7.71 are determined from Figures 7.84–7.89, while the values in Table 7.72 are determined from Figures 7.90–7.95.

These k and k_v values of different pollutants are obtained from boundary conditions of inflow discharge varies between 1.188 and 5.119 m³/d and influent concentration of pollutant ranges from 162 to 209 mg/l for BOD, 242 to 337 mg/l for COD, and 131 to 180 mg/l for TSS. Omitting C^* values

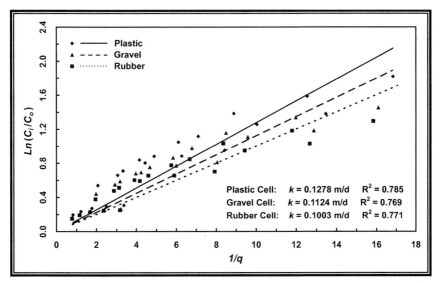

FIGURE 7.84 Average removal rate constants for BOD pollutant (plug flow model, Eq. 4.16).

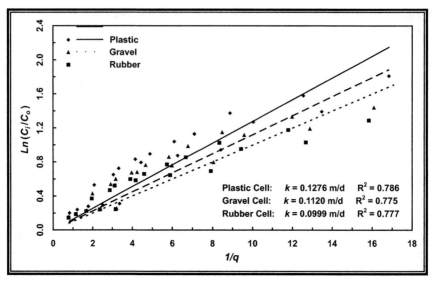

FIGURE 7.85 Average removal rate constants for COD pollutant (plug flow model, Eq. 4.16).

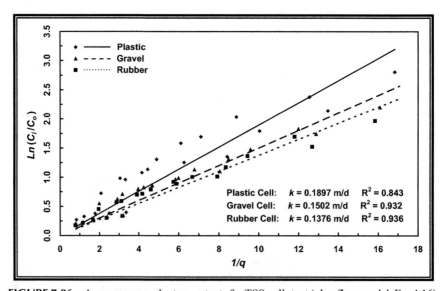

FIGURE 7.86 Average removal rate constants for TSS pollutant (plug flow model, Eq. 4.16).

in the plug flow models produce k and k_v values smaller than the models with C^* by:

- 19.41 & 19.48%; 15.93 & 15.91%; and 14.36 & 14.39% for plastic, gravel, and rubber cells, respectively for BOD.

Steady Stage Analysis

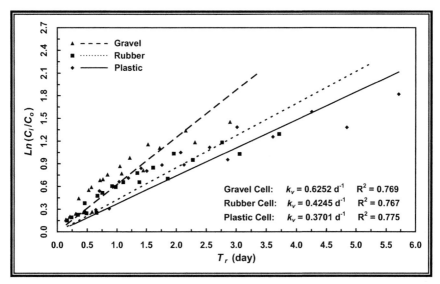

FIGURE 7.87 Average volumetric removal rate constants for BOD pollutant (plug flow model, Eq. 4.17).

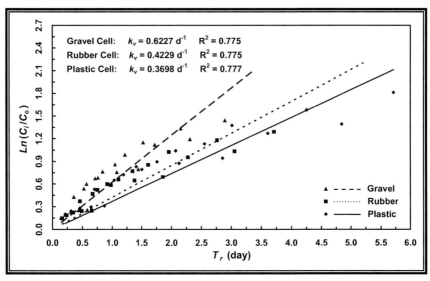

FIGURE 7.88 Average volumetric removal rate constants for COD pollutant (plug flow model, Eq. 4.17).

- 17.16 & 17.17%; 14.11 & 14.13%; and 12.81 & 12.82% for plastic, gravel, and rubber cells, respectively for COD.
- 39.27 & 40.06%; 69.04 & 69.03%; and 40.25 & 40.53% for plastic, gravel, and rubber cells, respectively for TSS.

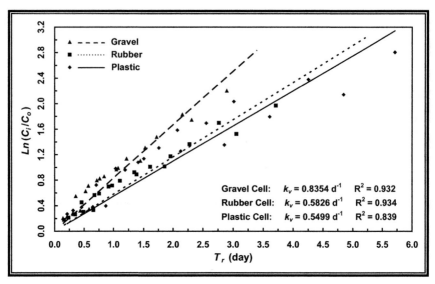

FIGURE 7.89 Average volumetric removal rate constants for TSS pollutant (plug flow model, Eq. 4.17).

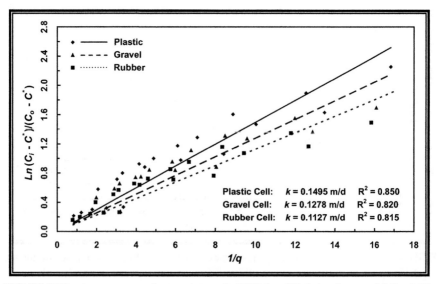

FIGURE 7.90 Average removal rate constants for BOD (modified plug flow model, Eq. 4.18).

Steady Stage Analysis

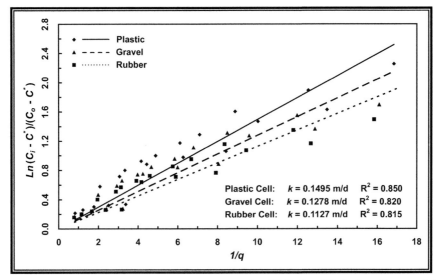

FIGURE 7.91 Average removal rate constants for COD (modified plug flow model, Eq. 4.18).

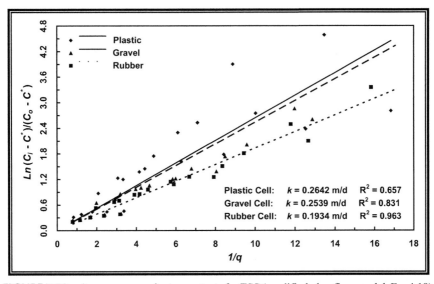

FIGURE 7.92 Average removal rate constants for TSS (modified plug flow model, Eq. 4.18).

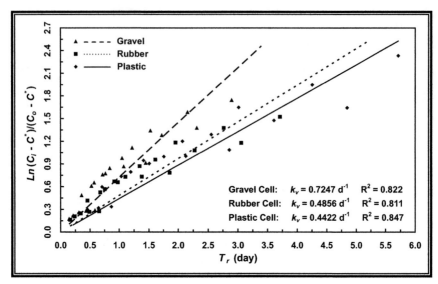

FIGURE 7.93 Average volumetric removal rate constants for BOD (modified plug flow, Eq. 4.19).

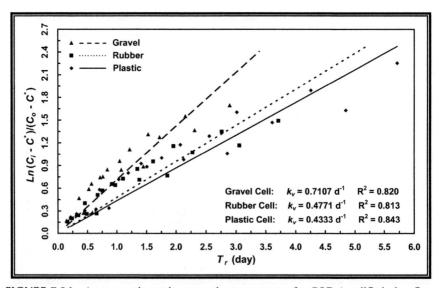

FIGURE 7.94 Average volumetric removal rate constants for COD (modified plug flow, Eq. 4.19).

Steady Stage Analysis

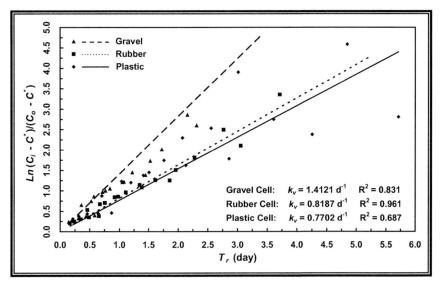

FIGURE 7.95 Average volumetric removal rate constants for TSS (modified plug flow, Eq. 4.19).

TABLE 7.71 Removal Rate Constants for BOD, COD, and TSS (Eqs. 4.16–4.17)

Pollutant	Plastic Cell		Gravel Cell		Rubber Cell	
	k(m/d)	kv(d^{-1})	k(m/d)	kv(d^{-1})	k(m/d)	kv(d^{-1})
BOD	0.1278	0.3701	0.1124	0.6252	0.1003	0.4245
COD	0.1276	0.3698	0.1120	0.6227	0.0999	0.4229
TSS	0.1897	0.5499	0.1502	0.8354	0.1379	0.5826

TABLE 7.72 Removal Rate Constants for BOD, COD, and TSS (Eqs. 4.18–4.19)

Pollutant	Plastic Cell		Gravel Cell		Rubber Cell	
	k(m/d)	kv(d^{-1})	k(m/d)	kv(d^{-1})	k(m/d)	kv(d^{-1})
BOD	0.1526	0.4422	0.1303	0.7247	0.1147	0.4856
COD	0.1495	0.4333	0.1278	0.7107	0.1127	0.4771
TSS	0.2642	0.7702	0.2539	1.4121	0.1934	0.8187

7.14.2 ESTIMATING K, K_V BY MIXED FLOW MODEL (BOD – COD – TSS)

Table 7.73 gives the average removal and volumetric rate constants by applying mixed flow models with C^* equal zero (Eqs. 4.20 and 4.21) for

TABLE 7.73 Removal Rate Constants for BOD, COD, and TSS (Eqs. 4.20–4.21)

Pollutant	Plastic Cell		Gravel Cell		Rubber Cell	
	k(m/d)	kv(d^{-1})	k(m/d)	kv(d^{-1})	k(m/d)	kv(d^{-1})
BOD	0.1803	0.5621	0.1567	0.870	0.1398	0.6118
COD	0.1806	0.5638	0.1544	0.8571	0.1382	0.6043
TSS	0.2693	0.8348	0.1980	1.0993	0.1779	0.7767

TABLE 7.74 Removal Rate Constants for BOD, COD, and TSS (Eqs. 4.22–4.23)

Pollutant	Plastic Cell		Gravel Cell		Rubber Cell	
	k(m/d)	kv(d^{-1})	k(m/d)	kv(d^{-1})	k(m/d)	kv(d^{-1})
BOD	0.2084	0.6473	0.1779	0.9876	0.1571	0.6867
COD	0.2056	0.6396	0.1730	0.9607	0.1535	0.6706
TSS	0.3891	1.1879	0.2885	1.6021	0.2307	1.0039

BOD, COD, and TSS pollutants. Table 7.74 gives these rate constants by applying modified mixed flow models with C^* values (Eqs. 4.22 and 4.23).

Neglecting C^* values in the mixed flow models, produces smaller k and k_v values than the corresponding ones obtained from these models which take C^* into consideration by:

- 15.59 & 15.16%; 13.53 & 13.52%; and 12.37 & 12.24% for plastic, gravel, and rubber cells, respectively for BOD.
- 13.84 & 13.44%; 12.05 & 12.09%; and 11.07 & 10.97% for plastic, gravel, and rubber cells, respectively for COD.
- 44.49 & 42.30%; 45.71 & 45.74%; and 29.68 & 29.25% for plastic, gravel, and rubber cells, respectively for TSS.

Generally the k and k_v values obtained from the mixed flow models are about 1.3 to 1.4 times the values given by the corresponding plug flow models for BOD, COD, and TSS pollutants. The plastic media produces higher k values followed by gravel then rubber media.

7.14.3 ESTIMATING K, K$_V$ FOR NH$_3$, PO$_4$, AND FC

Table 7.75 gives the calculated average removal and volumetric removal rate constants by applying plug flow models (Eqs. 4.16 and 4.17) for ammonia, phosphate, and fecal coliforms pollutants. Table 7.76 gives these removal rate constants by applying mixed flow models (Eqs. 4.20 and 4.21).

Steady Stage Analysis

TABLE 7.75 Removal Rate Constants for NH_3, PO_4, and FC (Eqs. 4.16–4.17)

Pollutant	Plastic Cell		Gravel Cell		Rubber Cell	
	k (m/d)	kv (d⁻¹)	k (m/d)	kv (d⁻¹)	k (m/d)	kv (d⁻¹)
Ammonia	0.1319	0.3886	0.1098	0.6101	0.1024	0.4367
Phosphate	0.1852	0.5455	0.1443	0.8016	0.1226	0.5228
Fecal Coliforms	0.5349	1.5759	0.5496	3.0530	0.5503	2.3471

TABLE 7.76 Removal Rate Constants for NH_3, PO_4, and FC (Eqs. 4.20–4.21)

Pollutant	Plastic Cell		Gravel Cell		Rubber Cell	
	k (m/d)	kv (d⁻¹)	k (m/d)	kv (d⁻¹)	k (m/d)	kv (d⁻¹)
Ammonia	0.1510	0.4449	0.1220	0.6778	0.1127	0.4805
Phosphate	0.2241	0.6603	0.1658	0.9210	0.1380	0.5887
Fecal Coliforms	0.9345	2.753	0.9492	5.2729	0.9420	4.018

These k and k_v values of different pollutants are obtained from boundary conditions of influent concentrations of pollutant ranges; which are 15.28 to 26.16 mg/l for ammonia, 2.05 to 3.54 mg/l for phosphate, and 200,000 to 270,000 MPN/100 ml for fecal coliforms. The k and k_v values obtained from the mixed flow models are bigger than those given by the corresponding plug flow models by about 10 to 70% for ammonia, phosphate, and fecal coliforms pollutants.

7.14.4 ESTIMATING K, K_V FOR HEAVY METALS

Figures 7.96–7.100 give the relationship between $ln(C_i/C_o)$ and $1/q$ to plastic, gravel, and rubber cells for zinc, iron, manganese, lead, and cadmium elements. Figures 7.101–7.105 show the relationship between $ln(C_i/C_o)$ and the retention time for these elements.

Table 7.77 summarizes the average removal and volumetric removal rate constants by applying the plug flow models for heavy metals. The values of average removal rate constants in this table are determined from Figures 7.96–7.100, while the values of average volumetric removal rate constants are shown in Figures 7.101–7.105.

Table 7.78 gives the calculated average removal and volumetric removal rate constants for plastic, gravel, and rubber media applying mixed flow models for heavy metals elements.

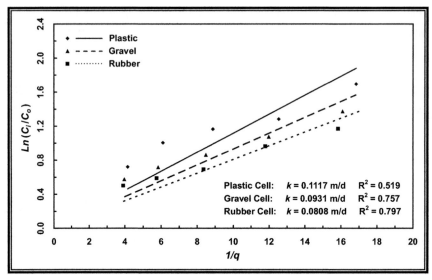

FIGURE 7.96 Average removal rate constants for treating Zn element (plug flow model, Eq. 4.16).

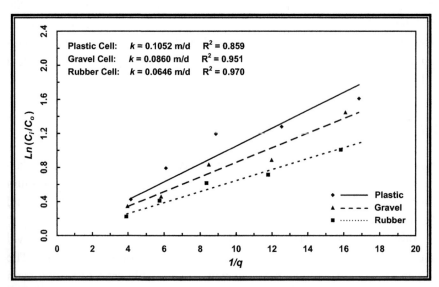

FIGURE 7.97 Average removal rate constants for treating Fe element (plug flow model, Eq. 4.16).

Steady Stage Analysis

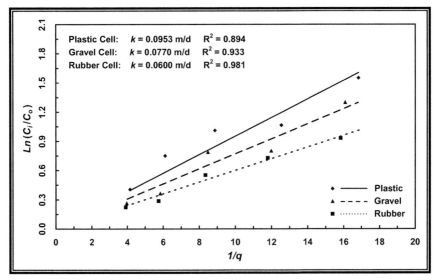

FIGURE 7.98 Average removal rate constants for treating Mn element (plug flow model, Eq. 4.16).

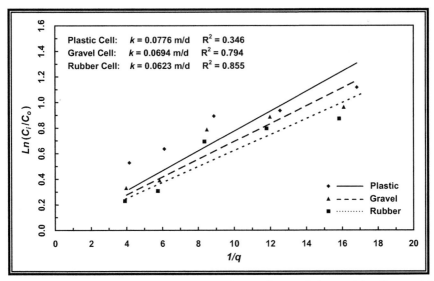

FIGURE 7.99 Average removal rate constants for treating Lead element (plug flow model, Eq. 4.16).

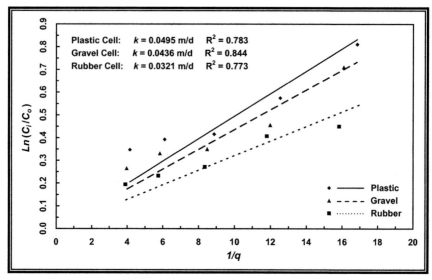

FIGURE 7.100 Average removal rate constants for treating Cadmium element (plug flow model, Eq. 4.16).

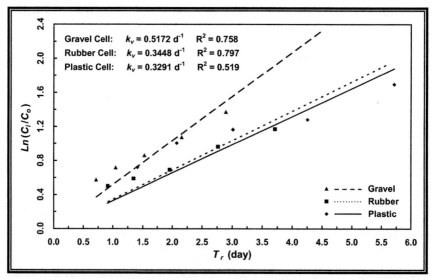

FIGURE 7.101 Average volumetric removal rate constants for treating Zn (plug flow model, Eq. 4.17).

Steady Stage Analysis

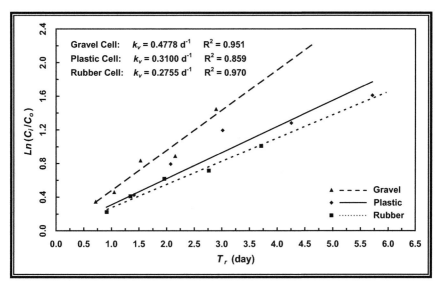

FIGURE 7.102 Average volumetric removal rate constants for treating Fe (plug flow model, Eq. 4.17).

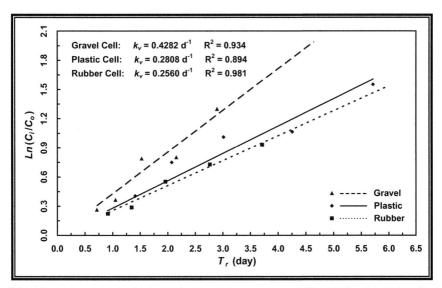

FIGURE 7.103 Average volumetric removal rate constants for treating Mn (plug flow model, Eq. 4.17).

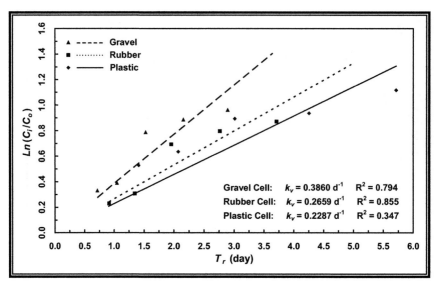

FIGURE 7.104 Average volumetric removal rate constants for treating Pb (plug flow model, Eq. 4.17).

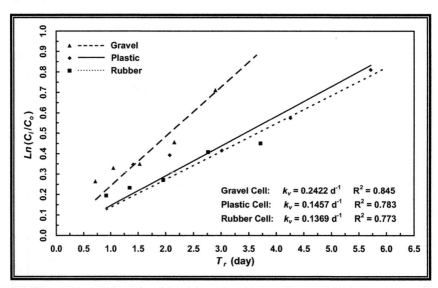

FIGURE 7.105 Average volumetric removal rate constants for treating Cd (plug flow model, Eq. 4.17).

These k and k_v values of different pollutants are obtained from boundary conditions of inlet concentrations of heavy metals having ranges of 1.28 to 2.16 mg/l for zinc, 0.63 to 1.19 mg/l for iron, 0.15 to 0.50 mg/l for manganese, 0.021 to 0.072 mg/l for lead, and 1.26 to 3.06 μg/l for cadmium.

TABLE 7.77 Removal Rate Constants for Heavy Metals Elements (Eqs. 4.16–4.17)

Parameter	Plastic Cell		Gravel Cell		Rubber Cell	
	k (m/d)	kv (d^{-1})	k (m/d)	kv (d^{-1})	k (m/d)	kv (d^{-1})
Zinc	0.1117	0.3291	0.0931	0.5172	0.0808	0.3448
Iron	0.1052	0.310	0.0860	0.4778	0.0646	0.2755
Manganese	0.0953	0.2808	0.0770	0.4282	0.0600	0.2560
Lead	0.0776	0.2287	0.0694	0.3860	0.0623	0.2659
Cadmium	0.0495	0.1457	0.0436	0.2422	0.0321	0.1369

TABLE 7.78 Removal Rate Constants for Heavy Metals Elements (Eqs. 4.20–4.21)

Parameter	Plastic Cell		Gravel Cell		Rubber Cell	
	k (m/d)	kv (d^{-1})	k (m/d)	kv (d^{-1})	k (m/d)	kv (d^{-1})
Zinc	0.1552	0.4571	0.1219	0.6769	0.1033	0.4405
Iron	0.1294	0.3811	0.0951	0.5285	0.0706	0.3011
Manganese	0.1155	0.3402	0.0814	0.4522	0.0631	0.2692
Lead	0.1045	0.3078	0.0822	0.4564	0.0686	0.2927
Cadmium	0.0614	0.1808	0.052	0.2886	0.0386	0.1646

The k and k_v values obtained from the mixed flow models are bigger than those given by the plug flow models by about 10 to 30% for heavy metals pollutants. The plastic cell produces higher k values followed by gravel cell.

KEYWORDS

- plastic media
- porosity
- removal rate constants
- rubber media
- steady stage
- treatment efficiency

CHAPTER 8

ANNs MODELING AND SPSS ANALYSIS

CONTENTS

8.1 Introduction 357
8.2 ANNs Modeling for Set Up Stage 358
8.3 ANNs Modeling – Steady Stage (BOD – COD – TSS) 374
8.4 ANNs Modeling – Steady Stage (NH_3 – PO_4 – DO – FC) 388
8.5 ANNs Modeling – Steady Stage (Heavy Metals) 407
8.6 Statistical Modeling for Set Up Stage 430
8.7 Statistical Modeling – Steady (BOD – COD – TSS) 438
8.8 Statistical Modeling – Steady (NH_3 – PO_4 – DO – FC) 447
8.9 Statistical Modeling – Steady (Heavy Metals) 457
8.10 Limits of Regression Equations 470
8.11 Comparison Between (Exp. – ANN – SPSS) Results 470
8.12 Significance TESTs 511
Keywords 514

8.1 INTRODUCTION

Horizontal subsurface flow constructed wetlands have a variety of complex and interrelated physical, chemical, and biological processes, so the mathematical representation for these processes is difficult. For this reason it was decided to use artificial neural networks (ANNs) and statistical analysis (SPSS) in this study. This chapter is divided into three main parts as follows:

Parts of this chapter have been reprinted from Zidan, A. A., Rashed, A. A., Hatata, A. Y., and Abdel Hadi, M. A., "Artificial Neural Networks to Predict Wastewater Treatment in Different Meida HSSG Constructed Wetlands," International Water Technology Journal, Vol. 5, No. 1, pp. 32-42, March 2015. Reprinted with permission.

- Artificial neural networks were used for modeling the input variables and forecast the output concentrations (the design and training ANNs models "3 Matlab programs" for data of set up and steady stages).
- Statistical analysis for the two stages was modeled using stochastic package for social science (SPSS software version 17). Several regression equations (linear and nonlinear) were tested and the best ones were chosen. The best equation which gave the good convergence of the target about the line of perfect agreement was applied.
- A comparison between the experimental measured data and both ANNs and SPSS results are presented and the best modeling is chosen to represent these data.

The input variables for the models are influent concentration (C_i), loading rate (q), media surface area (A_s), actual velocity (v), and time from start of operation (T_o) in set up stage and C_i, q, A_s, and v in steady stage. The output result for all models is the effluent concentration (C_o).

8.2 ANNs MODELING FOR SET UP STAGE

It is important to model the pollutants in set up stage as the porosity of all cells decreases with time from start of operation and the wetland plants and attached biofilm are growing. This stage is one of the main wetland system operation processes that has little research in literature. In this section BOD, COD, and TSS pollutants are modeled. A general ANN program was designed and built in Matlab to represent each pollutant in the set up stage, Program No.1, Appendix III.

8.2.1 ANNs STRUCTURE

Different networks with one and two hidden layers were considered and their performance was evaluated. It was found that the networks with reasonable number of neurons in one hidden layer cannot cover all used data. On the other hand, networks with two hidden layers provided better results without having high number of neurons in their hidden layers. The number of neurons in hidden layers was taken as two or three and was increased till desired results were obtained by testing the neural network. Various transfer functions were tested and the functions that had the ability to represent the data were chosen. *Tansig* or *logsig* transfer functions were used between input and first hidden layer (FHL), also between FHL and second hidden layer (SHL). Whereas the linear transfer function was used for SHL and output layers.

The software used for implementing the algorithm is the Matlab neural network toolbox. The Matlab programs and the list of orders and definitions of the transfer functions used in these programs, in addition to the weight matrixes and biases vectors for the selected networks are listed in Appendix III (Demuth et al., 2009). In this study, many different neural network structures having 5-input variables and one output value for set up stage were designed. Two hidden layers with different number of neurons were considered and trained. Some of the multifeed forward neural networks (MFFNNs) were tested, and a comparison between these different networks is listed in Tables 8.1–8.3 for pollutants under study.

The network structure 5–5–3–1 means 5 input variables, 5 neurons in the first hidden layer, 3 neurons in the second hidden layers, and only one pollutant concentration in the output layer. The number of epoch 106 means that the program performs 106 trials to reach the best mean square error (Table 8.1). The gradient and the momentum factor (Mu) are the values by which weights modified each iteration. The networks shown in Figures 8.1 and 8.2 exhibit satisfactory results having a structure of (5–5–3–1), (5–4–3–1), and (5–5–3–1) for BOD, COD, and TSS pollutants in set up stage, respectively.

TABLE 8.1 Comparison Between Different Tested Networks for BOD (Set Up Stage)

Network Structure	Number of Epoch	Mean Square Error	Gradient	Mu
5–3–3–1	41	3.50	165	2.11
5–4–3–1	34	3.99	55.8	0.452
5–5–3–1	106	0.0883	29.1	0.0175
5–5–4–1	59	1.24	37.6	0.374

TABLE 8.2 Comparison Between Different Tested Networks for COD (Set Up Stage)

Network Structure	Number of Epoch	Mean Square Error	Gradient	Mu
5–3–2–1	42	6.09	195	3.06
5–3–3–1	49	1.74	79.0	1.04
5–4–3–1	81	0.653	92.3	0.492
5–4–4–1	50	1.11	284	0.526
5–5–3–1	40	1.11	96.8	0.302

TABLE 8.3 Comparison Between Different Tested Networks for TSS (Set Up Stage)

Network Structure	Number of Epoch	Mean Square Error	Gradient	Mu
5–3–2–1	104	0.404	25.6	5.88
5–3–3–1	103	0.383	56.2	1.09
5–4–3–1	46	0.330	8.52	0.120
5–5–3–1	58	0.0572	12.4	0.00505
5–5–5–1	45	0.190	1.85	0.0132

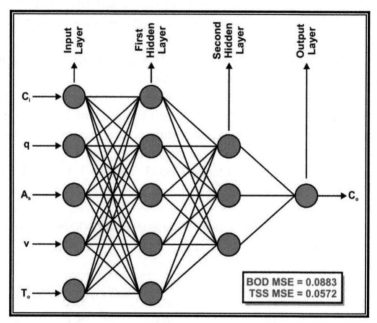

FIGURE 8.1 Optimum structure of ANNs for BOD and TSS.

The developed ANNs models are capable to minimize the mean square error (MSE) to the final values of 0.0883, 0.653, and 0.0572, for BOD, COD, and TSS pollutants, respectively. The corresponding neural network training tools for these pollutants are presented in Figures 8.3–8.5.

These networks use Marquardt-Levenberg algorithm for training. The algorithm randomly divides input and target vectors into three sets, 60% for training, 20% for validate the network, and 20% for test the network.

During training, the Figures 8.3–8.5 display the training progress and allow for interrupting this progress at any point by clicking stop

ANNs Modeling and SPSS Analysis

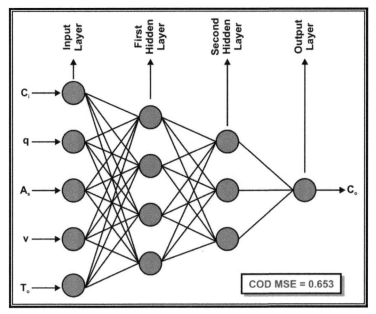

FIGURE 8.2 Optimum model structure of ANN for COD.

training button. The performance click in the training window, gives a plot of the training errors, validation errors, and test errors appears, as shown in Figures 8.6–8.8 for BOD, COD, and TSS, respectively (the mean square error training convergence diagrams for the selected ANNs using "trainlm" training function).

Regression click in the training window performs some analysis of the selected network response as the linear regression between the network outputs and the corresponding targets. If even more accurate results were required, the following approaches could be tried by modifying:

- The number of neurons in hidden layers.
- The number of training vectors.
- The number of input values, if more relevant information is available.

Figures 8.6–8.8 show reasonable results because of the following considerations:

- The final mean square error is small and; the test and validation set errors have similar characteristics.
- No significant over fitting has occurred by iterations 86, 61, and 38 (where the best validation performance occurs). The training procedure

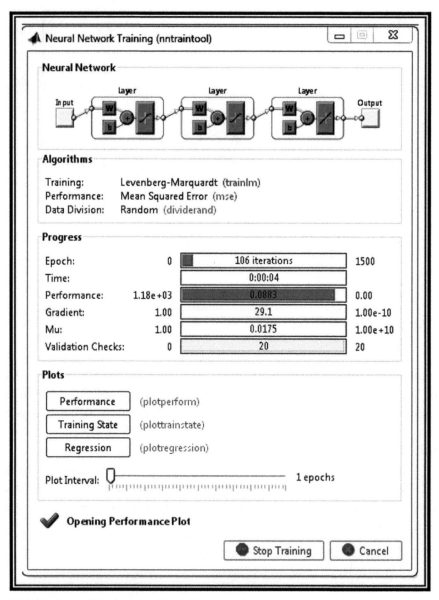

FIGURE 8.3 ANN training tool for BOD (printed screen – set up stage).

stops after an additional number of 20 iterations in order to ensure that the minimum value of square error (*MSE*) has been reached. In this case the total number of iterations is 106, 81, and 58, for BOD, COD, and TSS, respectively.

ANNs Modeling and SPSS Analysis

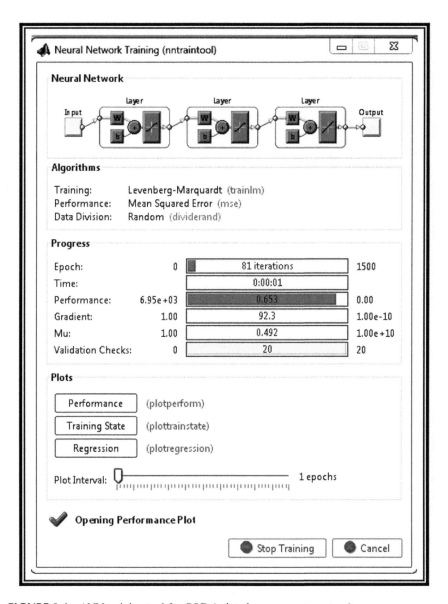

FIGURE 8.4 ANN training tool for COD (printed screen – set up stage).

8.2.2 CALIBRATION PROCESS FOR NETWORKS

About 120 patterns for input variables were used to calibrate the designed networks for BOD, COD, and TSS pollutants in set up stage. The training performance which means the determination of difference between the experimental

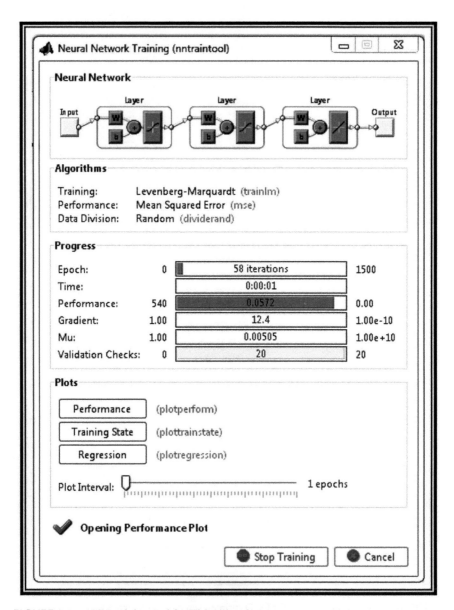

FIGURE 8.5 ANN training tool for TSS (printed screen – set up stage).

and the artificial neural network output concentrations. This performance and error values for these pollutants are shown in Figures 8.9–8.11. The error and the percentage error are computed using the following formulae:

ANNs Modeling and SPSS Analysis

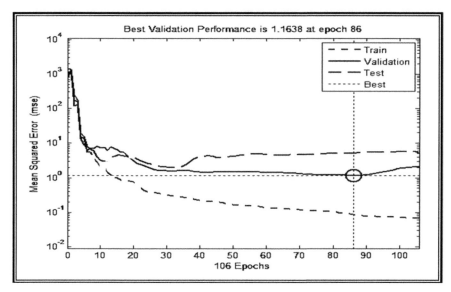

FIGURE 8.6 MSE training convergence of the ANN (BOD – set up stage – printed screen).

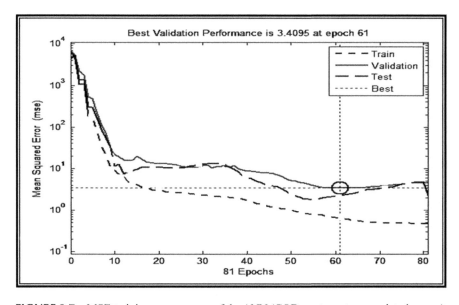

FIGURE 8.7 MSE training convergence of the ANN (COD – set up stage – printed screen).

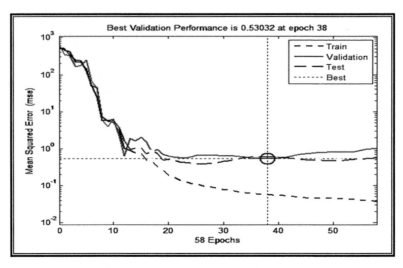

FIGURE 8.8 MSE training convergence of the ANN (TSS – set up stage – printed screen).

$$Error = C_{Exp} - C_{ANN} \qquad (4.30)$$

$$E_n = \left(\frac{C_{Exp} - C_{ANN}}{C_{Exp}}\right) \times 100 \qquad (4.31)$$

where: E_n = artificial neural network percentage error, %; C_{Exp} = experimental measured output concentration, mg/l; C_{ANN} = artificial neural network output concentration, mg/l.

From the analysis of results and referring to Figures 8.9–8.11, it is noticed that, the average error between the experimental effluent concentration and the ANNs model output for 120 patterns varies between −0.44 and +0.88 mg/l for BOD. For COD the average error ranges from −0.92 to +0.93 mg/l, whereas, for TSS the average error varies between −0.36 and +0.34 mg/l.

Figures 8.12–8.14 show the equality diagrams of the 120 patterns for BOD, COD, and TSS pollutants, respectively. The dotted line in these diagrams represents the line of perfect agreement (the slope of the line is 45 degree). All the equality diagrams in this chapter declare the number of patterns (points) that have a percentage error (E_n) less than ±5%, between ±5 and ±10%, and more than ±10%.

ANNs Modeling and SPSS Analysis

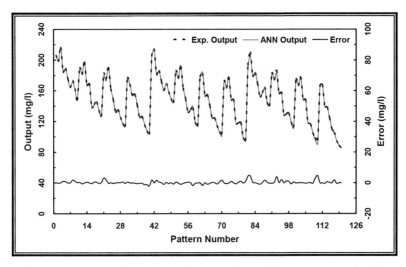

FIGURE 8.9 Training performance and error values of 120 patterns for BOD outputs (Exp. and ANN).

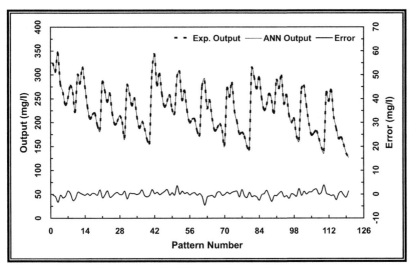

FIGURE 8.10 Training performance and error values of 120 patterns for COD outputs (Exp. and ANN).

The Figures 8.12–8.14 demonstrate accurate results which are close to the experimental values. All percentage errors between experimental and ANN model outputs for pollutants under study are less than ±5% for 120 patterns (calibration process).

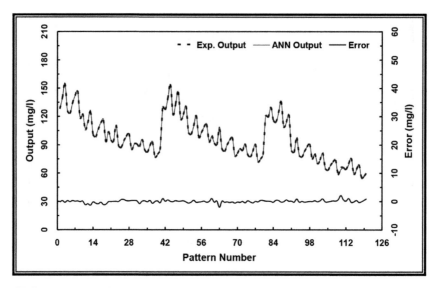

FIGURE 8.11 Training performance and error values of 120 patterns for TSS outputs (Exp. and ANN).

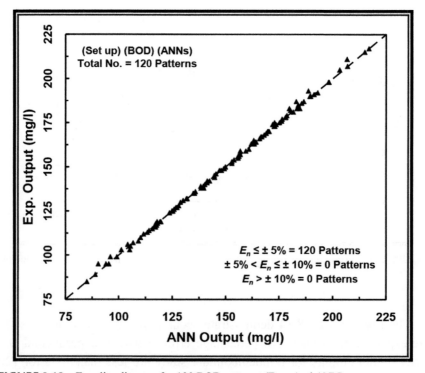

FIGURE 8.12 Equality diagram for 120 BOD patterns (Exp. And ANN).

ANNs Modeling and SPSS Analysis 369

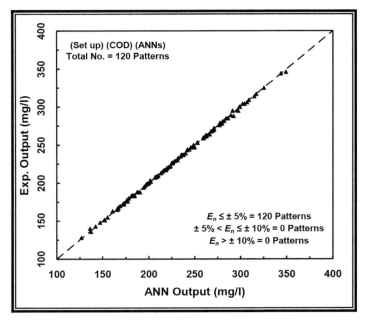

FIGURE 8.13 Equality diagram for 120 COD patterns (Exp. and ANN).

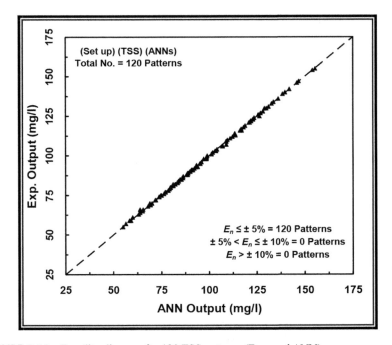

FIGURE 8.14 Equality diagram for 120 TSS patterns (Exp. and ANN).

8.2.3 VALIDATION PROCESS FOR NETWORKS

About 24 random patterns for input variables were used to validate the designed networks for the studied pollutants. These random 24 patterns for input variables were entered to the Matlab Program No.1 for set up stage and the program was run with the best structure of each pollutant (5–3 for BOD, 4–3 for COD, and 5–3 for TSS as the number of neurons in the first and second hidden layers) then the model output was obtained.

The validation performance and error values of 24 patterns for BOD, COD, and TSS pollutants are illustrated in Figures 8.15–8.17.

From the analysis of results and referring to Figures 8.15–8.17, it can be observed that, the average error between the experimental and the ANN model output concentrations for 24 patterns varies between:

- −0.87 and +1.34 mg/l for BOD pollutant.
- −2.02 and +1.60 mg/l for COD pollutant.
- −0.61 and +0.53 mg/l for TSS pollutant.

Figures 8.18–8.20 show the equality diagrams of 24 patterns for BOD, COD, and TSS pollutants, respectively. The 1:1 dotted line in these diagrams represents the line of perfect agreement.

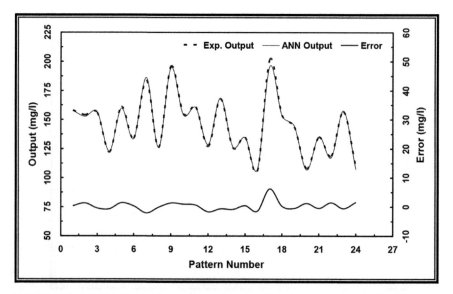

FIGURE 8.15 Validation performance and error values of 24 patterns for BOD outputs (Exp. and ANN).

ANNs Modeling and SPSS Analysis

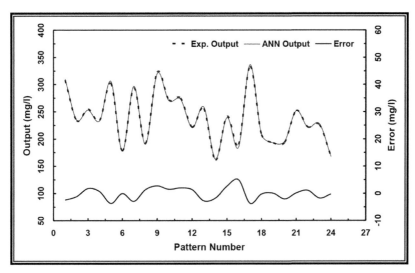

FIGURE 8.16 Validation performance and error values of 24 patterns for COD outputs (Exp. and ANN).

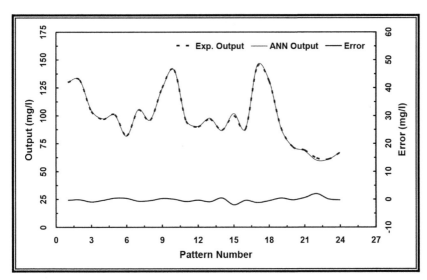

FIGURE 8.17 Validation performance and error values of 24 patterns for TSS outputs (Exp. and ANN).

From Figures 8.20–8.22, it can be noticed that:
- All percentage errors between experimental and ANN outputs for BOD, COD, and TSS pollutants are less than ±5% for the 24 patterns which are used in validation process of set up stage.

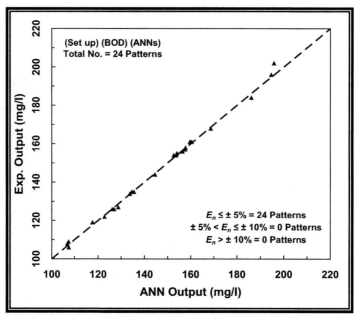

FIGURE 8.18 Equality diagram for 24 BOD patterns (Exp. And ANN).

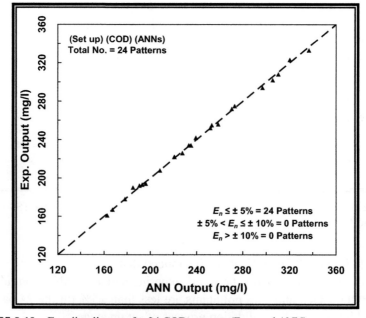

FIGURE 8.19 Equality diagram for 24 COD patterns (Exp. and ANN).

ANNs Modeling and SPSS Analysis

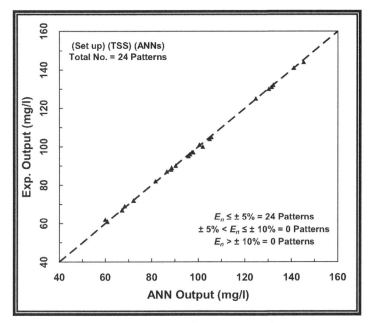

FIGURE 8.20 Equality diagram for 24 TSS patterns (Exp. and ANN).

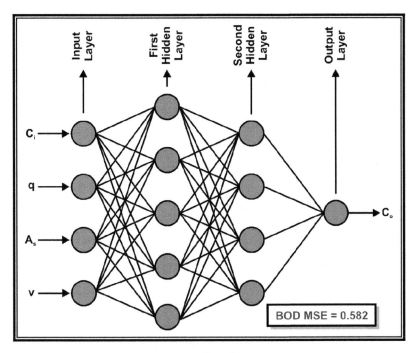

FIGURE 8.21 Optimum structure of ANN for BOD.

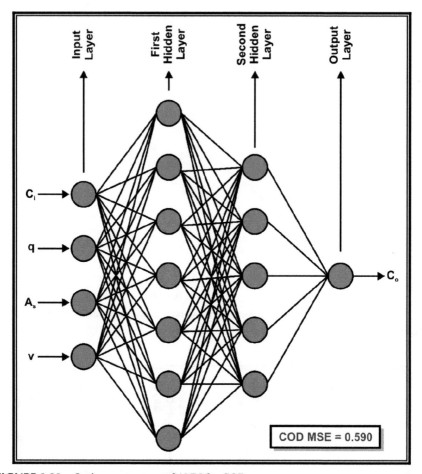

FIGURE 8.22 Optimum structure of ANN for COD.

- The model results are very close to the experimental output concentrations for TSS pollutant.
- For BOD and COD, the model results are matching with the experimental effluents value with the few peculiar points.

8.3 ANNs MODELING – STEADY STAGE (BOD – COD – TSS)

It is important to model the pollutants in the steady stage as media porosity, plants, and attached biofilm are reached the steady state. Through this stage, behavior of the horizontal subsurface flow constructed wetland systems treatment performance is studied, simulated, and determined. In this section BOD, COD, and TSS pollutants in steady stage are modeled. A general ANN program was designed and built in Matlab to represent each pollutant in this stage (Program No.2).

8.3.1 STRUCTURE OF ANNS

Many different neural networks structure having 4-input variables (C_i, q, A_s, and v) and one output value (C_o) for steady stage was designed. Two hidden layers with different number of neurons were considered and trained. Some of the MFFNNs were tested, and the comparison between these different networks is listed in Tables 8.4–8.6 for studied pollutants. The weight matrixes and biases vectors for the selected networks are listed in Appendix III.

TABLE 8.4 Comparison Between Different Tested Networks for BOD (Steady Stage)

Network Structure	Number of Epoch	Mean Square Error	Gradient	Mu
4–3–3–1	236	2.99	26.2	0.160
4–4–3–1	320	2.07	23.4	0.000260
4–4–4–1	53	3.09	209	13.0
4–5–3–1	119	0.880	38.1	0.445
4–5–4–1	77	0.582	10.7	0.145
4–5–5–1	42	3.03	60.4	8.74

TABLE 8.5 Comparison Between Different Tested Networks for COD (Steady Stage)

Network Structure	Number of Epoch	Mean Square Error	Gradient	Mu
4–3–3–1	53	7.96	0.000638	1.16E-10
4–4–3–1	89	4.76	8.68	20.9
4–5–4–1	147	2.99	216	0.0811
4–6–5–1	54	0.879	96.4	1.78
4–7–5–1	103	0.590	78.8	0.226
4–7–6–1	61	1.89	221	2.91

TABLE 8.6 Comparison Between Different Tested Networks for TSS (Steady Stage)

Network Structure	Number of Epoch	Mean Square Error	Gradient	Mu
4–3–3–1	222	4.60	0.0908	0.00433
4–4–3–1	47	2.19	54.1	2.39
4–4–4–1	121	1.28	61.5	2.05
4–5–3–1	102	1.76	158	0.981
4–5–4–1	546	0.962	1.29	0.233
4–6–5–1	60	0.347	0.750	2.01
4–7–5–1	52	1.18	130	0.502

The artificial neural networks which show satisfactory results for BOD, COD, and TSS pollutants having a structure of (4–5–4–1), (4–7–5–1), and (4–6–5–1), respectively. Figures 8.21–8.23 illustrate these networks structures for the studied pollutants in steady stage. The best network which gives the minimum mean square error and also minimum percentage error between experimental and ANNs outputs is chosen.

The developed ANNs models are capable to minimize the MSE to final values of 0.582, 0.590, and 0.347, for BOD, COD, and TSS, respectively. The corresponding neural network training tools for these pollutants are presented in Figures 8.24–8.26.

The mean square error training convergence diagrams for the selected ANNs using "trainlm" training function are shown in Figures 8.27–8.29.

From Figures 8.27–8.29, it is obtained that, the best validation performance occurs by a number of iterations 57, 83, and 40 for BOD, COD, and TSS pollutants in steady stage, respectively. The training stops when the validation error increases by additional 20 iterations, which occurs at a number of iteration 77, 103, and 60 for the studied pollutants.

8.3.2 CALIBRATION PROCESS FOR NETWORKS

A number of 240 patterns for input variables were used to calibrate the designed networks for the studied pollutants. The training performance and error values for BOD, COD, and TSS pollutants are illustrated in Figures 8.30–8.32, respectively.

From the analysis of results and referring to Figures 8.30–8.32, it is obtained that, the error between the experimental effluent concentration and

ANNs Modeling and SPSS Analysis

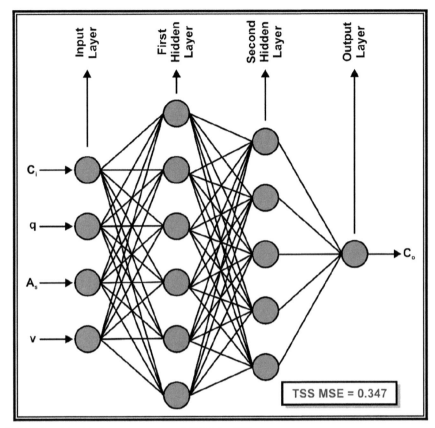

FIGURE 8.23 Optimum structure of ANN for TSS.

the ANNs model outputs varies between −0.85 and +0.66 mg/l for BOD pollutant. For COD this error ranges from −1.08 to +1.06 mg/l, whereas, for TSS the error varies between −0.63 and +0.84 mg/l. Figures 8.33–8.35 show the corresponding equality diagrams for these pollutants.

From Figures 8.33–8.35, it can be observed that, the results of calibration process are very encouraging and match accurately with the target values. Few peculiar points are observed in these equality diagrams. For BOD pollutant 236 points give percentage error (E_n) less than ±5% and 4 points gave E_n between ±5 and ±10%. For COD pollutant 238 points give E_n less than ±5% and 2 points give E_n between ±5 and ±10%. For TSS pollutant 227 points give E_n less than ±5% and 11 points give E_n between ±5 and ±10%, and 2 points give E_n more than ±10%.

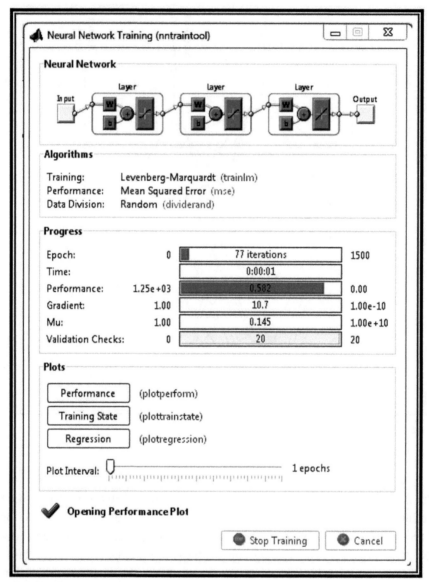

FIGURE 8.24 ANN training tool for BOD (printed screen – steady stage).

8.3.3 VALIDATION PROCESS FOR NETWORKS

The validation performance and error values for steady stage of the proposed artificial neural networks models for 60 patterns are presented in Figures 8.36–8.38 for BOD, COD, and TSS pollutants, respectively.

ANNs Modeling and SPSS Analysis

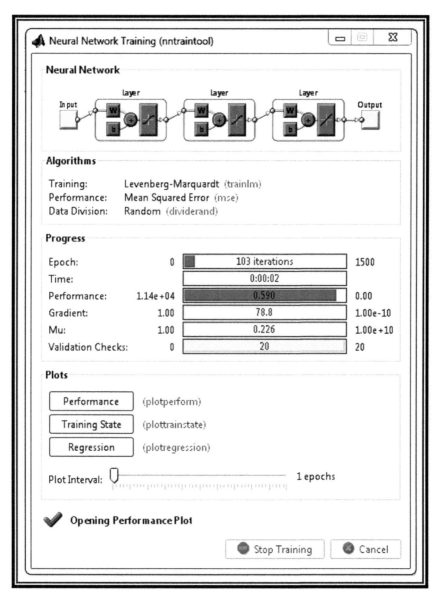

FIGURE 8.25 ANN training tool for COD (printed screen – steady stage).

From the analysis of results and referring to Figures 8.36–8.38, it can be observed that, the average error between the experimental and the ANN model output concentrations for 60 patterns varies between:

- −0.81 and +0.80 mg/l for BOD pollutant.

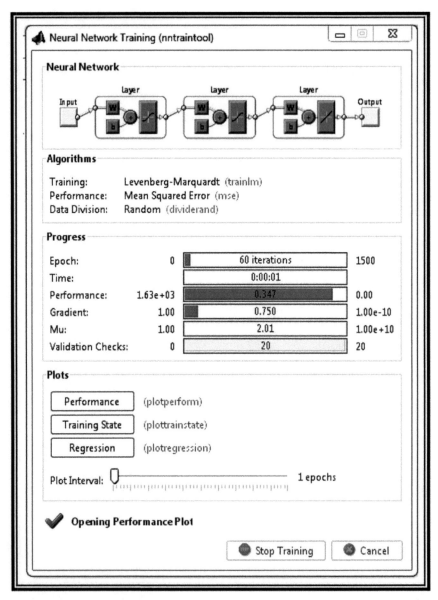

FIGURE 8.26 ANN training tool for TSS (printed screen – steady stage).

- −1.60 and +1.61 mg/l for COD pollutant.
- −0.70 and +0.99 mg/l for TSS pollutant.

Figures 8.39–8.41 present the equality diagrams of 60 patterns for BOD, COD, and TSS pollutants, respectively. The dotted line in these diagrams represents the line of perfect agreement.

ANNs Modeling and SPSS Analysis

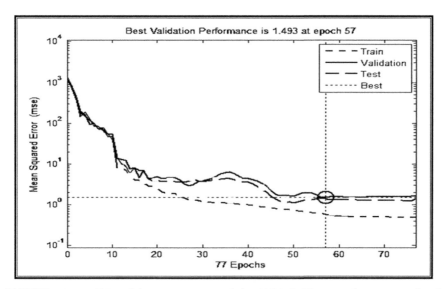

FIGURE 8.27 MSE training convergence of the ANN (BOD – steady stage – printed screen).

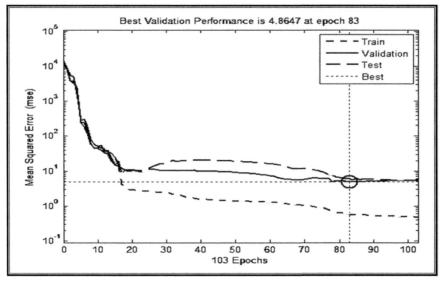

FIGURE 8.28 MSE training convergence of the ANN (COD – steady stage – printed screen).

From Figures 8.39–8.41, it can be noticed that:

- The model results are very close to the experimental output concentration for BOD and COD pollutants.

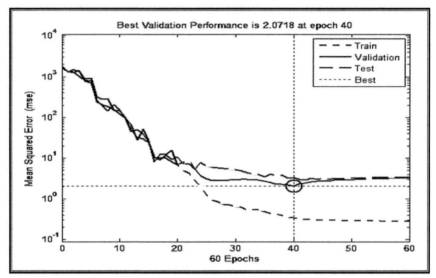

FIGURE 8.29 MSE training convergence of the ANN (TSS – steady stage – printed screen).

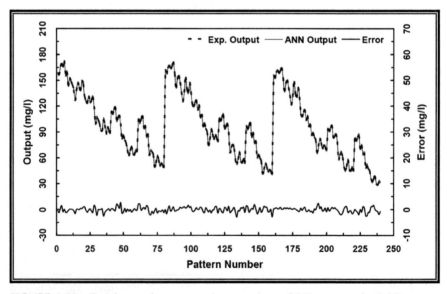

FIGURE 8.30 Training performance and error values of 240 patterns for BOD outputs (Exp. and ANN).

ANNs Modeling and SPSS Analysis

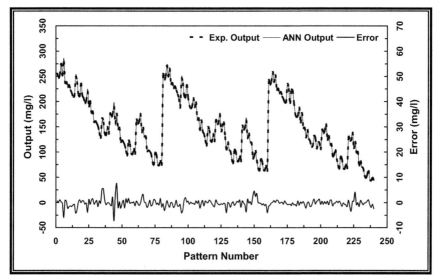

FIGURE 8.31 Training performance and error values of 240 patterns for COD outputs (Exp. and ANN).

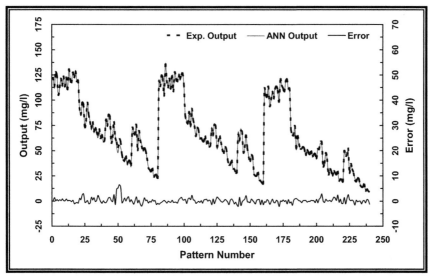

FIGURE 8.32 Training performance and error values of 240 patterns for TSS outputs (Exp. and ANN).

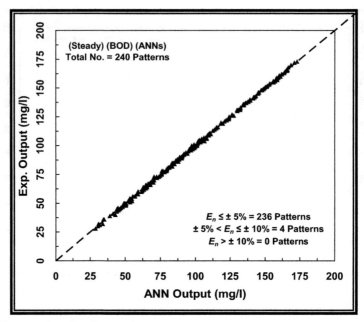

FIGURE 8.33 Equality diagram for 240 BOD patterns (Exp. and ANN).

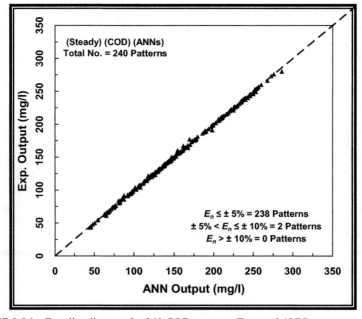

FIGURE 8.34 Equality diagram for 240 COD patterns (Exp. and ANN).

ANNs Modeling and SPSS Analysis

385

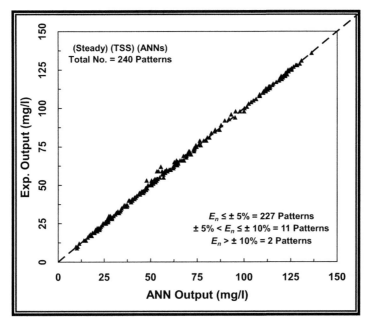

FIGURE 8.35 Equality diagram for 240 TSS patterns (Exp. and ANN).

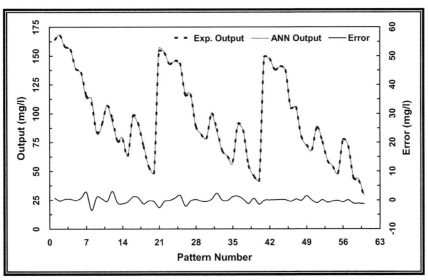

FIGURE 8.36 Validation performance and error values of 60 patterns for BOD outputs (Exp. and ANN).

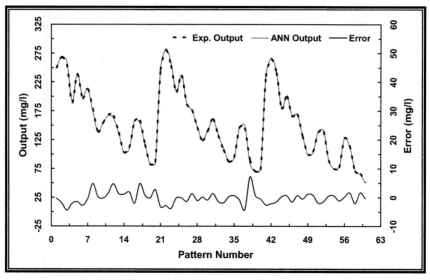

FIGURE 8.37 Validation performance and error values of 60 patterns for COD outputs (Exp. and ANN).

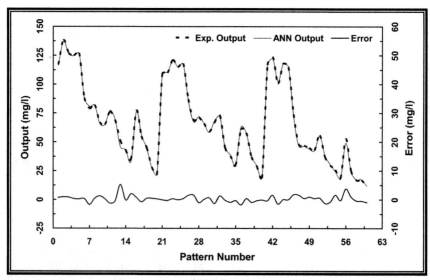

FIGURE 8.38 Validation performance and error values of 60 patterns for TSS outputs (Exp. and ANN).

ANNs Modeling and SPSS Analysis 387

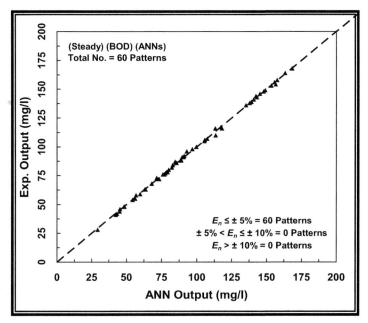

FIGURE 8.39 Equality diagram for 60 BOD patterns (Exp. and ANN).

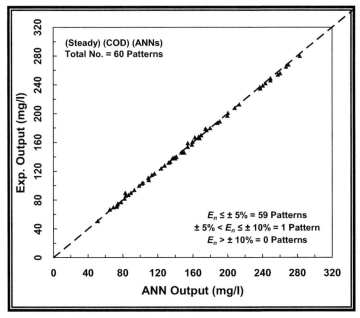

FIGURE 8.40 Equality diagram for 60 COD patterns (Exp. and ANN).

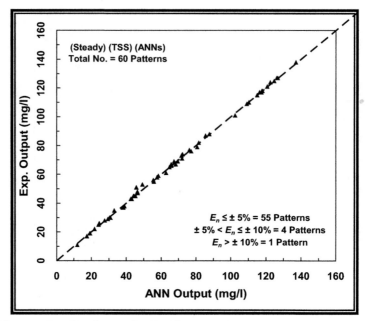

FIGURE 8.41 Equality diagram for 60 TSS patterns (Exp. and ANN).

- For TSS, the model results are matching with the experimental effluent value except few peculiar points.
- For BOD pollutant all points (60) give percentage error (E_n) less than ±5%.
- For COD pollutant 59 points give E_n less than ±5% and one point give E_n between ±5 and ±10%.
- For TSS pollutant 55 points give E_n less than ±5% and 4 points gave E_n between ±5 and ±10%, and one point give E_n more than ±10%.

8.4 ANNs MODELING – STEADY STAGE (NH_3 – PO_4 – DO – FC)

This section deals with modeling ammonia (NH_3), phosphate (PO_4), dissolved oxygen (DO), and fecal coliforms (FC) parameters. A general ANN program was designed and trained in Matlab toolbox to represent each parameter (Program No.3). This program is suitable for the parameters under study with the change of the axes limits and number of hidden layers. The influent and effluent samples at the end of wetland cells were only analyzed.

8.4.1 STRUCTURE OF ANNs

Different network structures having four input variables (C_i, q, A_s, and v) and one output value (C_o) are designed. Except for the dissolved oxygen networks which has three input variables (influent concentrations equal zero) and one output. Two hidden layers with different number of neurons were trained. A comparison between these networks is listed in Tables 8.7–8.10 for parameters under study. The weight matrixes and biases vectors for the best networks are written in Appendix III.

TABLE 8.7 Comparison Between Different Tested Networks for NH_3 (Steady Stage)

Network Structure	Number of Epoch	Mean Square Error	Gradient	Mu
4–2–2–1	82	0.0485	0.431	0.0139
4–3–3–1	25	0.167	0.283	0.00258
4–4–3–1	33	0.112	0.883	0.0174
4–4–4–1	28	0.0694	0.380	0.00125
4–5–3–1	34	0.117	3.40	0.00523
4–5–4–1	31	0.0768	0.758	0.00831

TABLE 8.8 Comparison Between Different Tested Networks for PO_4 (Steady Stage)

Network Structure	Number of Epoch	Mean Square Error	Gradient	Mu
4–3–3–1	46	0.000179	0.00833	0.000221
4–4–3–1	40	2.75E-05	0.00105	1.03E-06
4–4–4–1	30	0.000442	0.00347	5.10E-05
4–5–4–1	48	9.47E-05	0.00956	7.37E-05
4–6–5–1	27	0.000250	0.00505	3.69E-05

TABLE 8.9 Comparison Between Different Tested Networks for DO (Steady Stage)

Network Structure	Number of Epoch	Mean Square Error	Gradient	Mu
3–3–3–1	21	0.00784	3.20E-11	2.91E-07
3–4–3–1	39	0.00578	4.67E-11	1.20E-06
3–4–4–1	18	0.00537	4.94E-14	4.62E-07
3–5–3–1	16	0.00732	7.18E-14	2.86E-07
3–5–4–1	16	0.00778	4.15E-13	8.68E-06

TABLE 8.10 Comparison Between Different Tested Networks for FC (Steady Stage)

Network Structure	Number of Epoch	Mean Square Error	Gradient	Mu
4–3–2–1	50	2.57E+04	1.99E+06	107
4–3–3–1	56	9.94E+03	3.74E+05	60.8
4–4–3–1	76	1.26E+04	1.58E+05	1.85E+03
4–4–4–1	32	1.42E+04	2.21E+05	3.42
4–5–4–1	51	5.78E+04	1.53E+04	1.39E+03

The networks show satisfactory results for ammonia, phosphate, dissolved oxygen, and fecal coliforms having the structure of (4–2–2–1), (4–4–3–1), (3–4–4–1), and (4–3–3–1), respectively. Figures 8.42–8.45 show these network structures for the parameters under study.

The developed ANNs models are capable to minimize the mean square error (MSE) to final values of 0.0485, 2.75E-05, 0.00537, and 9.94E+03, for NH_3, PO_4, DO, and FC, respectively. The corresponding network training tools for these parameters are shown in Figures 8.46–8.49.

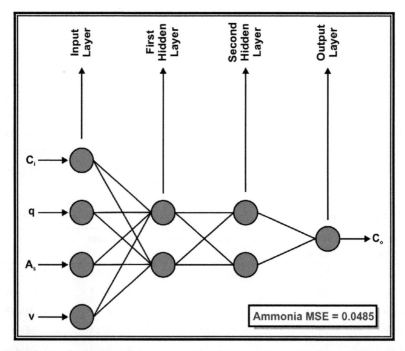

FIGURE 8.42 Optimum model structure of artificial neural network for ammonia pollutant.

ANNs Modeling and SPSS Analysis

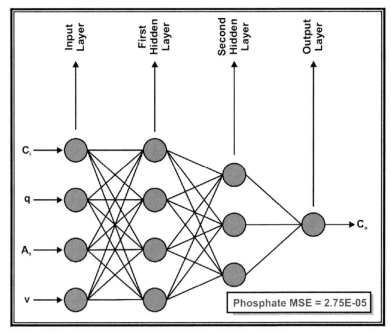

FIGURE 8.43 Optimum model structure of artificial neural network for phosphate pollutant.

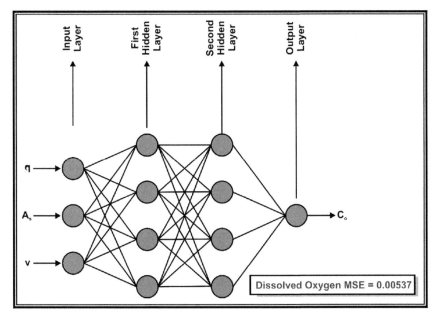

FIGURE 8.44 Optimum model structure of artificial neural network for dissolved oxygen parameter.

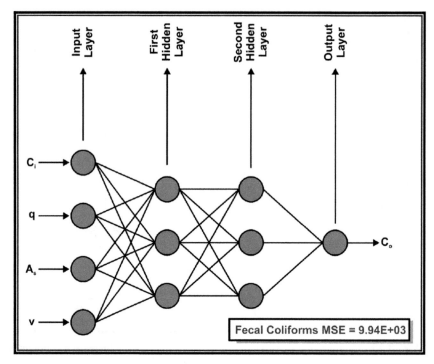

FIGURE 8.45 Optimum structure of ANN for fecal coliforms.

The mean square error training convergence diagrams for the selected networks using "trainlm" training function are illustrated in Figures 8.50–8.53 for ammonia, phosphate, dissolved oxygen, and fecal coliforms, respectively.

From Figures 8.50–8.53, it is obtained that, the best calibration performance occurs by iterations 62, 20, 14 and 36 for ammonia, phosphate, dissolved oxygen, and fecal coliforms parameters, respectively. The training stops when the validation error increased for 20 iterations, which occurred at iteration 82, 40, and 56 for NH_3, PO_4, and FC. For DO the training stops when the validation error increased for 4 iterations only (18) because of the stability of the value of validation error.

8.4.2 CALIBRATION PROCESS FOR NETWORKS

There are 60 patterns for input variables used to calibrate the designed neural networks for studied parameters. The training performance and error values for ammonia, phosphate, dissolved oxygen, and fecal coliforms are illustrated in Figures 8.54–8.57.

ANNs Modeling and SPSS Analysis 393

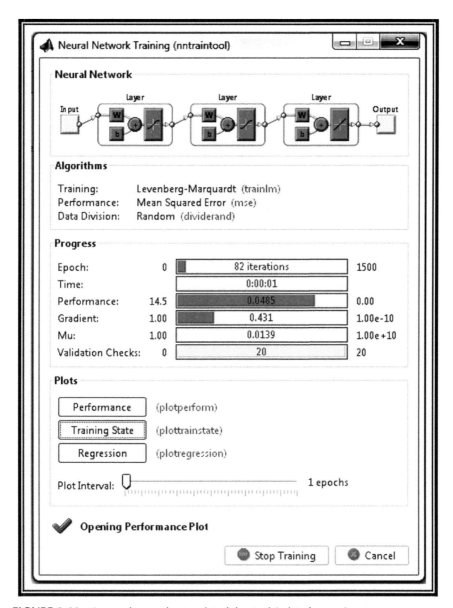

FIGURE 8.46 Ammonia neural network training tool (printed screen).

From the analysis of results and referring to Figures 8.54–8.57, it is noticed that, the error between the experimental effluent values and the ANNs model outputs vary between −0.209 and +0.180 mg/l for ammonia pollutant. For phosphate this error ranges from −0.031 to +0.013 mg/l.

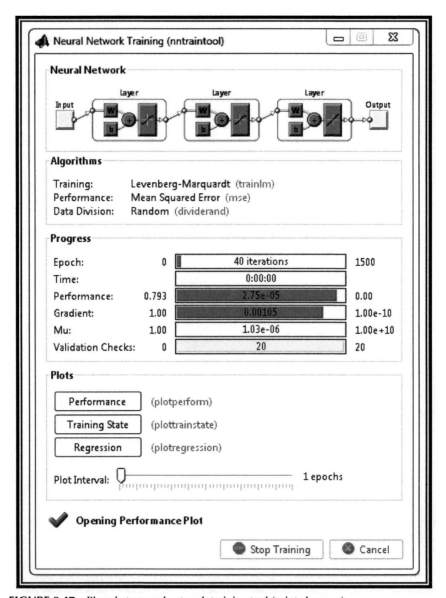

FIGURE 8.47 Phosphate neural network training tool (printed screen).

For dissolved oxygen the error ranges from −0.10 to +0.08 mg/l, whereas, the error for fecal coliforms varies between −72 and +95 MPN/100 ml.

Figures 8.58–8.61 illustrate the equality diagrams for ammonia, phosphate, dissolved oxygen, and fecal coliforms parameters of steady stage, respectively.

ANNs Modeling and SPSS Analysis

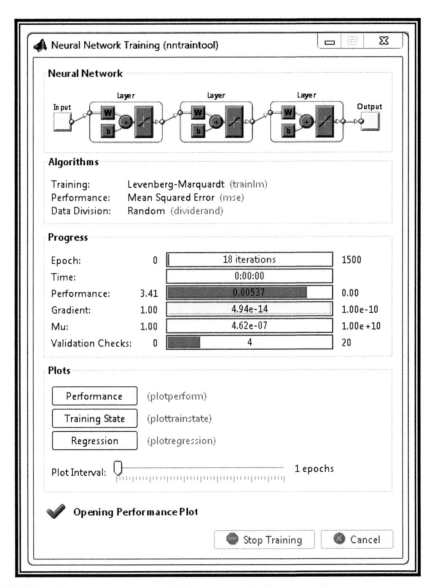

FIGURE 8.48 Dissolved oxygen ANN training tool (printed screen).

From Figures 8.58–8.61, it can be noticed that:
- The model results are close to the experimental measured output concentration for fecal coliforms pollutant.
- For ammonia, phosphate, and dissolved oxygen the model results are matching with the experimental effluent concentration with the few peculiar points.

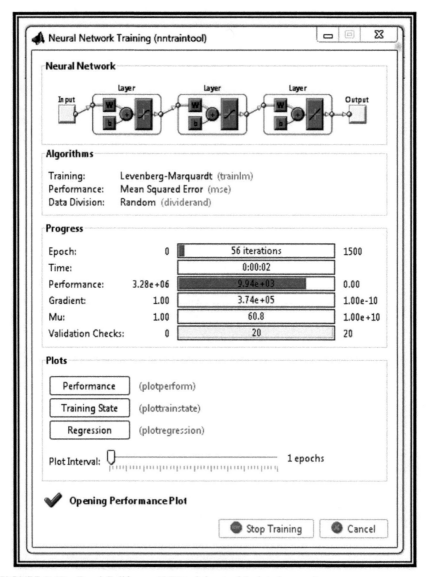

FIGURE 8.49 Fecal Coliforms ANN training tool (printed screen).

- For NH_3 52 points give E_n less than ±5% and 4 points give E_n between ±5 and ±10%, and 4 points give E_n more than ±10%.
- For DO 51 points give E_n less than ±5% and 6 points give E_n between ±5 and ±10%, and 3 points give E_n more than ±10%.
- For FC 54 points give E_n less than ±5% and 6 points give E_n between ±5 and ±10%.

ANNs Modeling and SPSS Analysis

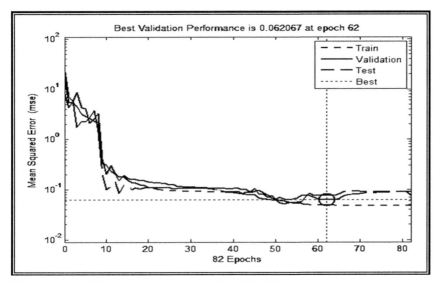

FIGURE 8.50 MSE training convergence of the ANN (NH_3 – steady stage – printed screen).

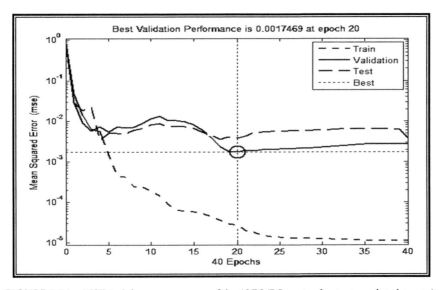

FIGURE 8.51 MSE training convergence of the ANN (PO_4 – steady stage – printed screen).

8.4.3 VALIDATION PROCESS FOR NETWORKS

The validation performance and error values of the proposed artificial neural networks models for 15 patterns are presented in Figures 8.62–8.65

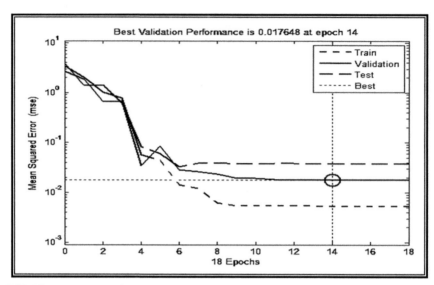

FIGURE 8.52 MSE training convergence of the ANN (DO – steady stage – printed screen).

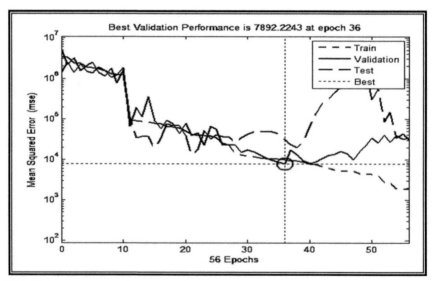

FIGURE 8.53 MSE training convergence of the ANN (FC – steady stage – printed screen).

for ammonia, phosphate, dissolved oxygen, and fecal coliforms parameters, respectively.

From the analysis of results and referring to Figures 8.62–8.65, it can be observed that, the average error between the experimental and the ANN model output concentrations for 15 patterns varies between:

ANNs Modeling and SPSS Analysis

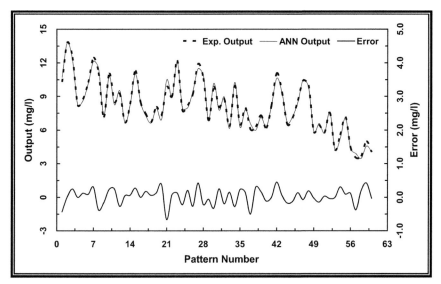

FIGURE 8.54 Training performance and error values of 60 patterns for NH_3 outputs (Exp. and ANN).

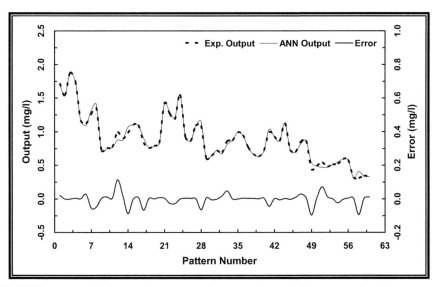

FIGURE 8.55 Training performance and error values of 60 patterns for PO_4 outputs (Exp. and ANN).

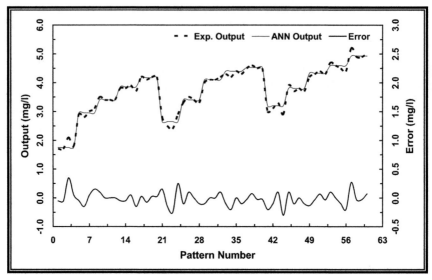

FIGURE 8.56 Training performance and error values of 60 patterns for DO outputs (Exp. and ANN).

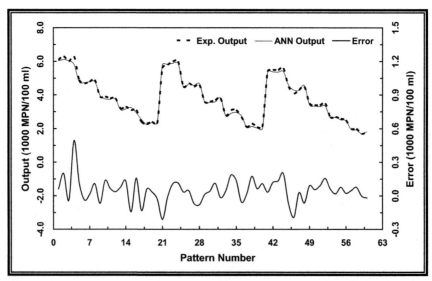

FIGURE 8.57 Training performance and error values of 60 patterns for FC outputs (Exp. and ANN).

ANNs Modeling and SPSS Analysis

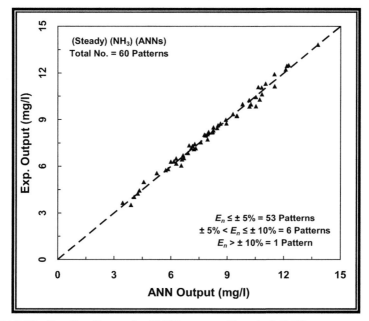

FIGURE 8.58 Equality diagram for 60 NH_3 patterns (Exp. And ANN).

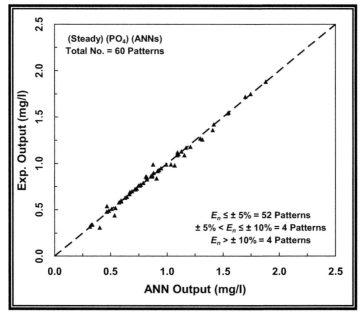

FIGURE 8.59 Equality diagram for 60 PO_4 patterns (Exp. and ANN).

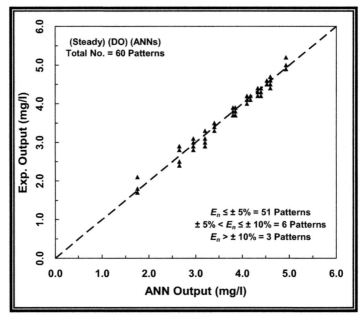

FIGURE 8.60 Equality diagram for 60 DO patterns (Exp. and ANN).

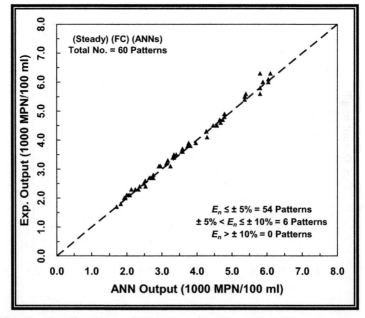

FIGURE 8.61 Equality diagram for 60 FC patterns (Exp. and ANN).

ANNs Modeling and SPSS Analysis

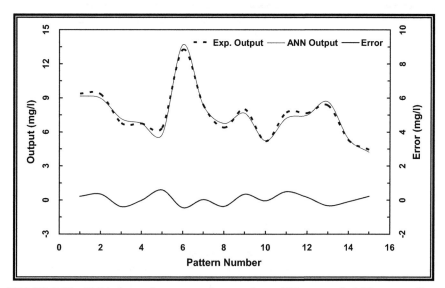

FIGURE 8.62 Validation performance and error values of 15 patterns for NH_3 outputs (Exp. and ANN).

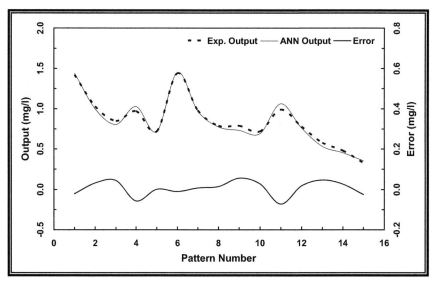

FIGURE 8.63 Validation performance and error values of 15 patterns for PO_4 outputs (Exp. and ANN).

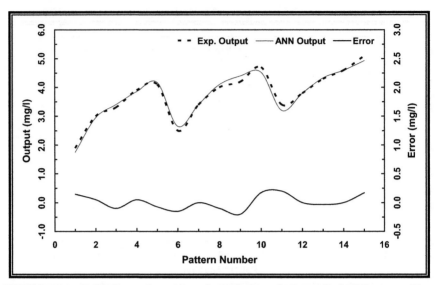

FIGURE 8.64 Validation performance and error values of 15 patterns for DO outputs (Exp. and ANN).

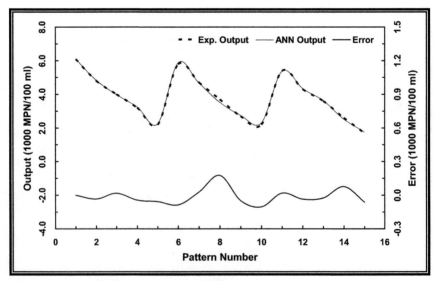

FIGURE 8.65 Validation performance and error values of 15 patterns for FC outputs (Exp. and ANN).

- −0.236 and +0.306 mg/l for ammonia pollutant.
- −0.037 and +0.027 mg/l for phosphate pollutant.
- −0.093 and +0.099 mg/l for dissolved oxygen parameter.
- −49 and +64 MPN/100 ml for fecal coliforms pollutant.

Figures 8.66–8.69 present the equality diagrams of 15 patterns for ammonia, phosphate, dissolved oxygen, and fecal coliforms parameters, respectively.

From Figures 8.66–8.69, it can be noticed that:
- The model results are very close to experimental output values for fecal coliforms pollutant.
- For NH_3, PO_4, and DO parameters, the model results are matching with the experimental effluent concentration.
- For NH_3 10 points give E_n less than ±5% and 5 points give E_n between ±5 and ±10%.
- For PO_4 8 points give E_n less than ±5% and 7 points give E_n between ±5 and ±10%.
- For DO 12 points give E_n less than ±5% and 3 points give E_n between ±5 and ±10%.
- For FC 15 points give E_n less than ±5%.

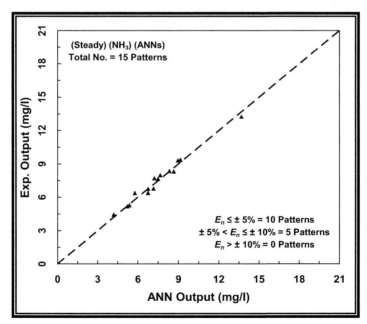

FIGURE 8.66 Equality diagram for 15 NH_3 patterns (Exp. and ANN).

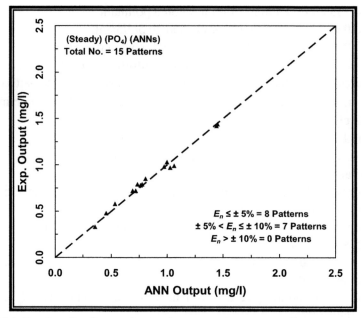

FIGURE 8.67 Equality diagram for 15 PO_4 patterns (Exp. and ANN).

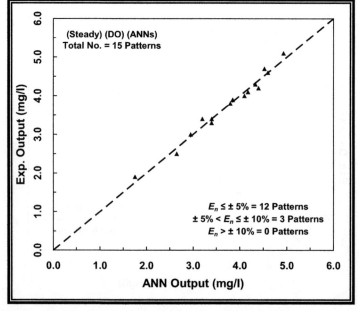

FIGURE 8.68 Equality diagram for 15 DO patterns (Exp. and ANN).

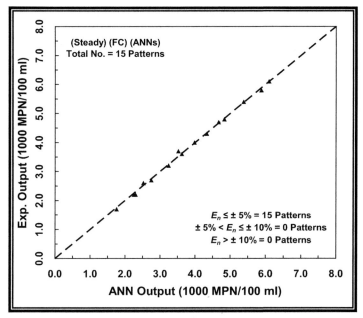

FIGURE 8.69 Equality diagram for 15 FC patterns (Exp. and ANN).

8.5 ANNs MODELING – STEADY STAGE (HEAVY METALS)

This section deals with the modeling of some heavy metals. A general artificial neural network program was designed and tested in Matlab to represent each pollutant (Program No.3). This program is convenient for the studied pollutants with the change of the axes limits and number of hidden layers. The inlet and outlet samples at the end of cells were only analyzed for these heavy metals "Zinc (Zn), Iron (Fe), Manganese (Mn), Lead (Pb), and Cadmium (Cd)."

8.5.1 STRUCTURE OF ANNs

Different network structures having 4 input variables (C_i, q, A_s, and v) and one output value (C_o) are designed. Two hidden layers with different number of neurons are tested. The comparison between these networks is listed in Tables 8.11–8.15 for all parameters. The weight matrixes and biases vectors for the best-selected networks are written in Appendix III.

TABLE 8.11 Comparison Between Different Tested Networks for Zn (Steady Stage)

Network Structure	Number of Epoch	Mean Square Error	Gradient	Mu
4–3–3–1	50	9.78E-05	0.00293	1.79E-06
4–4–3–1	58	7.92E-06	0.000207	1.14E-06
4–4–4–1	39	3.89E-05	0.000321	5.45E-07
4–5–3–1	43	8.52E-05	0.00102	1.96E-05
4–5–4–1	43	1.73E-05	0.000556	2.40E-06
4–5–5–1	35	5.76E-05	0.000140	1.32E-06

TABLE 8.12 Comparison Between Different Tested Networks for Fe (Steady Stage)

Network Structure	Number of Epoch	Mean Square Error	Gradient	Mu
4–3–3–1	31	0.000116	0.000635	1.53E-05
4–4–3–1	59	3.03E-05	0.00561	7.97E-06
4–4–4–1	41	3.27E-05	0.000444	1.58E-05
4–5–3–1	34	0.000147	0.00145	2.12E-05
4–5–4–1	31	7.64E-05	0.000235	1.18E-05

TABLE 8.13 Comparison Between Different Tested Networks for Mn (Steady Stage)

Network Structure	Number of Epoch	Mean Square Error	Gradient	Mu
4–3–3–1	38	6.70E-05	0.000315	2.36E-06
4–4–3–1	32	5.68E-05	0.000792	1.01E-05
4–4–4–1	40	1.60E-05	0.000502	1.84E-05
4–5–3–1	32	2.66E-05	0.000195	7.76E-06
4–5–4–1	32	5.81E-05	0.000139	1.24E-06
4–5–5–1	29	8.74E-05	0.000219	8.95E-07

TABLE 8.14 Comparison Between Different Tested Networks for Pb (Steady Stage)

Network Structure	Number of Epoch	Mean Square Error	Gradient	Mu
4–3–2–1	32	1.73E-06	7.78E-06	2.42E-08
4–3–3–1	32	4.61E-06	2.45E-06	2.42E-08
4–4–3–1	35	3.39E-06	4.77E-06	1.98E-08
4–4–4–1	31	2.47E-06	4.40E-06	4.50E-09
4–5–3–1	31	2.64E-06	7.73E-06	6.20E-08

TABLE 8.15 Comparison Between Different Tested Networks for Cd (Steady Stage)

Network Structure	Number of Epoch	Mean Square Error	Gradient	Mu
4–3–3–1	52	6.89E-10	7.93E-08	8.46E-10
4–4–3–1	65	4.21E-11	4.43E-09	8.23E-12
4–4–4–1	41	3.20E-10	2.74E-09	6.96E-11
4–5–3–1	47	3.42E-10	7.65E-09	1.02E-10

The networks which showed satisfactory results for the studied heavy metals having a structure of (4–4–3–1), (4–4–3–1), (4–4–4–1), (4–3–2–1), and (4–4–3–1) for Zinc, Iron, Manganese, Lead, and Cadmium elements, respectively. Figure 8.70 shows the network structure for Zn, Fe, and Cd pollutants. Figures 8.71 and 8.72 illustrate these neural network structures for Mn and Pb elements, respectively.

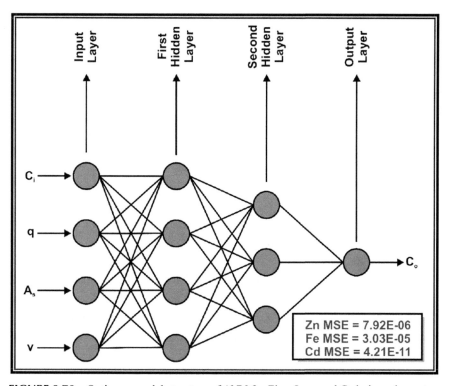

FIGURE 8.70 Optimum model structure of ANN for Zinc, Iron, and Cadmium elements.

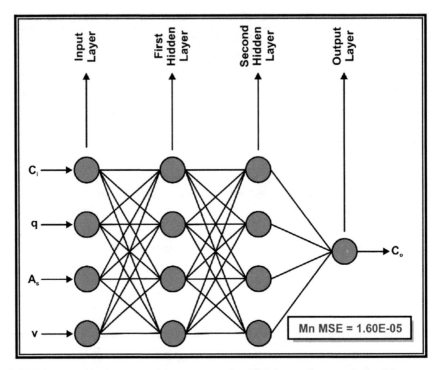

FIGURE 8.71 Optimum model structure of artificial neural network for Manganese element.

The developed ANNs models are capable to minimize the mean square error to the final values of 7.92E-06, 3.03E-05, 1.60E-05, 1.73E-06, and 4.21E-11, for Zinc, Iron, Manganese, Lead, and Cadmium, respectively. The corresponding network training tools for these pollutants are presented in Figures 8.73–8.77.

The mean square error training convergence diagrams for the selected networks using "trainlm" training function are shown in Figures 8.78–8.82 for heavy metals under study.

From Figures 8.78–8.82 it is concluded that, the best validation performance occurs by a number of iterations 38, 39, 20, 12, and 45 for Zinc, Iron, Manganese, Lead, and Cadmium elements in steady stage, respectively. The training stops when the error increases after an additional of 20 iterations, which occurred at iteration 58, 59, 40, 32, and 65 for these heavy metals.

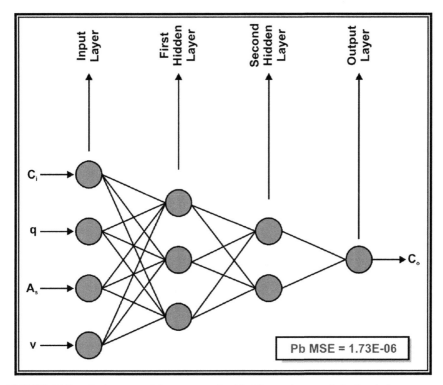

FIGURE 8.72 Optimum model structure of artificial neural network for Lead element.

8.5.2 CALIBRATION PROCESS FOR NETWORKS

A number of 60 patterns for input variables were used to calibrate the designed neural networks for heavy metals under study. The training performance and error values for Zn, Fe, Mn, Pb, and Cd are illustrated in Figures 8.83–8.87, respectively.

From the analysis of results and referring to Figures 8.83–8.87, it is concluded that, the error between the experimental effluent concentration and the ANNs model outputs varies between −8.8 and +8.4 μg/l for Zinc. For Iron this error ranges from −9.0 to +6.0 μg/l. For Manganese the error ranges from −3.7 to +3.8 μg/l. For Lead the error ranges from −1.2 to +0.5.0 μg/l, whereas, for Cadmium the error varies between −0.008 and +0.023 μg/l.

Figures 8.88–8.92 present the equality diagrams of 60 patterns for Zinc, Iron, Manganese, Lead, and Cadmium elements in steady stage, respectively.

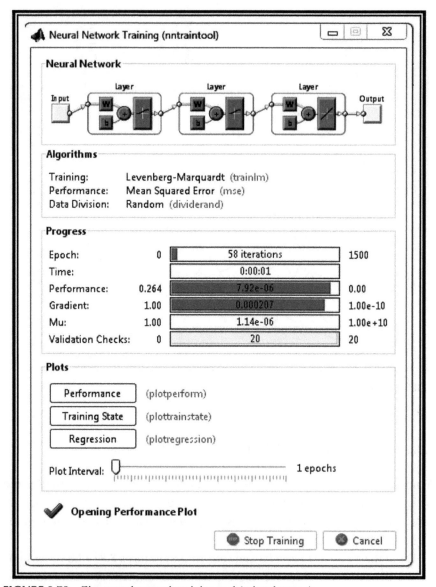

FIGURE 8.73 Zinc neural network training tool (printed screen).

From Figures 8.88–8.92, it can be obtained that:
- The model results are close to the experimental measured output concentration for Zinc and Cadmium pollutants.
- For Iron, Manganese, and Lead elements, the model results are matching with the experimental effluent concentration with few peculiar points.

ANNs Modeling and SPSS Analysis

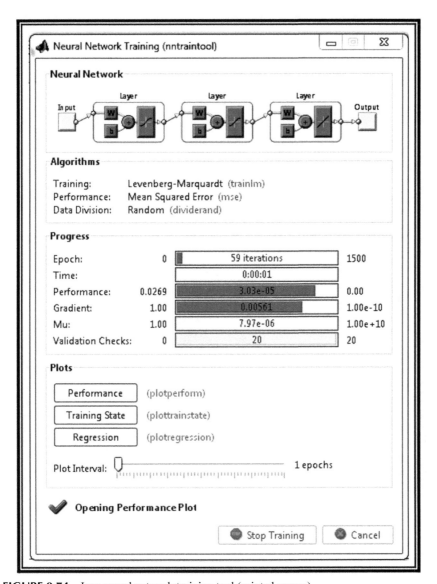

FIGURE 8.74 Iron neural network training tool (printed screen).

- For Zn, 55 points give E_n less than ±5% and 5 points give E_n between ±5 and ±10%.
- For Iron, 51 points give E_n less than ±5% and 6 points give E_n between ±5 and ±10%, and 3 points give E_n more than ±10%.
- For Manganese, 52 points give E_n less than ±5% and 7 points gave E_n between ±5 and ±10%, and one point gives E_n more than ±10%.

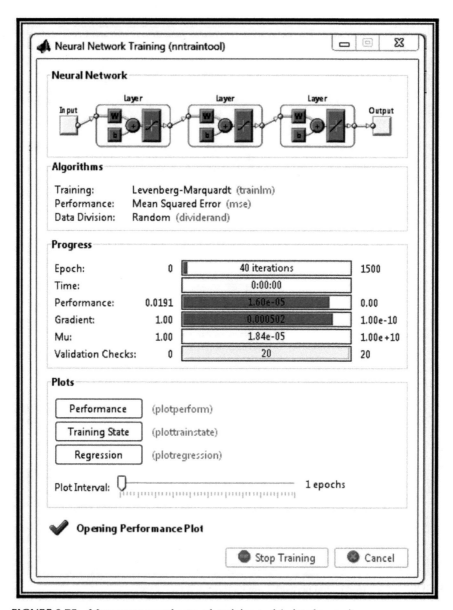

FIGURE 8.75 Manganese neural network training tool (printed screen).

- For Lead, 42 points give E_n less than ±5% and 17 points give E_n between ±5 and ±10%, and one point give E_n more than ±10%.
- For Cadmium, 59 points give E_n less than ±5% and one point give E_n between ±5 and ±10%.

ANNs Modeling and SPSS Analysis

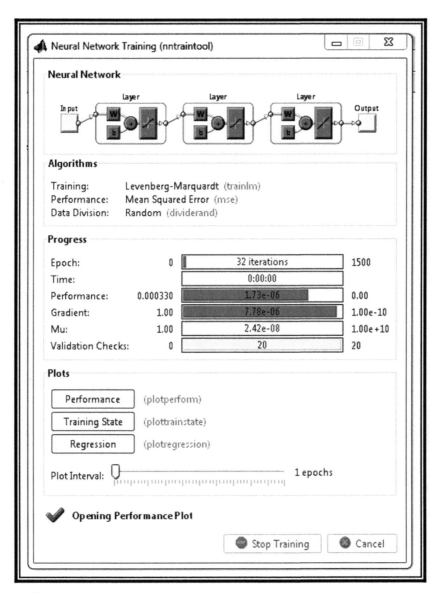

FIGURE 8.76 Lead neural network training tool (printed screen).

8.5.3 VALIDATION PROCESS FOR NETWORKS

The validation performance and error values of the proposed artificial neural networks models for 15 patterns are presented in Figures 8.93–8.97 for Zinc, Iron, Manganese, Lead, and Cadmium, respectively.

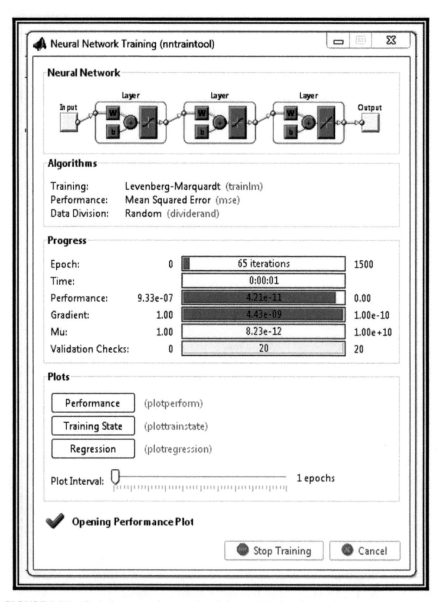

FIGURE 8.77 Cadmium neural network training tool (printed screen).

From the analysis of results and referring to Figures 8.93–8.97, it can be observed that, the error between the experimental effluent concentration and the ANNs model outputs varies between −21.8 and +16.2 µg/l for Zinc. For Iron this error ranges from −22.3 to +14.8 µg/l. For Manganese the error

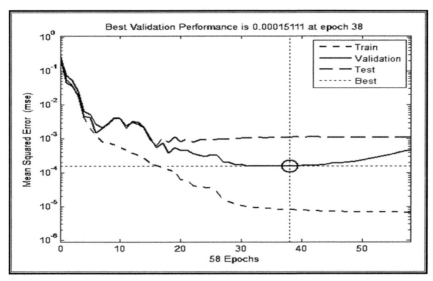

FIGURE 8.78 MSE training convergence of the ANN (Zn – steady stage – printed screen).

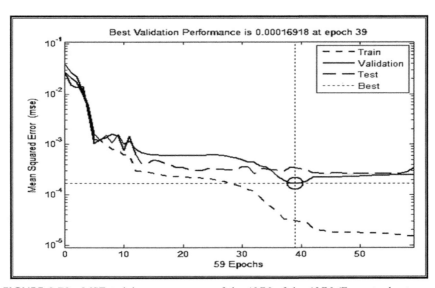

FIGURE 8.79 MSE training convergence of the ANN of the ANN (Fe – steady stage – printed screen).

ranges from −6.0 to +2.6 µg/l. For Lead the error ranges from −1.1 to +0.4 µg/l, whereas, for Cadmium the error varies between −0.009 and +0.032 µg/l.

Figures 8.98–8.102 present the equality diagrams of 15 patterns for Zinc, Iron, Manganese, Lead, and Cadmium elements in steady stage, respectively.

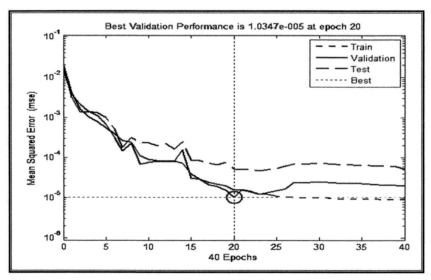

FIGURE 8.80 MSE training convergence of the ANN of the ANN (Mn – steady stage – printed screen).

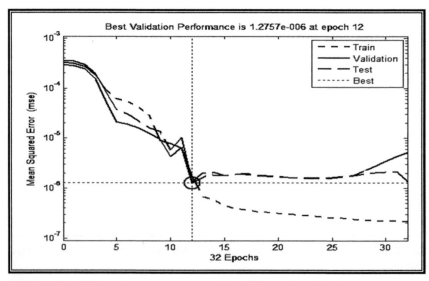

FIGURE 8.81 MSE training convergence of the ANN of the ANN (Pb – steady stage – printed screen).

ANNs Modeling and SPSS Analysis 419

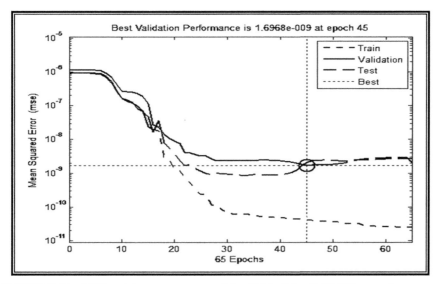

FIGURE 8.82 MSE training convergence of the ANN of the ANN (Cd – steady stage – printed screen).

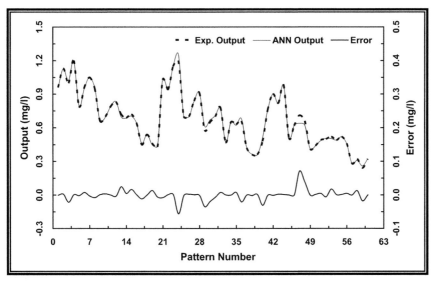

FIGURE 8.83 Training performance and error values of 60 patterns for Zn outputs (Exp. and ANN).

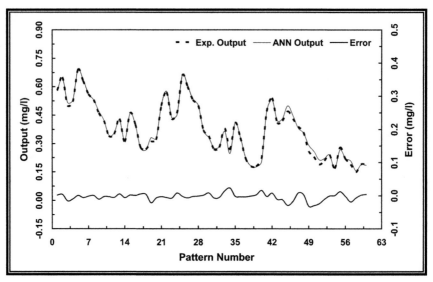

FIGURE 8.84 Training performance and error values of 60 patterns for Fe outputs (Exp. and ANN).

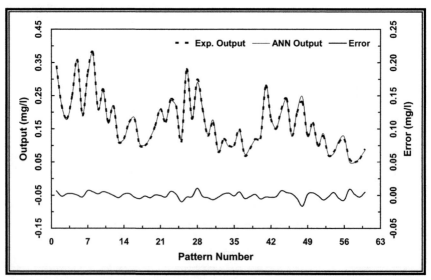

FIGURE 8.85 Training performance and error values of 60 patterns for Mn outputs (Exp. and ANN).

ANNs Modeling and SPSS Analysis

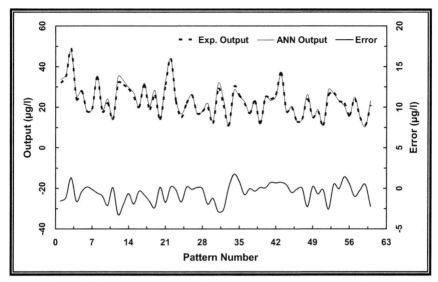

FIGURE 8.86 Training performance and error values of 60 patterns for Pb outputs (Exp. and ANN).

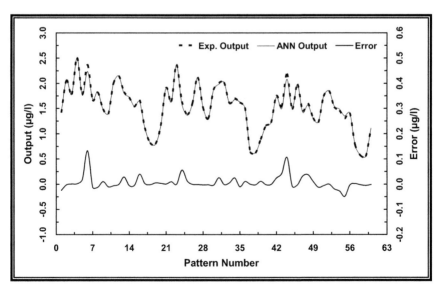

FIGURE 8.87 Training performance and error values of 60 patterns for Cd outputs (Exp. and ANN).

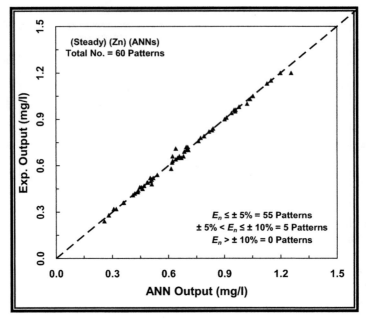

FIGURE 8.88 Equality diagram for 60 Zn patterns (Exp. and ANN).

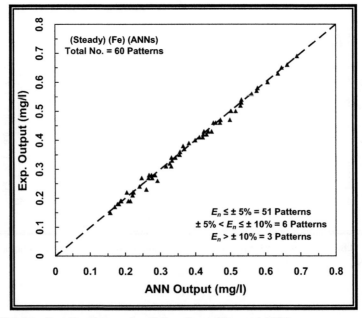

FIGURE 8.89 Equality diagram for 60 Fe patterns (Exp. and ANN).

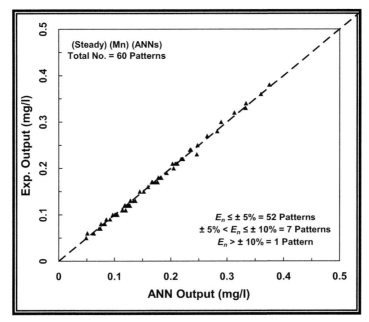

FIGURE 8.90 Equality diagram for 60 Mn patterns (Exp. and ANN).

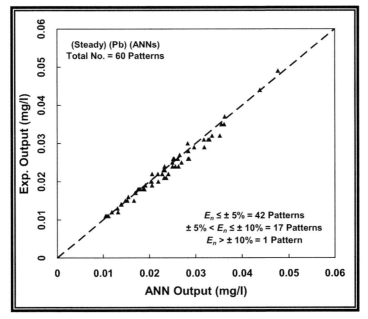

FIGURE 8.91 Equality diagram for 60 Pb patterns (Exp. and ANN).

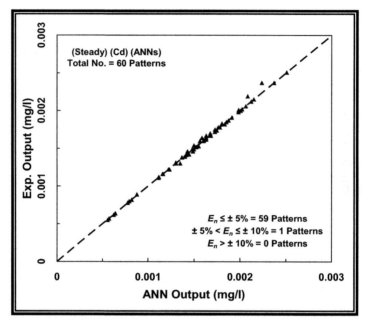

FIGURE 8.92 Equality diagram for 60 Cd patterns (Exp. and ANN).

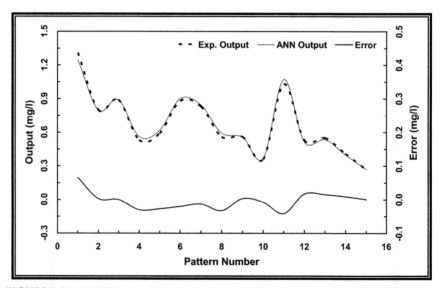

FIGURE 8.93 Validation performance and error values of 15 patterns for Zn outputs (Exp. and ANN).

ANNs Modeling and SPSS Analysis 425

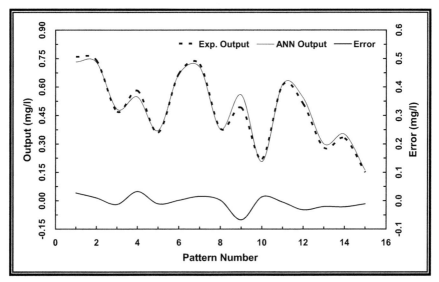

FIGURE 8.94 Validation performance and error values of 15 patterns for Fe outputs (Exp. and ANN).

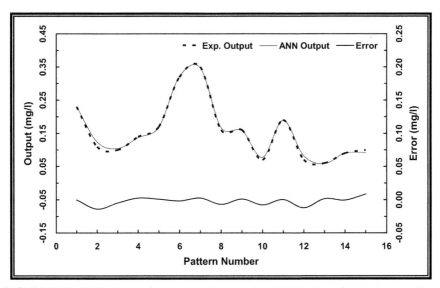

FIGURE 8.95 Validation performance and error values of 15 patterns for Mn outputs (Exp. and ANN).

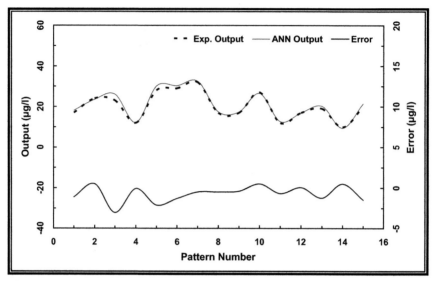

FIGURE 8.96 Validation performance and error values of 15 patterns for Pb outputs (Exp. and ANN).

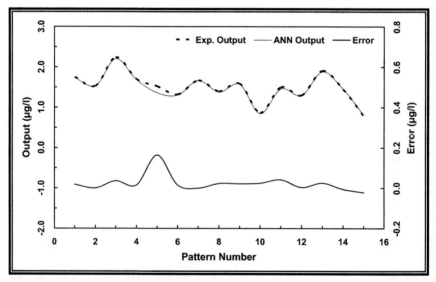

FIGURE 8.97 Validation performance and error values of 15 patterns for Cd outputs (Exp. and ANN).

ANNs Modeling and SPSS Analysis 427

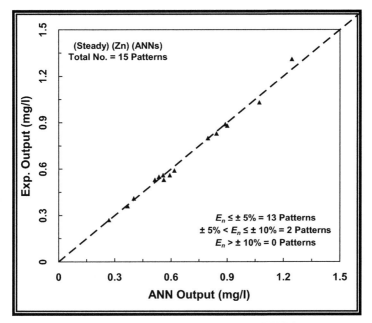

FIGURE 8.98 Equality diagram for 15 Zn patterns (Exp. and ANN).

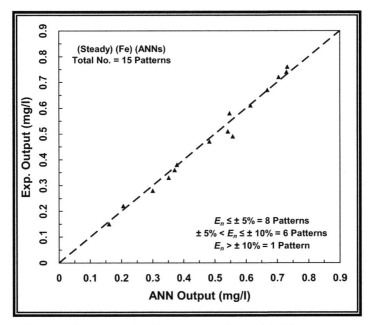

FIGURE 8.99 Equality diagram for 15 Fe patterns (Exp. and ANN).

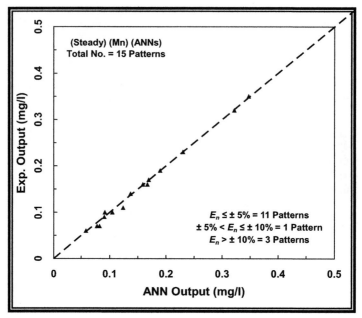

FIGURE 8.100 Equality diagram for 15 Mn patterns (Exp. and ANN).

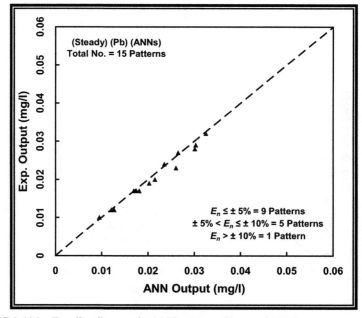

FIGURE 8.101 Equality diagram for 15 Pb patterns (Exp. and ANN).

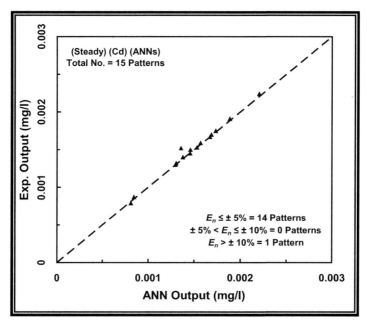

FIGURE 8.102 Equality diagram for 15 Cd patterns (Exp. and ANN).

From Figures 8.98–8.102, it can be obtained that:

- The model results are very close to the experimental measured output concentration for Zinc pollutant.
- For Iron, Manganese, Lead, and Cadmium elements, the model results are matching with the experimental effluent concentration even with few peculiar points. The peculiar points may be excluded to decrease the average error values for the pollutants (E_n more than ±10%).
- For Zinc 13 points give E_n less than ±5% and 2 points give E_n between ±5 and ±10%.
- For Iron 8 points give E_n less than ±5% and 6 points give E_n between ±5 and ±10%, and one point give E_n more than ±10%.
- For Manganese 11 points give E_n less than ±5% and one point gives E_n between ±5 and ±10%, and 3 points give E_n more than ±10%.
- For Lead 9 points give E_n less than ±5% and 5 points give E_n between ±5 and ±10%, and one point give E_n more than ±10%.
- For Cd 14 points give E_n less than ±5% and one point gives E_n more than ±10%.

8.6 STATISTICAL MODELING FOR SET UP STAGE

Statistical procedures were carried out using the SPSS software. The linear and nonlinear regressions were tested but linear regression gave the least mean square error and best convergence about the line of perfect agreement. Three linear regression equations were obtained to represent the experimental data for the studied parameters in set up stage. The same data used in artificial neural networks with the same excluded ones for calibration and validation were undergone to the SPSS analysis. This symmetry was used to make a comparison between ANNs and SPSS results.

8.6.1 REGRESSION EQUATIONS USING SPSS

The same five input variables used in artificial neural network program, which are C_i, q, A_s, v, and T_o are used in SPSS as independent variables to get the dependent output concentration. The following equation (Eq. 8.1) represents the general form of the deduced regression equations.

Table 8.16 shows the constant values in this equation and the determination coefficients (R^2) for BOD, COD, and TSS pollutants. Application limits of such equation are summarized later in Table 8.20.

$$C_o = b_1 + b_2 \times C_i + b_3 \times q + b_4 \times A_s + b_5 \times v + b_6 \times T_o \qquad (8.1)$$

where: C_i, C_o = input and output concentrations, mg/l; q = hydraulic loading rate, m/d; A_s = surface area of used media, m²; v = actual velocity, m/d; T_o = time from starting operation, d; b_i = constant i = 1, 2, ... 6.

8.6.2 CALIBRATION PROCESS FOR REGRESSION EQUATIONS

Substituting input variables in the proposed regression equations, it gives the output concentration. The performance of these equations for 120 patterns

TABLE 8.16 Constants of Eq. (8.1) and R^2 Values for Set Up Stage Pollutants

Pollutant	b_1	b_2	b_3	b_4	b_5	b_6	R^2
BOD	47.678	0.723	15.370	−0.015	−0.122	−0.251	0.983
COD	78.853	0.721	21.633	−0.025	−0.204	−0.391	0.982
TSS	−1.724	0.616	18.936	−0.018	0.905	−0.045	0.984

ANNs Modeling and SPSS Analysis 431

with the error values are presented in Figures 8.103–8.105 for BOD, COD, and TSS pollutants, respectively. The error and the percentage error are computed using the following formulae:

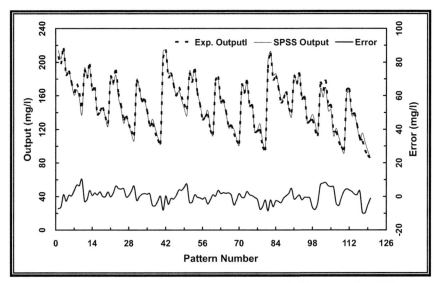

FIGURE 8.103 Training performance and error values of 120 patterns for BOD outputs (Exp. and SPSS).

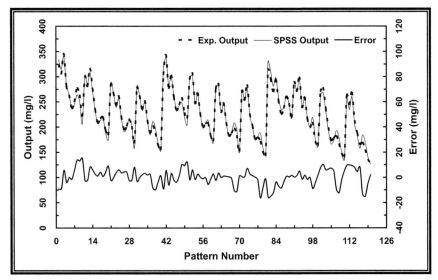

FIGURE 8.104 Training performance and error values of 120 patterns for COD outputs (Exp. and SPSS).

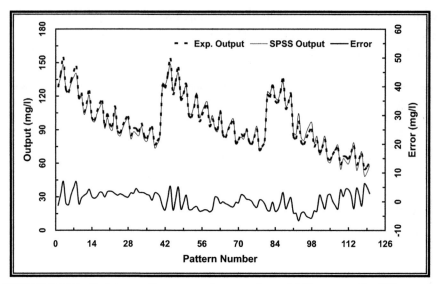

FIGURE 8.105 Training performance and error values of 120 patterns for TSS outputs (Exp. and SPSS).

$$Error = C_{Exp} - C_{SPSS} \qquad (4.32)$$

$$E_s = \left(\frac{C_{Exp} - C_{SPSS}}{C_{Exp}}\right) \times 100 \qquad (4.33)$$

where: E_s = SPSS percentage error, %; C_{Exp} = experimental measured output concentration, mg/l; C_{spss} = regression equation output concentration, mg/l.

From the analysis of results and referring to Figures 8.103–8.105, the error between the regression equation output and the experimental concentration varies between −3.16 and +3.08 mg/l for BOD. For COD pollutant this error ranges from −5.41 to +4.63 mg/l, whereas, for TSS the error varies between −2.26 and +2.66 mg/l.

Figures 8.106–8.108 show the equality diagrams of the 120 patterns for BOD, COD, and TSS pollutants, respectively (experimental output concentration versus the output of the deduced regression equation).

From Figures 8.106–8.108, it can be observed that:

- The deduced regression equations results are matching with the target values for COD pollutant.

ANNs Modeling and SPSS Analysis 433

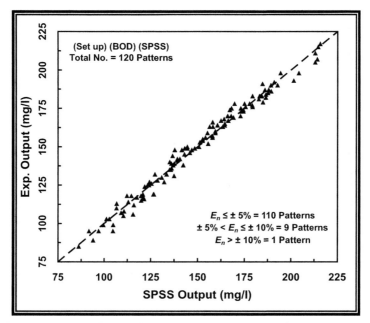

FIGURE 8.106 Equality diagram for 120 BOD patterns (Exp. and SPSS).

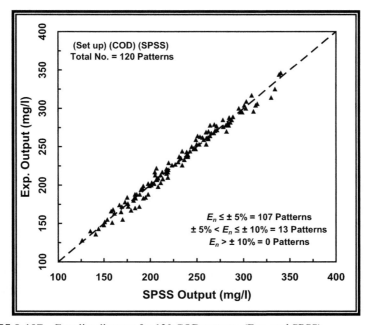

FIGURE 8.107 Equality diagram for 120 COD patterns (Exp. and SPSS).

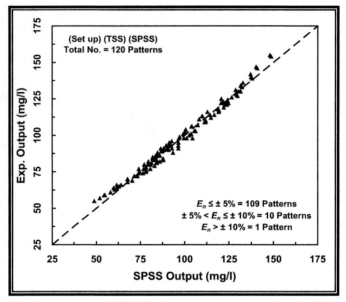

FIGURE 8.108 Equality diagram for 120 TSS patterns (Exp. and SPSS).

- For BOD and TSS pollutants, the model results are acceptable with the experimental output values except a few peculiar points.
- The difference between experimental output concentrations and SPSS results are higher than the corresponding results of ANNs model outputs for calibration process.
- For BOD, 110 points give percentage error (E_s) less than ±5% and 9 points between ±5:10%, while one point gives E_s higher than ±10%.
- For COD, 107 points give E_s less than ±5% and 13 points between ±5:10%.
- For TSS, 109 points give E_s less than ±5% and 10 points between ±5:10%, while one point gives E_s higher than ±10%.

8.6.3 VALIDATION PROCESS FOR REGRESSION EQUATIONS

The same 24 random patterns for input variables that used to validate the designed artificial neural networks for BOD, COD, and TSS pollutants were used also in validation process for the deduced regression equations. These random 24 patterns for input variables were substituted in the regression equations to get the output concentrations.

The validation performance and error values of the proposed equations for 24 patterns are presented in Figures 8.109–8.111 for BOD, COD, and TSS parameters, respectively.

From the analysis of results and referring to Figures 8.109–8.111, it can be obtained that:

- The error between the experimental results and SPSS model output varies between −4.18 and +3.10 mg/l for BOD.
- The error between experimental results and SPSS model output varies between −5.67 and +5.28 mg/l for COD.
- The error between experimental results and SPSS model output varies between −2.17 and +2.60 mg/l for TSS.

Figures 8.112–8.114 show the equality diagrams of the 24 patterns for BOD, COD, and TSS pollutants, respectively.

From Figures 8.106–8.108, it can be observed that:

- The SPSS model validation results are matching with the experimental output values for the three studied pollutants in set up stage.
- The difference between experimental output concentrations and SPSS results are higher than the corresponding results of ANNs model outputs for validation process.

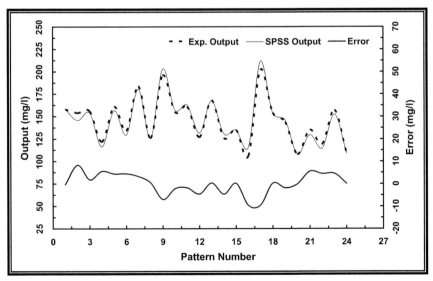

FIGURE 8.109 Validation performance and error values of 24 patterns for BOD outputs (Exp. and SPSS).

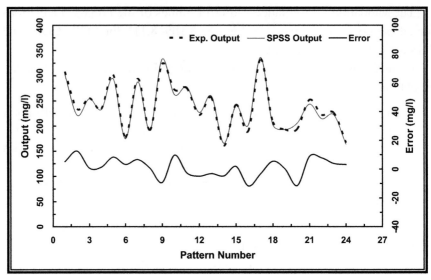

FIGURE 8.110 Validation performance and error values of 24 patterns for COD outputs (Exp. and SPSS).

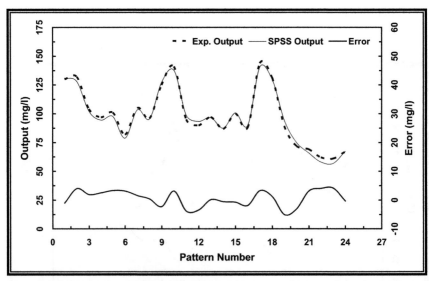

FIGURE 8.111 Validation performance and error values of 24 patterns for TSS outputs (Exp. and SPSS).

ANNs Modeling and SPSS Analysis

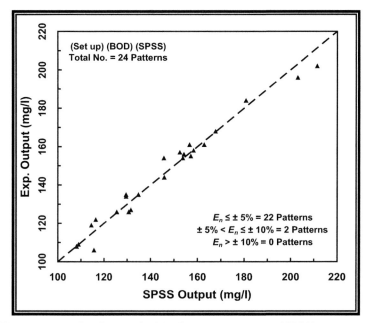

FIGURE 8.112 Equality diagram for 24 BOD patterns (Exp. and SPSS).

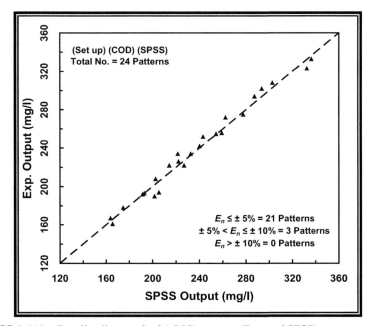

FIGURE 8.113 Equality diagram for 24 COD patterns (Exp. and SPSS).

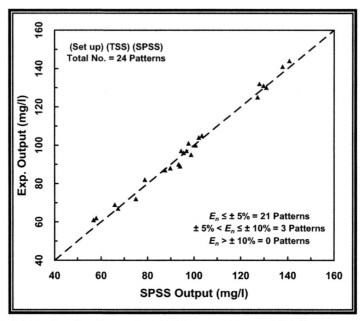

FIGURE 8.114 Equality diagram for 24 TSS patterns (Exp. and SPSS).

- For BOD, 22 points give E_s less than ±5%, and two points lies between ±5:10%.
- For COD and TSS, 21 points give E_s less than ±5%, and three points lies between ±5:10%.

8.7 STATISTICAL MODELING – STEADY (BOD – COD – TSS)

Statistical procedures were carried out using the SPSS program. Three non-linear regression equations were obtained to represent the experimental data for the studied parameters in steady stage.

All the data used for calibration and validation processes in the artificial neural networks models were underwent to SPSS analysis. This symmetry was conducted to make a comparison between ANNs and SPSS results. Many different types of regression equations were tested and the selected one that gives higher value of the determination coefficient (R^2) and consequently a good convergence about the line of perfect agreement.

8.7.1 REGRESSION EQUATIONS USING SPSS

Equation (8.2) represents the general form of the deduced regression equations for BOD and COD, whereas Eq. (8.3) is the obtained one for TSS parameter ($R^2 = 0.977$). Table 8.17 shows the constant values in Eq. (8.2) in addition to the determination coefficients for BOD and COD pollutants. Limits of such equations are summarized later in Table 8.21.

$$C_o = b_1 \times C_i^{Z_1} + b_2 \times q^{Z_2} + b_3 \times A_s^{Z_3} + b_4 \times v^{Z_4} \tag{8.2}$$

$$C_o = C_i^{1.172}\left(0.599 - 0.033q^{-0.971}\right)\left(1.74 - 0.295 A_s^{0.196}\right)\left(0.558 + 0.026 v^{0.683}\right) \tag{8.3}$$

where: C_o = output concentration, mg/l; C_i = input concentration, mg/l; q = hydraulic loading rate, m/d; A_s = surface area of used media, m²; v = actual velocity, m/d; b_i = constant i = 1, 2, … 4; Z_i = constant i = 1, 2, … 4.

8.7.2 CALIBRATION PROCESS FOR REGRESSION EQUATIONS

The performance of the proposed regression equations for 240 patterns is presented in Figures 8.115–8.117 for BOD, COD, and TSS.

From the analysis of results and referring to Figures 8.115–8.117, it is concluded that, the error between the measured concentration and SPSS model outputs varies between −3.81 and +3.45 mg/l for BOD pollutant. For COD this error ranges from −6.07 to +5.56 mg/l, whereas, for TSS the error

TABLE 8.17A Constants Value of Eq. 8.2 for BOD and COD Pollutants

Pollutant	b_1	b_2	b_3	b_4
BOD	127.809	−38.910	−151.226	0.001
COD	129.130	−60.037	−149.542	0.002

TABLE 8.17B Constants Value of Eq. (8.2) for BOD and COD Pollutants

Pollutant	Z_1	Z_2	Z_3	Z_4	R^2
BOD	0.235	−0.370	0.087	3.329	0.986
COD	0.265	−0.366	0.115	3.287	0.986

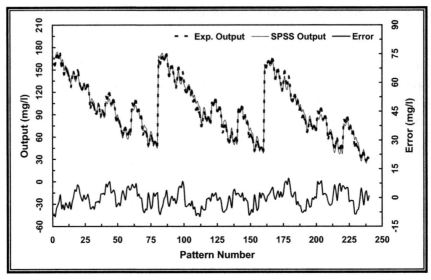

FIGURE 8.115 Training performance and error values of 240 patterns for BOD outputs (Exp. and SPSS).

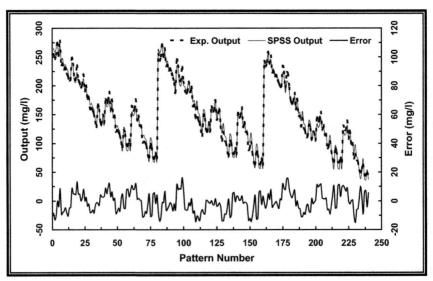

FIGURE 8.116 Training performance and error values of 240 patterns for COD outputs (Exp. and SPSS).

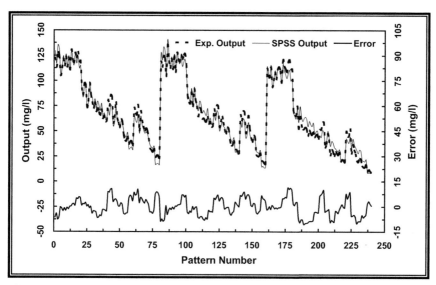

FIGURE 8.117 Training performance and error values of 240 patterns for TSS outputs (Exp. and SPSS).

varies between −4.25 and +4.21 mg/l. Figures 8.118–8.120 show the equality diagrams for BOD, COD, and TSS pollutants in steady stage, respectively.

From Figures 8.118–8.120, it can be observed that:

- The SPSS model calibration results are acceptable with the experimental output values for BOD and COD pollutants except some peculiar points. For TSS pollutant the number of peculiar points is relatively big.
- The difference between experimental and SPSS results are higher than the corresponding results of ANNs model outputs.
- For BOD pollutant, 154 points give E_s less than ±5% and 66 points give E_s between ±5 and ±10%, and 20 points give E_s more than ±10%.
- For COD pollutant, 146 points give E_s less than ±5% and 67 points give E_s between ±5 and ±10%, and 27 points give E_s more than ±10%.
- For TSS pollutant, 96 points give E_s less than ±5% and 67 points give E_s between ±5 and ±10%, and 77 points give $E_s \geq \pm 10\%$.

8.7.3 VALIDATION PROCESS FOR REGRESSION EQUATIONS

The validation performance and error values of the proposed equations for 60 patterns are presented in Figures 8.121–8.123 for BOD, COD, and TSS parameters, respectively.

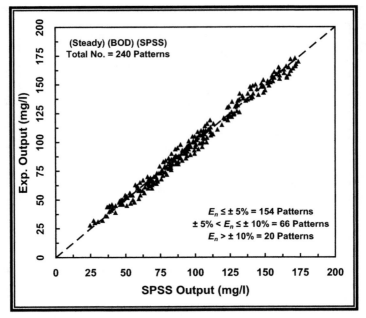

FIGURE 8.118 Equality diagram for 240 BOD patterns (Exp. and SPSS).

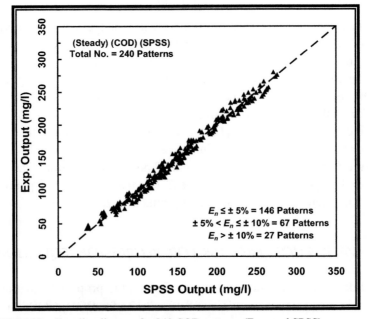

FIGURE 8.119 Equality diagram for 240 COD patterns (Exp. and SPSS).

ANNs Modeling and SPSS Analysis

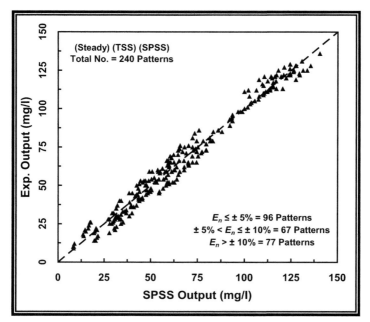

FIGURE 8.120 Equality diagram for 240 TSS patterns (Exp. and SPSS).

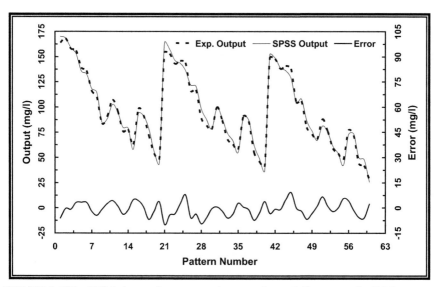

FIGURE 8.121 Validation performance and error values of 60 patterns for BOD outputs (Exp. and SPSS).

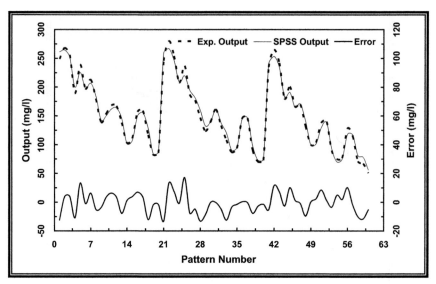

FIGURE 8.122 Validation performance and error values of 60 patterns for COD outputs (Exp. and SPSS).

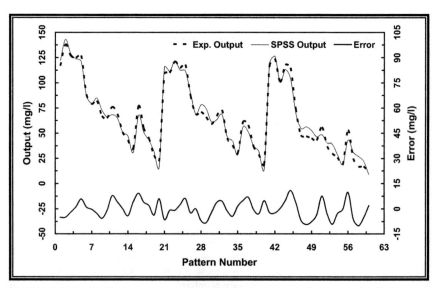

FIGURE 8.123 Validation performance and error values of 60 patterns for TSS outputs (Exp. and SPSS).

ANNs Modeling and SPSS Analysis

From the analysis of results and referring to Figures 8.121–8.123, it can be obtained that:

- The error between experimental results and SPSS model outputs varies between −3.56 and +3.49 mg/l for BOD pollutant.
- The error between experimental results and SPSS model outputs varies between −5.51 and +5.94 mg/l for COD pollutant.
- The error between experimental results and SPSS model outputs varies between −4.12 and +4.41 mg/l for TSS pollutant.

The equality diagrams for BOD, COD, and TSS parameters in steady stage calibration process are presented in Figures 7.124–7.126.

From Figures 8.124–8.126, it can be found that:

- The SPSS model validation results are acceptable with the experimental output values for BOD and COD pollutants except some peculiar points. For TSS pollutant the number of peculiar points is relatively big.
- The difference between experimental and SPSS results are higher than the corresponding results of ANNs model outputs.
- For BOD pollutant 38 points give E_s less than ±5% and 17 points give E_s between ±5 and ±10%, and 5 points give E_s more than ±10%.

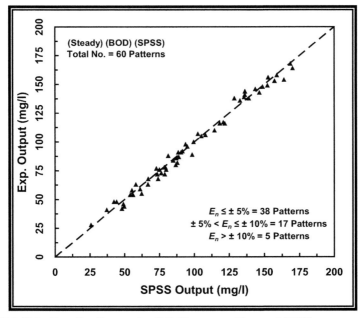

FIGURE 8.124 Equality diagram for 60 BOD patterns (Exp. and SPSS).

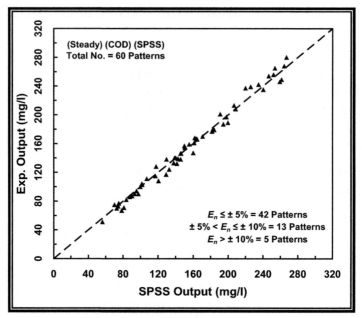

FIGURE 8.125 Equality diagram for 60 COD patterns (Exp. and SPSS).

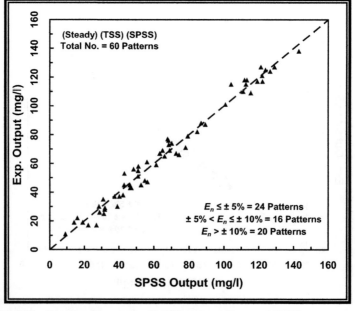

FIGURE 8.126 Equality diagram for 60 TSS patterns (Exp. and SPSS).

- For COD pollutant 42 points give E_s less than ±5% and 13 points give E_s between ±5 and ±10%, and 5 points give E_s more than ±10%.
- For TSS pollutant 24 points give E_s less than ±5% and 16 points give E_s between ±5 and ±10%, and 20 points give $E_s \geq \pm 10\%$.

8.8 STATISTICAL MODELING – STEADY (NH_3 – PO_4 – DO – FC)

Statistical procedures were carried out using the SPSS software. Four nonlinear regression equations were obtained to represent the experimental data for ammonia, phosphate, dissolved oxygen, and fecal coliforms.

The same data used in artificial neural networks with the same excluded ones for calibration and validation were undergone to the SPSS analysis. This symmetry was conducted to make a comparison between ANNs and SPSS results. Many different types of regression equations were tested and the select one was that gave the best convergence about the line of perfect agreement and least mean square error.

8.8.1 REGRESSION EQUATIONS USING SPSS

Equation 8.4 represents the general form of the deduced regression equations for the four studied parameters. Table 8.18 shows the constants value

TABLE 8.18A Constants Value of Eq. (8.4) for the Studied Parameters

Pollutant	b	b_1	b_2	b_3	b_4
Ammonia	2.479	0.023	–0.544	–0.006	7.243
Phosphate	3.134	0.006	2.355	–1.138	0.581
D. Oxygen	–253.494	0.0	201.386	41.688	1.00
F. Coliforms	12635.701	32.208	–5398.275	–294.716	0.143

TABLE 8.18B Constants Value of Eq. (8.4) for the Studied Parameters

Pollutant	Z_1	Z_2	Z_3	Z_4	R^2
Ammonia	1.781	–0.798	0.858	0.024	0.969
Phosphate	3.598	0.556	0.167	0.001	0.826
D. Oxygen	1.00	–0.009	0.026	0.233	0.938
F. Coliforms	0.345	–0.259	0.242	3.045	0.984

in Eq. (8.4) in addition to the determination coefficients. The input concentration for DO was not taken into consideration as its value equal to zero. Also the unit of input and output concentrations for fecal coliforms was MPN/100 ml and mg/l for the remaining parameters. Application limits of such equation are summarized later in Table 8.21.

$$C_o = b + b_1 \times C_i^{Z_1} + b_2 \times q^{Z_2} + b_3 \times A_s^{Z_3} + b_4 \times v^{Z_4} \quad (8.4)$$

8.8.2 CALIBRATION PROCESS FOR REGRESSION EQUATIONS

The performance and the error values of the proposed regression equations for 60 patterns are presented in Figures 8.127–8.130 for ammonia, phosphate, dissolved oxygen, and fecal coliforms, respectively.

From the analysis of results and referring to Figures 8.127–8.130 it is demonstrated that, the error between the experimental values and the SPSS model outputs varies between −0.407 and +0.248 mg/l for ammonia. For phosphate this error ranges from −0.043 to +0.072 mg/l. For dissolved oxygen the error ranges from −0.179 to +0.170 mg/l, whereas, for fecal coliforms this error varies between −130 and +122 MPN/100 ml. Figures 8.131–8.134 show the equality diagrams for these four parameters.

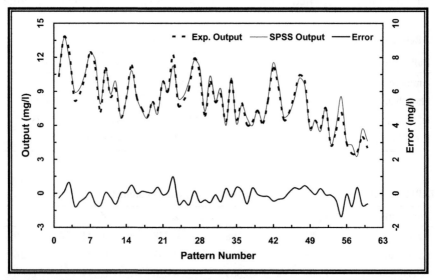

FIGURE 8.127 Training performance and error values of 60 patterns for NH_3 outputs (Exp. and SPSS).

ANNs Modeling and SPSS Analysis 449

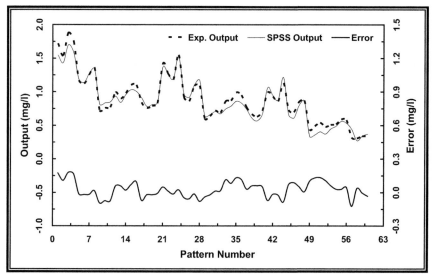

FIGURE 8.128 Training performance and error values of 60 patterns for PO_4 outputs (Exp. and SPSS).

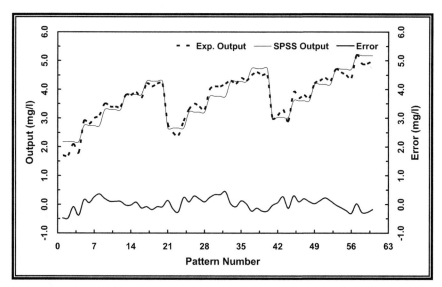

FIGURE 8.129 Training performance and error values of 60 patterns for DO outputs (Exp. and SPSS).

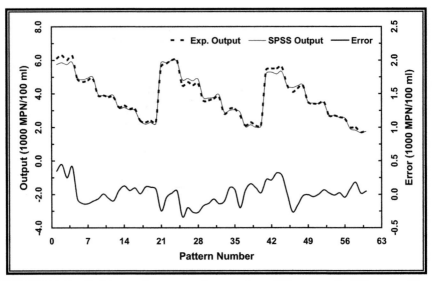

FIGURE 8.130 Training performance and error values of 60 patterns for FC outputs (Exp. and SPSS).

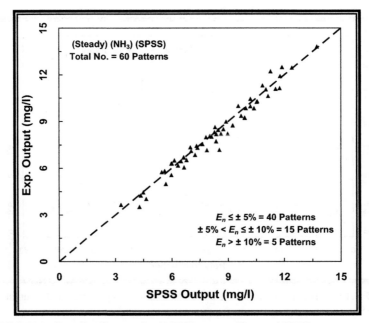

FIGURE 8.131 Equality diagram for 60 NH_3 patterns (Exp. and SPSS).

ANNs Modeling and SPSS Analysis 451

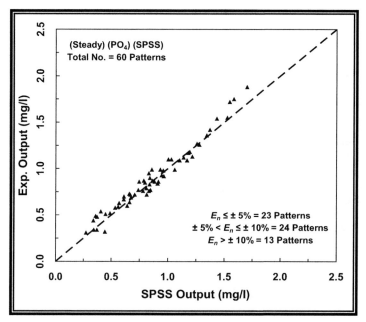

FIGURE 8.132 Equality diagram for 60 PO$_4$ patterns (Exp. and SPSS).

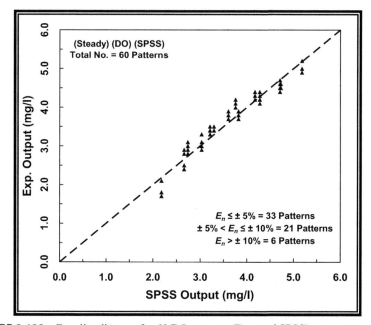

FIGURE 8.133 Equality diagram for 60 DO patterns (Exp. and SPSS).

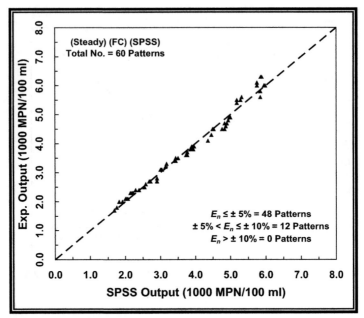

FIGURE 8.134 Equality diagram for 60 FC patterns (Exp. and SPSS).

Figures 8.131–8.134 highlight that:

- The SPSS model calibration results are matching with the experimental output values for FC pollutant. For NH_3, PO_4, and DO, the results are acceptable except some peculiar points.
- The difference between experimental and SPSS results are higher than the corresponding results of ANNs model outputs.
- For ammonia, 40 points give E_s less than ±5% and 15 points give E_s between ±5 and ±10%, and 5 points give $E_s \geq \pm 10\%$.
- For phosphate, 23 points give E_s less than ±5% and 24 points give E_s between ±5 and ±10%, and 13 points gave $E_s \geq \pm 10\%$.
- For DO, 33 points give E_s less than ±5% and 21 points give E_s between ±5 and ±10%, and 6 points give $E_s \geq \pm 10\%$.
- For fecal coliforms, 48 points give E_s less than ±5% and 12 points give E_s between ±5 and ±10%.

8.8.3 VALIDATION PROCESS FOR REGRESSION EQUATIONS

The validation performance and error values of the proposed equations for 15 patterns are presented in Figures 8.135–8.138 for ammonia, phosphate, dissolved oxygen, and fecal coliforms, respectively.

ANNs Modeling and SPSS Analysis

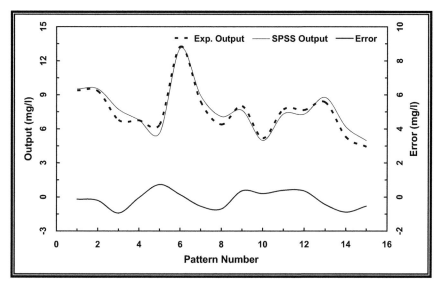

FIGURE 8.135 Validation performance and error values of 15 patterns for NH_3 outputs (Exp. and SPSS).

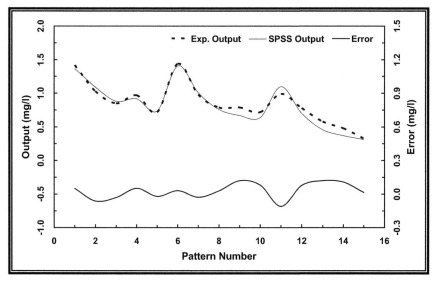

FIGURE 8.136 Validation performance and error values of 15 patterns for PO_4 outputs (Exp. and SPSS).

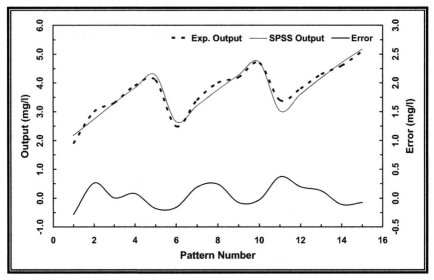

FIGURE 8.137 Validation performance and error values of 15 patterns for DO outputs (Exp. and SPSS).

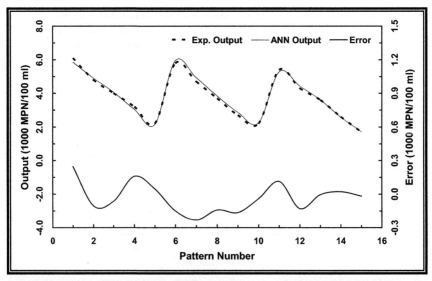

FIGURE 8.138 Validation performance and error values of 15 patterns for FC outputs (Exp. and SPSS).

From the analysis of results and referring to Figures 8.135–8.138, it can be obtained that:

- The error between experimental results and SPSS model outputs varies between −0.487 and +0.366 mg/l for ammonia.
- The error between experimental results and SPSS model outputs varies between −0.049 and +0.069 mg/l for phosphate.
- The error between experimental results and SPSS model outputs varies between −0.130 and +0.184 mg/l for dissolved oxygen.
- The error between experimental results and SPSS model outputs varies between −104 and +117 MPN/100 ml for fecal coliforms.

Figures 8.139–8.142 show the equality diagrams of validation process for NH_3, PO_4, DO, and FC parameters, respectively.

From Figures 8.139–8.142, it can be concluded that:

- The SPSS model validation results are matching with the experimental outputs for FC. For NH_3, PO_4, and DO the results are acceptable with some peculiar points.
- The difference between experimental and SPSS results is higher than the corresponding results of ANNs model outputs.

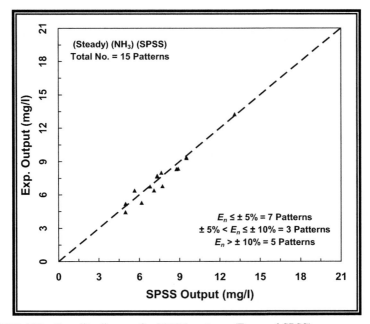

FIGURE 8.139 Equality diagram for 15 NH_3 patterns (Exp. and SPSS).

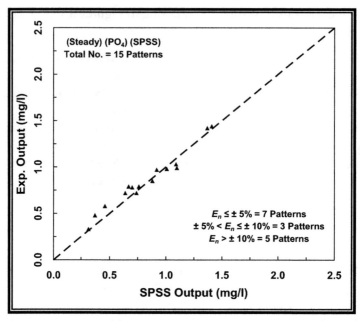

FIGURE 8.140 Equality diagram for 15 PO_4 patterns (Exp. and SPSS).

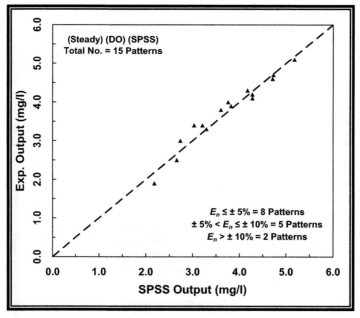

FIGURE 8.141 Equality diagram for 15 DO patterns (Exp. and SPSS).

ANNs Modeling and SPSS Analysis

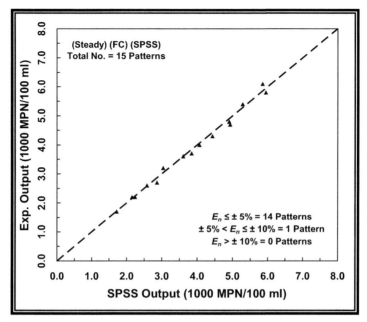

FIGURE 8.142 Equality diagram for 15 FC patterns (Exp. and SPSS).

- For NH_3 pollutant, 7 points give E_s less than ±5% and 3 points give E_s between ±5 and ±10%, and 5 points give $E_s \geq \pm 10\%$.
- For PO_4 pollutant, 7 points give E_s less than ±5% and 3 points give E_s between ±5 and ±10%, and 5 points give $E_s \geq \pm 10\%$.
- For DO parameter, 8 points give E_s less than ±5% and 5 points give E_s between ±5 and ±10%, and 2 points give $E_s \geq \pm 10\%$.
- For FC pollutant, 14 points give E_s less than ±5% and one point give E_s between ±5 and ±10%.

8.9 STATISTICAL MODELING – STEADY (HEAVY METALS)

Statistical procedures were carried out using the SPSS software. Four non-linear and one linear regression equations were obtained to represent the experimental data for the studied heavy metals parameters in steady stage. The same data used in artificial neural networks with the same excluded ones for calibration and validation were undergone to the SPSS analysis. This symmetry was conducted to make a comparison between ANNs and SPSS results. Many different types of regression equations were tested and the selected one which gave the least mean square error and the best convergence about the line of perfect agreement.

8.9.1 REGRESSION EQUATIONS USING SPSS

Equation (8.5) represents the general form of the deduced regression equations for Zinc, Iron, Manganese, and Lead, whereas Eq. (8.6) is the obtained one for Cadmium element ($R^2 = 0.959$). Table 8.19 shows the constant values in Eq. (8.5) in addition to the determination coefficient. Application limits of such equations are summarized later in Table 8.21.

$$C_o = b + b_1 \times C_i^{Z_1} + b_2 \times q^{Z_2} + b_3 \times A_s^{Z_3} + b_4 \times v^{Z_4} \qquad (8.5)$$

$$\begin{aligned}C_o = &\ 2.5E-06 + 0.721 \times C_i + 0.003 \times q - 2.529E-07 \\ &\times A_s - 2.133E-05 \times v\end{aligned} \qquad (8.6)$$

8.9.2 CALIBRATION PROCESS FOR REGRESSION EQUATIONS

The performance and error value of the proposed regression equations for 60 patterns are presented in Figures 8.143–8.147 for Zinc, Iron, Manganese, Lead, Cadmium elements, respectively.

TABLE 8.19A Constants of Eq. (8.5) for Zn, Fe, Mn, and Pb Elements

Pollutant	b	b_1	b_2	b_3	b_4
Zinc	−72.867	0.065	73.392	32.690	0.014
Iron	0.00	0.642	10.736	−9.679	0.899
Manganese	−16.160	0.620	19.131	−2.036	0.145
Lead	−0.079	0.182	0.060	1.033	1.553E-05

TABLE 8.19B Constants of Eq. (8.5) for Zn, Fe, Mn, and Pb Elements

Pollutant	Z_1	Z_2	Z_3	Z_4	R^2
Zinc	2.645	0.005	−0.548	0.522	0.983
Iron	0.786	0.027	0.019	−0.028	0.950
Manganese	1.468	0.006	0.038	−2.495	0.934
Lead	0.291	0.655	−0.604	1.721	0.939

ANNs Modeling and SPSS Analysis

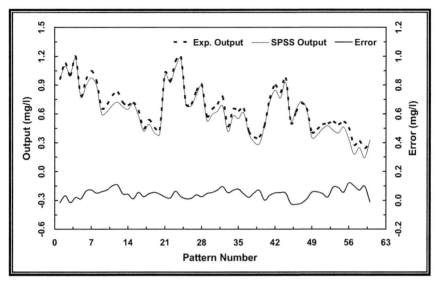

FIGURE 8.143 Training performance and error values of 60 patterns for Zn outputs (Exp. and SPSS).

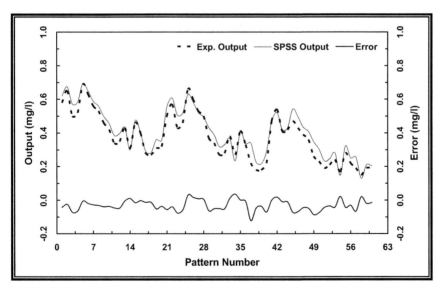

FIGURE 8.144 Training performance and error values of 60 patterns for Fe outputs (Exp. and SPSS).

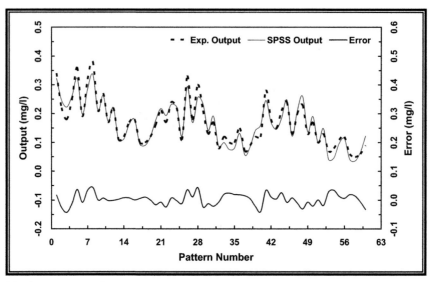

FIGURE 8.145 Training performance and error values of 60 patterns for Mn outputs (Exp. and SPSS).

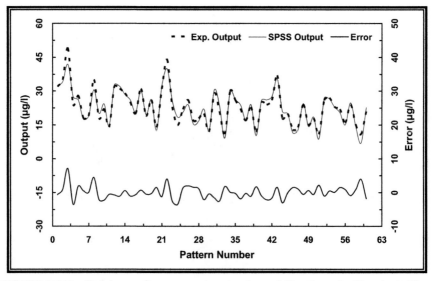

FIGURE 8.146 Training performance and error values of 60 patterns for Pb outputs (Exp. and SPSS).

ANNs Modeling and SPSS Analysis 461

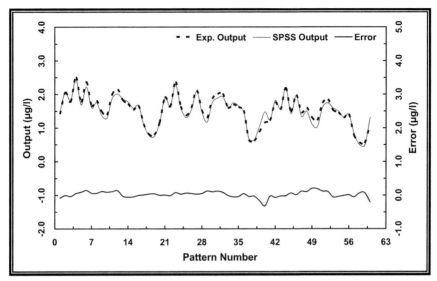

FIGURE 8.147 Training performance and error values of 60 patterns for Cd outputs (Exp. and SPSS).

From the analysis of results and referring to Figures 8.143–8.147, it is remarked that, the error between the experimental values and the SPSS model outputs varies between −0.0177 and +0.0524 mg/l for Zinc element. For Iron this error ranges from −0.0418 to +0.0154 mg/l, whereas, for Manganese the error varies between −0.0164 and +0.0163 mg/l.

For Lead the error ranges from −1.3 to +1.7 µg/l, whereas, for Cadmium the error varies between −0.06 µg/l and +0.08 µg/l. Figures 8.148–8.152 illustrate the equality diagrams for Zinc, Iron, Manganese, Lead, Cadmium elements, respectively.

From Figures 8.148–8.152, it can be obtained that:

- The SPSS model calibration results are acceptable with the experimental output values for cadmium element with some peculiar points. For the remaining elements the number of peculiar points is relatively big.
- The difference between experimental and SPSS results is higher than the corresponding results of ANNs model outputs.
- For Zinc element, 21 points give E_s less than ±5% and 18 points give E_s between ±5 and ±10%, and 21 points give $E_s \geq \pm 10\%$.
- For Iron element, 21 points give E_s less than ±5% and 6 points give E_s between ±5 and ±10%, and 33 points give $E_s \geq \pm 10\%$.
- For Manganese element, 16 points give E_s less than ±5% and 13 points give E_s between ±5 and ±10%, and 31 points give E_s more than ±10%.

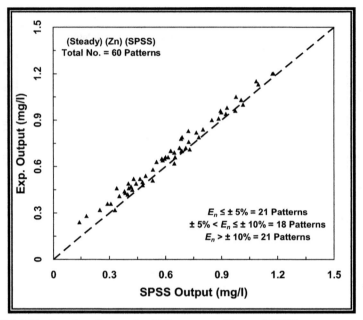

FIGURE 8.148 Equality diagram for 60 Zn patterns (Exp. and SPSS).

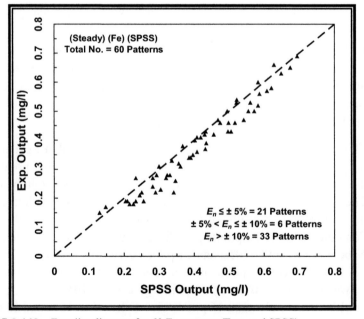

FIGURE 8.149 Equality diagram for 60 Fe patterns (Exp. and SPSS).

ANNs Modeling and SPSS Analysis 463

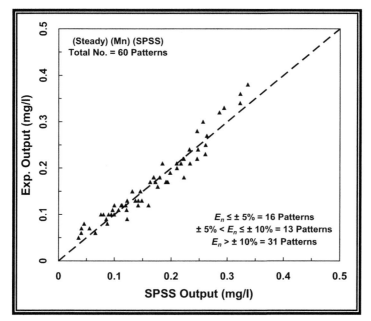

FIGURE 8.150 Equality diagram for 60 Mn patterns (Exp. and SPSS).

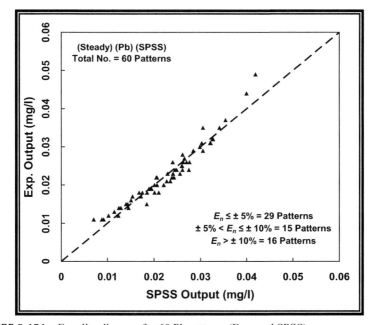

FIGURE 8.151 Equality diagram for 60 Pb patterns (Exp. and SPSS).

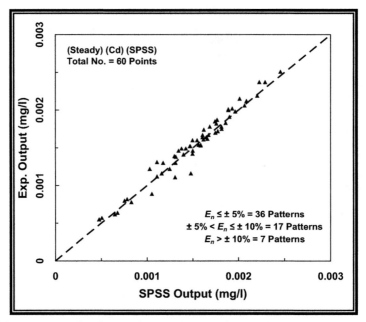

FIGURE 8.152 Equality diagram for 60 Cd patterns (Exp. and SPSS).

- For Lead element, 29 points give E_s less than ±5% and 15 points give E_s between ±5 and ±10%, and 16 points give E_s more than ±10%.
- For Cadmium element, 36 points give E_s less than ±5% and 17 points give E_s between ±5 and ±10%, and 7 points give E_s more than ±10%.

8.9.3 VALIDATION PROCESS FOR REGRESSION EQUATIONS

The validation performance and error values of the proposed equations for 15 patterns are presented in Figures 8.153–8.157 for Zinc, Iron, Manganese, Lead, Cadmium elements, respectively.

From the analysis of results and referring to Figures 8.153–8.157, it can be obtained that, the error between the experimental results and SPSS model outputs varies between:

- −0.0029 and +0.0524 mg/l for Zinc element.
- −0.0507 and +0.023 mg/l for Iron element.
- −0.0241 and +0.0194 mg/l for Manganese element.
- −1.50 and +2.0 µg/l for Lead element.
- −0.05 and +0.1 µg/l for Cadmium element.

ANNs Modeling and SPSS Analysis

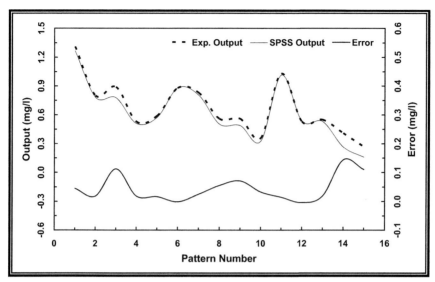

FIGURE 8.153 Validation performance and error values of 15 patterns for Zn outputs (Exp. and SPSS).

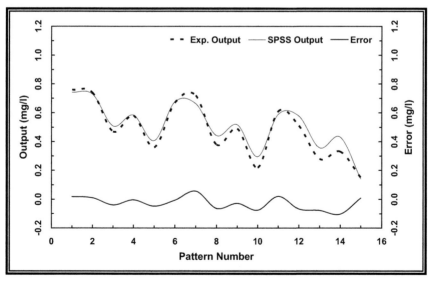

FIGURE 8.154 Validation performance and error values of 15 patterns for Fe outputs (Exp. and SPSS).

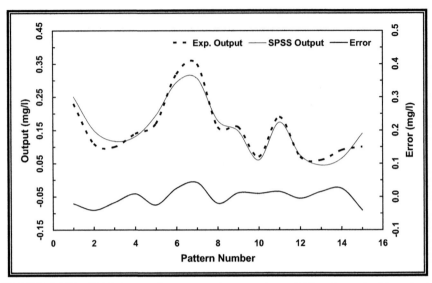

FIGURE 8.155 Validation performance and error values of 15 patterns for Mn outputs (Exp. and SPSS).

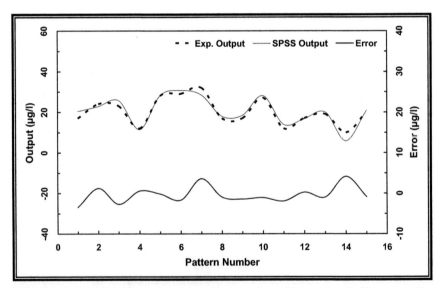

FIGURE 8.156 Validation performance and error values of 15 patterns for Pb outputs (Exp. and SPSS).

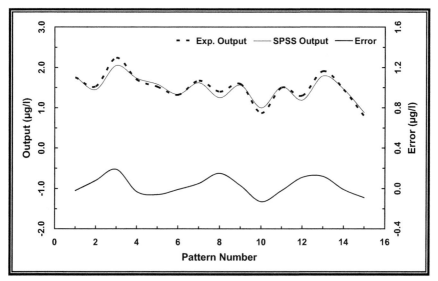

FIGURE 8.157 Validation performance and error values of 15 patterns for Cd outputs (Exp. and SPSS).

The equality diagrams for Zinc, Iron, Manganese, Lead, Cadmium elements, respectively are presented in Figures 8.158–8.162, respectively.

From Figures 8.158–8.162, it can be noticed that:

- For Zinc element, 9 points give E_s less than ±5% and 2 points give E_s between ±5 and ±10%, and 4 points give $E_s \geq \pm 10\%$.
- For Iron element, 7 points give E_s less than ±5% and 3 points give E_s between ±5 and ±10%, and 5 points give $E_s \geq \pm 10\%$.
- For Manganese element, 6 points give E_s between ±5 and ±10%, and 9 points give E_s more than ±10%.
- For Lead element, 5 points give E_s less than ±5% and 5 points give E_s between ±5 and ±10%, and 5 points give E_s more than ±10%.
- For Cadmium element, 9 points give E_s less than ±5% and 3 points give E_s between ±5 and ±10%, and 3 points give E_s more than ±10%.
- The SPSS model validation results are acceptable compared with the experimental output values for Cd element with some peculiar points. For Zn, Fe, and Pb elements the number of peculiar points is relatively big. As for Mn element the results are not good compared with the other elements.
- The difference between experimental and SPSS results is higher than the corresponding results of ANNs model outputs.

468 Constructed Subsurface Wetlands

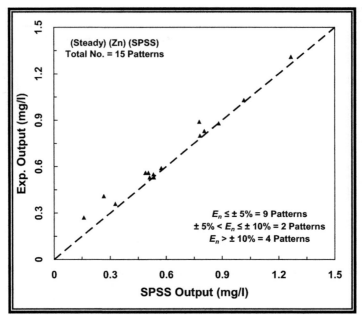

FIGURE 8.158 Equality diagram for 15 Zn patterns (Exp. and SPSS).

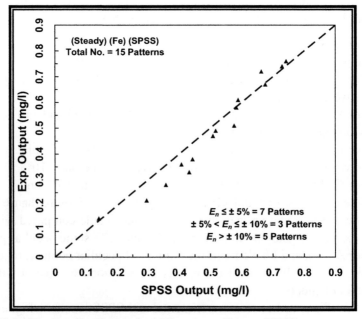

FIGURE 8.159 Equality diagram for 15 Fe patterns (Exp. and SPSS).

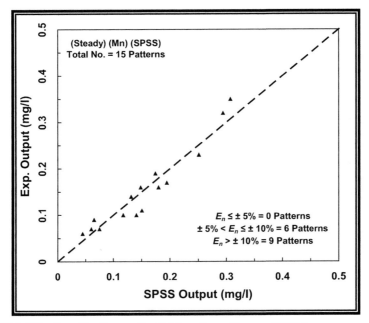

FIGURE 8.160 Equality diagram for 15 Mn patterns (Exp. and SPSS).

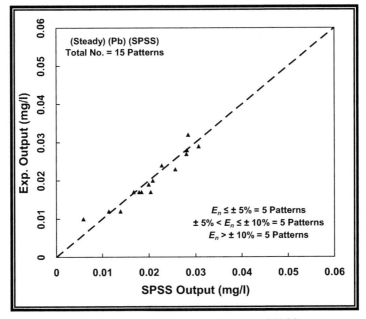

FIGURE 8.161 Equality diagram for 15 Pb patterns (Exp. and SPSS).

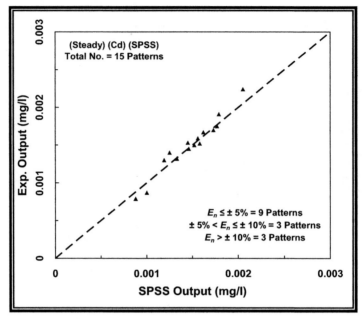

FIGURE 8.162 Equality diagram for 15 Cd patterns (Exp. and SPSS).

8.10 LIMITS OF REGRESSION EQUATIONS

Tables 8.20 presents the limits of independent variables for set up stage pollutants (Eq. 8.1). Whereas Table 8.21 gives these limits for steady stage parameters (Eqs. 8.2–8.6).

8.11 COMPARISON BETWEEN (EXP. – ANN – SPSS) RESULTS

This section discusses a comparison process between each of artificial neural network outputs and SPSS outputs with the experimental measured values for both set up and steady stages.

8.11.1 SET-UP STAGE

The input variables and the results of the validation process of 24 patterns for the obtained models (ANNs and SPSS) and the experimental output concentrations are shown in Tables 8.22–8.24 for BOD, COD, and TSS pollutants, respectively. The total experimental data (144 patterns for construction and test) used in artificial neural networks and SPSS analysis are listed in Appendix III.

ANNs Modeling and SPSS Analysis

TABLE 8.20 Limits of Independent Variables for Set Up Stage Pollutants

Pollutant	C_i(mg/l)	q (m/d)	A_s(m²)	v(m/d)	T_o(d)
BOD	$168 < C_i < 232$	$0.319 < q < 1.695$	$177 < A_s < 2265$	$8.43 < v < 21.10$	$56 < T_o < 210$
COD	$251 < C_i < 368$	$0.319 < q < 1.695$	$177 < A_s < 2265$	$8.43 < v < 21.10$	$56 < T_o < 210$
TSS	$138 < C_i < 180$	$0.319 < q < 1.695$	$177 < A_s < 2265$	$8.43 < v < 21.10$	$56 < T_o < 210$

TABLE 8.21A Limits of Independent Variables for Steady Stage

Pollutant	C_i(mg/l)	q (m/d)	A_s(m²)	v(m/d)
BOD	$162 < C_i < 209$	$0.059 < q < 1.280$	$177 < A_s < 2265$	$1.65 < v < 14.03$
COD	$242 < C_i < 337$	$0.059 < q < 1.280$	$177 < A_s < 2265$	$1.65 < v < 14.03$
TSS	$131 < C_i < 180$	$0.059 < q < 1.280$	$177 < A_s < 2265$	$1.65 < v < 14.03$

TABLE 8.21B Limits of Independent Variables for Steady Stage

Pollutant	C_i(mg/l)	q (m/d)	A_s(m²)	v(m/d)
Ammonia	$15.28 < C_i < 26.16$	$0.059 < q < 0.256$	$1138 < A_s < 2265$	$1.75 < v < 13.99$
Phosphate	$1.89 < C_i < 3.54$	$0.059 < q < 0.256$	$1138 < A_s < 2265$	$1.75 < v < 13.99$
Dissolved Oxygen	Equal zero	$0.059 < q < 0.256$	$1138 < A_s < 2265$	$1.75 < v < 13.99$
Fecal Coliforms	$200 < C_i < 270$	$0.059 < q < 0.256$	$1138 < A_s < 2265$	$1.75 < v < 13.99$

*Fecal coliforms concentrations in units 1000 MPN/100 ml.

TABLE 8.21C Limits of Independent Variables for Steady Stage

Pollutant	C_i(mg/l)	q (m/d)	A_s(m²)	v(m/d)
Zinc	$1.28 < C_i < 2.16$	$0.059 < q < 0.256$	$1138 < A_s < 2265$	$1.75 < v < 13.99$
Iron	$0.63 < C_i < 1.05$	$0.059 < q < 0.256$	$1138 < A_s < 2265$	$1.75 < v < 13.99$
Manganese	$0.15 < C_i < 0.50$	$0.059 < q < 0.256$	$1138 < A_s < 2265$	$1.75 < v < 13.99$
Lead	$0.021 < C_i < 0.072$	$0.059 < q < 0.256$	$1138 < A_s < 2265$	$1.75 < v < 13.99$
Cadmium	$1.26 < C_i < 3.06$	$0.059 < q < 0.256$	$1138 < A_s < 2265$	$1.75 < v < 13.99$

*Cadmium concentrations in units μg/l.

Table 8.25 summarizes the average percentage errors (negative and positive) between experimental output concentration and both ANNs (E_n) and SPSS (E_s) outputs for the 120 and 24 training performance patterns

TABLE 8.22 Comparison Between Experimental, ANNs, and SPSS for BOD Outputs

Media Type		C_i	q	A_s	v	T_o	C_{Exp}	C_{ANN}	C_{SPSS}	E_n	E_s
		\multicolumn{5}{c}{BOD Input Data}					
1	Rubber Media	172	1.695	177.5	16.07	140	158	157.70	158.33	0.19	−0.21
2		174	1.695	177.5	16.28	196	154	152.70	145.69	0.85	5.39
3		191	0.678	568.9	14.93	126	156	156.43	154.21	−0.28	1.15
4		168	0.678	568.9	15.17	210	122	122.73	116.47	−0.60	4.53
5		198	0.424	960.3	14.09	98	161	159.61	156.62	0.87	2.72
6		180	0.424	960.3	14.29	154	134	133.71	129.53	0.22	3.34
7		232	0.339	1137.8	13.90	84	184	186.11	180.78	−1.15	1.75
8		180	0.339	1137.8	14.26	154	126	126.13	125.57	−0.10	0.34
9	Gravel Media	211	1.668	210.3	17.36	70	196	194.67	203.03	0.68	−3.58
10		172	1.668	210.3	18.09	140	155	154.05	157.17	0.61	−1.40
11		198	0.667	741.0	20.58	98	161	160.36	162.86	0.40	−1.16
12		184	0.667	741.0	20.98	182	127	128.68	131.61	−1.32	−3.63
13		211	0.417	1271.6	19.14	70	168	168.66	167.66	−0.39	0.20
14		194	0.417	1271.6	19.93	168	126	126.85	130.68	−0.68	−3.71
15		191	0.334	1481.9	18.20	126	135	134.58	134.82	0.31	0.13
16		184	0.334	1481.9	18.43	182	106	107.42	115.68	−1.34	−9.13

ANNs Modeling and SPSS Analysis

TABLE 8.22 (Continued)

Media Type		C_i	q	A_s	v	T_o	C_{Exp}	C_{ANN}	C_{SPSS}	E_n	E_s
17	Plastic Media	220	1.594	285.2	11.75	56	202	195.80	211.47	3.07	−4.69
18		184	1.594	285.2	12.31	182	154	153.73	153.75	0.17	0.16
19		191	0.638	1132.6	9.05	126	144	144.50	145.85	−0.34	−1.29
20		168	0.638	1132.6	9.15	210	108	106.78	108.13	1.13	−0.12
21		191	0.399	1980.1	8.62	126	135	135.48	129.52	−0.36	4.06
22		180	0.399	1980.1	8.64	154	119	117.56	114.53	1.21	3.75
23		211	0.319	2265.3	9.06	70	157	157.56	152.48	−0.36	2.88
24		180	0.319	2265.3	9.29	154	109	107.37	108.95	1.50	0.04

TABLE 8.23 Comparison Between Experimental, ANNs, and SPSS for COD Outputs

Media Type		C_i	q	A_s	v	T_o	C_{Exp}	C_{ANN}	C_{SPSS}	E_n	E_s
1	Rubber Media	331	1.695	177.5	15.97	112	308	310.39	302.69	−0.78	1.72
2		264	1.695	177.5	16.28	196	234	235.14	221.48	−0.49	5.35
3		300	0.678	568.9	14.79	98	255	253.15	254.27	0.72	0.29
4		308	0.678	568.9	15.05	168	234	233.29	232.61	0.30	0.59
5		368	0.424	960.3	13.96	84	302	305.62	293.65	−1.20	2.76
6		264	0.424	960.3	14.43	196	178	178.00	174.78	0.00	1.81
7		368	0.339	1137.8	13.90	84	294	296.89	287.39	−0.98	2.25
8		273	0.339	1137.8	14.26	154	192	190.58	191.45	0.74	0.28

TABLE 8.23 (Continued)

Media Type		C_i	q	A_s	v	T_o	C_{Exp}	C_{ANN}	C_{SPSS}	E_n	E_s
Gravel Media	9	344	1.668	210.3	17.21	56	323	320.23	332.30	0.86	−2.88
	10	308	1.668	210.3	18.19	168	272	270.47	262.35	0.56	3.55
	11	325	0.667	741.0	20.09	70	275	272.95	277.62	0.75	−0.95
	12	308	0.667	741.0	20.92	168	222	220.60	226.87	0.63	−2.20
	13	325	0.417	1271.6	19.14	70	256	258.70	259.14	−1.06	−1.22
	14	264	0.417	1271.6	20.05	196	161	162.70	165.70	−1.05	−2.92
	15	331	0.334	1481.9	18.15	112	242	239.31	240.18	1.11	0.75
	16	308	0.334	1481.9	18.36	168	190	185.02	201.66	2.62	−6.13
Plastic Media	17	368	1.594	285.2	11.93	84	333	336.62	336.26	−1.09	−0.98
	18	251	1.594	285.2	12.36	210	208	208.21	202.55	−0.10	2.62
	19	257	0.638	1132.6	9.06	140	193	192.92	193.04	0.04	−0.02
	20	297	0.638	1132.6	9.11	182	194	196.04	205.45	−1.05	−5.90
	21	325	0.399	1980.1	8.46	70	252	251.86	243.20	0.05	3.49
	22	300	0.399	1980.1	8.57	98	222	220.88	214.21	0.50	3.51
	23	331	0.319	2265.3	9.25	112	226	227.72	222.09	−0.76	1.73
	24	273	0.319	2265.3	9.29	154	167	167.23	163.84	−0.14	1.89

ANNs Modeling and SPSS Analysis

TABLE 8.24 Comparison Between Experimental, ANNs, and SPSS for TSS Outputs

Media Type		C_i	q	A_s	v	T_o	C_{Exp}	C_{ANN}	C_{SPSS}	E_n	E_s
Rubber Media	1	150	1.695	177.5	15.36	56	130	130.28	130.96	−0.22	−0.74
	2	155	1.695	177.5	16.32	210	132	132.14	127.98	−0.10	3.05
	3	148	0.678	568.9	14.54	70	104	104.93	102.05	−0.89	1.87
	4	144	0.678	568.9	15.14	196	97	97.33	94.46	−0.34	2.62
	5	162	0.424	960.3	13.96	84	101	100.54	97.66	0.45	3.31
	6	138	0.424	960.3	14.37	182	82	81.64	78.84	0.44	3.86
	7	180	0.339	1137.8	14.04	98	105	105.65	103.40	−0.62	1.53
	8	172	0.339	1137.8	14.31	168	96	96.43	95.56	−0.45	0.46
Gravel Media	9	146	1.668	210.3	17.96	112	125	124.70	127.23	0.24	−1.78
	10	166	1.668	210.3	18.14	154	141	140.98	137.82	0.02	2.26
	11	143	0.667	741.0	20.75	126	95	95.79	98.77	−0.83	−3.97
	12	138	0.667	741.0	20.98	182	90	90.30	93.38	−0.33	−3.75
	13	162	0.417	1271.6	19.35	84	97	97.87	96.80	−0.89	0.20
	14	155	0.417	1271.6	20.10	210	87	86.53	87.51	0.54	−0.58
	15	180	0.334	1481.9	18.03	98	100	101.97	100.70	−1.97	−0.70
	16	166	0.334	1481.9	18.31	154	88	88.37	89.82	−0.42	−2.07

TABLE 8.24 (Continued)

Media Type		C_i	q	A_s	v	T_o	C_{Exp}	C_{ANN}	C_{SPSS}	E_n	E_s
				TSS Input Data							
17	Plastic Media	180	1.594	285.2	12.05	98	144	145.18	140.71	−0.82	2.29
18		166	1.594	285.2	12.23	154	131	131.49	129.72	−0.38	0.97
19		162	0.638	1132.6	8.92	84	89	88.54	94.05	0.51	−5.68
20		138	0.638	1132.6	9.11	182	72	72.17	75.03	−0.24	−4.21
21		148	0.399	1980.1	8.46	70	69	68.38	65.86	0.89	4.56
22		144	0.399	1980.1	8.70	196	62	59.99	57.93	3.24	6.56
23		146	0.319	2265.3	9.25	112	61	60.91	56.80	0.15	6.88
24		166	0.319	2265.3	9.29	154	67	67.24	67.27	−0.36	−0.41

TABLE 8.25 Average Errors Between Experimental and Both ANN and SPSS (Set Up Stage)

Parameter	Average E_n (%)(ANNs)		Average E_s (%)(SPSS)	
	120 Patterns	24 Patterns	120 Patterns	24 Patterns
BOD	−0.32 : +0.60	−0.63 : +0.86	−2.36 : +2.06	−2.89 : +2.18
COD	−0.39 : +0.45	−0.79 : +0.68	−2.56 : +2.02	−2.58 : +2.17
TSS	−0.36 : +0.40	−0.59 : +0.72	−2.39 : +2.98	−2.39 : +2.89

for pollutants under study. Figures 8.163–8.165 illustrate the percentage errors E_n and E_s for 120 calibration patterns for BOD, COD, and TSS, respectively.

From Table 8.25 and Figures 8.163–8.165, it can be obtained that:

- For 120 calibration patterns, the BOD average percentage error between the experimental output concentrations and the SPSS model output concentrations varies between −2.36% and +2.06% but this percentage error ranges from −0.32% to +0.60% using ANNs model.
- For 120 calibration patterns, the COD average percentage error between the experimental output concentrations and the SPSS model output concentrations varies between −2.56% and +2.02% but this percentage error ranges from −0.39% to +0.45% using ANNs model.
- For 120 calibration patterns, the TSS average percentage error between the experimental output concentrations and the SPSS model output concentrations varies between −2.39% and +2.98% but this percentage error ranges from −0.36% to +0.40% using ANNs model.
- The artificial neural network models represent the experimental data in calibration process which are more accurate than the SPSS models. This is due to the ability of ANNs models to construct a variety of complex relationships between variables to represent it precisely.

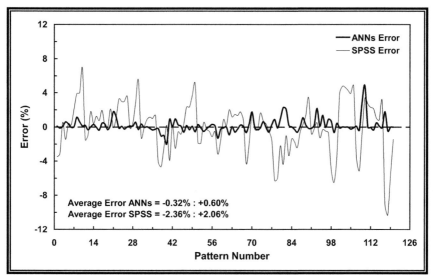

FIGURE 8.163 Percentage errors between Experimental and both ANNs and SPSS outputs (BOD – 120 Patterns).

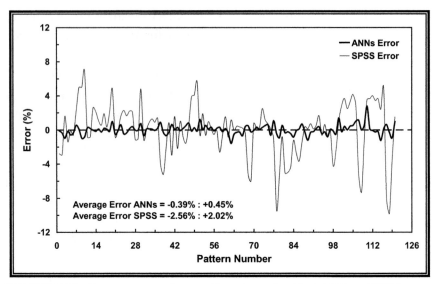

FIGURE 8.164 Percentage errors between Experimental and both ANNs and SPSS outputs (COD – 120 Patterns).

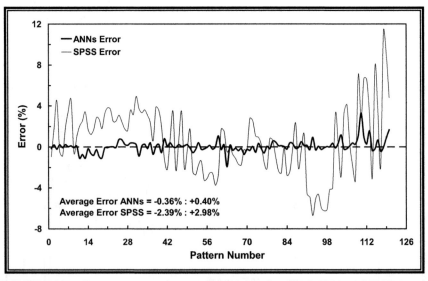

FIGURE 8.165 Percentage errors between Experimental and both ANNs and SPSS outputs (TSS – 120 Patterns).

Figures 8.166–8.168 illustrate the percentage errors E_n and E_s for 24 validation patterns for BOD, COD, and TSS pollutants, respectively. These figures are provided with the average negative and positive percentage errors for these two used models.

ANNs Modeling and SPSS Analysis

From Table 8.25 and Figures 8.166–8.168, it can be found that:

- For 24 validation patterns, the BOD average percentage error between the experimental measured output and SPSS model outputs varies between −2.89% and +2.18% but this percentage error ranges from −0.63% to +0.86% using ANNs.
- For 24 validation patterns, the COD average percentage error between the experimental measured output and SPSS model outputs varies between −2.58% and +2.17% but this percentage error ranges from −0.79% to +0.68% using ANNs.
- For 24 validation patterns, the TSS average percentage error between the experimental measured output and SPSS model outputs varies between −2.39% and +2.89% but this percentage error ranges from −0.59% to +0.72% using ANNs.
- The ANNs models represent the experimental data in validation process more accurately than the SPSS models.

8.11.2 STEADY STAGE (BOD, COD, AND TSS)

The results of the validation process of 60 patterns for the obtained models and the percentage errors for BOD, COD, and TSS are shown in Tables 8.26–8.34.

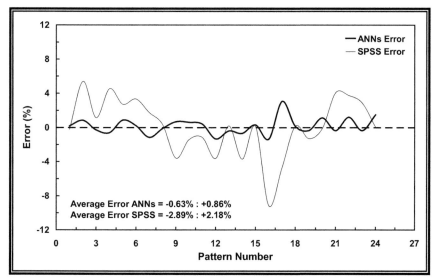

FIGURE 8.166 Percentage errors between Experimental and both ANNs and SPSS outputs (BOD – 24 Patterns).

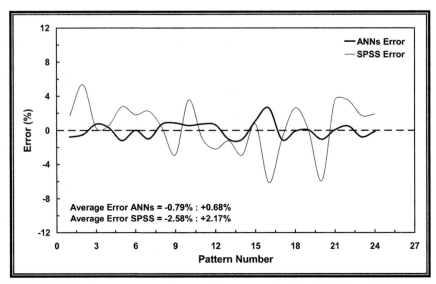

FIGURE 8.167 Percentage errors between Experimental and both ANNs and SPSS outputs (COD – 24 Patterns).

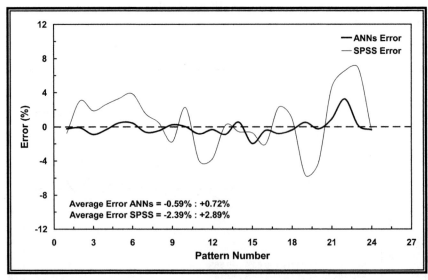

FIGURE 8.168 Percentage errors between Experimental and both ANNs and SPSS outputs (TSS – 24 Patterns).

TABLE 8.26 Experimental, ANN, and SPSS Output Concentrations (BOD – Rubber – Steady)

No.	C_i	q	A_s	v	C_{Exp}	C_{ANN}	C_{SPSS}	E_n	E_s
1	190	1.280	177.4	12.33	164	163.33	170.08	0.41	−3.70
2	203	0.871	177.4	8.39	168	168.28	168.41	−0.17	−0.24
3	198	0.599	177.4	5.77	158	157.65	158.88	0.22	−0.56
4	199	0.424	177.4	4.09	156	155.66	152.76	0.22	2.08
5	179	0.316	177.4	3.04	139	139.08	135.63	−0.06	2.42
6	197	0.512	568.9	10.91	136	135.23	132.74	0.57	2.40
7	188	0.348	568.9	7.42	116	113.22	118.20	2.39	−1.90
8	198	0.240	568.9	5.11	110	113.53	114.47	−3.21	−4.07
9	162	0.170	568.9	3.62	84	82.91	84.91	1.30	−1.09
10	183	0.126	568.9	2.69	91	90.58	88.50	0.46	2.75
11	182	0.320	960.3	10.62	107	107.39	102.63	−0.37	4.09
12	188	0.218	960.3	7.22	96	92.91	94.98	3.22	1.07
13	180	0.150	960.3	4.97	76	76.92	79.87	−1.21	−5.09
14	199	0.106	960.3	3.52	78	79.25	79.34	−1.61	−1.72
15	179	0.079	960.3	2.62	63	63.58	58.12	−0.92	7.75
16	182	0.256	1137.8	10.91	98	96.83	93.68	1.19	4.40
17	200	0.174	1137.8	7.42	92	91.10	91.48	0.98	0.57
18	198	0.120	1137.8	5.11	72	73.28	78.85	−1.78	−9.51
19	178	0.085	1137.8	3.62	54	54.27	56.11	−0.51	−3.91
20	177	0.063	1137.8	2.69	48	48.52	44.32	−1.07	7.66

TABLE 8.27 Experimental, ANN, and SPSS Output Concentrations (BOD – Gravel – Steady)

No.	C_i	q	A_s	v	C_{Exp}	C_{ANN}	C_{SPSS}	E_n	E_s
1	182	1.259	210.3	13.88	154	156.60	163.97	−1.69	−6.48
2	188	0.857	210.3	9.44	153	153.40	157.23	−0.26	−2.76
3	182	0.589	210.3	6.49	143	143.11	146.54	−0.08	−2.48
4	189	0.417	210.3	4.60	146	145.27	143.62	0.50	1.63
5	187	0.311	210.3	3.42	144	142.37	136.23	1.13	5.40

TABLE 8.27 (Continued)

No.	\multicolumn{4}{c}{BOD Input Data}	C_{Exp}	C_{ANN}	C_{SPSS}	E_n	E_s
	C_i	q	A_s	v					
6	182	0.504	740.9	13.99	116	118.06	121.85	−1.78	−5.04
7	203	0.343	740.9	9.52	117	117.25	120.73	−0.22	−3.19
8	180	0.236	740.9	6.55	89	88.78	98.45	0.24	−10.61
9	178	0.167	740.9	4.64	82	82.31	87.91	−0.38	−7.20
10	177	0.124	740.9	3.45	78	78.49	78.54	−0.62	−0.69
11	182	0.315	1271.6	14.03	100	99.68	99.45	0.32	0.55
12	188	0.214	1271.6	9.54	87	84.61	88.87	2.74	−2.15
13	182	0.147	1271.6	6.57	68	67.96	74.03	0.05	−8.87
14	189	0.104	1271.6	4.65	63	63.10	66.78	−0.16	−5.99
15	187	0.078	1271.6	3.46	58	56.79	55.23	2.08	4.77
16	182	0.252	1481.9	13.99	91	89.60	90.48	1.54	0.57
17	203	0.171	1481.9	9.52	86	85.65	87.12	0.41	−1.30
18	180	0.118	1481.9	6.55	55	56.22	62.33	−2.22	−13.32
19	178	0.083	1481.9	4.64	46	45.37	49.14	1.38	−6.83
20	177	0.062	1481.9	3.45	41	42.42	37.23	−3.46	9.19

TABLE 8.28 Experimental, ANN, and SPSS Output Concentrations (BOD – Plastic – Steady)

No.	\multicolumn{4}{c}{BOD Input Data}	C_{Exp}	C_{ANN}	C_{SPSS}	E_n	E_s
	C_i	q	A_s	v					
1	182	1.204	285.2	9.35	149	149.04	152.26	−0.03	−2.19
2	188	0.819	285.2	6.36	148	147.94	148.79	0.04	−0.53
3	182	0.563	285.2	4.37	138	137.88	138.91	0.09	−0.66
4	189	0.399	285.2	3.10	141	140.80	136.12	0.14	3.46
5	187	0.297	285.2	2.31	138	137.88	128.71	0.09	6.73
6	182	0.481	1132.6	7.09	105	105.42	105.05	−0.40	−0.04
7	203	0.328	1132.6	4.82	106	105.64	108.03	0.34	−1.91
8	180	0.225	1132.6	3.32	80	80.09	86.75	−0.11	−8.44
9	178	0.160	1132.6	2.35	73	71.52	76.38	2.03	−4.62
10	177	0.119	1132.6	1.75	68	67.98	66.95	0.03	1.54
11	182	0.301	1980.1	6.69	88	88.95	81.35	−1.08	7.55

TABLE 8.28 (Continued)

No.	\multicolumn{4}{c	}{BOD Input Data}	C_{Exp}	C_{ANN}	C_{SPSS}	E_n	E_s
	C_i	q	A_s	v					
12	188	0.205	1980.1	4.55	76	75.88	74.98	0.16	1.35
13	182	0.141	1980.1	3.13	59	59.80	61.16	−1.36	−3.66
14	189	0.100	1980.1	2.22	54	54.35	54.04	−0.65	−0.07
15	187	0.074	1980.1	1.65	48	48.16	42.42	−0.33	11.62
16	182	0.241	2265.3	7.09	77	77.64	72.81	−0.83	5.45
17	203	0.164	2265.3	4.82	72	71.93	73.51	0.10	−2.09
18	180	0.113	2265.3	3.32	44	45.13	49.68	−2.58	−12.91
19	199	0.080	2265.3	2.35	42	43.19	48.08	−2.84	−14.47
20	179	0.059	2265.3	1.75	28	29.31	25.74	−4.68	8.06

TABLE 8.29 Experimental, ANN, and SPSS Output Concentrations (COD – Rubber – Steady)

No.	\multicolumn{4}{c	}{COD Input Data}	C_{Exp}	C_{ANN}	C_{SPSS}	E_n	E_s
	C_i	q	A_s	v					
1	289	1.280	177.4	12.33	249	249.30	261.24	−0.12	−4.92
2	322	0.871	177.4	8.39	268	269.86	264.24	−0.69	1.40
3	319	0.599	177.4	5.77	256	260.41	251.96	−1.72	1.58
4	242	0.424	177.4	4.09	189	191.05	199.76	−1.08	−5.69
5	306	0.316	177.4	3.04	239	240.47	225.75	−0.61	5.54
6	289	0.512	568.9	10.91	197	199.79	197.94	−1.42	−0.48
7	337	0.348	568.9	7.42	213	213.45	206.70	−0.21	2.96
8	319	0.240	568.9	5.11	179	174.24	184.00	2.66	−2.80
9	267	0.170	568.9	3.62	139	138.75	142.65	0.18	−2.63
10	306	0.126	568.9	2.69	154	154.22	150.36	−0.14	2.37
11	285	0.320	960.3	10.62	168	166.60	161.71	0.83	3.74
12	322	0.218	960.3	7.22	167	162.29	163.51	2.82	2.09
13	319	0.150	960.3	4.97	138	136.72	145.71	0.93	−5.58
14	267	0.106	960.3	3.52	103	101.89	101.82	1.07	1.15
15	306	0.079	960.3	2.62	111	109.18	107.07	1.64	3.54
16	289	0.256	1137.8	10.91	157	159.00	150.05	−1.28	4.43

TABLE 8.29 (Continued)

No.	COD Input Data				C_{Exp}	C_{ANN}	C_{SPSS}	E_n	E_s
	C_i	q	A_s	v					
17	337	0.174	1137.8	7.42	159	154.18	155.46	3.03	2.23
18	319	0.120	1137.8	5.11	117	116.29	129.02	0.61	−10.27
19	267	0.085	1137.8	3.62	82	82.14	83.73	−0.17	−2.11
20	306	0.063	1137.8	2.69	86	83.33	87.65	3.11	−1.92

TABLE 8.30 Experimental, ANN, and SPSS Output Concentrations (COD – Gravel – Steady)

No.	COD Input Data				C_{Exp}	C_{ANN}	C_{SPSS}	E_n	E_s
	C_i	q	A_s	v					
1	289	1.259	210.3	13.88	246	249.20	259.21	−1.30	−5.37
2	337	0.857	210.3	9.44	280	282.81	266.77	−1.00	4.72
3	319	0.589	210.3	6.49	254	257.81	246.47	−1.50	2.97
4	267	0.417	210.3	4.60	208	208.19	208.62	−0.09	−0.30
5	306	0.311	210.3	3.42	237	237.16	219.90	−0.07	7.22
6	289	0.504	740.9	13.99	187	188.35	194.43	−0.72	−3.97
7	303	0.343	740.9	9.52	177	175.70	181.66	0.73	−2.63
8	291	0.236	740.9	6.55	147	148.04	160.05	−0.71	−8.88
9	267	0.167	740.9	4.64	124	123.87	132.58	0.10	−6.92
10	306	0.124	740.9	3.45	140	140.87	140.10	−0.62	−0.07
11	292	0.315	1271.6	14.03	161	159.63	161.15	0.85	−0.10
12	290	0.214	1271.6	9.54	133	134.39	137.74	−1.04	−3.57
13	291	0.147	1271.6	6.57	108	109.82	120.44	−1.68	−11.52
14	267	0.104	1271.6	4.65	87	86.69	90.39	0.36	−3.90
15	306	0.078	1271.6	3.46	94	93.36	95.44	0.69	−1.53
16	289	0.252	1481.9	13.99	146	146.87	145.62	−0.60	0.26
17	337	0.171	1481.9	9.52	146	150.19	146.25	−2.87	−0.17
18	279	0.118	1481.9	6.55	90	82.84	97.65	7.95	−8.50
19	267	0.083	1481.9	4.64	70	69.28	72.66	1.03	−3.80
20	302	0.062	1481.9	3.45	73	73.52	74.33	−0.71	−1.83

TABLE 8.31 Experimental, ANN, and SPSS Output Concentrations (COD – Plastic – Steady)

No.	COD Input Data				C_{Exp}	C_{ANN}	C_{SPSS}	E_n	E_s
	C_i	q	A_s	v					
1	289	1.204	285.2	9.35	235	237.54	240.17	−1.08	−2.20
2	337	0.819	285.2	6.36	265	267.13	253.53	−0.80	4.33
3	319	0.563	285.2	4.37	242	243.52	234.72	−0.63	3.01
4	242	0.399	285.2	3.10	180	179.71	182.56	0.16	−1.42
5	273	0.297	285.2	2.31	201	200.33	190.89	0.33	5.03
6	285	0.481	1132.6	7.09	166	167.52	164.58	−0.92	0.85
7	322	0.328	1132.6	4.82	170	169.29	170.79	0.42	−0.47
8	291	0.225	1132.6	3.32	132	132.49	141.50	−0.37	−7.20
9	242	0.160	1132.6	2.35	100	98.84	99.78	1.16	0.22
10	268	0.119	1132.6	1.75	104	103.19	101.53	0.78	2.37
11	289	0.301	1980.1	6.69	138	139.94	129.50	−1.40	6.16
12	337	0.205	1980.1	4.55	140	141.37	138.74	−0.98	0.90
13	279	0.141	1980.1	3.13	90	89.41	93.32	0.65	−3.69
14	267	0.100	1980.1	2.22	75	74.38	70.03	0.83	6.63
15	306	0.074	1980.1	1.65	77	78.11	75.00	−1.44	2.60
16	292	0.241	2265.3	7.09	128	127.78	117.79	0.17	7.97
17	322	0.164	2265.3	4.82	115	113.33	116.85	1.46	−1.61
18	286	0.113	2265.3	3.32	71	73.06	81.07	−2.90	−14.18
19	316	0.080	2265.3	2.35	67	65.33	78.49	2.49	−17.15
20	306	0.059	2265.3	1.75	51	51.30	56.19	−0.58	−10.17

TABLE 8.32 Experimental, ANN, and SPSS Output Concentrations (TSS – Rubber – Steady)

No.	TSS Input Data				C_{Exp}	C_{ANN}	C_{SPSS}	E_n	E_s
	C_i	q	A_s	v					
1	140	1.280	177.4	12.33	117	116.28	122.11	0.61	−4.37
2	170	0.871	177.4	8.39	138	137.02	143.01	0.71	−3.63
3	165	0.599	177.4	5.77	127	126.15	129.01	0.67	−1.58
4	169	0.424	177.4	4.09	125	124.66	123.83	0.27	0.93
5	176	0.316	177.4	3.04	127	126.75	121.17	0.20	4.59

TABLE 8.32 (Continued)

No.	TSS Input Data				C_{Exp}	C_{ANN}	C_{SPSS}	E_n	E_s
	C_i	q	A_s	v					
6	140	0.512	568.9	10.91	88	87.66	86.97	0.39	1.17
7	141	0.348	568.9	7.42	79	80.68	79.30	−2.13	−0.38
8	165	0.240	568.9	5.11	82	81.56	84.72	0.54	−3.32
9	164	0.170	568.9	3.62	67	65.73	72.68	1.89	−8.47
10	176	0.126	568.9	2.69	65	64.72	66.04	0.43	−1.59
11	140	0.320	960.3	10.62	76	77.24	68.30	−1.63	10.13
12	151	0.218	960.3	7.22	69	69.49	64.88	−0.71	5.97
13	144	0.150	960.3	4.97	51	45.78	50.97	10.25	0.05
14	169	0.106	960.3	3.52	43	43.15	47.16	−0.36	−9.66
15	165	0.079	960.3	2.62	35	33.00	30.86	5.71	11.83
16	153	0.256	1137.8	10.91	77	76.22	67.81	1.02	11.93
17	139	0.174	1137.8	7.42	55	55.74	51.04	−1.35	7.19
18	148	0.120	1137.8	5.11	45	44.53	43.04	1.04	4.36
19	159	0.085	1137.8	3.62	28	27.51	31.74	1.74	−13.37
20	165	0.063	1137.8	2.69	22	21.75	16.04	1.15	27.10

TABLE 8.33 Experimental, ANN, and SPSS Output Concentrations (TSS – Gravel – Steady)

No.	TSS Input Data				C_{Exp}	C_{ANN}	C_{SPSS}	E_n	E_s
	C_i	q	A_s	v					
1	135	1.259	210.3	13.88	109	109.06	115.40	−0.06	−5.88
2	139	0.857	210.3	9.44	110	110.27	110.99	−0.24	−0.90
3	160	0.589	210.3	6.49	121	120.69	121.89	0.26	−0.73
4	159	0.417	210.3	4.60	115	115.01	112.61	−0.01	2.07
5	168	0.311	210.3	3.42	118	117.65	111.81	0.30	5.24
6	149	0.504	740.9	13.99	87	85.51	89.37	1.72	−2.73
7	131	0.343	740.9	9.52	69	67.39	68.96	2.33	0.06
8	162	0.236	740.9	6.55	71	72.02	78.03	−1.44	−9.90
9	177	0.167	740.9	4.64	66	65.89	74.29	0.16	−12.56
10	176	0.124	740.9	3.45	59	58.34	61.30	1.11	−3.90
11	140	0.315	1271.6	14.03	67	68.32	63.31	−1.97	5.51

TABLE 8.33 (Continued)

No.	TSS Input Data				C_{Exp}	C_{ANN}	C_{SPSS}	E_n	E_s
	C_i	q	A_s	v					
12	170	0.214	1271.6	9.54	73	71.71	68.45	1.76	6.23
13	144	0.147	1271.6	6.57	45	45.21	46.35	−0.46	−3.01
14	169	0.104	1271.6	4.65	38	38.90	42.41	−2.36	−11.59
15	171	0.078	1271.6	3.46	30	30.32	28.43	−1.08	5.23
16	140	0.252	1481.9	13.99	61	62.79	56.12	−2.93	8.01
17	151	0.171	1481.9	9.52	58	57.61	51.16	0.68	11.79
18	144	0.118	1481.9	6.55	37	37.90	37.50	−2.43	−1.35
19	177	0.083	1481.9	4.64	29	29.32	31.86	−1.10	−9.87
20	171	0.062	1481.9	3.45	19	19.10	14.15	−0.52	25.55

TABLE 8.34 Experimental, ANN, and SPSS Output Concentrations (TSS – Plastic – Steady)

No.	TSS Input Data				C_{Exp}	C_{ANN}	C_{SPSS}	E_n	E_s
	C_i	q	A_s	v					
1	153	1.204	285.2	9.35	117	117.37	119.16	−0.32	−1.85
2	170	0.819	285.2	6.36	124	122.44	126.49	1.26	−2.01
3	148	0.563	285.2	4.37	101	102.47	100.83	−1.46	0.17
4	173	0.399	285.2	3.10	118	117.97	113.17	0.02	4.09
5	171	0.297	285.2	2.31	115	114.96	104.12	0.04	9.46
6	149	0.481	1132.6	7.09	74	72.24	70.11	2.37	5.25
7	131	0.328	1132.6	4.82	48	46.39	54.86	3.35	−14.29
8	148	0.225	1132.6	3.32	47	46.55	56.33	0.95	−19.86
9	159	0.160	1132.6	2.35	45	44.07	52.72	2.07	−17.16
10	168	0.119	1132.6	1.75	43	42.57	46.51	0.99	−8.15
11	147	0.301	1980.1	6.69	56	55.44	48.47	1.00	13.45
12	141	0.205	1980.1	4.55	37	38.28	40.25	−3.45	−8.80
13	162	0.141	1980.1	3.13	30	30.79	39.14	−2.65	−30.47
14	159	0.100	1980.1	2.22	26	24.39	28.78	6.20	−10.68
15	165	0.074	1980.1	1.65	19	19.25	19.14	−1.30	−0.75
16	149	0.241	2265.3	7.09	53	49.23	43.18	7.11	18.52
17	131	0.164	2265.3	4.82	25	24.06	31.26	3.77	−25.05

TABLE 8.34 (Continued)

No.	TSS Input Data				C_{Exp}	C_{ANN}	C_{SPSS}	E_n	E_s
	C_i	q	A_s	v					
18	144	0.113	2265.3	3.32	17	17.40	27.00	−2.35	−58.85
19	177	0.080	2265.3	2.35	17	17.48	22.31	−2.85	−31.21
20	180	0.059	2265.3	1.75	11	11.94	9.11	−8.54	17.17

The total experimental data (300 patterns for construction and test) used in ANNs and SPSS analysis are listed in Appendix III.

Table 8.35 summarizes the average percentage errors (average negative and average positive) between experimental output concentration and both ANNs and SPSS outputs (E_n and E_s), respectively for the 240 and 60 training performance patterns for BOD, COD, and TSS pollutants in steady stage. Figures 8.169–8.174 illustrate the percentage errors E_n and E_s for 240 calibration patterns and 60 validation patterns.

Table 8.35 and Figures 8.169–8.174, show that:

- For 240 calibration patterns, the BOD average percentage error between the experimental values and the SPSS model results varies between −4.61% and +4.10% but this percentage error ranges from −1.25% to +0.77% using ANNs model.
- For 60 validation patterns, the BOD average percentage error between the experimental values and the SPSS model results varies between −4.45% and +4.29% but this percentage error ranges from −1.18% to +0.84% using ANNs model.
- For 240 calibration patterns, the COD average percentage error between the experimental values and the SPSS model results varies between −4.83% and +4.54% but this percentage error ranges from −0.85% to +0.81% using ANNs model.

TABLE 8.35 Average Errors Between Experimental and Both ANN and SPSS (Steady Stage)

Parameter	Average E_n(%) (ANNs)		Average E_s(%) (SPSS)	
	240 Patterns	60 Patterns	240 Patterns	60 Patterns
BOD	−1.25 : +0.77	−1.18 : +0.84	−4.61 : +4.10	−4.45 : +4.29
COD	−0.85 : +0.81	−0.95 : +1.37	−4.83 : +4.54	−4.63 : +3.32
TSS	−1.55 : +1.54	−1.75 : +1.83	−10.09 : +7.91	−10.24 : +8.26

ANNs Modeling and SPSS Analysis

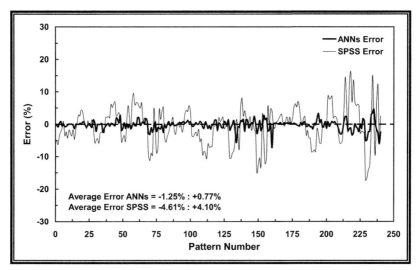

FIGURE 8.169 Percentage errors between Experimental and both ANNs and SPSS outputs (BOD – 240 Patterns).

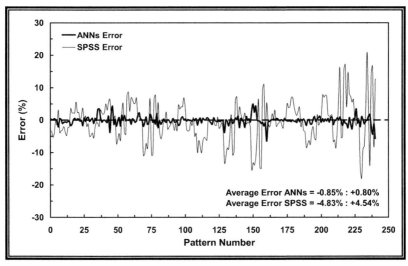

FIGURE 8.170 Percentage errors between Experimental and both ANNs and SPSS outputs (COD – 240 Patterns).

- For 60 validation patterns, the COD average percentage error between the experimental values and the SPSS model results varies between −4.63% and +3.32% but this percentage error ranges from −0.95% to +1.37% using ANNs model.

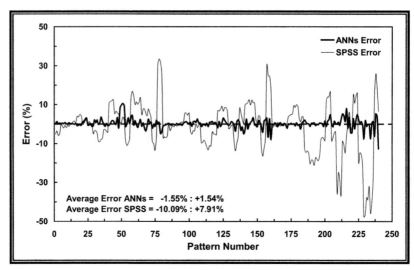

FIGURE 8.171 Percentage errors between Experimental and both ANNs and SPSS outputs (TSS – 240 Patterns).

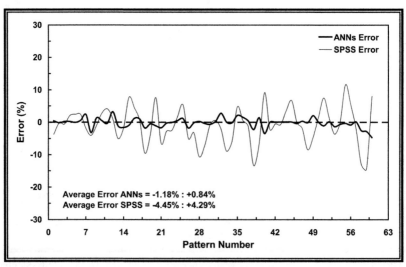

FIGURE 8.172 Percentage errors between Experimental and both ANNs and SPSS outputs (BOD – 60 Patterns).

- For 240 calibration patterns, the TSS average percentage error between the experimental values and the SPSS model results varies between −10.09% and +7.91% but this percentage error ranges from −1.55% to +1.54% using ANNs model.

ANNs Modeling and SPSS Analysis

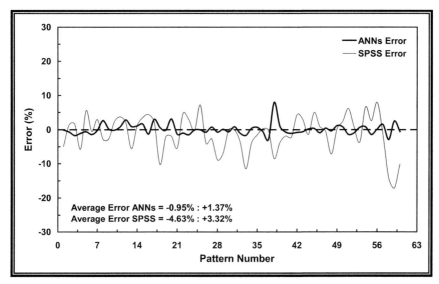

FIGURE 8.173 Percentage errors between Experimental and both ANNs and SPSS outputs (COD – 60 Patterns).

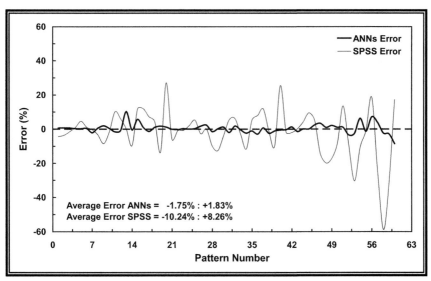

FIGURE 8.174 Percentage errors between Experimental and both ANNs and SPSS outputs (TSS – 60 Patterns).

- For 60 validation patterns, the TSS average percentage error between the experimental values and the SPSS model results varies between −10.24% and +8.26% but this percentage error ranges from −1.75% to +1.83% using ANNs model.

- The ANNs model represents the experimental data more accurately than the SPSS model for calibration and validation process.

8.11.3 STEADY STAGE (NH$_3$, PO$_4$, DO, AND FC)

The results of the validation process of 15 patterns for the obtained models and the percentage errors are shown in Tables 8.36–8.39 for ammonia, phosphate, dissolved oxygen, and fecal coliforms parameters, respectively.

The total experimental data (75 patterns for construction and test) used in artificial neural networks and SPSS analysis are listed in Appendix III. Table 8.40 summarizes the average percentage errors (average negative and average positive) between experimental output concentration and both ANNs and SPSS outputs (E_n and E_s), respectively for the 60 and 15 training performance patterns for NH$_3$, PO$_4$, DO, and FC parameters.

Figures 8.175–8.182 illustrate the percentage error between ANNs and SPSS and the experimental output values for 60 calibration and 15 validation patterns for NH$_3$, PO$_4$, DO, and FC parameters, respectively.

TABLE 8.36A Experimental, ANNs, and SPSS Effluents (NH$_3$ – Steady)

	Media Type	Ammonia Input Data				C_{Exp}	C_{ANN}	C_{SPSS}	E_n	E_s
		C_i	q	A_s	v					
1	Rubber cell	16.76	0.256	1137.8	10.91	9.38	9.158	9.507	2.36	−1.35
2		18.48	0.174	1137.8	7.42	9.32	8.964	9.518	3.81	−2.12
3		15.92	0.120	1137.8	5.11	6.78	7.160	7.720	−5.60	−13.86
4		16.11	0.085	1137.8	3.62	6.75	6.762	6.786	−0.18	−0.54
5		15.96	0.063	1137.8	2.69	6.38	5.781	5.646	9.39	11.50

TABLE 8.36B Experimental, ANNs, and SPSS Effluents (NH$_3$ – Steady)

	Media Type	Ammonia Input Data				C_{Exp}	C_{ANN}	C_{SPSS}	E_n	E_s
		C_i	q	A_s	v					
6	Gravel cell	26.16	0.252	1481.9	13.99	13.25	13.693	13.109	−3.34	1.07
7		18.48	0.171	1481.9	9.52	8.36	8.329	8.895	0.37	−6.40
8		15.92	0.118	1481.9	6.55	6.39	6.755	7.086	−5.72	−10.89
9		19.92	0.083	1481.9	4.64	8.00	7.655	7.633	4.31	4.58
10		15.96	0.062	1481.9	3.45	5.19	5.226	4.988	−0.69	3.89

ANNs Modeling and SPSS Analysis

TABLE 8.36C Experimental, ANNs, and SPSS Effluents (NH_3 – Steady)

Media Type		\multicolumn{4}{c}{Ammonia Input Data}	C_{Exp}	C_{ANN}	C_{SPSS}	E_n	E_s
		C_i	q	A_s	v					
11	Plastic	16.76	0.241	2265.3	7.09	7.72	7.217	7.322	6.51	5.15
12	cell	18.48	0.164	2265.3	4.82	7.66	7.491	7.305	2.21	4.64
13		23.74	0.113	2265.3	3.32	8.33	8.655	8.767	−3.90	−5.25
14		20.37	0.080	2265.3	2.35	5.29	5.379	6.174	−1.69	−16.70
15		20.22	0.059	2265.3	1.75	4.44	4.214	4.972	5.09	−11.99

TABLE 8.37A Experimental, ANNs, and SPSS Effluents (PO_4 – Steady)

Media Type		\multicolumn{4}{c}{Phosphate Input Data}	C_{Exp}	C_{ANN}	C_{SPSS}	E_n	E_s
		C_i	q	A_s	v					
1	Rubber	2.77	0.256	1137.8	10.91	1.420	1.441	1.369	−1.45	3.57
2	cell	2.53	0.174	1137.8	7.42	1.030	0.998	1.091	3.12	−5.91
3		2.33	0.120	1137.8	5.11	0.850	0.806	0.880	5.20	−3.53
4		2.94	0.085	1137.8	3.62	0.970	1.027	0.918	−5.86	5.36
5		2.66	0.063	1137.8	2.69	0.720	0.719	0.740	0.13	−2.74

TABLE 8.37B Experimental, ANNs, and SPSS Effluents (PO_4 – Steady)

Media Type		\multicolumn{4}{c}{Phosphate Input Data}	C_{Exp}	C_{ANN}	C_{SPSS}	E_n	E_s
		C_i	q	A_s	v					
6	Gravel	3.32	0.252	1481.9	13.99	1.440	1.450	1.409	−0.72	2.17
7	cell	2.85	0.171	1481.9	9.52	0.980	0.972	1.007	0.79	−2.79
8		2.57	0.118	1481.9	6.55	0.790	0.775	0.761	1.86	3.73
9		2.7	0.083	1481.9	4.64	0.790	0.734	0.670	7.09	15.20
10		2.89	0.062	1481.9	3.45	0.720	0.692	0.640	3.87	11.17

TABLE 8.37C Experimental, ANNs, and SPSS Effluents (PO_4 – Steady)

Media Type		\multicolumn{4}{c}{Phosphate Input Data}	C_{Exp}	C_{ANN}	C_{SPSS}	E_n	E_s
		C_i	q	A_s	v					
11	Plastic	3.32	0.241	2265.3	7.09	0.990	1.062	1.098	−7.27	−10.93
12	cell	2.85	0.164	2265.3	4.82	0.780	0.762	0.702	2.31	9.98
13		2.57	0.113	2265.3	3.32	0.580	0.533	0.460	8.10	20.78
14		2.7	0.080	2265.3	2.35	0.480	0.454	0.372	5.47	22.52
15		2.8	0.059	2265.3	1.75	0.330	0.355	0.314	−7.59	4.73

TABLE 8.38A Experimental, ANNs, and SPSS Effluents (DO – Steady)

	Media Type	DO Input Data			C_{Exp}	C_{ANN}	C_{SPSS}	E_n	E_s
		q	A_s	v					
1	Rubber	0.256	1137.8	10.91	1.90	1.75	2.18	7.70	−14.73
2	cell	0.174	1137.8	7.42	3.00	2.95	2.74	1.73	8.72
3		0.120	1137.8	5.11	3.30	3.40	3.29	−3.01	0.16
4		0.085	1137.8	3.62	3.90	3.85	3.82	1.28	2.03
5		0.063	1137.8	2.69	4.10	4.17	4.28	−1.79	−4.35

TABLE 8.38B Experimental, ANNs, and SPSS Effluents (DO – Steady)

	Media Type	DO Input Data			C_{Exp}	C_{ANN}	C_{SPSS}	E_n	E_s
		q	A_s	v					
6	Gravel	0.252	1481.9	13.99	2.50	2.65	2.66	−5.98	−6.34
7	cell	0.171	1481.9	9.52	3.40	3.40	3.21	−0.02	5.64
8		0.118	1481.9	6.55	4.00	4.10	3.76	−2.46	6.07
9		0.083	1481.9	4.64	4.20	4.40	4.28	−4.75	−1.81
10		0.062	1481.9	3.45	4.70	4.52	4.73	3.73	−0.60

TABLE 8.38C Experimental, ANNs, and SPSS Effluents (DO – Steady)

	Media Type	DO Input Data			C_{Exp}	C_{ANN}	C_{SPSS}	E_n	E_s
		q	A_s	v					
11	Plastic	0.241	2265.3	7.09	3.40	3.20	3.03	5.86	10.88
12	cell	0.164	2265.3	4.82	3.80	3.80	3.60	0.01	5.19
13		0.113	2265.3	3.32	4.30	4.33	4.17	−0.74	2.96
14		0.080	2265.3	2.35	4.60	4.60	4.71	0.04	−2.39
15		0.059	2265.3	1.75	5.10	4.93	5.18	3.35	−1.49

TABLE 8.39A Experimental, ANNs, and SPSS Effluents (FC – Steady)

	Media Type	Fecal Coliforms Input Data				C_{Exp}	C_{ANN}	C_{SPSS}	E_n	E_s
		C_i	q	A_s	v					
1	Rubber	240,000	0.256	1137.8	10.91	6100	6100	5854	−0.01	4.03
2	cell	240,000	0.174	1137.8	7.42	4800	4833	4905	−0.69	−2.19
3		260,000	0.120	1137.8	5.11	4000	3983	4063	0.43	−1.58
4		220,000	0.085	1137.8	3.62	3200	3243	3041	−1.36	4.97
5		200,000	0.063	1137.8	2.69	2200	2257	2153	−2.57	2.15

TABLE 8.39B Experimental, ANNs, and SPSS Effluents (FC – Steady)

Media Type		Fecal Coliforms Input Data				C_{Exp}	C_{ANN}	C_{SPSS}	E_n	E_s
		C_i	q	A_s	v					
6	Gravel	240,000	0.252	1481.9	13.99	5800	5885	5949	−1.47	−2.57
7	cell	270,000	0.171	1481.9	9.52	4700	4671	4931	0.62	−4.91
8		230,000	0.118	1481.9	6.55	3700	3524	3841	4.77	−3.82
9		210,000	0.083	1481.9	4.64	2700	2749	2864	−1.80	−6.07
10		270,000	0.062	1481.9	3.45	2200	2304	2240	−4.71	−1.83

TABLE 8.39C Experimental, ANNs, and SPSS Effluents (FC – Steady)

Media Type		Fecal Coliforms Input Data				C_{Exp}	C_{ANN}	C_{SPSS}	E_n	E_s
		C_i	q	A_s	v					
11	Plastic	240,000	0.241	2265.3	7.09	5400	5380	5286	0.36	2.10
12	cell	240,000	0.164	2265.3	4.82	4300	4334	4429	−0.80	−3.00
13		260,000	0.113	2265.3	3.32	3600	3624	3604	−0.68	−0.12
14		220,000	0.080	2265.3	2.35	2600	2523	2578	2.98	0.83
15		210,000	0.059	2265.3	1.75	1700	1761	1718	−3.59	−1.04

TABLE 8.40 Average Errors Between Experimental and Both ANN and SPSS (Steady Stage)

Parameter	Average E_n(%)(ANNs)		Average E_s(%)(SPSS)	
	60 Patterns	15 Patterns	60 Patterns	15 Patterns
Ammonia	−2.74 : +2.33	−3.02 : +4.26	−5.77 : +3.06	−7.68 : +5.14
Phosphate	−4.69 : +1.66	−4.58 : +3.79	−6.26 : +8.97	−5.18 : +9.92
D. Oxygen	−3.00 : +2.45	−2.68 : +2.96	−6.19 : +4.73	−4.53 : +5.21
F. Coliforms	−1.94 : +2.58	−1.77 : +1.83	−3.17 : +3.18	−2.71 : +2.81

Table 8.40 and Figures 8.175–8.182, exhibit that:

- For 60 calibration patterns, the NH_3 average percentage error between the experimental values and the SPSS model results varies between −5.77% and +3.06% but this percentage error ranges from −2.74% to +2.33% using ANNs model.
- For 15 validation patterns, the NH_3 average percentage error between the experimental values and the SPSS model results varies between −7.68% and +5.14% but this percentage error ranges from −3.02% to +4.26% using ANNs model.

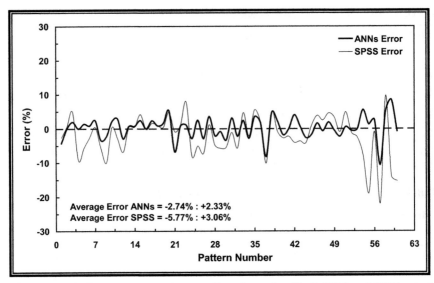

FIGURE 8.175 Percentage errors between Experimental and both ANNs and SPSS outputs (NH_3 – 60 Patterns).

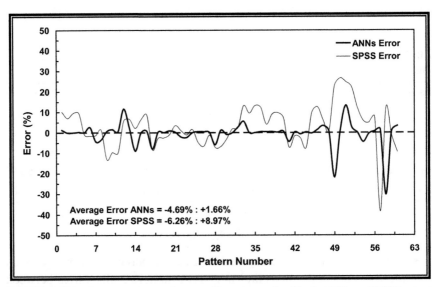

FIGURE 8.176 Percentage errors between Experimental and both ANNs and SPSS outputs (PO_4 – 60 Patterns).

- For 60 calibration patterns, the PO_4 average percentage error between the experimental values and the SPSS model results varies between −6.26% and +8.97% but this percentage error ranges from −4.69% to +1.66% using ANNs model.

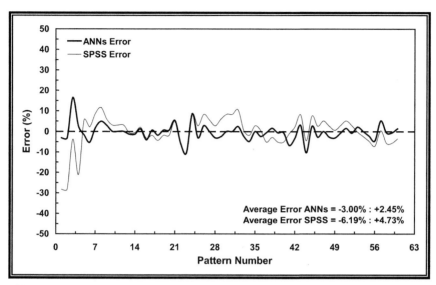

FIGURE 8.177 Percentage errors between Experimental and both ANNs and SPSS outputs (DO – 60 Patterns).

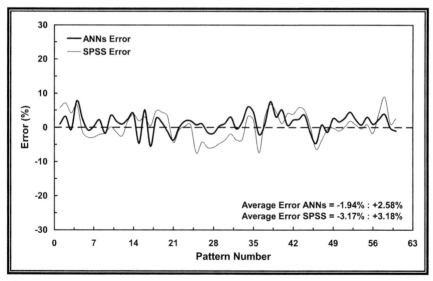

FIGURE 8.178 Percentage errors between Experimental and both ANNs and SPSS outputs (FC – 60 Patterns).

- For 15 validation patterns, the PO_4 average percentage error between the experimental values and the SPSS model results varies between −5.18% and +9.92% but this percentage error ranges from −4.58% to +3.79% using ANNs model.

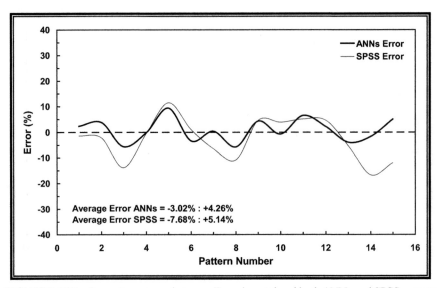

FIGURE 8.179 Percentage errors between Experimental and both ANNs and SPSS outputs (NH_3 – 15 Patterns).

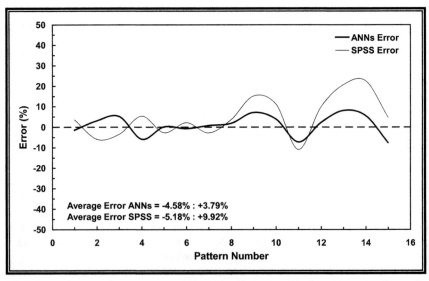

FIGURE 8.180 Percentage errors between Experimental and both ANNs and SPSS outputs (PO_4 – 15 Patterns).

- For 60 calibration patterns, the DO average percentage error between the experimental values and the SPSS model results varies between −6.19% and +4.73% but this percentage error ranges from −3.00% to +2.45% using ANNs model.

ANNs Modeling and SPSS Analysis

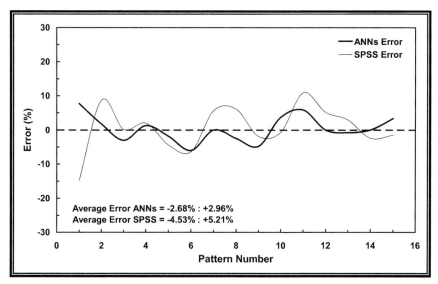

FIGURE 8.181 Percentage errors between Experimental and both ANNs and SPSS outputs (DO – 15 Patterns).

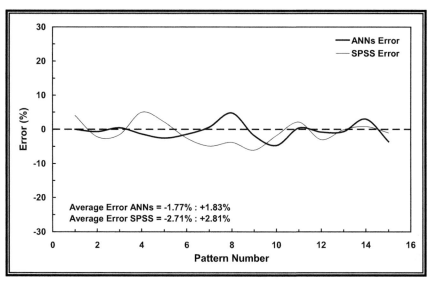

FIGURE 8.182 Percentage errors between Experimental and both ANNs and SPSS outputs (FC – 15 Patterns).

- For 15 validation patterns, the DO average percentage error between the experimental values and the SPSS model results varies between −4.53% and +5.21% but this percentage error ranges from −2.68% to +2.96% using ANNs model.

- For 60 calibration patterns, the FC average percentage error between the experimental values and the SPSS model results varies between −3.17% and +3.18% but this percentage error ranges from −1.94% to +2.58% using ANNs model.
- For 15 validation patterns, the FC average percentage error between the experimental values and the SPSS model results varies between −2.71% and +2.81% but this percentage error ranges from −1.77% to +1.83% using ANNs model.
- The ANNs model represents the experimental data more accurately than the SPSS model. This is due to the ability of ANNs model to construct a variety of complex relationships between variables to represent it precisely.

8.11.4 STEADY STAGE (HEAVY METALS)

The results of the validation process of 15 patterns for the obtained models and the percentage errors are shown in Tables 8.41–8.45 for Zn, Fe, Mn, Pb, and Cd elements, respectively. The total experimental data (75 patterns for construction and test) used in the artificial neural networks and SPSS analysis are listed in Appendix III.

Table 8.46 summarizes the average percentage errors (average negative and average positive) between experimental output concentration and both ANNs and SPSS outputs (E_n and E_s), respectively for the 60 and 15 training performance patterns for heavy metals parameters. Figures 8.183–8.187 show the percentage error between ANNs and SPSS and the experimental output for 60 calibration patterns for Zn, Fe, Mn, Pb, and Cd. These percentage errors for 15 validation patterns for the studied parameters are presented in Figures 8.188–8.192.

TABLE 8.41A Experimental, ANNs, and SPSS Effluents (Zn – Steady)

	Media Type	\multicolumn{4}{c}{Zinc Input Data}	C_{Exp}	C_{ANN}	C_{SPSS}	E_n	E_s
		C_i	q	A_s	v					
1	Rubber cell	2.16	0.256	1137.8	10.91	1.31	1.245	1.265	4.99	3.43
2		1.42	0.174	1137.8	7.42	0.80	0.797	0.782	0.35	2.27
3		1.79	0.120	1137.8	5.11	0.89	0.890	0.778	−0.01	12.61
4		1.43	0.085	1137.8	3.62	0.53	0.561	0.511	−5.80	3.54
5		1.87	0.063	1137.8	2.69	0.59	0.617	0.574	−4.66	2.79

TABLE 8.41B Experimental, ANNs, and SPSS Effluents (Zn – Steady)

Media Type		Zinc Input Data				C_{Exp}	C_{ANN}	C_{SPSS}	E_n	E_s
		C_i	q	A_s	v					
6	Gravel	1.55	0.252	1481.9	13.99	0.88	0.900	0.882	–2.32	–0.18
7	cell	1.74	0.171	1481.9	9.52	0.83	0.843	0.805	–1.60	2.98
8		1.28	0.118	1481.9	6.55	0.56	0.593	0.505	–5.89	9.85
9		1.64	0.083	1481.9	4.64	0.56	0.557	0.489	0.54	12.65
10		1.5	0.062	1481.9	3.45	0.36	0.368	0.327	–2.21	9.05

TABLE 8.41C Experimental, ANNs, and SPSS Effluents (Zn – Steady)

Media Type		Zinc Input Data				C_{Exp}	C_{ANN}	C_{SPSS}	E_n	E_s
		C_i	q	A_s	v					
11	Plastic	2.16	0.241	2265.3	7.09	1.03	1.072	1.016	–4.03	1.40
12	cell	1.42	0.164	2265.3	4.82	0.53	0.513	0.534	3.19	–0.79
13		1.79	0.113	2265.3	3.32	0.55	0.535	0.532	2.75	3.36
14		1.43	0.080	2265.3	2.35	0.41	0.401	0.266	2.24	35.08
15		1.44	0.059	2265.3	1.75	0.27	0.269	0.159	0.24	40.93

TABLE 8.42A Experimental, ANNs, and SPSS Effluents (Fe – Steady)

Media Type		Iron Input Data				C_{Exp}	C_{ANN}	C_{SPSS}	E_n	E_s
		C_i	q	A_s	v					
1	Rubber	0.95	0.256	1137.8	10.91	0.76	0.732	0.742	3.63	2.37
2	cell	1.12	0.174	1137.8	7.42	0.74	0.730	0.729	1.38	1.46
3		0.87	0.120	1137.8	5.11	0.47	0.483	0.509	–2.67	–8.28
4		1.19	0.085	1137.8	3.62	0.58	0.547	0.584	5.64	–0.65
5		0.98	0.063	1137.8	2.69	0.36	0.371	0.407	–2.92	–13.10

TABLE 8.42B Experimental, ANNs, and SPSS Effluents (Fe – Steady)

Media Type		Iron Input Data				C_{Exp}	C_{ANN}	C_{SPSS}	E_n	E_s
		C_i	q	A_s	v					
6	Gravel	0.95	0.252	1481.9	13.99	0.67	0.668	0.676	0.25	–0.89
7	cell	1.12	0.171	1481.9	9.52	0.72	0.704	0.663	2.16	7.90
8		0.87	0.118	1481.9	6.55	0.38	0.378	0.443	0.55	–16.53
9		1.19	0.083	1481.9	4.64	0.49	0.557	0.518	–13.76	–5.66
10		0.89	0.062	1481.9	3.45	0.22	0.206	0.295	6.33	–34.10

TABLE 8.42C Experimental, ANNs, and SPSS Effluents (Fe – Steady)

Media Type		Iron Input Data				C_{Exp}	C_{ANN}	C_{SPSS}	E_n	E_s
		C_i	q	A_s	v					
11	Plastic	0.95	0.241	2265.3	7.09	0.61	0.614	0.589	−0.71	3.38
12	cell	1.12	0.164	2265.3	4.82	0.51	0.542	0.577	−6.21	−13.12
13		0.87	0.113	2265.3	3.32	0.28	0.300	0.357	−7.19	−27.44
14		1.19	0.080	2265.3	2.35	0.33	0.351	0.432	−6.43	−30.90
15		0.76	0.059	2265.3	1.75	0.15	0.161	0.141	−7.16	5.90

TABLE 8.43A Experimental, ANNs, and SPSS Effluents (Mn – Steady)

Media Type		Manganese Input Data				C_{Exp}	C_{ANN}	C_{SPSS}	E_n	E_s
		C_i	q	A_s	v					
1	Rubber	0.28	0.256	1137.8	10.91	0.23	0.230	0.251	−0.12	−9.17
2	cell	0.15	0.174	1137.8	7.42	0.11	0.124	0.150	−12.74	−36.80
3		0.17	0.120	1137.8	5.11	0.10	0.105	0.117	−4.70	−17.29
4		0.28	0.085	1137.8	3.62	0.14	0.137	0.131	1.93	6.25
5		0.44	0.063	1137.8	2.69	0.17	0.169	0.194	0.53	−14.35

TABLE 8.43B Experimental, ANNs, and SPSS Effluents (Mn – Steady)

Media Type		Manganese Input Data				C_{Exp}	C_{ANN}	C_{SPSS}	E_n	E_s
		C_i	q	A_s	v					
6	Gravel	0.41	0.252	1481.9	13.99	0.32	0.322	0.294	−0.52	8.11
7	cell	0.5	0.171	1481.9	9.52	0.35	0.347	0.307	0.75	12.23
8		0.36	0.118	1481.9	6.55	0.16	0.167	0.180	−4.30	−12.38
9		0.37	0.083	1481.9	4.64	0.16	0.159	0.148	0.73	7.35
10		0.26	0.062	1481.9	3.45	0.07	0.078	0.060	−10.88	14.08

TABLE 8.43C Experimental, ANNs, and SPSS Effluents (Mn – Steady)

Media Type		Manganese Input Data				C_{Exp}	C_{ANN}	C_{SPSS}	E_n	E_s
		C_i	q	A_s	v					
11	Plastic	0.28	0.241	2265.3	7.09	0.19	0.190	0.174	0.21	8.26
12	cell	0.15	0.164	2265.3	4.82	0.07	0.082	0.075	−17.17	−6.96
13		0.17	0.113	2265.3	3.32	0.06	0.058	0.045	2.68	25.70
14		0.28	0.080	2265.3	2.35	0.09	0.090	0.065	−0.52	27.64
15		0.44	0.059	2265.3	1.75	0.10	0.091	0.141	8.59	−40.63

ANNs Modeling and SPSS Analysis

TABLE 8.44A Experimental, ANNs, and SPSS Effluents (Pb – Steady)

	Media Type	\multicolumn{4}{c}{Lead Input Data}	C_{Exp}	C_{ANN}	C_{SPSS}	E_n	E_s
		C_i	q	A_s	v					
1	Rubber	0.021	0.256	1137.8	10.91	0.017	0.0181	0.0204	−6.16	−19.94
2	cell	0.033	0.174	1137.8	7.42	0.024	0.0235	0.0228	2.05	5.17
3		0.047	0.120	1137.8	5.11	0.023	0.0260	0.0257	−11.69	−11.70
4		0.027	0.085	1137.8	3.62	0.012	0.0121	0.0114	−0.82	4.89
5		0.066	0.063	1137.8	2.69	0.028	0.0301	0.0282	−7.06	−0.58

TABLE 8.44B Experimental, ANNs, and SPSS Effluents (Pb – Steady)

	Media Type	C_i	q	A_s	v	C_{Exp}	C_{ANN}	C_{SPSS}	E_n	E_s
6	Gravel	0.040	0.252	1481.9	13.99	0.029	0.0303	0.0307	−4.30	−5.73
7	cell	0.048	0.171	1481.9	9.52	0.032	0.0325	0.0284	−1.55	11.20
8		0.036	0.118	1481.9	6.55	0.017	0.0175	0.0179	−2.87	−5.38
9		0.043	0.083	1481.9	4.64	0.017	0.0174	0.0184	−2.34	−8.34
10		0.072	0.062	1481.9	3.45	0.027	0.0265	0.0280	1.83	−3.89

TABLE 8.44C Experimental, ANNs, and SPSS Effluents (Pb – Steady)

	Media Type	C_i	q	A_s	v	C_{Exp}	C_{ANN}	C_{SPSS}	E_n	E_s
11	Plastic	0.021	0.241	2265.3	7.09	0.012	0.0127	0.0139	−5.49	−15.92
12	cell	0.033	0.164	2265.3	4.82	0.017	0.0170	0.0167	0.23	1.53
13		0.047	0.113	2265.3	3.32	0.019	0.0203	0.0200	−6.17	−5.02
14		0.027	0.080	2265.3	2.35	0.010	0.0095	0.0059	4.97	41.44
15		0.061	0.059	2265.3	1.75	0.020	0.0215	0.0208	−6.88	−4.25

TABLE 8.45A Experimental, ANNs, and SPSS Effluents (Cd – Steady)

	Media Type	C_i	q	A_s	v	C_{Exp}	C_{ANN}	C_{SPSS}	E_n	E_s
1	Rubber	0.00211	0.256	1137.8	10.91	0.00175	0.0017	0.0018	1.03	−1.14
2	cell	0.00191	0.174	1137.8	7.42	0.00153	0.0015	0.0015	0.01	4.92
3		0.00290	0.120	1137.8	5.11	0.00224	0.0022	0.0021	1.59	8.26
4		0.00255	0.085	1137.8	3.62	0.00170	0.0017	0.0017	0.85	−1.73
5		0.00241	0.063	1137.8	2.69	0.00152	0.0014	0.0016	10.74	−4.16

TABLE 8.45B Experimental, ANNs, and SPSS Effluents (Cd – Steady)

Media Type		Cadmium Input Data				C_{Exp}	C_{ANN}	C_{SPSS}	E_n	E_s
		C_i	q	A_s	v					
6	Gravel	0.00173	0.252	1481.9	13.99	0.00132	0.0013	0.0013	1.18	−0.81
7	cell	0.00233	0.171	1481.9	9.52	0.00167	0.0017	0.0016	−0.01	3.17
8		0.00196	0.118	1481.9	6.55	0.00140	0.0014	0.0013	1.63	10.48
9		0.00247	0.083	1481.9	4.64	0.00159	0.0016	0.0016	1.36	1.98
10		0.00175	0.062	1481.9	3.45	0.00087	0.0008	0.0010	2.82	−15.03

TABLE 8.45C Experimental, ANNs, and SPSS Effluents (Cd – Steady)

Media Type		Cadmium Input Data				C_{Exp}	C_{ANN}	C_{SPSS}	E_n	E_s
		C_i	q	A_s	v					
11	Plastic	0.00211	0.241	2265.3	7.09	0.00150	0.0015	0.0015	2.74	−1.30
12	cell	0.00191	0.164	2265.3	4.82	0.00130	0.0013	0.0012	0.28	8.24
13		0.00290	0.113	2265.3	3.32	0.00191	0.0019	0.0018	1.30	6.52
14		0.00255	0.080	2265.3	2.35	0.00145	0.0015	0.0015	−0.41	−0.35
15		0.00182	0.059	2265.3	1.75	0.00079	0.0008	0.0009	−2.76	−11.45

TABLE 8.46 Average Errors Between Experimental and Both ANN and SPSS (Steady Stage)

Parameter	Average E_n(%) (ANNs)		Average E_s(%) (SPSS)	
	60 Patterns	15 Patterns	60 Patterns	15 Patterns
Zinc	−1.43 : +1.35	−3.32 : +2.04	−2.86 : +9.48	−0.48 : +10.77
Iron	−2.97 : +1.94	−5.88 : +2.85	−13.57 : +5.03	−15.07 : +4.20
Manganese	−2.83 : +2.51	−6.37 : +2.20	−11.11 : +11.57	−19.65 : +13.71
Lead	−4.63 : +2.23	−5.03 : +2.27	−6.23 : +8.20	−8.07 : +12.84
Cadmium	−0.57 : +1.28	−1.06 : +2.13	−4.82 : +5.45	−4.50 : +6.22

Table 8.46 and Figures 8.183–8.192 show that:

- For 60 calibration patterns, the Zn average percentage error between the experimental values and the SPSS model results varies between −2.86% and +9.48% but this percentage error ranges from −1.43% to +1.35% using the ANNs model.

ANNs Modeling and SPSS Analysis

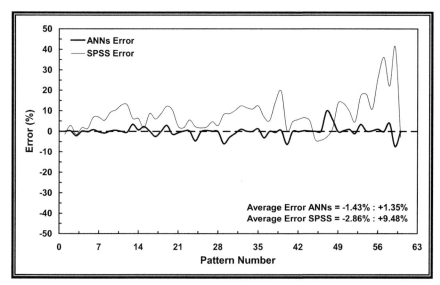

FIGURE 8.183 Percentage errors between Experimental and both ANNs and SPSS outputs (Zn – 60 Patterns).

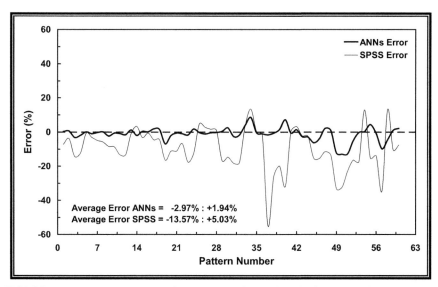

FIGURE 8.184 Percentage errors between Experimental and both ANNs and SPSS outputs (Fe – 60 Patterns).

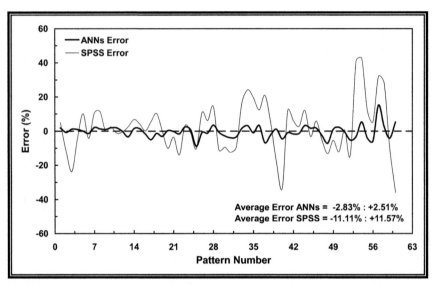

FIGURE 8.185 Percentage errors between Experimental and both ANNs and SPSS outputs (Mn – 60 Patterns).

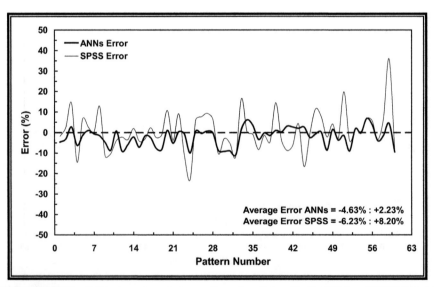

FIGURE 8.186 Percentage errors between Experimental and both ANNs and SPSS outputs (Pb – 60 Patterns).

ANNs Modeling and SPSS Analysis

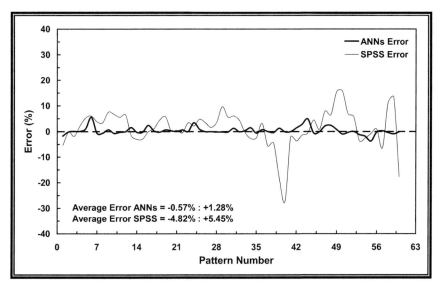

FIGURE 8.187 Percentage errors between Experimental and both ANNs and SPSS outputs (Cd – 60 Patterns).

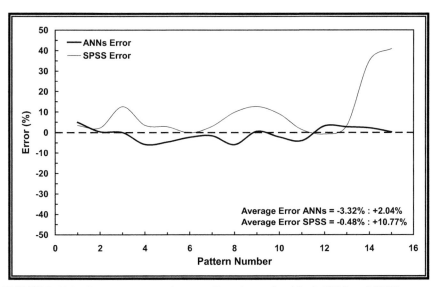

FIGURE 8.188 Percentage errors between Experimental and both ANNs and SPSS outputs (Zn – 15 Patterns).

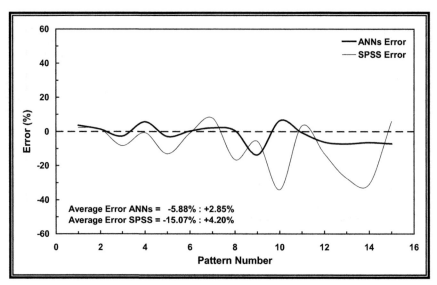

FIGURE 8.189 Percentage errors between Experimental and both ANNs and SPSS outputs (Fe – 15 Patterns).

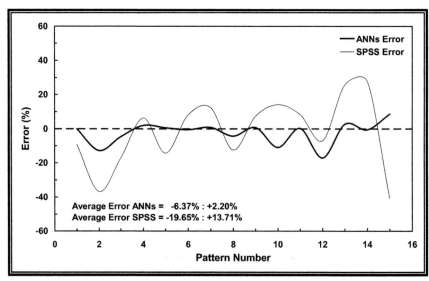

FIGURE 8.190 Percentage errors between Experimental and both ANNs and SPSS outputs (Mn – 15 Patterns).

ANNs Modeling and SPSS Analysis

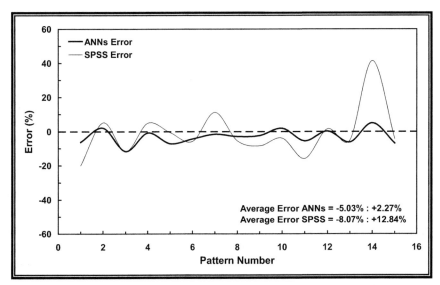

FIGURE 8.191 Percentage errors between Experimental and both ANNs and SPSS outputs (Pb – 15 Patterns).

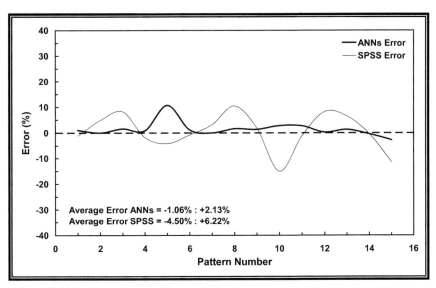

FIGURE 8.192 Percentage errors between Experimental and both ANNs and SPSS outputs (Cd – 15 Patterns).

- For 15 validation patterns, the Zn average percentage error between the experimental values and the SPSS model results varies between −0.48% and +10.77% but this percentage error ranges from −3.32% to +2.04% using the ANNs.
- For 60 calibration patterns, the Fe average percentage error between the experimental values and the SPSS model results varies between −13.57% and +5.03% but this percentage error ranges from −2.97% to +1.94% using the ANNs.
- For 15 validation patterns, the Fe average percentage error between the experimental values and the SPSS model results varies between −15.07% and +4.20% but this percentage error ranges from −5.88% to +2.85% using the ANNs.
- For 60 calibration patterns, the Mn average percentage error between the experimental values and the SPSS model results varies between −11.11% and +11.57% but this percentage error ranges from −2.83% to +2.51% using the ANNs.
- For 15 validation patterns, the Mn average percentage error between the experimental values and the SPSS model results varies between −19.65% and +13.71% but this percentage error ranges from −6.37% to +2.20% using the ANNs.
- For 60 calibration patterns, the Pb average percentage error between the experimental values and the SPSS model results varies between −6.23% and +8.20% but this percentage error ranges from −4.63% to +2.23% using the ANNs model.
- For 15 validation patterns, the Pb average percentage error between the experimental values and the SPSS model results varies between −8.07% and +12.84% but this percentage error ranges from −5.03% to +2.27% using the ANNs.
- For 60 calibration patterns, the Cd average percentage error between the experimental values and the SPSS model results varies between −4.82% and +5.45% but this percentage error ranges from −0.57% to +1.28% using the ANNs model.
- For 15 validation patterns, the Cd average percentage error between the experimental values and the SPSS model results varies between −4.50% and +6.22% but this percentage error ranges from −1.06% to +2.13% using the ANNs model.

- The ANNs models represent the experimental data more accurately than those given by the SPSS model. This is due to the ability of ANNs model to construct precise complex relationships between the different layers of solution process.

8.12 SIGNIFICANCE TESTs

This part deals with two types of data; the first one is the measured concentrations of parameters at the end of wetland cells as given in Chapters 6 and 7. The second type is the verified output data used in ANNs and SPSS models. The aim of the following statistical tests is to estimate the significant difference between the media under study, in addition to the difference between the experimental data and both ANNs and SPSS model outputs, and also between SPSS and ANNs results. The normality test was carried out for all groups of data. One-way ANOVA is used for the normal distribution. For case of abnormal distribution, the nonparametric test (Chi-square) is applied. All tests were performed by SPSS software at 5% level of significance and the results are listed in details in Appendix (IV).

8.12.1 NORMALITY TEST

Table 8.47 presents the significance test results to show if each group of the data follows the normal distribution or not, for both set up and steady stages parameters.

8.12.2 ONE WAY ANOVA AND CHI-SQUARE TESTS

Tables 8.48 and 8.49 exhibit comparisons between results using the one way ANOVA and Chi-square tests at 5% level for different groups of the used data. The tested groups were rubber, gravel, and plastic media; rubber and gravel; rubber and plastic; and gravel and plastic media (Table 8.48), in addition to the comparisons between the experimental outputs and both ANNs and SPSS; and SPSS and ANNs groups (Table 8.49).

TABLE 8.47 Normality Test for Groups of Set Up and Steady Stages Parameters

Stage	Parameter	Rubber	Gravel	Plastic	Exp.	ANNs	SPSS
Set-up	BOD	Normal	Normal	Normal	Normal	Normal	Normal
	COD	Normal	Normal	Normal	Normal	Normal	Normal
	TSS	Normal	Normal	Normal	Normal	Normal	Normal
Steady	BOD	Normal	Normal	Normal	Abnormal	Abnormal	Normal
	COD	Abnormal	Abnormal	Abnormal	Normal	Normal	Normal
	TSS	Normal	Normal	Normal	Abnormal	Abnormal	Normal
	Ammonia	Normal	Normal	Normal	Normal	Normal	Normal
	Phosphate	Normal	Normal	Normal	Normal	Normal	Normal
	DO	Normal	Abnormal	Normal	Normal	Normal	Normal
	FC	Normal	Normal	Normal	Normal	Normal	Normal
	Zinc	Normal	Normal	Abnormal	Normal	Normal	Normal
	Iron	Normal	Normal	Normal	Normal	Normal	Normal
	Manganese	Normal	Normal	Normal	Normal	Normal	Normal
	Lead	Normal	Normal	Normal	Normal	Normal	Normal
	Cadmium	Normal	Normal	Normal	Normal	Normal	Normal

TABLE 8.48 Significant Difference Between the Different Media

Stage	Parameter	Significant difference between the following groups:			
		R & G & P	R & G	R & P	G & P
Set-up (12)[n]	BOD	Non Sig.[a]	Non Sig.[a]	Non Sig.[a]	Non Sig.[a]
	COD	Non Sig.[a]	Non Sig.[a]	Non Sig.[a]	Non Sig.[a]
	TSS	Sig.[a]	Non Sig.[a]	Sig.[a]	Sig.[a]
Steady (25)[n]	BOD	Sig.[a]	Non Sig.[a]	Sig.[a]	Sig.[a]
	COD	Sig.[c]	Non Sig.[c]	Sig.[c]	Sig.[c]
	TSS	Sig.[a]	Non Sig.[a]	Sig.[a]	Sig.[a]
	Ammonia	Sig.[a]	Non Sig.[a]	Sig.[a]	Sig.[a]
	Phosphate	Sig.[a]	Non Sig.[a]	Sig.[a]	Sig.[a]
	DO	Sig.[c]	Sig.[c]	Sig.[a]	Non Sig.[c]
	FC	Non Sig.[a]	Non Sig.[a]	Non Sig.[a]	Non Sig.[a]
	Zinc	Sig.[c]	Non Sig.[a]	Sig.[c]	Sig.[c]
	Iron	Sig.[a]	Non Sig.[a]	Sig.[a]	Sig.[a]
	Manganese	Sig.[a]	Non Sig.[a]	Sig.[a]	Non Sig.[a]
	Lead	Sig.[a]	Non Sig.[a]	Sig.[a]	Non Sig.[a]
	Cadmium	Non Sig.[a]	Non Sig.[a]	Sig.[a]	Non Sig.[a]

*R = Rubber media; G = Gravel media; P = Plastic media.

TABLE 8.49 Significant Difference Between Pairs of Groups

Stage	Parameter	No. of Data	Significant difference between the following groups:		
			Exp. & ANNs	Exp. & SPSS	SPSS & ANNs
Set-up	BOD	24	Non Sig.[a]	Non Sig.[a]	Non Sig.[a]
	COD	24	Non Sig.[a]	Non Sig.[a]	Non Sig.[a]
	TSS	24	Non Sig.[a]	Non Sig.[a]	Non Sig.[a]
Steady	BOD	60	Non Sig.[c]	Non Sig.[c]	Non Sig.[c]
	COD	60	Non Sig.[a]	Non Sig.[a]	Non Sig.[a]
	TSS	60	Non Sig.[c]	Non Sig.[c]	Non Sig.[c]
	Ammonia	15	Non Sig.[a]	Non Sig.[a]	Non Sig.[a]
	Phosphate	15	Non Sig.[a]	Non Sig.[a]	Non Sig.[a]
	DO	15	Non Sig.[a]	Non Sig.[a]	Non Sig.[a]
	FC	15	Non Sig.[a]	Non Sig.[a]	Non Sig.[a]

TABLE 8.49 (Continued)

Stage	Parameter	No. of Data	Significant difference between the following groups: Exp. & ANNs	Exp. & SPSS	SPSS & ANNs
	Zinc	15	Non Sig.ᵃ	Non Sig.ᵃ	Non Sig.ᵃ
	Iron	15	Non Sig.ᵃ	Non Sig.ᵃ	Non Sig.ᵃ
	Manganese	15	Non Sig.ᵃ	Non Sig.ᵃ	Non Sig.ᵃ
	Lead	15	Non Sig.ᵃ	Non Sig.ᵃ	Non Sig.ᵃ
	Cadmium	15	Non Sig.ᵃ	Non Sig.ᵃ	Non Sig.ᵃ

ᵃUsing one way ANOVA test; ᶜUsing Chi-square test; ⁿNumber of used data.

KEYWORDS

- ANNs
- artificial neural networks modeling
- calibration
- HSSF
- regression equations
- SPSS
- statistical modeling
- validation

CHAPTER 9

SUMMARY, CONCLUSIONS, AND RECOMMENDATIONS

CONTENTS

9.1 Summary 515
9.2 General Conclusions 524
9.3 Recommendations 525
Keywords 526

9.1 SUMMARY

The following items are the outcomes of constructing the horizontal subsurface flow wetlands (HSSF) model using three different media (plastic, gravel, and rubber) and reeds plant species (*Phragmites Australis*). For loading load ranged from 5.94 to 169.50 cm/d, discharge ranged from 1.19 to 6.78 m^3/d, and retention time varied between 2.62 and 137.11 hr.

9.1.1 POROSITY OF TREATMENT MEDIA

- The porosity values reached the steady region after six months from start of operation for the wetland system (short-term effect) in this study.
- Through 218 days (tested period for porosity) from start of operation, the porosity decreases by 19.43% for cell entrance coarse gravel, 16.94% for gravel media, 12.33 for rubber media, and by 9.01% for plastic media. This indicates that the plastic and rubber media have a clogging ability smaller than the gravel one.
- At the end of set up stage, the values of porosity were 0.365 for cell entrance coarse gravel, 0.358 for gravel media, 0.505 for rubber media, and 0.788 for plastic media.

- During set up stage, the porosity varied with time according to a polynomial function of third degree for the three used media as well as for coarse gravel at the inlet and outlet zones.

9.1.2 SURFACE AREA OF TREATMENT MEDIA

- The surface area per unit bed volume (specific surface area) was calculated as 66 and 177 m^2/m^3 for cell entrance coarse gravel and gravel media, respectively.
- The specific surface area for rubber and plastic pieces was calculated to be 130 and 283 m^2/m^3, respectively.
- The higher specific surface area of plastic followed by gravel media led to have a wider area for the attached bacterial biofilm and better pollutants treatment comparing with rubber media.

9.1.3 SET-UP STAGE TREATMENT

- At outlets, BOD removal efficiency increased from 24.09 to 49.40% for plastic; from 21.36 to 43.45% for gravel; and from 19.55 to 38.69% for rubber cell at start and end of the stage, respectively.
- At outlets, plastic media gave more reduction of BOD than rubber and gravel media by average values of 8.66 and 4.83%, in the same sequence.
- At cell outlet, COD removal efficiency raised from 23.55 to 49.0% in plastic, from 20.93 to 43.03% in gravel, and 19.19 to 38.25% in rubber, at start and end of the stage, in the same order.
- The rubber media gave lower reduction of COD than the gravel and plastic media by average values of 3.80 and 8.52% at cells outlets, respectively.
- At outlets, TSS removal efficiency improved from 56 to 60.65% in plastic, from 42.67 to 48.39% in gravel, and 39.33 to 45.16% in rubber at start and end of the stage, in the same sequence.
- The plastic media gave more reduction of TSS than the gravel and rubber media by average values of 12.79 and 16.03%, in the same order at outlets. Gravel and rubber media had approximately the same effect for TSS performance.
- Plastic media had the best treatment performance for BOD, COD, and TSS followed by gravel then rubber. This is compatible with surface area measures of the three media types as plastic occupied the top followed by gravel and rubber.

Summary, Conclusions, and Recommendations 517

- The BOD and COD removal efficiencies for the three media were directly proportional to the time from start of sampling and follows a third order polynomial function, and in case of TSS followed a second order polynomial function.
- The three wetland systems have shown an exponential increase in the pollutant removal rate with the decrease in the retention time.
- During the set up stage (porosity steadiness period), the growth of planted reeds roots and the increase of bacterial biofilm attached to the media surface and plant parts, enhanced the pollutants accumulation, biodegradation, and the treatment efficiency.

9.1.4 STEADY STAGE TREATMENT

- At outlet, BOD removal efficiency enhanced from 56.83 to 83.8% in plastic cell after moving from Q_{max} to Q_{min}, from 49.43 to 76.62% for gravel, and from 45.31 to 72.59% for rubber.
- At cells outlets, plastic media gave an average BOD removal efficiency more than gravel media by about 6.75% and by about 10.88% for rubber media.
- For plastic, gravel, and rubber media, the BOD concentration (at outlet) was in the allowable limit (60 mg/l) at loading rate lower than 11.27, 11.79, and 8.48 cm/d, respectively.
- At outlet, COD removal efficiency improved from 56.45 to 83.68% in plastic cell after moving from Q_{max} to Q_{min}, from 49.02 to 76.37% in gravel, and from 44.97 to 72.5% in rubber.
- At cells outlets, rubber media gave an average COD removal efficiency lower than gravel and plastic media by 4.10 and 10.87%, respectively.
- For plastic, gravel, and rubber media, the COD concentration (at outlet) was in the permitted limit (80 mg/l) at loading rate lower than 11.27, 8.35, and 6.32 cm/d, respectively.
- At outlet, TSS removal efficiency improved from 66.06 to 93.96 in plastic cell after moving from Q_{max} to Q_{min}, from 55.01 to 88.96% in gravel, and from 50.31 to 86.06% in rubber.
- At cells outlets, rubber media gave an average TSS removal efficiency lower than gravel and plastic media by 3.12 and 13.95%, respectively.
- For plastic, gravel, and rubber media, the TSS concentration (at outlet) was in the admissible limit (50 mg/l) at loading rate lower than 24.07, 11.79, and 11.98 cm/d, respectively.

- Dissolved oxygen for plastic media was developed by an average value of 0.92 mg/l higher than gravel media and 0.38 mg/l higher than rubber media.
- For plastic, gravel, and rubber media, the DO concentration at cells outlets was in the allowable limit (4.0 mg/l) at loading rate lower than 11.27, 11.79, and 6.32 cm/d, in the same sequence.
- At cells outlets, ammonia removal efficiency improved from 57 to 78.53% in plastic cell after moving from Q_{max} to Q_{min}, from 47.97 to 67.69% in gravel cell, and from 45.75 to 62.48% in rubber cell.
- At cells outlets, plastic media gave average ammonia removal efficiency lower than gravel and rubber media by 8.56 and 11.84%, respectively.
- For plastic, gravel, and rubber media, the ammonia concentration (at outlet) was in the permitted limit (less than 10:12 mg/l) at loading rate lower than 24.07, 25.19, and 25.6 cm/d, respectively.
- At cells outlets, phosphate removal efficiency enhanced from 68.98 to 88.62% in plastic cell after moving from Q_{max} to Q_{min}, from 56.77 to 75.80% in gravel cell, and from 47.56 to 72.77% in rubber cell.
- The phosphate removal efficiency for rubber media decreased by an average value of 5.75% lower than gravel media and 16.19% lower than rubber media.
- For plastic, gravel, and rubber media, the phosphate concentration (at outlet) was in the acceptable limit (less than 5.0 mg/l) at loading rate lower than 24.07, 25.19, and 25.6 cm/d, respectively.
- The treatment performance of fecal coliforms varied from 97.1 to 99.3% with an average value of 98.3%. This high performance may be a result of microbial degradation, die-off, and predation.
- The fecal coliforms removal efficiency for rubber media decreased by an average value of 1.05% than gravel media and 2.27% than plastic media.
- For plastic, gravel, and rubber media, FC concentration (at outlet) was in the allowable limit (less than 5000 MPN/10 ml) at loading rate lower than 16.38, 17.13, and 17.41 cm/d, respectively.
- At cells outlets, Zinc removal efficiency enhanced from 51.32 to 81.83% in plastic cell after moving from Q_{max} to Q_{min}, from 43.54 to 74.91% in gravel cell, and from 39.35 to 68.65% in rubber cell.
- The Zinc removal efficiency for plastic media increased by an average value of 8.87 and 14.70% than gravel and rubber media, respectively.
- At cells outlets, Iron removal efficiency enhanced from 34.78 to 79.63% in plastic cell after moving from Q_{max} to Q_{min}, from 30.09 to 76.07% in gravel cell, and from 20.27 to 63.70% in rubber cell.
- The Iron removal efficiency for rubber media decreased by an average value of 8.51 and 19.03% lower than gravel and plastic media, respectively.

- At cells outlets, Manganese removal efficiency enhanced from 31.67 to 78.69% in plastic cell after moving from Q_{max} to Q_{min}, from 21.85 to 72.02% in gravel cell, and from 17.72 to 60.70% in rubber cell.
- The Manganese removal efficiency for plastic media increased by an average value of 11.64 and 19.49% higher than gravel and rubber media, respectively.
- At cells outlets, Lead removal efficiency enhanced from 40.30 to 66.91% in plastic cell after moving from Q_{max} to Q_{min}, from 27.32 to 62.58% in gravel cell, and from 20.16 to 57.41% in rubber cell.
- The Lead removal efficiency for plastic media increased by an average value of 8.14 and 13.21% higher than gravel and rubber media, respectively.
- At outlets, Cadmium removal efficiency enhanced from 29.25 to 55.73% in plastic cell after moving from Q_{max} to Q_{min}, from 23.35 to 50.81% in gravel cell, and from 17.76 to 36.40% in rubber cell.
- The Cadmium removal efficiency for rubber media decreased by an average value of 7.25 and 12.55% lower than gravel and plastic media, respectively.
- For plastic, gravel, and rubber media, the studied heavy metals concentration at cells outlets were reduced significantly in these systems and typically met effluent standards at loading rate lower than 24.07, 25.19, and 25.6 cm/d, respectively.
- The variation of all studied heavy metals removal efficiencies vary with retention time and discharge according to logarithmic functions for the used media.
- The studied wetland systems showed a logarithmic increase in the dissolved oxygen concentration level with the increase of the hydraulic retention time.
- The variation of BOD, COD, TSS, NH_3, PO_4, and FC removal efficiencies varied with retention time and discharge according to power and logarithmic functions for the used media, respectively.

9.1.5 POLLUTANTS REMOVAL RATE CONSTANTS

- The removal rate constants (k) for BOD pollutant using plug flow model were 0.1278, 0.1124, and 0.1003 m/d (k_v = 0.3701, 0.6252, and 0.4245 d^{-1}) for plastic, gravel, and rubber media, respectively; and 0.1526, 0.1303, and 0.1147 m/d (k_v = 0.4422, 0.7247, and 0.4856 d^{-1}) in case of modified plug flow.
- The removal rate constants for BOD using mixed flow model were 0.1803, 0.1567, and 0.1398 m/d (k_v = 0.5621, 0.870, and 0.6118 d^{-1}) for plastic,

gravel, and rubber media, in the same sequence; and 0.2084, 0.1779, and 0.1571 m/d (k_v = 0.6473, 0.9876, and 0.6867 d^{-1}) in case of modified mixed flow.
- The removal rate constants for COD using plug flow model were 0.1276, 0.1120, and 0.0999 m/d (k_v = 0.3698, 0.6227, and 0.4229 d^{-1}) for plastic, gravel, and rubber media; and 0.1495, 0.1278, and 0.1127 m/d (k_v = 0.4333, 0.7107, and 0.4771 d^{-1}) using modified plug flow.
- The removal rate constants for COD using mixed flow model were 0.1806, 0.1544, and 0.1382 m/d (k_v = 0.5638, 0.8571, and 0.6043 d^{-1}) for plastic, gravel, and rubber cells; and 0.2056, 0.173, and 0.1535 m/d (k_v = 0.6396, 0.9607, and 0.6706 d^{-1}) using modified mixed flow.
- The removal rate constants for TSS using plug flow model were 0.1897, 0.1502, and 0.1379 m/d (k_v = 0.5499, 0.8354, and 0.5826 d^{-1}) for plastic, gravel, and rubber media; and 0.2642, 0.2539, and 0.1934 m/d (k_v = 0.7702, 1.4121, and 0.8187 d^{-1}) in case of modified plug flow.
- The removal rate constants for TSS using mixed flow model were 0.2693, 0.1980, and 0.1779 m/d (k_v = 0.8348, 1.0993, and 0.7767 d^{-1}) for plastic, gravel, and rubber cells; and 0.3891, 0.2885, and 0.2307 m/d (k_v = 1.1879, 1.6021, and 1.0039 d^{-1}) using modified mixed flow.
- The removal rate constants for ammonia using plug flow model were 0.1319, 0.1098, and 0.1024 m/d (k_v = 0.3886, 0.6101, and 0.4367 d^{-1}) for plastic, gravel, and rubber media; and 0.151, 0.122, and 0.1127 m/d (k_v = 0.4449, 0.6778, and 0.4805 d^{-1}) in case of mixed flow, in the same sequence.
- The removal rate constants for phosphate using plug flow model were 0.1852, 0.1443, and 0.1226 m/d (k_v = 0.5455, 0.8016, and 0.5228 d^{-1}) for plastic, gravel, and rubber media; and 0.2241, 0.1658, and 0.138 m/d (k_v = 0.6603, 0.921, and 0.5887 d^{-1}) in case of mixed flow.
- The removal rate constants for fecal coliforms using plug flow model were 0.5349, 0.5496, and 0.5503 m/d (k_v = 1.5759, 3.053, and 2.3471 d^{-1}) for plastic, gravel, and rubber media; and 0.9345, 0.9492, and 0.942 m/d (k_v = 2.753, 5.2729, and 4.018 d^{-1}) in case of mixed flow, in the same order.
- The removal rate constants for Zinc using plug flow model were 0.1117, 0.0931, and 0.0808 m/d (k_v = 0.3291, 0.5172, and 0.3448 d^{-1}) for plastic, gravel, and rubber media; and 0.1552, 0.1219, and 0.1033 m/d (k_v = 0.4571, 0.6769, and 0.4405 d^{-1}) in case of mixed flow.
- The removal rate constants for Iron using plug flow model were 0.1052, 0.086, and 0.0646 m/d (k_v = 0.31, 0.4778, and 0.2755 d^{-1}) for plastic, gravel, and rubber media; and 0.1294, 0.0951, and 0.0706 m/d (k_v = 0.3811, 0.5285, and 0.3011 d^{-1}) in case of mixed flow.

- The removal rate constants for Manganese using plug flow model were 0.0953, 0.077, and 0.060 m/d (k_v = 0.2808, 0.4282, and 0.256 d⁻¹) for plastic, gravel, and rubber media; and 0.1155, 0.0814, and 0.0631 m/d (k_v = 0.3402, 0.4522, and 0.2682 d⁻¹) using mixed flow.
- The removal rate constants for Lead using plug flow model were 0.0776, 0.0694, and 0.0623 m/d (k_v = 0.2287, 0.3860, and 0.2659 d⁻¹) for plastic, gravel, and rubber media; and 0.1045, 0.0822, and 0.0686 m/d (k_v = 0.3078, 0.4564, and 0.2927 d⁻¹) in case of mixed flow.
- The removal rate constants for Cadmium using plug flow model were 0.0495, 0.0436, and 0.0321 m/d (k_v = 0.1457, 0.2422, and 0.1369 d⁻¹) for plastic, gravel, and rubber media; and 0.0614, 0.052, and 0.0386 m/d (k_v = 0.1808, 0.2886, and 0.1646 d⁻¹) in case of mixed flow.

The k and k_v values obtained from the mixed flow models are about 1.3 to 1.4 times the corresponding ones given by the plug flow models for BOD, COD, and TSS; about 1.1 to 1.7 times for ammonia, phosphate, and fecal coliforms; and about 1.1 to 1.3 times for heavy metals.

9.1.6 THE ANNs MODELING OF HSSF WETLANDS

The experimental data of HSSF constructed wetlands were simulated using the artificial neural networks using five input variables (influent concentration, loading rate, media surface area, actual velocity, and time from start of sampling) for set up stage pollutants, four input variables for steady stage pollutants (C_i, q, A_s, and v), and three input variables for DO parameter. The output of the model was the effluent concentration.

The total number of data for set up stage was 144 patterns, 120 patterns of these data used for model calibration and 24 patterns were used for model verification. For BOD, COD, and TSS pollutants in steady stage these numbers were 300 (total), 240 (calibration), and 60 (verification) patterns. For the remaining parameters in steady stage, 75 patterns were used, 60 (calibration) and 15 (verification).

For BOD, COD, and TSS pollutants in set up stage, the networks that exhibit satisfactory results having a structure of 5–5–3–1 (5 input variables, 5 number of neurons in the first hidden layer, 3 number of neurons in the second hidden layer, and 1 output variable), 5–4–3–1, and 5–5–3–1, respectively and the output layer was capable to minimize the mean square error of the network to the final values of 0.0883, 0.653, and 0.0572.

The average percentage error between the experimental effluent and the ANNs model output for the 120 patterns varied between −0.32 and +0.60% for BOD; −0.39 and +0.45% for COD; −0.36 and +0.40% for TSS. For 24 patterns these error ranged from −0.63 to +0.86% for BOD; −0.79 to +0.68% for COD; −0.59 to +0.72% for TSS.

For BOD, COD, and TSS pollutants in steady stage, the networks that show the good results having a structure of 4–5–4–1, 4–7–5–1, and 4–6–5–1, respectively and the output layer was capable to minimize the mean square error of the network to the final values of 0.582, 0.590, and 0.347.

The average percentage error between the experimental effluent and the ANNs model output for the 240 patterns varied between −1.25 and +0.77% for BOD; −0.85 and +0.81% for COD; −1.55 and +1.54% for TSS. For 60 patterns these error ranged from −1.18 to +0.84% for BOD; −0.95 to +1.37% for COD; −1.75 to +1.83% for TSS.

For NH_3, PO_4, DO, and FC parameters in steady stage, the networks that give satisfactory results having a structure of 4–2–2–1, 4–4–3–1, 3–4–4–1, and 4–3–3–1, respectively and the output layer was capable to minimize the mean square error of the network to the final values of 0.0485, 2.75E-05, 0.00537, and 9.94E+03.

The average percentage error between the experimental effluent and the ANNs model output for the 60 patterns varied between −2.74 and +2.33% for NH_3; −4.69 and +1.66% for PO_4; −3.0 and +2.45% for DO; −1.94 and +2.58% for FC. For 15 patterns these error ranged from −3.02 to +4.26% for NH_3; −4.58 to +3.79% for PO_4; −2.68 to +2.96% for DO; −1.77 to +1.83% for FC.

For Zn, Fe, Mn, Pb, and Cd elements in steady stage, the networks that give satisfactory results having a structure of 4–4–3–1, 4–4–3–1, 4–4–4–1, 4–3–2–1, and 4–4–3–1, respectively and the output layer is capable to minimize the mean square error of the network to the final values of 7.92E-06, 3.03E-05, 1.60E-05, 1.73E-06, and 4.21E-11.

The average percentage error between the experimental effluent and the ANNs model output for the 60 patterns varied between −1.43 and +1.35% for Zn; −2.97 and +1.94% for Fe; −2.83 and +2.51% for Mn; −4.63 and +2.23% for Pb; −0.57 and +1.28% for Cd. For 15 patterns these error ranged from −3.32 to +2.04% for Zn; −5.88 to +2.85% for Fe; −6.37 to +2.20% for Mn; −5.03 to +2.27% for Pb; −1.06 and +2.13% for Cd.

It could be said that models based on artificial neural networks have been successfully used in wastewater treatment systems and are very effective at capturing the nonlinear relationships between variables (multi-input) in complex systems.

9.1.7 THE SPSS MODELING OF HSSF WETLANDS

The same input and output variables presented in ANNs modeling were used in SPSS modeling with the same number of data for calibration and verification processes.

The average percentage error between the experimental effluent and the SPSS model output for the 120 patterns varied between −2.36 and +2.06% for BOD; −2.56 and +2.02% for COD; −2.39 and +2.98% for TSS. For 24 patterns these error ranged from −2.89 to +2.18% for BOD; −2.58 to +2.17% for COD; −2.39 to +2.89% for TSS.

The average percentage error between the experimental effluent and the SPSS model output for the 240 patterns varied between −4.61 and +4.10% for BOD; −4.83 and +4.54% for COD; −10.09 and +7.91% for TSS. For 60 patterns these error ranged from −4.45 to +4.29% for BOD; −4.63 to +3.32% for COD; −10.24 to +8.26% for TSS.

The average percentage error between the experimental effluent and the SPSS model output for the 60 patterns varied between −5.77 and +3.06% for NH_3; −6.26 and +8.97% for PO_4; −6.19 and +4.73% for DO; −3.17 and +3.18% for FC. For 15 patterns these error ranged from −7.68 to +5.14% for NH_3; −5.18 to +9.92% for PO_4; −4.53 to +5.21% for DO; −2.71 to +2.81% for FC.

The average percentage error between the experimental effluent and the SPSS model output for the 60 patterns varied between −2.86 and +9.48% for Zn; −13.57 and +5.03% for Fe; −11.11 and +11.57% for Mn; −6.23 and +8.20% for Pb; −4.82 and +5.45% for Cd. For 15 patterns these error ranged from −0.48 to +10.77% for Zn; −15.07 to +4.2% for Fe; −19.65 to +13.71% for Mn; −8.07 to +12.84% for Pb; −4.50 and +6.22% for Cd.

9.1.8 MEDIA CHARACTERISTICS

For plastic media, the porosity is higher than the porosities of both gravel and rubber media which minimize the clogging problem. Besides, the density is low for the plastic material, the plastic pieces tend to float creating more space at the bottom of the bed. Its specific surface area is about 1.6 and 2.2 times those given by gravel and rubber media, respectively.

For rubber media, a suitable fill medium for wetlands is provided, if the waste tires are chipped into small pieces about 40:50 mm length, 30:50 mm width, and 5:15 mm thickness. Tire chips are less dense and less expensive than gravel media. Because of their lower density, they are cheaper to haul. Tire chips have greater bulk porosity than gravel, so the wastewater is detained longer in the cell.

The use of tires in the constructed wetlands gives environmentally suitable solutions for disposal problems, which end up accumulating in rivers and public designations. Others burned tires, releasing gases into the atmosphere. The disadvantages of the rubber media are the lower specific surface area than the gravel media and the grease which produced by the tires during the first month of operation.

For gravel media, the mining, transportation, and handling of this conventional media are expensive and the clogging problem may be appeared after short time from operation. However, the gravel media is suitable for cultivation and root penetration.

9.1.9 SIGNIFICANCE TESTS

The outcomes of the significance tests declared the following points:

- Rubber, gravel, and plastic media; BOD and COD (set up stage) and fecal coliforms and Cadmium gave non-significant difference. The remaining parameters showed a significant difference.
- Rubber and gravel media; all the tested parameters in the two stages gave non-significant difference except dissolved oxygen.
- Rubber and plastic media; BOD and COD (set up stage) and fecal coliforms gave non-significant difference and the remaining parameters showed a significant difference.
- Gravel and plastic media; BOD and COD (set up stage) and DO, FC, Mn, Pb, and Cd gave non-significant difference. The remaining parameters showed a significant difference.
- Experimental and both ANNs and SPSS models, all the tested parameters in the two stages gave non-significant difference. Also, the comparison between SPSS and ANNs models.

9.2 GENERAL CONCLUSIONS

- During set up and steady stages, the plastic media was the best for treating all pollutants. This may be due to the highest surface area and biggest bacterial biofilm layer than both other media. Gravel media took the second order followed by rubber media in the third degree.

- The Samaha village experimental HSSF constructed wetland system managed to treat primary treated municipal wastewater with various efficiencies depends on pollutant loads and influent discharge in the range of 1.188 to 6.781 m^3/d.
- The characteristics of the wastewater improved significantly as the wastewater flowed through the wetland cell and the quality of the effluent water enhanced along the treatment path of flow.
- The better relative removal efficiency happened at the cells middle part (5 to 8 m) comparing with the other parts. This means that pollutants treatment mainly existed at the cell middle zone.
- The used HSSF wetland systems have shown an exponential decrease in pollutants concentrations with distance from inlet towards outlet. While these systems have shown a power increase in pollutant removal rate with treatment distance.
- The studied systems have shown a logarithmic increase in the pollutant concentration level with the increase of loading rate.
- The three wetland cells have shown an exponential increase in the pollutant effluent concentration level (BOD, COD, TSS, and NH$_3$) with the increase of pollutant influent concentration.
- The gravel cell has the lowest retention time (T_r) followed by the rubber and then the plastic cell. This is compatible with the porosity of the three media types. The gravel media has the smallest porosity followed by rubber media, while plastic media has the greatest porosity.
- The empirical relationships developed in this study can be used in the rational design of horizontal subsurface flow wetlands for conditions similar to the ones under which this study was conducted.
- Both ANNs and statistical linear and nonlinear equations through the SPSS package managed to mimic the HSSF wetlands for all used parameters with an acceptable accuracy. The artificial neural network models represent the experimental data in calibration and validation processes more accurately than the SPSS models. This is due to the ability of ANNs models to construct a variety of complex relationships between variables to represent them precisely.

9.3 RECOMMENDATIONS

The obtained results in this work encourage the development of future studies to increase the performance of these wastewater systems based on a better knowledge of the influence of hydraulic parameters, like flow, retention time, and hydraulic loading rate, on the pollutants removal efficiencies.

Based on the results and analyzes of this book, the following recommendations can be considered for further research:

- Initiating long-term (multi-year) studies to monitor the cumulative effects of dissolved and particulate organic matter on sediment accumulation, hydraulic conductivity and wastewater treatment efficiency.
- Check of the performance for wetland systems against high organic loads since Samaha wastewater have moderate loads.
- Appling different media, plants, and pollutants, to evaluate the performance ability of pollutant treatment.
- Examining other aspect ratio in treatment performance and media behavior.
- Studies are required for selecting the efficient engineering shape for constructed wetland instead of the rectangular shape. This selected shape must relay on the best pollutant removal, avoiding dead zones and water stagnations.
- Studying the use of mixed media in the same basin on treatment of pollutants either in sequence or in overlapping layers. Also, the use of different plants in the same cell on treatment performance.
- Evaluating the side effects to use the shredded tires on treatment and the effluent water quality and the carbon concentration.
- Checking plastic media with different diameters and lengths and choosing the best ones for performing the highest treatment.

KEYWORDS

- gravel
- HSSF
- plastic
- pollutants removal rate constants
- porosity
- rubber
- treatment media
- water quality

REFERENCES

Abidi, S.; Kallali, H.; Jedidi, N.; Bouzaiane, O.; Hassen, A. Comparative Pilot Study of the Performances of Two Constructed Wetland Wastewater Treatment Hybrid Systems. *Journal of Desalination*, **2009**, *248,* 49–56.

Akratos, C.S.; Papaspyros, J.N.E.; Tsihrintzis, V.A. An Artificial Neural Network Model and Design Equations for BOD and COD Removal Prediction in Horizontal Subsurface Flow Constructed Wetlands. *Chemical Engineering Journal*, **2008**, *143,* 96–110.

Akratos, C.S.; Papaspyros, J.N.E.; Tsihrintzis, V.A. Total Nitrogen and Ammonia Removal Prediction in Horizontal Subsurface Flow Constructed Wetlands: Use of Artificial Neural Networks and Development of a Design Equation. *Bioresource Technology J.*, **2009**, *100(2),* 586–596.

Al-Omari, A.; Fayyad, M. Treatment of Domestic Wastewater by Subsurface Flow Constructed Wetlands in Jordan. *Journal of Desalination*, **2003**, *155,* 27–39.

Albuquerque, A.; Bandeiras, R. Analysis of Hydrodynamic Characteristics of a Horizontal Subsurface Flow Constructed Wetland. International Conference on Water Pollution in natural Porous Media at Different Scales, **2007**, 329–338.

Albuquerque, A.; Oliveira, J.; Semitela, S.; Amaral, L. Influence of Bed Media Characteristics on Ammonia and Nitrate Removal in Shallow Horizontal Subsurface flow Constructed Wetlands. *Journal of Bioresource Tech.*, **2009**, *100,* 6269–6277.

Allen, W.C.; Hook, P.B.; Beiderman, J.A.; Stein, O.R. Wetlands and Aquatic Processes: Temperature and Wetland Plant Species Effects on Wastewater Treatment and Root Zone Oxidation. *Journal of Environmental Quality*, **2002**, *31,* 1010–1016.

Amos, P.W.; Younger, P.L. Substrate Characterization for a Subsurface Reactive Barrier to Treat Colliery Spoil Leachate. *Journal of Water Research*, **2003**, *37,* 108–120.

Anurita. Design of Research Facilities for Graywater Treatment Using Wetlands. Department of Civil Engineering Colorado State University, Fort Collins, Co., 2005.

APHA. Standard Methods for the Examination of Water and Wastewater. 20th Edition, American Public Health Association (APHA), American Water Works Association (AWWA) and the Water Environment Federation (WEF): Washington D.C., U.S.A., 1998.

Araújo, A.; Sousa, E.; Albuquerque, A. Longitudinal Dispersion in a Horizontal Subsurface Flow Constructed Wetland: a Numerical Solution. *Anziam Journal*, **2008**, *50,* 339–353.

Arias, C.A.; Brix, H.; Garza, M.F. Alternatives for Phosphorus Removal in Subsurface Flow Constructed Wetlands. International Meeting on Phytodepuration, University of Aarhus, Denmark, **2005**, 73–79.

Attiogbe, F.K.; Glover Amengor, M.; Nyadziehe, K.T. Correlating Biochemical and Chemical Oxygen Demand of Effluents – A Case Study of Selected Industries in Kumasi, Ghana. University of Science and Technology, Kumasi, Ghana, 1999.

Awad, A.M.; Saleh, H.I. Evaluating Contaminants Removal Rates in Subsurface Flow Constructed Wetland in Egypt. American Society of Civil Engineers Conference Proceedings of Wetlands Engineering and River Restoration, doi: 10.1061/40581, 2001.

Babatunde, A.O.; Zhao, Y.Q.; Doyle, R.J.; Rackard, S.M.; Kumar, J.L.G.; Hu, Y.S. Performance Evaluation and Prediction for a Pilot Two-Stage On-Site Constructed Wetland

System Employing Dewatered Alum Sludge as Main Substrate. *Bioresource Technology*, **2011**, *102(10),* 5645–5652.

Barton, C.D.; Karathanasis, A.D. Renovation of a Failed Constructed Wetland Treating Acid Mine Drainage. *Journal of Environmental Geology*, **1999**, *39(1),* 39–50.

Baskar, G.; Deeptha, V.T.; Rahman, A.A. Treatment of Wastewater from Kitchen in an Institution Hostel Mess Using Constructed Wetland. *International Journal of Recent Trends in Engineering*, **2009**, *1(6),* 54–58.

Borges, A.C.; Calijuri, M.C.; Matos, A.T.; Queiroz, M.E.L.R. Horizontal Subsurface Flow Constructed Wetlands for Mitigation of Ametryn-Contaminated Water. *Water SA Journal*, **2009**, *35(4),* 441–446.

Borkar, R.P.; Mahatme, P.S. Wastewater Treatment with Vertical Flow Constructed Wetland. *International Journal of Environmental Sciences*, **2011**, *2(2),* 590–603.

Botch, S.L.; Light, R. J. Constructed Wetlands for Municipal Wastewater Treatment. Institute of Local Government Affairs, A Joint Project of the Office of the Comptroller, State of Illinois and Southern Illinois University, 1994.

Brisson, J.; Chazarenc, F. Maximizing Pollutant Removal in Constructed Wetlands: Should We Pay More Attention to Macrophyte Species Selection? *Science of the Total Environment*, **2009**, *407,* 3923–3930.

Brix, H. Do Macrophytes Play a Role in Constructed Treatment Wetlands. *Journal of Water Science and Technology*, **1997**, *35(5),* 11–17.

Brix, H. Plants Used in Constructed Wetlands and Their Functions. Proceedings of Conference ICN and INAG, Lisbon, **2003**, 81–109.

Brix, H. Use of Constructed Wetlands in Water Pollution Control: Historical Development, Present Status, and Future Perspectives. *Journal of Water Science and Technology*, **1994**, *30(8),* 209–223.

Brix, H.; Arias, C.A.; Bubba, M.D. Media Selection for Sustainable Phosphorus Removal in Subsurface Flow Constructed Wetlands. *Journal of Water Science and Technology*, **2001**, *44,* No. 11, 47–54.

Brown, D.S.; Reed, S.C. Inventory of Constructed Wetlands in the United States. *Journal of Water Science and Technology*, **1994**, *29(4),* 309–318.

Bubba, M.D.; Arias, C.A.; Brix, H. Phosphorus Adsorption Maximum of Sands for Use as Media in Subsurface Flow Constructed Reed Beds as Measured by the Langmuir Isotherm. *Journal of Water Research*, **2003**, *37,* 3390–3400.

Butler, J.; Williams, J. Gravel Bed Hydroponic Wetlands for Wastewater Treatment. DFID Engineering Division Theme W4 Summary, University of Portsmouth, England, 1997.

Butler, J.E.; Loveridge, R.F.; Dewedar, A. Gravel Bed Hydroponics: an Artificial Wetland Capable of Sewage Treatment, 1991.

Calheiros, C.S.C.; Rangel, A.O.S.S.; Castro, P.M.L. Evaluation of Different Substrates to Support the Growth of Typha Latifolia in Constructed Wetlands Treating Tannery Wastewater over Long-Term Operation. Universidade Católica, Portugal, 2008.

Caselles-Osorio, A. Influence of the Characteristics of Organic Matter on the Efficiency of Horizontal Subsurface-Flow Constructed Wetlands. Ph.D. Thesis, University of Catalonia, Barcelona, Spain, **2006**, p. 147.

Chan, T.P.; Shah, N.R.; Cooper, T.J.; Alleman, J.E.; Govindaraju, R.S. Enhancing Oxygen Transfer in Subsurface-Flow Constructed Wetlands. American Society of Civil Engineers Conference Proceedings of World Water and Environmental Resources, doi: 10.1061/40792, **2004**, *173,* 307.

References

Chazarenc, F.; Merlin, G.; Gonthier, Y. Hydrodynamics of Horizontal Subsurface Flow Constructed Wetlands. *Journal of Ecological Engineering*, **2003**, *21,* No. 2–3, 165–173.

Chen, M.; Tang, Y.; Li, X.; Yu, Z. Study on the Heavy Metals Removal Efficiencies of Constructed Wetlands with Different substrates. *Journal of Water Resource and Protection*, **2009,** *1,* 22–28.

Chen, Z. M.; Chen, B.; Zhou, J.B.; Zhou, Z.; Li Xi, Y.X.R.; Lin, C.; Chen, G.Q. A Vertical Subsurface Flow Constructed Wetland in Beijing. *Communications in Nonlinear Science and Numerical Simulation*, **2008,** *13,* 1986–1997.

Chervek, D. The Use of Wetlands as Water Treatment Systems. BAE 558, 2005.

Collaço, A.B.; Roston, D.M. Use of Shredded Tires as Support Medium for Subsurface Flow Constructed Wetland. *Engenharia Ambiental – Espírito Santo do Pinhal*, **2006,** *3(1),* 21–31.

Cooke, A.J.; Rowe, R.K. Extension of Porosity and Surface Area Models for Uniform Porous Media. *Journal of Environmental Engineering*, **1999,** *125(2),* 126–136.

Cordesius, H.; Hedström, S. A Feasibility Study on Sustainable Wastewater Treatment Using Constructed Wetlands, an Example from Cochabamba, Bolivia. M.Sc. Thesis, Division of Water Resources Engineering, Department of Building and Environmental Technology, Lund University, Sweden, 2009.

Crites, R.W.; Middlebrooks, J.; Reed, S.C. *Natural Wastewater Treatment Systems, Chapter 7: Subsurface and Vertical Flow Constructed Wetlands.* Taylor and Francis Group, 2006.

Czech, A. Constructed Wetlands for Wastewater Treatment, Wetlands Conservation and Water Recycling Kraków, 25–27 October 2005. Invitation to International, Practical Workshop, Jagiellonian University, 2005.

Davis, L. *A Handbook of Constructed Wetlands.* A Guide to Creating Wetlands for Agricultural Wastewater, Domestic Wastewater, Coal Mine Drainage, and Stormwater, Volume No. 1, General Considerations. United States Department of Agriclture-Natural Resources Conservation Service (USDA-NRCS) and the United States Environmental Protection Agency: Washington D.C., 1995a.

Davis, L. *A Handbook of Constructed Wetlands.* A Guide to Creating Wetlands for Agricultural Wastewater, Domestic Wastewater, Coal Mine Drainage, and Stormwater, Volume No. 2, Domestic Wastewater. United States Department of Agriculture-Natural Resources Conservation Service (USDA-NRCS) and the United States Environmental Protection Agency: Washington D.C., 1995b.

Davis, M.L.; Cornwell, D.A. *Introduction to Environmental Engineering.* Third Edition, McGraw Hill, 1998.

Davison, L.; Bayley, M.; Kohlenberg, T.; Craven, J. Performance of Reed Beds and Single Pass Sand Filters with Characterization of Domestic Effluent: NSW North Coast. A Research Report, The NSW Department of Local Government's Septic Safe Scheme, 2001.

Demuth, H.; Beale, M.; Hagan, M. *Neural Network Toolbox™ 6 User's Guide.* Copyright 1992–2009 by The MathWorks Inc., Natick, MA 01760-2098, U.S.A., 2009.

Dreyfus, G. *Neural Networks: Methodology and Applications.* Springer-Verlag Berlin Heidelberg, 2005.

Drizo, A.; Frost, C.A.; Grace, J.; Smith, K.A. Physico-Chemical Screening of Phosphate-Removing Substrates for Use in Constructed Wetland Systems. *Journal of Water Research*, **1999,** *33(17),* 3595–3602.

Du, K.L.; Swamy, M.N.S. *Neural Networks in a Soft Computing Framework*. Springer-Verlag London Ltd., 2006.

Duarte, A.A.; Canais-Seco, T.; Peres, J.A.; Bentes, I.; Pinto, J. Sustainability Indicators of Subsurface Flow Constructed Wetlands in Portuguese Small Communities. *Wseas Transactions on Environment and Development*, **2010**, *9(6)*, 625–634.

El-Gammal, H.A.A. Water Quality Protection in Rural Areas of Egypt. *International Water Technology Journal*, IWTJ, **2012**, *1(3)*, 230–238.

El-Hamouri, B.; Nazih, J.; Lahjouj, J. Subsurface Horizontal Flow Constructed Wetland for Sewage Treatment under Moroccan Climate Conditions. *Journal of Desalination*, **2007**, *215*, 153–158.

El-Khateeb, M.A.; El-Gohary, F.A. Combining UASB Technology and Constructed Wetland for Domestic Wastewater Reclamation and Reuse. *Water Science and Technology: Water Supply*, **2003**, *3(4)*, 201–208.

El-Refaie, G.G.; Rashed, A.A.; Abdul Khaleq, M.A. Microcosm Constructed Wetland for Improving Mixed Irrigation Water of El-Salam Canal, Egypt. The 7th INTECOL International Wetlands Conference, Utrecht, Nederland, 2004.

El-Zoghby, M.R. Constructed Wetland is a Low Cost Technology for Water Saving. 1st International Conference and Exhibition on Sustainable Water Supply and Sanitation, Cairo, Egypt, 2010.

Elmitwalli, T.A.; Al-Sarawey, A.; El-Sherbiny, M.F.; Zeeman, G.; Lettinga, G. Anaerobic Biodegradability and Treatment of Egyptian Domestic Sewage. 7th International Water Technology Conference Egypt, **2002a**, 263–273.

Ewemoje, O.E.; Sangodoyin, A.Y. *Developing a Pilot Scale Horizontal Subsurface Flow Constructed Wetlands for Phytoremediation of Primary Lagoon Effluents*. 11th Edition of the World Wide Workshop for Young Environmental Scientists, Resource or Risks, Arcueil, France, 2011.

Farooqi, I.H.; Basheer, F.; Ghaudhari, R.J. Constructed Wetland System (CWS) for Wastewater Treatment. The 12th World Lake Conference, **2008**, 1004–1009.

Ferro, A.M.; Kadlec, R.H.; Deschamp, J. Constructed Wetland System to Treat Wastewater at the BP Amoco Former Casper Refinery: Pilot Scale Project. The 9th International Petroleum Environmental Conference, 2002.

Fisher, J.; Acreman, M.C. Wetland Nutrient Removal: A Review of the Evidence. *Journal of Hydrology and Earth System Sciences*, **2004**, *8(4)*, 6703–685.

Fraser, L.H.; Carty, S.M.; Steer, D. A Test of Four Plant Species to Reduce Total Nitrogen and Total Phosphorus from Soil Leachate in Subsurface Wetland Microcosms. *Journal of Bio-resource Technology*, **2004**, *94*, 185–192.

Frazer-Williams, R.A.D. A Review of the Influence of Design Parameters on the Performance of Constructed Wetlands. *Journal of Chemical Engineering*, IEB, **2010**, Vol. ChE. *25(1)*, 29–42.

Freeman, J.A.; Skapura, D.M. *Neural Networks Algorithms, Applications, and Programming Techniques*. Addison-Wesley Publishing Company, Inc., U.S.A., 1991.

Garcia-Perez, A.; Jones, D.; Harrison, M. Recirculating Vertical Flow Constructed Wetlands for Treating Residential Wastewater. Rural Wastewater, Purdue University, 2007.

Gearheart, R.A. Constructed Wetlands for Natural Wastewater Treatment. South-west Hydrology, Humboldt State University, USA, 2006.

Graupe, D. *Principles of Artificial Neural Networks*. 2nd Edition, World Scientific Publishing Co., 2007.

References

Grismer; Mark, E.; Shepherd; Heather; L. Plants in Constructed Wetlands Help to Treat Agricultural Processing Wastewater. *California Agriculture, Division of Agriculture and Natural Resources Communication Services*, **2011**, *65(2)*, 72–79.

Grove, J.K.; Stein, O.R. Polar Organic Solvent Removal in Microcosm Constructed Wetlands. *Journal of Water Research*, **2005**, *39*, 4040–4050.

Halverson, N.V. Review of Constructed Subsurface Flow vs. Surface Flow Wetlands. U.S. Department of Energy under Contract No. DE-AC09–96SR18500, 2004.

Hammer, D.A. *Constructed Wetlands for Wastewater Treatment: Municipal, Industrial, and Agricultural*. Third Edition, Lewis Publishers, 1989.

Hanko, J.A.; Cundiff, H. Constructed Wetland Wastewater Treatment Facilities Guidance. Water-0001-NPD, Indiana Department of Environmental Management Nonrule Policy Document, 1997.

Hanson, A.; Swanson, L.; Ewing, D.; Grabas, G.; Meyer, S.; Ross, L.; Watmough, M.; Kirkby, J. Wetland Ecological Functions Assessment: An Overview of Approaches. Canadian Wildlife Service Technical report series, No. 497, Atlantic Region, 2008.

Hawkins, J. *Constructed Treatment Wetlands*. USDA-NRCS, University of Tennessee Extension Richmond, Virginia, U.S.A., 2002.

Hoddinott, B.C. Horizontal Subsurface Flow Constructed Wetlands for On-Site Wastewater Treatment. MD Candidate for Master in Public Health, Wright State University, Dayton Ohio, 2006.

Hodgson, C.J.; Perkins, J.; Labadz, J.C. The Use of Microbial Tracers to Monitor Seasonal Variations in Effluent Retention in a Constructed Wetland. *Journal of Water Research*, **2004**, *38*, 3833–3844.

Hoffmann, H.; Platzer, C.; Winker, M.; Muench, E.V. Technology Review of Constructed Wetlands, Subsurface Flow Constructed Wetlands for Grey-Water and Domestic Wastewater Treatment. Deutsche Gesellschaft Für Internationale Zusammenar-beit (GIZ) GmbH Sustainable Sanitation-Ecosan Program, Eschborn, Germany, 2011.

Hu, Y.H.; Hwang, J.-N. *Handbook of Neural Network Signal Processing*. CRC Press LLC, Boca Raton London New York Washington, D.C., 2002.

Huett, D.O.; Morris, S.G.; Smith, G.; Hunt, N. Nitrogen and Phosphorus Removal from Plant Nursery Runoff in Vegetated and Un-Vegetated Subsurface Flow Wetlands. *Journal of Water Research*, **2005**, *39*, 3259–3272.

Hussain, I.; Raschid, L.; Hanjra, M.A.; Marikar, F.; Van Der Hoek, W. Wastewater Use in Agriculture: Review of Impacts and Methodological Issues in Valuing Impacts. Working Paper 37, Colombo, Sri Lanka: International Water Management Institute, 2002.

Islam, A.M.; Tudor, T.; Bates, M. Evaluation of the Pollutant Removal Mechanisms of a Reed Bed System: Biochemical Parameters. 2nd International Workshop Advances in Cleaner Production, São Paulo, Brazil, 2009.

Jain, L.; Fanelli, A.M. *Recent Advances in Artificial Neural Networks Design and Applications*. CRC Press LLC, Boca Raton London New York Washington, D.C., 2000.

Jensen, L. Pilot Scale Constructed Wastewater Treatment Wetland. The Water Center, University of Washington, USA., 2001.

Kadlec, R.H.; Knight, R.L. *Treatment Wetlands*. CRC Lewis Publisher, New York, 1996.

Kadlec, R.H.; Wallace, S.D. *Treatment Wetlands*. Second Edition, CRC Press Taylor and Francis Group, 2009.

Kamarudzaman, A.N.; Abdul Aziz, R.; Ab Jalil, M.F. Removal of Heavy Metals from Landfill Leachate Using Horizontal and Vertical Subsurface Flow Constructed Wetland Planted with Limnocharis Flava. *International Journal of Civil and Environmental Engineering*, **2011**, *11(5)*, 85–91.

Karathanasis, A.D.; Potter, C.L.; Coyne, M.S. Vegetation Effects on Fecal Bacteria, BOD, and Suspended Solid Removal in Constructed Wetlands Treating Domestic Wastewater. *Ecological Engineering*, **2003**, *20*, 157–169.

Karu, C.; Pat, L.; Ken, L.; Marilyn, R.; Gary, S.; Larry, W. Guidelines for the Approval and Design of Natural and Constructed Treatment Wetlands for Water Quality Improvement. Municipal Program Development Branch, Environmental Sciences Division, Environmental Service, Alberta Environment, 2000.

Kayombo, S.; Mbwette, T.S.A.; Katima, J.H.Y.; Ladegaard, N.; Jørgensen, S.E. Waste Stabilization Ponds and Constructed Wetlands Design Manual. University of Dar Es-Salaam and Danish University, Copenhagen Denmark, 2004.

Kayranli, B.; Scholz, M.; Mustafa, A.; Hedmark, A. Carbon Storage and Fluxes within Freshwater Wetlands: a Critical Review. *Wetlands*, **2010**, *30(1)*, 111–124.

Khare, M.; Nagendra, S.M.S. *Artificial Neural Networks in Vehicular Pollution Modeling*. Springer-Verlag Berlin Heidelberg, 2007.

Kiracofe, B.D. Performance Evaluation of the Town of Monterey Wastewater Treatment Plant Utilizing Subsurface Flow Constructed Wetlands. M.Sc. Thesis, Faculty of Virginia Polytechnic Institute and State University, Blacksburg, Virginia, U.S.A., 2000.

Kopec, D.A. Small Subsurface Flow Constructed Wetlands with Soil Dispersal System. State of Ohio EPA, Division of Surface Water Guidance Document, 2007.

Kuschk, P.; Wießner, A.; Kappelmeyer, U.; Wießbrodt, E.; Kästner, M.; Stottmeister, U. Annual Cycle of Nitrogen Removal by a Pilot-Scale Subsurface Horizontal Flow in a Constructed Wetland under Moderate Climate. *Journal of Water Research*, **2003**, *37*, 4236–4242.

Kuschk, P.; Wießner, A.; Müller, R.; Kstner, M. Constructed Wetlands – Treating Wastewater with Cenoses of Plants and Microorganisms. Published by UFZ Centre for Environmental Research Leipzig-Halle, 2005.

Langergraber, G. Modeling of Processes in Subsurface Flow Constructed Wetlands: A Review. *Vadose Zone Journal*, Special Section, **2008**, *7(2)*, 830–842.

Langergraber, G.; Rousseau, D.P.L.; García, J.; Mena, J. CWM1 – A General Model to Describe Bio-kinetic Processes in Subsurface Flow Constructed Wetlands. *Journal of Water Science and Technology*, **2009**, *59(9)*, 1687–1697.

Lavrova, S.; Koumanova, B. Polishing of Aerobically Treated Wastewater in a Constructed Wetland System. *Journal of the University of Chemical Technology and Metallurgy*, **2007**, *42(2)*, 195–200.

Lee, R.E. Set-Wet: A Wetland Simulation Model to Optimize NPS Pollution Control. M.Sc. Thesis, Faculty of the Virginia Polytechnic Institute and State University, U.S.A., 1999.

Leonard, K.M. Analysis of Residential Subsurface Flow Constructed Wetlands Performance in Northern Alabama. *Small Flows Quarterly*, **2000**, *1(3)*, 34–39.

Leonard, K.M.; Swanson, G.W. Comparison of Operational Design Criteria for Subsurface Flow Constructed Wetlands for Wastewater Treatment. *Journal of Water Science and Technology*, **2001**, *43(11)*, 301–307.

Lesikar, B.; Weaver, R.; Richter, A.; O'Neill, C. On-Site Wastewater Treatment Systems: Constructed Wetland Media. AgriLIFE Extension, Texas A & M University, U.S.A, L-5459 (1), 2005.

Lin, X.; Lan, C.; Shu, W. Treatment of Landfill Leachate by Subsurface-Flow Constructed Wetland: A Microcosm Test. Sun Yatsen (Zhongshan) University, Guangzhou, China, 2000.

Liu, W. Subsurface Flow Constructed Wetlands Performance Evaluation, Modeling, and Statistical Analysis. Ph.D., University of Nebraska, U.S.A., 2002.

Mander, U.; Jenssen, P. (2003). Constructed Wetlands for Wastewater Treatment in Cold Climates. *Current Science*, **2003**, *84(10)*, 1367–1368.

Manios, T.; Stentiford, E.I.; Millner, P. Removal of Total Suspended Solids from Wastewater in Constructed Horizontal Flow Subsurface Wetlands. *Journal of Environmental Science and Health*, **2003**, *A38(6)*, 1073–1085.

Marchand, L.; Mench, M.; Jacob, D.L.; Otte, M.L. Metal and Metalloid Removal in Constructed Wetlands, with Emphasis on the Importance of Plants and Standardized Measurements: A Review. *Journal of Environmental Pollution*, **2010**, *158*, 3447–3461.

Marsili-Libelli, S.; Checchi, N. Identification of Dynamic Models for Horizontal Subsurface Constructed Wetlands. *Journal of Ecological Modeling*, **2005**, *187*, 201–218.

Masi, F.; Martinuzzi, N. Constructed Wetlands for the Mediterranean Countries: Hybrid Systems for Water Reuse and Sustainable Sanitation. *Desalination Journal*, **2007**, *215*, 44–55.

Mayo, A.W.; Bigambo, T. Nitrogen Transformation in Horizontal Subsurface Flow Constructed Wetlands I: Model Development. *Physics and Chemistry of the Earth*, **2005**, *30*, 658–667.

Mays, P.A.; Edwards, G.S. Comparison of Heavy Metals Accumulation in a Natural Wetland and Constructed wetlands Receiving Acid Mine Drainage. *Journal of Ecological Engineering*, **2001**, *16(4)*, 487–500.

Mbuligwe, S.E. Comparative Treatment of Dye-Rich Wastewater in Engineered Wetland Systems Vegetated with Different Plants. *Journal of Water Research*, **2005**, *39*, 271–280.

Mburu, N.; Thumbi, G.M.; Mayabi, A.O. Removal of Bacterial Pathogens from Domestic Wastewater in Tropical Subsurface Horizontal Flow Constructed Wetland. The 12th World Lake Conference, **2008**, 1010–1015.

McEntee, D. Constructed wetlands Opportunities for Local authorities. Chartered Engineer Dublin City Council, 2006.

McKenzie, C. Constructed Wetland. *Small Flows Quarterly*, U.S.A., **2004**, *5(4)*, 26–29.

Melton, R.H. BOD_5 Removal in Subsurface Flow Constructed Wetlands with Respect to Aspect Ratio and Influent loading. M.Sc. Thesis, Texas A & M University, USA, 2005.

Michele, F.; Valter, B.; Ermanno, Z. Pedo-Bio-Remediation by Constructed Wetlands: Wastewater Purification in Alpine Livestock Farming Zones Humides Artificielles. *Scientific Registration*, **1998**, *2115*, Symposium, No. 21.

Mink, L.R.; Heusser, C.; Brannon, E.L. Use of Surface and Subsurface Wetlands for Treatment of Municipal Wastewater. Research and Extension Regional Water Quality Conf., Russia, 2002.

Mitchell, C.; McNevin, D. Alternative Analysis of BOD Removal in Subsurface Flow Constructed Wetlands Employing Monod Kinetics. *Water Research*, **2001**, *35(5)*, 1295–1303.

Mitsch, W.J.; Gosselink, J.G. *Wetlands*. Second Edition, Van Nostrand Reinhold Company: New York, 1993.

Mitsch, W.J.; Wise, K.M. Water Quality, Fate of Metals, and Predictive Model Validation of a Constructed Wetland Treating Acid Mine Drainage. *Journal of Water Research*, **1998**, *32(6)*, 1888–1900.

Moore, J.A. Using Constructed Wetlands to Improve Water Quality. Water Quality, EC 1408, Oregon State University, 1993.

Mowjood, M.I.; Herath, G.B.B.; Jinadasa, K.B.S.N.; Weerakoon, G.M.P.R. Performance Evaluation of Subsurface Flow Constructed Wetland Systems under Variable Hydraulic

Loading Rates. International Conference on Sustainable Built Environment (ICSBE), Kandy, University of Peradeniya, Sri Lanka, 2010.

Murray-Gulde, C.; Heatley, J.E.; Karanfil, T.; Rodgers, J.H.; Myers, J.E. Performance of a Hybrid Reverse Osmosis-Constructed Wetland Treatment System for Brackish Oil Field Produced Water. *Journal of Water Research*, **2003**, *37*, 705–713.

Myers, J.E.; Jackson, L.M. Evaluation of Subsurface Flow and Free Water Surface Wetlands Treating NPR-3 Produced Water – Year No. 1. Doe/Rmotc-020144, Prepared for the United States Department of Energy/Rocky Mountain Oilfield Testing Center, 2001.

NAWQAM Drainage Water Reuse and Pilot Schemes: Drainage Wastewater Treatment Technologies and Approaches. Report No. DR-TE-0209-007-fbFN, 2002.

Nelson, M.; Alling, A.; Dempster, W.F.; Thillo, M.V.; Allen, J. Advantages of Using Subsurface Flow Constructed Wetlands for Wastewater Treatment in Space Applications: Ground-Based Mars Base Prototype. *Advances Space Research*, **2003**, *31(7)*, 1799–1804.

Nikolić, V.; Milićević, D.; Milenković, S. Wetlands, Constructed Wetlands and Their Role in Wastewater Treatment with Principles and Examples of Using It in Serbia. *Architecture and Civil Engineering*, **2009**, *7(1)*, 65–82.

Nilsson, E.; Sha, L.; Qian, W.; Leedo, M. Constructed Wetlands "Wastewater Treatment." VVA No.1 Decentralized water and wastewater treatment, 2012.

Nitisoravut, S.; Klomjek P. Inhibition Kinetics of Salt-Affected Wetland for Municipal Wastewater Treatment. *Journal of Water Research*, **2005**, *39*, 4413–4419.

Noor, A.M.; Shiam, L.C.; Hong, F.W.; Soetardjo, S.; Abdul Khalil, H.P.S. Application of Vegetated Constructed Wetland with Different Filter Media for Removal of Ammoniacal Nitrogen and Total Phosphorus in Landfill Leachate. *International Journal of Chemical Engineering and Applications*, **2010**, *1(3)*, 270–275.

Odong, J. Evaluation of Empirical Formulae for Determination of Hydraulic Conductivity Based on Grain-Size Analysis. *Journal of American Science*, **2007**, *3(3)*, 54–60.

Ouellet-Plamondon, C.; Chazarenc, F.; Comeau, Y.; Brisson, J. Artificial Aeration to Increase Pollutant Removal Efficiency of Constructed Wetlands in Cold Climate. *Journal of Ecological Engineering*, **2006**, *27(3)*, 258–264.

Pant, H.K.; Reddy, K.R.; Lemon, E. Phosphorus Retention Capacity of Root Bed Media of Subsurface Flow Constructed Wetlands. *Journal Ecological Engineering*, **2001**, *17*, 345–355.

Pendleton, C.H.; Morris, J.W.F.; Goldemund, H.; Rozema, L.R.; Mallamo, M.S.; Agricola, L. Leachate Treatment Using Vertical Subsurface Flow Wetland Systems-Findings from Two Pilot Studies. International Waste Management and Landfill Symposium, Sardinia, 2005.

Pescod, M.B. Wastewater Treatment and Use in Agriculture-FAO Irrigation and Drainage Paper 47. Food and Agriculture Organization of the United Nations, Rome, Italy, 1992.

Powell, J.; Homer, J.; Glassmeyer, C.; Sauer, N. Alternative Wastewater Treatment: On-Site Biotreatment Wetlands at the Fernald Preserve Visitors Center. WM Conference, Phoenix, Arizona, 2009.

Ragusa, S.R.; McNevin, D.; Qasem, S.; Mitchell, C. Indicators of Biofilm Development and Activity in Constructed Wetlands Microcosms. *Journal of Water Research*, **2004**, *38*, 2865–2873.

Rani, S.H.C.; Din, M.F.M.; Yusof, M.B.M.; Chelliapan, S. Overview of Subsurface Constructed Wetlands Application in Tropical Climates. *Universal Journal of Environmental Research and Technology*, **2011**, *1(2)*, 103–114.

Rashed, A.A. Reciprocating Subsurface Wetlands for Drainage Water Treatment: A Case Study in Egypt. *El-Mansoura Engineering Journal*, **2007**, *32(3)*.

Rashed, A.A. Treatment of Municipal Pollution through Re-Engineered Drains: A Case Study, Edfina Drain, West Nile Delta. The 11th Iced International Drainage Workshop (IDW) 23–27 September, Cairo, Egypt, 2012.

Rashed, A.A.; Abdel-Rashid, A. Polluted Drainage Water Natural On-Stream Remediation. The 26th Water Conference, Alexandria, 2008.

Rashed, A.A.; El-Quosy, D.; Abdel-Gawad, S.T.; Bayoumi, M.N. Environmental Protection of Lake Manzala, Egypt and Reuse of Treated Water by a Constructed Wetland. International Workshop on Development and Management of Flood Plains and Wetlands (IWFW) Beijing, China, 2000.

Reinoso, R.; Torresa, L.A.; Bécaresb, E. Efficiency of Natural Systems for Removal of Bacteria and Pathogenic Parasites from Wastewater. *Science of the Total Environment*, **2008**, *395*, 80–86.

Rousseau, D.; Geenens, D.; Vanrolleghem, P.A.; De Pauw, N. Short-Term Behavior of Constructed Reed Beds: Pilot Plant Experiments under Different Temperature Conditions. Ghent University, Belgium, 2001.

Rousseau, D.P.L.; Vanrolleghem, P.A.; De Pauw, N. Model-Based Design of Horizontal Subsurface Flow Constructed Treatment Wetlands: A Review. *Journal of Water Research*, **2004**, *38*, 1484–1493.

RTM. Regional Technical Meeting on the Implementation of the Ramsar Convention in the Arab Region: Egypt Report, 2009.

Samarasinghe, S. *Neural Networks for Applied Sciences and Engineering*. Taylor & Francis Group, 2007.

Sarafraz, S.; Mohammad, T.A.; Noor, M.J.M.M.; Liaghat, A. Wastewater Treatment Using Horizontal Subsurface Flow Constructed Wetland. *American Journal of Environmental Sciences*, **2009**, *5(1)*, 99–105.

Shuib, N.; Davies, W.R.; Baskaran, K.; Muthukumaran, S. Effluent Quality Performance of Horizontal Subsurface Flow Constructed Wetlands Using Natural Zeolite (Escott). International Conference on Environment Science and Engineering IPCBEE, **2011**, *8*, IACSIT Press, Singapore.

Sirianuntapiboon, S.; Jitvimolnimit, S. Effect of Plantation Pattern on the Efficiency of Subsurface Flow Constructed Wetland (SFCW) for Sewage Treatment. *African Journal of Agricultural Research*, **2007**, *2(9)*, 447–454.

Sirianuntapiboon, S.; Kongchum, M.; Jitmaikasem, W. (2006). Effects of Hydraulic Retention Time and Media of Constructed Wetland for Treatment of Domestic Wastewater. *African Journal of Agricultural Research*, **2006**, *1(2)*, 27–37.

Smith, R.D.; Ammann, A.; Barlodus, C.; Brinson, M.M. An Approach for Assessing Wetland Functions Using Hydrogeomorphic Classification, Reference wetlands, and Functional Indices. U.S. Army Corps of Engineers, Waterways Experiment Station, 1995.

Snow, A.; Anderson, B.C.; Wootton, B.; Hellebust, A. Efficacy of a Hybrid Subsurface Flow Constructed Wetland for the Treatment of Aquaculture Wastewater. 44th Central Canadian Symposium on Water Quality Research Burlington, 2009.

Song, K.-Y.; Zoh, K.-D.; Kang, H. Release of Phosphate in a Wetland by Changes in Hydrological Regime. *Science of the Total Environment*, **2007**, *380*, 13–18.

Steer, D.N.; Fraser, L.H.; Seibert, B.A. Cell-to-Cell Pollution Reduction Effectiveness of Subsurface Domestic Treatment Wetlands. *Journal of Bioresource Technology*, **2005**, *96*, 969–976.

Tanner, C.; Kadlec, R.H. Oxygen Flux Implications of Observed Nitrogen Rates in Subsurface-Flow Treatment Wetlands. *Water Science and Technology*, **2003,** *48(5),* 191–198.

Tanner, C.; Sukias, J.P. Status of Wastewater Treatment Wetlands in New Zealand. National Institute of Water and Atmospheric Research (NIWA), Hamilton, New Zealand, 2002.

Thiyagarajan, G.; Ranghaswami, M.V.; Selvakumar, S.; Muralidharan, J. Constructed Wetlands. Science Technology Entrepreneur, Tamil Nadu Agricultural University, Coimbatore–641003, Tamil Nadu, 2006.

Tomenko, V.; Ahmed, S.; Popov, V. Modeling Constructed Wetland Treatment System Performance. *Journal of Ecological Modeling,* **2007,** *205,* No. 3–4, 355–364.

Toscano, A.; Langergraber, G.; Consoli, S.; Cirelli, G.L. Modeling Pollutant Removal in a Pilot-Scale Two-Stage Subsurface Flow Constructed Wetlands. *Journal of Ecological Engineering,* **2009,** *35,* 281–289.

U.S. EPA. Constructed Wetlands for Wastewater Treatment and Wildlife Habitat: 17 Case Studies. EPA 832-R-93–005, Office of Research and Development, 1993a.

U.S. EPA. Functions and Values of Wetlands. Office of Water, Office of Wetlands, Oceans and Watersheds, EPA 843-F-01-002C, 2001.

U.S. EPA. R.E.D. Facts: Ametryn. Prevention, Pesticides and Toxic Substances (7508C), EPA-738-F-05-007, 2005.

U.S. EPA. Subsurface Flow Constructed Wetlands for Wastewater Treatment: A Technology Assessment. EPA 832-R-93-008, Office of Water, 1993b.

Ulrich, H.; Klaus, D.; Irmgard, F.; Annette, H.; Juan, P.L.; Regine, S. Microbiological Investigations for Sanitary Assessment of Wastewater Treated in Constructed Wetlands. *Journal of Water Research,* **2005,** *39,* 4849–4858.

Urbanc-Bercic, O. Investigation into the Use of Constructed Reed beds for Municipal Waste Dump Leachate Treatment. *Journal of Water Science and Technology,* **1994,** *29(4),* 289–294.

USAID. Integrated Water Resource Management II: Feasibility of Wastewater Reuse-Report No. 14. The American People through the United States Agency for International Development (USAID), International Resources Group (IRG), 2010.

USDA. *Chapter 3 Constructed Wetland.* Part 637 Environmental Engineering, National Engineering Handbook, 210-VI-NEH, 2002.

Vacca, G.; Wand, H.; Nikolausz, M.; Kuschk, P.; Kastner, M. (2005). Effect of Plants and Filter Materials on Bacteria Removal in Pilot-Scale Constructed Wetlands. *Journal of Water Research,* **2005,** *39(7),* 1361–1373.

Vrhovšek, D.; Kukanja, V.; Bulc, T. Constructed Wetland (CW) for Industrial Wastewater Treatment. *Journal of Water Research,* **1996,** *30(10),* 2287–2292.

Vymazal, J. Removal of Nutrients in Various Types of Constructed Wetlands. *Science of the Total Environment,* **2007,** *380,* 48–65.

Vymazal, J. Review Constructed Wetlands for Wastewater Treatment. *Water Journal,* **2010,** *2,* 530–549.

Vymazal, J.; Kröpfelová, L. Removal of Organics in Constructed Wetlands with Horizontal Subsurface Flow: A Review of the Field Experience. *Science of the Total Environment,* **2009,** *407,* 3911–3922.

Wiessner, A.; Kappelmeyer, U.; Kuschk, P.; Kästner, M. Sulphate Reduction and the Removal of Carbon and Ammonia in a Laboratory-Scale Constructed Wetland. *Journal of Water Research,* **2005,** *39,* 4643–4650.

Wieβner, A.; Kappelmeyer, U.; Kuschk, P.; Kästner, M. Influence of the Redox Condition Dynamics on the Removal Efficiency of a Laboratory-Scale Constructed Wetland. *Journal of Water Research,* **2005,** *39,* 248–256.

Wießner, A.; Kuschk, P.; Stottmeister, U.; Struchmann, D.; Jank, M. Treating a Lignite Pyrolysis Wastewater in a Constructed Subsurface Flow Wetland. *Journal of Water Research*, **1999,** *33(5),* 1296–1302.

Wynn, J.; Belfi, H.; Cantino, H.; Ajamian, M. Ecological Wastewater Treatment for Appalachia: Constructed Wetlands and Related Innovations. An Emerging Issue Paper Prepared for the Appalachian Regional Commission by Rural Action, 1997.

Wynn, T.M.; Liehr, S.K. Development of a Constructed Subsurface-Flow Wetland Simulation Model. *Journal of Ecological Engineering*, **2001,** *16(4),* 519–536.

Xing, A. Recent Developments in Wetland Technology for Wastewater Treatment. M.Sc. Thesis in Applied Environmental Science, School of Business and Engineering, Halmstad University, 2012.

Yalcuk, A. Modeling Different Types of Constructed Wetlands Treating Phenol from Olive Mill Wastewater by Using Artificial Neural Network. Abant Izzet Beysal University, Faculty of Engineering and Architecture, Environmental Engineering Department, Gölkoy-Bolu/Turkey, 2012.

Yamagiwa, K.; Ong, S. Up-flow Constructed Wetland for On-Site Industrial Wastewater Treatment. Niigata University, Japan, 2007.

Yeh, T.Y.; Chou, C.C.; Pan, C.T. Heavy Metal Removal within Pilot-Scale Constructed Wetlands Receiving River Water Contaminated by Confined Swine Operations. *Journal of Desalination,* **2009,** *249,* 368–373.

Zhang, L.; Zhang, L.; Liu, Y.; Shen, Y.; Liu, H.; Xiong, Y. Effect of Limited Artificial Aeration on Constructed Wetland Treatment of Domestic Wastewater. *Journal of Desalination,* **2010,** *250,* 915–920.

Zidan, A.A.; El-Gamal, M.A.; Rashed, A.A.; Abd El-Hady, M.A. BOD Treatment in HSSF Constructed Wetlands Using Different Media (Set-up Stage). *Mansoura Engineering. Journal (MEJ),* **2013,** *38(3)*.

Zurita, F.; Anda, J.D.; Belmont, M.A. Treatment of Domestic Wastewater and Production of Commercial Flowers in Vertical and Horizontal Subsurface-Flow Constructed Wetlands. *Journal of Ecological Engineering,* **2009,** *35,* 861–869.

APPENDICES

APPENDIX I

POROSITY MEASUREMENTS

POROSITY DETERMINATION

The porosity was measured at three layers each layer thickness equal to 16.67 cm (media depth/3). The calculated values of the porosity are given in Tables I-1 to I-8.

The average of these nine readings was taken as the porosity of media at specified date. Due to the difficulty of implementation a porosity bucket

TABLE I-1A Calculated Values for Media Porosity (Date 5/9/2009)

Layer	Vp	Vb	Vd	Vw	Vv	n
		Porosity Measurements for Gravel Media				
$L_a d_1$	3040.85	1443.46	1597.39	2220	622.61	0.431
$L_a d_2$	3040.85	1613.14	1427.71	2210	782.29	0.485
$L_a d_3$	3040.85	1584.20	1456.65	2160	703.35	0.444
$L_b d_1$	6079.88	2886.05	3193.83	4310	1116.17	0.387
$L_b d_2$	6079.88	3225.31	2854.57	4280	1425.43	0.442
$L_b d_3$	6079.88	3167.46	2912.42	4230	1317.58	0.416
$L_c d_1$	9120.74	4329.51	4791.23	6500	1708.77	0.395
$L_c d_2$	9120.74	4838.45	4282.29	6450	2167.71	0.448
$L_c d_3$	9120.74	4751.66	4369.08	6410	2040.92	0.430
Porosity of Coarse Gravel Layer = 0.453						0.431

TABLE I-1B Calculated Values for Media Porosity (Date 5/9/2009)

Layer	V_p	V_b	V_d	V_w	V_v	n
\multicolumn{7}{c}{**Porosity Measurements for Rubber Media**}					
$L_a d_1$	3040.85	1671.79	1369.06	2460	1090.94	0.653
$L_a d_2$	3040.85	1388.99	1651.86	2380	728.14	0.524
$L_a d_3$	3040.85	1527.12	1513.73	2370	856.27	0.561
$L_b d_1$	6079.88	3342.58	2737.30	4850	2112.70	0.632
$L_b d_2$	6079.88	2777.15	3302.73	4760	1457.27	0.525
$L_b d_3$	6079.88	3053.32	3026.56	4700	1673.44	0.548
$L_c d_1$	9120.74	5014.38	4106.36	7270	3163.64	0.631
$L_c d_2$	9120.74	4166.15	4954.59	7240	2285.41	0.549
$L_c d_3$	9120.74	4580.44	4540.30	7110	2569.70	0.561
The average value						**0.576**

TABLE I-1C Calculated Values for Media Porosity (Date 5/9/2009)

Layer	V_p	V_b	V_d	V_w	V_v	n
\multicolumn{7}{c}{**Porosity Measurements for Plastic Media**}					
$L_a d_1$	3040.85	1584.20	1456.65	2890	1433.35	0.905
$L_a d_2$	3040.85	1513.01	1527.84	2850	1322.16	0.874
$L_a d_3$	3040.85	1443.46	1597.39	2880	1282.61	0.889
$L_b d_1$	6079.88	3167.46	2912.42	5660	2747.58	0.867
$L_b d_2$	6079.88	3025.12	3054.76	5640	2585.24	0.855
$L_b d_3$	6079.88	2886.05	3193.83	5660	2466.17	0.855
$L_c d_1$	9120.74	4751.66	4369.08	8440	4070.92	0.857
$L_c d_2$	9120.74	4538.13	4582.61	8490	3907.39	0.861
$L_c d_3$	9120.74	4329.51	4791.23	8410	3618.77	0.836
The average value						**0.866**

TABLE I-2A Calculated Values for Media Porosity (Date 19/9/2009)

Layer	V_p	V_b	V_d	V_w	V_v	n
\multicolumn{7}{c}{**Porosity Measurements for Gravel Media**}					
$L_a d_1$	3040.85	1443.46	1597.39	2170	572.61	0.397
$L_a d_2$	3040.85	1613.14	1427.71	2180	752.29	0.466
$L_a d_3$	3040.85	1584.20	1456.65	2140	683.35	0.431
$L_b d_1$	6079.88	2886.05	3193.83	4260	1066.17	0.369

Appendices

TABLE I-2A (Continued)

Layer	V_p	V_b	V_d	V_w	V_v	n
\multicolumn{7}{c}{Porosity Measurements for Gravel Media}
$L_b\,d_2$	6079.88	3225.31	2854.57	4140	1285.43	0.399
$L_b\,d_3$	6079.88	3167.46	2912.42	4130	1217.58	0.384
$L_c\,d_1$	9120.74	4329.51	4791.23	6390	1598.77	0.369
$L_c\,d_2$	9120.74	4838.45	4282.29	6350	2067.71	0.427
$L_c\,d_3$	9120.74	4751.66	4369.08	6250	1880.92	0.396
Porosity of Coarse Gravel Layer = 0.431						0.404

TABLE I-2B Calculated Values for Media Porosity (Date 19/9/2009)

Layer	V_p	V_b	V_d	V_w	V_v	n
\multicolumn{7}{c}{Porosity Measurements for Rubber Media}
$L_a\,d_1$	3040.85	1671.79	1369.06	2420	1050.94	0.629
$L_a\,d_2$	3040.85	1388.99	1651.86	2330	678.14	0.488
$L_a\,d_3$	3040.85	1527.12	1513.73	2340	826.27	0.541
$L_b\,d_1$	6079.88	3342.58	2737.30	4810	2072.70	0.620
$L_b\,d_2$	6079.88	2777.15	3302.73	4690	1387.27	0.500
$L_b\,d_3$	6079.88	3053.32	3026.56	4670	1643.44	0.538
$L_c\,d_1$	9120.74	5014.38	4106.36	7240	3133.64	0.625
$L_c\,d_2$	9120.74	4166.15	4954.59	7180	2225.41	0.534
$L_c\,d_3$	9120.74	4580.44	4540.30	7050	2509.70	0.548
The average value						0.558

TABLE I-2C Calculated Values for Media Porosity (Date 19/9/2009)

Layer	V_p	V_b	V_d	V_w	V_v	n
\multicolumn{7}{c}{Porosity Measurements for Plastic Media}
$L_a\,d_1$	3040.85	1584.20	1456.65	2840	1383.35	0.873
$L_a\,d_2$	3040.85	1513.01	1527.84	2820	1292.16	0.854
$L_a\,d_3$	3040.85	1443.46	1597.39	2850	1252.61	0.868
$L_b\,d_1$	6079.88	3167.46	2912.42	5550	2637.58	0.833
$L_b\,d_2$	6079.88	3025.12	3054.76	5570	2515.24	0.831

TABLE I-2C (Continued)

Layer	V_p	V_b	V_d	V_w	V_v	n
Porosity Measurements for Plastic Media							
$L_b\ d_3$	6079.88	2886.05	3193.83	5610	2416.17	0.837
$L_c\ d_1$	9120.74	4751.66	4369.08	8360	3990.92	0.840
$L_c\ d_2$	9120.74	4538.13	4582.61	8300	3717.39	0.819
$L_c\ d_3$	9120.74	4329.51	4791.23	8340	3548.77	0.820
The average value						0.842

TABLE I-3A Calculated Values for Media Porosity (Date 5/10/2009)

Layer	V_p	V_b	V_d	V_w	V_v	n
Porosity Measurements for Gravel Media							
$L_a\ d_1$	3040.85	1443.46	1597.39	2180	582.61	0.404
$L_a\ d_2$	3040.85	1613.14	1427.71	2110	682.29	0.423
$L_a\ d_3$	3040.85	1584.20	1456.65	2100	643.35	0.406
$L_b\ d_1$	6079.88	2886.05	3193.83	4250	1056.17	0.366
$L_b\ d_2$	6079.88	3225.31	2854.57	4120	1265.43	0.392
$L_b\ d_3$	6079.88	3167.46	2912.42	4070	1157.58	0.365
$L_c\ d_1$	9120.74	4329.51	4791.23	6440	1648.77	0.381
$L_c\ d_2$	9120.74	4838.45	4282.29	6250	1967.71	0.407
$L_c\ d_3$	9120.74	4751.66	4369.08	6230	1860.92	0.392
Porosity of Coarse Gravel Layer = 0.411						0.393

TABLE I-3B Calculated Values for Media Porosity (Date 5/10/2009)

Layer	V_p	V_b	V_d	V_w	V_v	n
Porosity Measurements for Rubber Media							
$L_a\ d_1$	3040.85	1671.79	1369.06	2380	1010.94	0.605
$L_a\ d_2$	3040.85	1388.99	1651.86	2330	678.14	0.488
$L_a\ d_3$	3040.85	1527.12	1513.73	2290	776.27	0.508
$L_b\ d_1$	6079.88	3342.58	2737.30	4740	2002.70	0.599
$L_b\ d_2$	6079.88	2777.15	3302.73	4690	1387.27	0.500
$L_b\ d_3$	6079.88	3053.32	3026.56	4610	1583.44	0.519
$L_c\ d_1$	9120.74	5014.38	4106.36	7240	3133.64	0.625
$L_c\ d_2$	9120.74	4166.15	4954.59	7180	2225.41	0.534
$L_c\ d_3$	9120.74	4580.44	4540.30	6920	2379.70	0.520
The average value						0.544

Appendices

TABLE I-3C Calculated Values for Media Porosity (Date 5/10/2009)

Layer	V_p	V_b	V_d	V_w	V_v	n
\multicolumn{7}{c}{**Porosity Measurements for Plastic Media**}
$L_a d_1$	3040.85	1584.20	1456.65	2720	1263.35	0.797
$L_a d_2$	3040.85	1513.01	1527.84	2810	1282.16	0.847
$L_a d_3$	3040.85	1443.46	1597.39	2800	1202.61	0.833
$L_b d_1$	6079.88	3167.46	2912.42	5510	2597.58	0.820
$L_b d_2$	6079.88	3025.12	3054.76	5640	2585.24	0.855
$L_b d_3$	6079.88	2886.05	3193.83	5580	2386.17	0.827
$L_c d_1$	9120.74	4751.66	4369.08	8290	3920.92	0.825
$L_c d_2$	9120.74	4538.13	4582.61	8280	3697.39	0.815
$L_c d_3$	9120.74	4329.51	4791.23	8370	3578.77	0.827
The average value						**0.827**

TABLE I-4A Calculated Values for Media Porosity (Date 14/10/2009)

Layer	V_p	V_b	V_d	V_w	V_v	n
\multicolumn{7}{c}{**Porosity Measurements for Gravel Media**}
$L_a d_1$	3040.85	1443.46	1597.39	2160	562.61	0.390
$L_a d_2$	3040.85	1613.14	1427.71	2100	672.29	0.417
$L_a d_3$	3040.85	1584.20	1456.65	2070	613.35	0.387
$L_b d_1$	6079.88	2886.05	3193.83	4220	1026.17	0.356
$L_b d_2$	6079.88	3225.31	2854.57	4110	1255.43	0.389
$L_b d_3$	6079.88	3167.46	2912.42	4060	1147.58	0.362
$L_c d_1$	9120.74	4329.51	4791.23	6300	1508.77	0.348
$L_c d_2$	9120.74	4838.45	4282.29	6230	1947.71	0.403
$L_c d_3$	9120.74	4751.66	4369.08	6150	1780.92	0.375
Porosity of Coarse Gravel Layer = 0.398						**0.381**

TABLE I-4B Calculated Values for Media Porosity (Date 14/10/2009)

Layer	V_p	V_b	V_d	V_w	V_v	n
\multicolumn{7}{c}{**Porosity Measurements for Rubber Media**}
$L_a d_1$	3040.85	1671.79	1369.06	2410	1040.94	0.623
$L_a d_2$	3040.85	1388.99	1651.86	2320	668.14	0.481
$L_a d_3$	3040.85	1527.12	1513.73	2330	816.27	0.535

TABLE I-4B (Continued)

Layer	V_p	V_b	V_d	V_w	V_v	n
Porosity Measurements for Rubber Media						
$L_b\ d_1$	6079.88	3342.58	2737.30	4730	1992.70	0.596
$L_b\ d_2$	6079.88	2777.15	3302.73	4460	1157.27	0.417
$L_b\ d_3$	6079.88	3053.32	3026.56	4650	1623.44	0.532
$L_c\ d_1$	9120.74	5014.38	4106.36	7120	3013.64	0.601
$L_c\ d_2$	9120.74	4166.15	4954.59	7050	2095.41	0.503
$L_c\ d_3$	9120.74	4580.44	4540.30	6860	2319.70	0.506
The average value						0.533

TABLE I-4C Calculated Values for Media Porosity (Date 14/10/2009)

Layer	V_p	V_b	V_d	V_w	V_v	n
Porosity Measurements for Plastic Media						
$L_a\ d_1$	3040.85	1584.20	1456.65	2820	1363.35	0.861
$L_a\ d_2$	3040.85	1513.01	1527.84	2770	1242.16	0.821
$L_a\ d_3$	3040.85	1443.46	1597.39	2750	1152.61	0.799
$L_b\ d_1$	6079.88	3167.46	2912.42	5530	2617.58	0.826
$L_b\ d_2$	6079.88	3025.12	3054.76	5610	2555.24	0.845
$L_b\ d_3$	6079.88	2886.05	3193.83	5540	2346.17	0.813
$L_c\ d_1$	9120.74	4751.66	4369.08	8270	3900.92	0.821
$L_c\ d_2$	9120.74	4538.13	4582.61	8120	3537.39	0.779
$L_c\ d_3$	9120.74	4329.51	4791.23	8270	3478.77	0.804
The average value						0.819

TABLE I-5A Calculated Values for Media Porosity (Date 20/11/2009)

Layer	V_p	V_b	V_d	V_w	V_v	n
Porosity Measurements for Gravel Media						
$L_a\ d_1$	3040.85	1443.46	1597.39	2140	542.61	0.376
$L_a\ d_2$	3040.85	1613.14	1427.71	2070	642.29	0.398
$L_a\ d_3$	3040.85	1584.20	1456.65	2070	613.35	0.387
$L_b\ d_1$	6079.88	2886.05	3193.83	4160	966.17	0.335
$L_b\ d_2$	6079.88	3225.31	2854.57	4130	1275.43	0.395

Appendix I

TABLE I-5A (Continued)

Layer	V_p	V_b	V_d	V_w	V_v	n
\multicolumn{7}{c}{**Porosity Measurements for Gravel Media**}
$L_b\,d_3$	6079.88	3167.46	2912.42	4060	1147.58	0.362
$L_c\,d_1$	9120.74	4329.51	4791.23	6310	1518.77	0.351
$L_c\,d_2$	9120.74	4838.45	4282.29	6180	1897.71	0.392
$L_c\,d_3$	9120.74	4751.66	4369.08	6110	1740.92	0.366
Porosity of Coarse Gravel Layer = 0.387						**0.374**

TABLE I-5B Calculated Values for Media Porosity (Date 20/11/2009)

Layer	V_p	V_b	V_d	V_w	V_v	n
\multicolumn{7}{c}{**Porosity Measurements for Rubber Media**}
$L_a\,d_1$	3040.85	1671.79	1369.06	2380	1010.94	0.605
$L_a\,d_2$	3040.85	1388.99	1651.86	2280	628.14	0.452
$L_a\,d_3$	3040.85	1527.12	1513.73	2320	806.27	0.528
$L_b\,d_1$	6079.88	3342.58	2737.30	4720	1982.70	0.593
$L_b\,d_2$	6079.88	2777.15	3302.73	4630	1327.27	0.478
$L_b\,d_3$	6079.88	3053.32	3026.56	4600	1573.44	0.515
$L_c\,d_1$	9120.74	5014.38	4106.36	7080	2973.64	0.593
$L_c\,d_2$	9120.74	4166.15	4954.59	7000	2045.41	0.491
$L_c\,d_3$	9120.74	4580.44	4540.30	6790	2249.70	0.491
The average value						**0.527**

TABLE I-5C Calculated Values for Media Porosity (Date 20/11/2009)

Layer	V_p	V_b	V_d	V_w	V_v	n
\multicolumn{7}{c}{**Porosity Measurements for Plastic Media**}
$L_a\,d_1$	3040.85	1584.20	1456.65	2740	1283.35	0.810
$L_a\,d_2$	3040.85	1513.01	1527.84	2720	1192.16	0.788
$L_a\,d_3$	3040.85	1443.46	1597.39	2800	1202.61	0.833
$L_b\,d_1$	6079.88	3167.46	2912.42	5490	2577.58	0.814
$L_b\,d_2$	6079.88	3025.12	3054.76	5520	2465.24	0.815
$L_b\,d_3$	6079.88	2886.05	3193.83	5590	2396.17	0.830
$L_c\,d_1$	9120.74	4751.66	4369.08	8360	3990.92	0.840
$L_c\,d_2$	9120.74	4538.13	4582.61	8110	3527.39	0.777
$L_c\,d_3$	9120.74	4329.51	4791.23	8250	3458.77	0.799
The average value						**0.812**

TABLE I-6A Calculated Values for Media Porosity (Date 20/12/2009)

Layer	V_p	V_b	V_d	V_w	V_v	n
\multicolumn{7}{c}{Porosity Measurements for Gravel Media}
$L_a d_1$	3040.85	1443.46	1597.39	2120	522.61	0.362
$L_a d_2$	3040.85	1613.14	1427.71	2090	662.29	0.411
$L_a d_3$	3040.85	1584.20	1456.65	2020	563.35	0.356
$L_b d_1$	6079.88	2886.05	3193.83	4160	966.17	0.335
$L_b d_2$	6079.88	3225.31	2854.57	4070	1215.43	0.377
$L_b d_3$	6079.88	3167.46	2912.42	4050	1137.58	0.359
$L_c d_1$	9120.74	4329.51	4791.23	6250	1458.77	0.337
$L_c d_2$	9120.74	4838.45	4282.29	6180	1897.71	0.392
$L_c d_3$	9120.74	4751.66	4369.08	6080	1710.92	0.360
Porosity of Coarse Gravel Layer = 0.376						**0.365**

TABLE I-6B Calculated Values for Media Porosity (Date 20/12/2009)

Layer	V_p	V_b	V_d	V_w	V_v	n
\multicolumn{7}{c}{Porosity Measurements for Rubber Media}
$L_a d_1$	3040.85	1671.79	1369.06	2340	970.94	0.581
$L_a d_2$	3040.85	1388.99	1651.86	2330	678.14	0.488
$L_a d_3$	3040.85	1527.12	1513.73	2270	756.27	0.495
$L_b d_1$	6079.88	3342.58	2737.30	4690	1952.70	0.584
$L_b d_2$	6079.88	2777.15	3302.73	4630	1327.27	0.478
$L_b d_3$	6079.88	3053.32	3026.56	4470	1443.44	0.473
$L_c d_1$	9120.74	5014.38	4106.36	7120	3013.64	0.601
$L_c d_2$	9120.74	4166.15	4954.59	6910	1955.41	0.469
$L_c d_3$	9120.74	4580.44	4540.30	6730	2189.70	0.478
The average value						**0.516**

TABLE I-6C Calculated Values for Media Porosity (Date 20/12/2009)

Layer	V_p	V_b	V_d	V_w	V_v	n
\multicolumn{7}{c}{Porosity Measurements for Plastic Media}
$L_a d_1$	3040.85	1584.20	1456.65	2740	1283.35	0.810
$L_a d_2$	3040.85	1513.01	1527.84	2780	1252.16	0.828
$L_a d_3$	3040.85	1443.46	1597.39	2710	1112.61	0.771

Appendices

TABLE I-6C (Continued)

Layer	V_p	V_b	V_d	V_w	V_v	n
\multicolumn{7}{c}{Porosity Measurements for Plastic Media}
$L_b\, d_1$	6079.88	3167.46	2912.42	5400	2487.58	0.785
$L_b\, d_2$	6079.88	3025.12	3054.76	5570	2515.24	0.831
$L_b\, d_3$	6079.88	2886.05	3193.83	5590	2396.17	0.830
$L_c\, d_1$	9120.74	4751.66	4369.08	8100	3730.92	0.785
$L_c\, d_2$	9120.74	4538.13	4582.61	8080	3497.39	0.771
$L_c\, d_3$	9120.74	4329.51	4791.23	8170	3378.77	0.780
The average value						**0.799**

TABLE I-7A Calculated Values for Media Porosity (Date 6/2/2010)

Layer	V_p	V_b	V_d	V_w	V_v	n
\multicolumn{7}{c}{Porosity Measurements for Gravel Media}
$L_a\, d_1$	3040.85	1443.46	1597.39	2090	492.61	0.341
$L_a\, d_2$	3040.85	1613.14	1427.71	2070	642.29	0.398
$L_a\, d_3$	3040.85	1584.20	1456.65	2050	593.35	0.375
$L_b\, d_1$	6079.88	2886.05	3193.83	4130	936.17	0.324
$L_b\, d_2$	6079.88	3225.31	2854.57	4030	1175.43	0.364
$L_b\, d_3$	6079.88	3167.46	2912.42	3990	1077.58	0.340
$L_c\, d_1$	9120.74	4329.51	4791.23	6190	1398.77	0.323
$L_c\, d_2$	9120.74	4838.45	4282.29	6250	1967.71	0.407
$L_c\, d_3$	9120.74	4751.66	4369.08	6210	1840.92	0.387
Porosity of Coarse Gravel Layer = 0.371						**0.362**

TABLE I-7B Calculated Values for Media Porosity (Date 6/2/2010)

Layer	V_p	V_b	V_d	V_w	V_v	n
\multicolumn{7}{c}{Porosity Measurements for Rubber Media}
$L_a\, d_1$	3040.85	1671.79	1369.06	2330	960.94	0.575
$L_a\, d_2$	3040.85	1388.99	1651.86	2270	618.14	0.445
$L_a\, d_3$	3040.85	1527.12	1513.73	2290	776.27	0.508
$L_b\, d_1$	6079.88	3342.58	2737.30	4690	1952.70	0.584
$L_b\, d_2$	6079.88	2777.15	3302.73	4570	1267.27	0.456

TABLE I-7B (Continued)

Layer	V_p	V_b	V_d	V_w	V_v	n
\multicolumn{7}{c}{Porosity Measurements for Rubber Media}
$L_b\,d_3$	6079.88	3053.32	3026.56	4570	1543.44	0.505
$L_c\,d_1$	9120.74	5014.38	4106.36	7060	2953.64	0.589
$L_c\,d_2$	9120.74	4166.15	4954.59	6920	1965.41	0.472
$L_c\,d_3$	9120.74	4580.44	4540.30	6710	2169.70	0.474
The average value						0.512

TABLE I-7C Calculated Values for Media Porosity (Date 6/2/2010)

Layer	V_p	V_b	V_d	V_w	V_v	n
\multicolumn{7}{c}{Porosity Measurements for Plastic Media}
$L_a\,d_1$	3040.85	1584.20	1456.65	2710	1253.35	0.791
$L_a\,d_2$	3040.85	1513.01	1527.84	2760	1232.16	0.814
$L_a\,d_3$	3040.85	1443.46	1597.39	2800	1202.61	0.833
$L_b\,d_1$	6079.88	3167.46	2912.42	5350	2437.58	0.770
$L_b\,d_2$	6079.88	3025.12	3054.76	5530	2475.24	0.818
$L_b\,d_3$	6079.88	2886.05	3193.83	5590	2396.17	0.830
$L_c\,d_1$	9120.74	4751.66	4369.08	7970	3600.92	0.758
$L_c\,d_2$	9120.74	4538.13	4582.61	8010	3427.39	0.755
$L_c\,d_3$	9120.74	4329.51	4791.23	8210	3418.77	0.790
The average value						0.795

TABLE I-8A Calculated Values for Media Porosity (Date 6/2/2010)

Layer	V_p	V_b	V_d	V_w	V_v	n
\multicolumn{7}{c}{Porosity Measurements for Gravel Media}
$L_a\,d_1$	3040.85	1443.46	1597.39	2090	492.61	0.341
$L_a\,d_2$	3040.85	1613.14	1427.71	2070	642.29	0.398
$L_a\,d_3$	3040.85	1584.20	1456.65	2020	563.35	0.356
$L_b\,d_1$	6079.88	2886.05	3193.83	4150	956.17	0.331
$L_b\,d_2$	6079.88	3225.31	2854.57	4070	1215.43	0.377
$L_b\,d_3$	6079.88	3167.46	2912.42	4010	1097.58	0.347
$L_c\,d_1$	9120.74	4329.51	4791.23	6240	1448.77	0.335
$L_c\,d_2$	9120.74	4838.45	4282.29	6140	1857.71	0.384
$L_c\,d_3$	9120.74	4751.66	4369.08	6050	1680.92	0.354
Porosity of Coarse Gravel Layer = 0.365						0.358

Appendix I

TABLE I-8B Calculated Values for Media Porosity (Date 10/4/2010)

Layer	V_p	V_b	V_d	V_w	V_v	n
\multicolumn{7}{c}{Porosity Measurements for Rubber Media}
$L_a\, d_1$	3040.85	1671.79	1369.06	2350	980.94	0.587
$L_a\, d_2$	3040.85	1388.99	1651.86	2260	608.14	0.438
$L_a\, d_3$	3040.85	1527.12	1513.73	2270	756.27	0.495
$L_b\, d_1$	6079.88	3342.58	2737.30	4770	2032.70	0.608
$L_b\, d_2$	6079.88	2777.15	3302.73	4550	1247.27	0.449
$L_b\, d_3$	6079.88	3053.32	3026.56	4460	1433.44	0.469
$L_c\, d_1$	9120.74	5014.38	4106.36	6990	2883.64	0.575
$L_c\, d_2$	9120.74	4166.15	4954.59	6880	1925.41	0.462
$L_c\, d_3$	9120.74	4580.44	4540.30	6650	2109.70	0.461
The average value						0.505

TABLE I-8C Calculated Values for Media Porosity (Date 10/4/2010)

Layer	V_p	V_b	V_d	V_w	V_v	n
\multicolumn{7}{c}{Porosity Measurements for Plastic Media}
$L_a\, d_1$	3040.85	1584.20	1456.65	2680	1223.35	0.772
$L_a\, d_2$	3040.85	1513.01	1527.84	2740	1212.16	0.801
$L_a\, d_3$	3040.85	1443.46	1597.39	2760	1162.61	0.805
$L_b\, d_1$	6079.88	3167.46	2912.42	5390	2477.58	0.782
$L_b\, d_2$	6079.88	3025.12	3054.76	5480	2425.24	0.802
$L_b\, d_3$	6079.88	2886.05	3193.83	5570	2376.17	0.823
$L_c\, d_1$	9120.74	4751.66	4369.08	8050	3680.92	0.775
$L_c\, d_2$	9120.74	4538.13	4582.61	7970	3387.39	0.746
$L_c\, d_3$	9120.74	4329.51	4791.23	8190	3398.77	0.785
The average value						0.788

in the first 2.0 m of the wetland cell, so the porosity of coarse gravel at inlet and outlet zones were taken as the average porosity of $L_a d_1$, $L_a d_2$, and $L_a d_3$ for gravel media, where this layer had the same gradate and degree of compaction.

$L_a d_1 - L_a d_2 - L_a d_3$ = porosity in first layer at distances 2.5, 5.5, and 7.5 m from inlet

$L_b d_1 - L_b d_2 - L_b d_3$ = porosity in first and second layers at distances 2.5, 5.5, and 7.5 m from inlet

$L_c d_1 - L_c d_2 - L_c d_3$ = porosity in the three layers at distances 2.5, 5.5, and 7.5 m from inlet

For computing the initial porosity of gravel cell at distances 2.5, 5.5, and 7.5 m from cell inlet (L_a-d_1, L_a-d_2, L_a-d_3, L_b-d_1, L_b-d_2, L_b-d_3, L_c-d_1, L_c-d_2, and L_c-d_3), the following calculations are carried out (Table I-2):

1. Porosity at L_a-d_1 (n_1)

$V_{p1} = \pi/4 \times 15.242 \times 16.67 = 3040.85\ cm^3$
$V_{b1} = \pi/4 \times 10.502 \times 16.67 = 1443.46\ cm^3$
$V_{d1} = V_{p1} - V_{b1} = 3040.85 - 1443.46 = 1597.39\ cm^3$
$V_{w1} = 2220\ cm^3$ (added water, their values in Appendix I)
$V_{v1} = V_{w1} - V_{d1} = 2220 - 1597.39 = 622.61\ cm^3$
$n_1 = V_{v1}/V_{b1} = 622.61/1443.46 = 0.431$

2. Porosity at L_a-d_2 (n_2)

$V_{p2} = \pi/4 \times 15.242 \times 16.67 = 3040.85\ cm^3$
$V_{b2} = \pi/4 \times 11.102 \times 16.67 = 1613.14\ cm^3$
$V_{d2} = V_{p2} - V_{b2} = 3040.85 - 1613.14 = 1427.71\ cm^3$
$V_{w2} = 2210\ cm^3$
$V_{v2} = V_{w2} - V_{d2} = 2210 - 1427.71 = 782.29\ cm^3$
$n_2 = V_{v2}/V_{b2} = 782.29/1613.14 = 0.485$

3. Porosity at L_a-d_3 (n_3)

$V_{p3} = \pi/4 \times 15.242 \times 16.67 = 3040.85\ cm^3$
$V_{b3} = \pi/4 \times 112 \times 16.67 = 1584.20\ cm^3$
$V_{d3} = V_{p3} - V_{b3} = 3040.85 - 1584.20 = 1456.65\ cm^3$
$V_{w3} = 2160\ cm^3$
$V_{v3} = V_{w3} - V_{d3} = 2160 - 1456.65 = 703.35\ cm^3$
$n_3 = V_{v3}/V_{b3} = 703.35/1584.20 = 0.444$

Appendices

4. Porosity at L_b-d_1 (n_4)
$V_{p4} = \pi/4 \times 15.24^2 \times 33.33 = 6079.88\ cm^3$
$V_{b4} = \pi/4 \times 10.50^2 \times 33.33 = 2886.05\ cm^3$
$V_{d4} = V_{p4} - V_{b4} = 6079.88 - 2886.05 = 3193.83\ cm^3$
$V_{w4} = 4310\ cm^3$
$V_{v4} = V_{w4} - V_{d4} = 4310 - 3193.83 = 1116.17\ cm^3$
$n_4 = V_{v4}/V_{b4} = 1116.17/2886.05 = 0.387$

5. Porosity at L_b-d_2 (n_5)
$V_{p5} = \pi/4 \times 15.24^2 \times 33.33 = 6079.88\ cm^3$
$V_{b5} = \pi/4 \times 11.10^2 \times 33.33 = 3225.31\ cm^3$
$V_{d5} = V_{p5} - V_{b5} = 6079.88 - 3225.31 = 2854.57\ cm^3$
$V_{w5} = 4280\ cm^3$
$V_{v5} = V_{w5} - V_{d5} = 4280 - 2854.57 = 1425.43\ cm^3$
$n_5 = V_{v5}/V_{b5} = 1425.43/3225.31 = 0.442$

6. Porosity at L_b-d_3 (n_6)
$V_{p6} = \pi/4 \times 15.24^2 \times 33.33 = 6079.88\ cm^3$
$V_{b6} = \pi/4 \times 11^2 \times 33.33 = 3167.46\ cm^3$
$V_{d6} = V_{p6} - V_{b6} = 6079.88 - 3167.46 = 2912.42\ cm^3$
$V_{w6} = 4230\ cm^3$
$V_{v6} = V_{w6} - V_{d6} = 4230 - 2912.42 = 1317.58\ cm^3$
$n_6 = V_{v6}/V_{b6} = 1317.58/3167.46 = 0.416$

7. Porosity at L_c-d_1 (n_7)
$V_{p7} = \pi/4 \times 15.24^2 \times 50 = 9120.74\ cm^3$
$V_{b7} = \pi/4 \times 10.50^2 \times 50 = 4329.51\ cm^3$
$V_{d7} = V_{p7} - V_{b7} = 9120.74 - 4329.51 = 4791.23\ cm^3$
$V_{w7} = 6500\ cm^3$
$V_{v7} = V_{w7} - V_{d7} = 6500 - 4791.23 = 1708.77\ cm^3$
$n_7 = V_{v7}/V_{b7} = 1708.77/4329.51 = 0.395$

8. Porosity at L_c-d_2 (n_8)
$V_{p8} = \pi/4 \times 15.24^2 \times 50 = 9120.74\ cm^3$
$V_{b8} = \pi/4 \times 11.10^2 \times 50 = 4838.45\ cm^3$
$V_{d8} = V_{p8} - V_{b8} = 9120.74 - 4838.45 = 4282.29\ cm^3$
$V_{w8} = 6450\ cm^3$
$V_{v8} = V_{w8} - V_{d8} = 6450 - 4282.29 = 2167.71\ cm^3$
$n_8 = V_{v8}/V_{b8} = 2167.71/4838.45 = 0.448$

9. Porosity at L_c-d_3 (n_9)

$V_{p9} = \pi/4 \times 15.242 \times 50 = 9120.74 \text{ cm}^3$
$V_{b9} = \pi/4 \times 112 \times 50 = 4751.66 \text{ cm}^3$
$V_{d9} = V_{p9} - V_{b9} = 9120.74 - 4751.66 = 4369.08 \text{ cm}^3$
$V_{w9} = 6410 \text{ cm}^3$
$V_{v9} = V_{w9} - V_{d9} = 6410 - 4369.08 = 2040.92 \text{ cm}^3$
$n_9 = V_{v9}/V_{b9} = 2040.92/4751.66 = 0.430$

The average initial porosity for gravel media (average of the nine readings) and coarse gravel (average of "L_a" readings) may be taken as:

$n = (n_1 + n_2 + n_3 + n_4 + n_5 + n_6 + n_7 + n_8 + n_9)/9$
$= (0.431 + 0.485 + 0.444 + 0.387 + 0.442 + 0.416 + 0.395 + 0.448 + 0.430)/9$
$= 0.431$

$n_{cg} = (n_1 + n_2 + n_3)/3$
$= (0.431 + 0.485 + 0.444)/3 = 0.453$

V_{pi} = volume of 15.24 cm diameter pipe, cm^3
V_{bi} = volume of bucket, filled with used media, cm^3
$V_{di} = V_{pi} - V_{bi}$ = volume between 15.24 cm pipe and bucket, cm^3
V_{wi} = volume of added water, cm^3
$V_{vi} = V_{wi} - V_{di}$ = volume of voids, cm^3
$n_i = V_{vi}/V_{bi}$ = media porosity
n = average porosity of gravel media
n_{cg} = average porosity of coarse gravel

APPENDIX II

TABLES OF WATER SAMPLES ANALYSIS

This appendix exhibits tables for the analysis of the experimental results. The order of appearance of these tables is arranged as given in Chapter 5.

1. Tables of Retention Time for set up stage

Tables from II-1 to II-3 represent the hydraulic calculations for water volumes inside cells and the corresponding actual retention times (T_r) for set up stage at different distances from inlet for plastic, gravel, and rubber media, respectively.

TABLE II-1 Hydraulic Retention Time for Plastic Media at Different Distances

T_s	n_{cg}	n_m	2 m Dis.		5 m Dis.		8 m Dis.		10 m Dis.	
			V_{w2}	T_r (d)	V_{w5}	T_r	V_{w8}	T_r	V_{w10}	T_r
0	0.393	0.816	1.085	0.170	3.533	0.554	5.981	0.938	7.067	1.108
14	0.389	0.813	1.078	0.169	3.517	0.552	5.956	0.934	7.035	1.103
28	0.384	0.809	1.069	0.168	3.496	0.548	5.923	0.929	6.992	1.096
42	0.379	0.803	1.058	0.166	3.467	0.544	5.876	0.921	6.935	1.087
56	0.375	0.799	1.050	0.165	3.447	0.541	5.844	0.916	6.895	1.081
70	0.374	0.798	1.048	0.164	3.442	0.540	5.836	0.915	6.885	1.080
84	0.372	0.797	1.045	0.164	3.436	0.539	5.827	0.914	6.872	1.078
98	0.371	0.796	1.043	0.164	3.431	0.538	5.819	0.912	6.862	1.076
112	0.370	0.794	1.040	0.163	3.422	0.537	5.804	0.910	6.845	1.073
126	0.368	0.792	1.036	0.162	3.412	0.535	5.788	0.908	6.825	1.070
140	0.367	0.791	1.034	0.162	3.407	0.534	5.780	0.906	6.815	1.069
154	0.366	0.789	1.031	0.162	3.398	0.533	5.765	0.904	6.797	1.066

TABLE II-2 Hydraulic Retention Time for Gravel Media at Different Distances

T_s	n_{cg}	n_m	2 m Dis.		5 m Dis.		8 m Dis.		10 m Dis.	
			V_{w2}	T_r	V_{w5}	T_r	V_{w8}	T_r	V_{w10}	T_r
0	0.393	0.378	0.775	0.116	1.909	0.286	3.043	0.456	3.819	0.572
14	0.389	0.376	0.769	0.115	1.897	0.284	3.025	0.453	3.794	0.569
28	0.384	0.372	0.760	0.114	1.876	0.281	2.992	0.448	3.751	0.562
42	0.379	0.367	0.750	0.112	1.851	0.277	2.952	0.442	3.701	0.555
56	0.375	0.365	0.743	0.111	1.838	0.275	2.933	0.440	3.676	0.551
70	0.374	0.364	0.741	0.111	1.833	0.275	2.925	0.438	3.666	0.549
84	0.372	0.363	0.738	0.111	1.827	0.274	2.916	0.437	3.653	0.548
98	0.371	0.362	0.736	0.110	1.822	0.273	2.908	0.436	3.643	0.546
112	0.370	0.361	0.734	0.110	1.817	0.272	2.900	0.435	3.633	0.545
126	0.368	0.360	0.730	0.109	1.810	0.271	2.890	0.433	3.621	0.543
140	0.367	0.359	0.728	0.109	1.805	0.271	2.882	0.432	3.611	0.541
154	0.366	0.358	0.726	0.109	1.800	0.270	2.874	0.431	3.601	0.540

TABLE II-3 Hydraulic Retention Time for Rubber Media at Different Distances

T_s	n_{cg}	n_m	2 m Dis.		5 m Dis.		8 m Dis.		10 m Dis.	
			V_{w2}	T_r	V_{w5}	T_r	V_{w8}	T_r	V_{w10}	T_r
0	0.393	0.530	0.883	0.130	2.473	0.365	4.063	0.599	4.946	0.729
14	0.389	0.528	0.876	0.129	2.460	0.363	4.044	0.596	4.921	0.726
28	0.384	0.524	0.867	0.128	2.439	0.360	4.011	0.592	4.878	0.719
42	0.379	0.519	0.857	0.126	2.414	0.356	3.971	0.586	4.828	0.712
56	0.375	0.515	0.849	0.125	2.394	0.353	3.939	0.581	4.788	0.706
70	0.374	0.514	0.847	0.125	2.389	0.352	3.931	0.580	4.778	0.705
84	0.372	0.513	0.844	0.124	2.383	0.351	3.922	0.578	4.766	0.703
98	0.371	0.512	0.842	0.124	2.378	0.351	3.914	0.577	4.756	0.701
112	0.370	0.510	0.839	0.124	2.369	0.349	3.899	0.575	4.738	0.699
126	0.368	0.509	0.836	0.123	2.363	0.348	3.890	0.574	4.726	0.697
140	0.367	0.507	0.833	0.123	2.354	0.347	3.875	0.571	4.708	0.694
154	0.366	0.506	0.831	0.123	2.349	0.346	3.867	0.570	4.698	0.693

$$V_{wx} = \left(V_{cg} \times n_{cg}\right) + \left(V_m \times n_m\right) \tag{4.2}$$

$$T_r = \frac{V_{wx}}{Q_i} \tag{4.3}$$

where: T_s = *time from start of sampling, day;* T_r = *hydraulic retention time, day;* n_{cg} = *porosity of coarse gravel;* n_m = *porosity of used media;* V_{wx} = *volume of water in wetland cell at distance x, m³;* Q_i = *discharge to wetland cell, m³/d;*

2. Tables of average difference

Tables II-4 to II-7 show the results of calculating the average difference between plastic removal efficiency and both gravel and rubber media and also the average difference between gravel and rubber media for BOD, COD, and TSS pollutants in set up stage at distances of 2, 5, 8, and 10 m from inlet, respectively. While Tables II-8 to II-11 show these results for BOD, COD, and TSS pollutants in the steady stage.

TABLE II-4 Average Difference for Set Up Stage Pollutants at 2 m from Cells Inlet

Run No.	BOD Rem. Eff. (%)			COD Rem. Eff. (%)			TSS Rem. Eff. (%)		
	P – G	P – R	G – R	P – G	P – R	G – R	P – G	P – R	G – R
1	1.36	2.27	0.91	2.62	3.20	0.58	6.00	6.00	0.00
2	1.42	2.37	0.95	2.46	3.07	0.61	5.40	6.08	0.68
3	1.72	2.58	0.86	2.99	3.53	0.54	5.55	6.17	0.62
4	1.52	3.03	1.51	2.67	4.00	1.33	5.56	6.11	0.55
5	1.46	2.92	1.46	2.72	3.93	1.21	6.17	6.85	0.68
6	1.57	3.14	1.57	2.46	3.86	1.40	5.59	6.29	0.70
7	1.75	3.49	1.74	2.72	4.28	1.56	5.77	6.41	0.64
8	1.67	3.89	2.22	2.57	4.77	2.20	6.02	6.62	0.60
9	2.58	4.64	2.06	3.57	5.52	1.95	5.81	6.40	0.59
10	2.71	4.89	2.18	4.05	5.73	1.68	5.79	6.52	0.73
11	2.88	5.18	2.30	3.79	5.69	1.90	5.55	6.25	0.70
12	2.38	4.77	2.39	3.58	5.58	2.00	5.81	6.45	0.64
Mean	1.92	3.60	1.68	3.02	4.43	1.41	5.75	6.35	0.59

TABLE II-5 Average Difference for Set Up Stage Pollutants at 5 m From Cells Inlet

Run No.	BOD Rem. Eff. (%)			COD Rem. Eff. (%)			TSS Rem. Eff. (%)		
	P – G	P – R	G – R	P – G	P – R	G – R	P – G	P – R	G – R
1	1.37	3.18	1.81	1.75	3.20	1.45	12.67	14.67	2.00
2	0.95	3.32	2.37	1.24	3.08	1.84	12.83	14.86	2.03
3	2.16	4.74	2.58	2.45	4.62	2.17	12.96	14.81	1.85

TABLE II-5 (Continued)

Run No.	BOD Rem. Eff. (%) P–G	P–R	G–R	COD Rem. Eff. (%) P–G	P–R	G–R	TSS Rem. Eff. (%) P–G	P–R	G–R
4	2.52	5.55	3.03	2.66	5.33	2.67	12.78	15.00	2.22
5	2.44	5.37	2.93	2.72	5.44	2.72	13.01	15.07	2.06
6	3.14	6.29	3.15	3.51	6.32	2.81	12.58	14.68	2.10
7	2.32	5.81	3.49	2.72	5.83	3.11	12.82	14.74	1.92
8	2.78	6.67	3.89	2.93	6.60	3.67	12.65	15.06	2.41
9	3.60	7.73	4.13	3.90	7.79	3.89	12.21	14.53	2.32
10	4.35	8.70	4.35	4.71	8.75	4.04	13.05	15.22	2.17
11	4.02	8.62	4.60	4.17	8.33	4.16	12.50	15.28	2.78
12	3.57	8.33	4.76	3.59	8.37	4.78	12.26	14.84	2.58
Mean	2.77	6.19	3.42	3.03	6.14	3.11	12.69	14.90	2.20

TABLE II-6 Average Difference for Set Up Stage Pollutants at 8 m From Cells Inlet

Run No.	BOD Rem. Eff. (%) P–G	P–R	G–R	COD Rem. Eff. (%) P–G	P–R	G–R	TSS Rem. Eff. (%) P–G	P–R	G–R
1	1.36	3.63	2.27	1.74	4.36	2.62	14.66	16.66	2.00
2	0.95	4.27	3.32	1.23	4.92	3.69	14.19	16.22	2.03
3	2.16	5.61	3.45	2.44	6.25	3.81	14.20	16.67	2.47
4	2.52	6.56	4.04	2.67	7.00	4.33	13.88	16.11	2.23
5	2.93	6.83	3.90	3.32	7.55	4.23	14.38	16.44	2.06
6	3.14	7.33	4.19	3.51	7.72	4.21	13.98	16.78	2.80
7	3.49	8.14	4.65	3.51	8.56	5.05	14.10	16.67	2.57
8	3.33	8.33	5.00	3.66	8.79	5.13	13.86	16.27	2.41
9	4.64	9.79	5.15	4.87	10.39	5.52	13.96	16.28	2.32
10	5.43	11.41	5.98	5.72	11.78	6.06	13.77	16.67	2.90
11	4.59	10.92	6.33	4.92	11.36	6.44	13.19	15.97	2.78
12	4.76	10.71	5.95	4.78	11.15	6.37	13.55	16.13	2.58
Mean	3.28	7.79	4.52	3.53	8.32	4.79	13.98	16.41	2.43

TABLE II-7 Average Difference for Set Up Stage Pollutants at 10 m From Cells Inlet

Run No.	BOD Rem. Eff. (%) P–G	P–R	G–R	COD Rem. Eff. (%) P–G	P–R	G–R	TSS Rem. Eff. (%) P–G	P–R	G–R
1	2.73	4.54	1.81	2.62	4.36	1.74	13.33	16.67	3.34
2	2.84	5.68	2.84	2.77	5.54	2.77	13.52	16.22	2.70

TABLE II-7 (Continued)

Run No.	BOD Rem. Eff. (%) P–G	P–R	G–R	COD Rem. Eff. (%) P–G	P–R	G–R	TSS Rem. Eff. (%) P–G	P–R	G–R
3	4.31	6.90	2.59	4.35	6.79	2.44	12.96	16.05	3.09
4	4.55	7.58	3.03	4.33	7.33	3.00	13.34	16.11	2.77
5	4.88	8.30	3.42	4.83	8.16	3.33	13.01	16.44	3.43
6	5.23	8.90	3.67	5.27	8.78	3.51	12.59	16.08	3.49
7	4.07	8.72	4.65	3.89	8.56	4.67	12.18	16.02	3.84
8	5.00	9.44	4.44	4.76	9.16	4.40	12.65	15.66	3.01
9	6.70	10.82	4.12	6.50	10.72	4.22	12.79	15.69	2.90
10	5.98	11.41	5.43	5.72	11.11	5.39	12.31	15.94	3.63
11	5.75	10.92	5.17	5.68	10.98	5.30	12.50	15.98	3.48
12	5.95	10.71	4.76	5.97	10.75	4.78	12.26	15.49	3.23
Mean	4.83	8.66	3.83	4.72	8.52	3.80	12.79	16.03	3.24

TABLE II-8 Average Difference for BOD, COD, and TSS (2 m Dis. – Steady Stage)

Cycle No.	BOD Rem. Eff. (%) P–G	P–R	G–R	COD Rem. Eff. (%) P–G	P–R	G–R	TSS Rem. Eff. (%) P–G	P–R	G–R
1	2.85	3.91	1.06	3.78	4.72	0.94	4.96	7.46	2.50
2	2.71	3.62	0.91	3.90	4.43	0.53	7.40	8.62	1.22
3	2.66	3.62	0.96	3.81	4.36	0.55	6.81	8.09	1.28
4	2.55	3.54	0.99	3.43	4.28	0.85	4.15	5.69	1.54
5	3.16	4.25	1.09	4.08	4.93	0.85	3.38	4.43	1.05
Mean	2.79	3.79	1.00	3.80	4.54	0.74	5.34	6.86	1.52

TABLE II-9 Average Difference for BOD, COD, and TSS (5 m Dis. – Steady Stage)

Cycle No.	BOD Rem. Eff. (%) P–G	P–R	G–R	COD Rem. Eff. (%) P–G	P–R	G–R	TSS Rem. Eff. (%) P–G	P–R	G–R
1	5.92	10.13	4.21	6.22	10.20	3.98	9.12	15.06	5.94
2	5.94	10.47	4.53	6.18	10.50	4.32	15.99	19.26	3.27
3	5.21	10.54	5.33	5.38	10.65	5.27	11.68	16.81	5.13
4	4.97	10.63	5.66	5.13	10.75	5.62	8.68	12.59	3.91
5	5.87	10.99	5.12	6.12	11.06	4.94	7.32	10.46	3.14
Mean	5.58	10.55	4.97	5.81	10.63	4.83	10.56	14.84	4.28

TABLE II-10 Average Difference for BOD, COD, and TSS (8 m Dis. – Steady Stage)

Cycle No.	BOD Rem. Eff. (%)			COD Rem. Eff. (%)			TSS Rem. Eff. (%)		
	P – G	P – R	G – R	P – G	P – R	G – R	P – G	P – R	G – R
1	6.34	10.66	4.32	6.42	10.94	4.52	10.76	17.12	6.36
2	5.64	10.57	4.93	5.72	10.87	5.15	15.33	18.32	2.99
3	4.90	10.12	5.22	4.98	10.30	5.32	13.75	18.10	4.35
4	4.53	10.30	5.77	4.58	10.49	5.91	6.18	8.91	2.73
5	5.43	10.55	5.12	5.48	10.76	5.28	5.69	9.99	4.30
Mean	5.37	10.44	5.07	5.44	10.67	5.24	10.34	14.49	4.15

TABLE II-11 Average Difference for BOD, COD, and TSS (10 m Dis. – Steady Stage)

Cycle No.	BOD Rem. Eff. (%)			COD Rem. Eff. (%)			TSS Rem. Eff. (%)		
	P – G	P – R	G – R	P – G	P – R	G – R	P – G	P – R	G – R
1	7.40	11.52	4.12	7.43	11.48	4.05	11.05	15.75	4.70
2	7.04	10.96	3.92	7.03	10.94	3.91	17.51	19.26	1.75
3	6.39	10.54	4.15	6.35	10.57	4.22	13.86	17.84	3.98
4	5.75	10.19	4.44	5.73	10.20	4.47	6.77	9.02	2.25
5	7.18	11.21	4.03	7.31	11.18	3.87	5.00	7.90	2.90
Mean	6.75	10.88	4.13	6.77	10.87	4.10	10.84	13.95	3.12

Tables II-12 to II-14 present the results of the average difference between plastic removal efficiency and both gravel and rubber media and also the average difference between gravel and rubber media for ammonia, phosphate, fecal coliforms, and the heavy metals in steady stage.

3. Tables of Water Samples (BOD – COD – TSS)

In set up stage all sampling points were considered (influent sample, 27 intermediate samples, and 3 effluents samples), Tables from II-15 to II-17 show the results of BOD concentrations at different sampling points for plastic, gravel, and rubber cells, respectively.

Tables from II-18 to II-20 show the results of chemical oxygen demand (COD) concentrations (mg/l) at different sampling points for plastic, gravel, and rubber cells, respectively.

Tables from II-21 to II-23 show the results of total suspended solids (TSS) concentrations (mg/l) at different sampling points for plastic, gravel, and rubber wetland cells, respectively.

Appendices

TABLE II-12 Average Difference for NH_3, PO_4, and FC (At Outlets – Steady Stage)

Cycle No.	NH_3 Rem. Eff. (%)			PO_4 Rem. Eff. (%)			FC Rem. Eff. (%)		
	P–G	P–R	G–R	P–G	P–R	G–R	P–G	P–R	G–R
1	9.03	11.25	2.22	12.21	21.42	9.21	0.16	0.30	0.14
2	4.29	7.78	3.49	7.04	13.96	6.92	0.11	0.20	0.09
3	6.18	9.05	2.87	8.33	14.02	5.69	0.09	0.18	0.09
4	12.46	15.09	2.63	11.78	15.68	3.90	0.11	0.25	0.14
5	10.84	16.05	5.21	12.82	15.85	3.03	0.15	0.22	0.07
Mean	8.56	11.84	3.28	10.44	16.19	5.75	0.12	0.23	0.11

TABLE II-13 Average Difference for Zn, Fe, and Mn (At Outlets – Steady Stage)

Cycle No.	Zn Rem. Eff. (%)			Fe Rem. Eff. (%)			Mn Rem. Eff. (%)		
	P–G	P–R	G–R	P–G	P–R	G–R	P–G	P–R	G–R
1	7.78	11.97	4.19	4.69	14.51	9.82	9.82	13.95	4.13
2	11.80	18.63	6.83	17.49	20.66	3.17	22.15	26.94	4.79
3	11.29	19.25	7.96	12.61	23.11	10.50	9.34	22.39	13.05
4	6.58	10.44	3.86	14.24	20.93	6.69	10.20	16.16	5.96
5	6.92	13.18	6.26	3.56	15.93	12.37	6.67	17.99	11.32
Mean	8.87	14.69	5.82	10.52	19.03	8.51	11.64	19.49	7.85

TABLE II-14 Average Difference for Pb and Cd (At Outlets – Steady Stage)

Cycle No.	Pb Rem. Eff. (%)			Cd Rem. Eff. (%)		
	P–G	P–R	G–R	P–G	P–R	G–R
1	12.98	20.14	7.16	5.90	11.49	5.59
2	15.95	20.61	4.66	3.91	11.48	7.57
3	4.27	8.56	4.29	4.61	10.05	5.44
4	3.17	7.26	4.09	7.13	10.38	3.25
5	4.33	9.50	5.17	4.92	19.33	14.41
Mean	8.14	13.21	5.07	5.29	12.55	7.25

TABLE II-15 BOD Concentrations for Plastic Cell Samples at Different Distances

T_s	S_i	2 m Dis.			5 m Dis.			8 m Dis.			S_e
		S_1	S_2	S_3	S_4	S_5	S_6	S_7	S_8	S_9	
0	220	202	207	198	184	178	186	181	171	173	167
14	211	200	188	191	180	166	176	166	170	163	157

TABLE II-15 (Continued)

T_s	S_i	2 m Dis.			5 m Dis.			8 m Dis.			S_e
		S_1	S_2	S_3	S_4	S_5	S_6	S_7	S_8	S_9	
28	232	204	215	215	190	181	190	179	173	181	168
42	198	182	169	186	152	160	156	153	141	149	139
56	205	189	180	183	154	162	162	151	144	152	139
70	191	173	175	163	146	136	150	132	139	135	125
84	172	152	149	155	131	125	128	114	120	120	110
98	180	158	153	160	132	134	124	122	116	119	109
112	194	171	158	166	131	129	133	119	120	112	106
126	184	156	149	157	123	116	118	105	103	107	95
140	174	142	149	145	112	115	110	103	97	98	89
154	168	145	137	138	112	105	107	97	90	99	85

TABLE II-16 BOD Concentrations for Gravel Cell Samples at Different Distances

T_s	S_i	2 m Dis.			5 m Dis.			8 m Dis.			S_e
		S_1	S_2	S_3	S_4	S_5	S_6	S_7	S_8	S_9	
0	220	199	201	215	190	175	193	174	183	176	173
14	211	196	194	198	174	180	174	161	173	170	163
28	232	213	219	213	198	190	188	179	188	181	178
42	198	180	187	178	166	163	155	148	156	155	148
56	205	189	194	178	164	162	166	146	166	153	149
70	191	178	175	166	153	152	146	148	137	138	135
84	172	158	157	150	131	129	136	126	119	126	117
98	180	154	163	163	144	134	127	130	128	118	118
112	194	173	160	177	134	135	145	125	125	129	119
126	184	157	156	164	122	130	130	118	114	113	106
140	174	161	149	141	118	123	117	105	106	110	99
154	168	148	143	141	115	119	108	98	104	107	95

TABLE II-17 BOD Concentrations for Rubber Cell Samples at Different Distances

T_s	S_i	2 m Dis.			5 m Dis.			8 m Dis.			S_e
		S_1	S_2	S_3	S_4	S_5	S_6	S_7	S_8	S_9	
0	220	202	206	212	187	186	197	184	177	188	177
14	211	188	201	205	184	179	180	180	171	174	169

Tables of Water Samples Analysis 563

TABLE II-17 (Continued)

T_s	S_i	2 m Dis.			5 m Dis.			8 m Dis.			S_e
		S_1	S_2	S_3	S_4	S_5	S_6	S_7	S_8	S_9	
28	232	208	225	218	197	194	203	186	196	190	184
42	198	188	179	188	166	172	163	164	159	160	154
56	205	195	189	185	173	173	164	162	159	167	156
70	191	170	181	177	152	160	155	148	153	145	142
84	172	155	162	157	142	135	137	134	134	128	125
98	180	169	160	163	147	135	144	125	139	137	126
112	194	181	179	163	147	140	151	134	133	141	127
126	184	159	169	161	136	131	139	124	123	131	116
140	174	152	153	156	119	132	130	119	113	122	108
154	168	145	153	146	120	119	126	117	107	115	103

TABLE II-18 COD Concentrations for Plastic Cell Samples at Different Distances

T_s	S_i	2 m Dis.			5 m Dis.			8 m Dis.			S_e
		S_1	S_2	S_3	S_4	S_5	S_6	S_7	S_8	S_9	
0	344	322	307	313	288	282	297	261	275	274	263
14	325	295	292	301	264	270	279	261	246	248	244
28	368	345	326	328	302	308	290	286	273	278	269
42	300	261	275	274	240	248	229	221	219	226	213
56	331	288	303	294	268	263	246	246	227	238	226
70	285	252	260	247	203	223	225	202	189	206	188
84	257	232	219	227	192	198	189	170	179	173	166
98	273	246	232	234	202	198	197	177	183	174	167
112	308	260	255	268	207	218	205	182	179	188	170
126	297	256	237	248	197	187	197	173	163	165	155
140	264	223	208	227	170	178	168	152	143	149	136
154	251	213	216	195	167	162	159	144	145	131	128

TABLE II-19 COD Concentrations for Gravel Cell Samples at Different Distances

T_s	S_i	2 m Dis.			5 m Dis.			8 m Dis.			S_e
		S_1	S_2	S_3	S_4	S_5	S_6	S_7	S_8	S_9	
0	344	322	316	331	288	294	303	275	270	284	272
14	325	303	312	296	265	280	280	265	245	258	253
28	368	350	350	332	320	305	302	287	296	282	285

TABLE II-19 (Continued)

T_s	S_i	2 m Dis.			5 m Dis.			8 m Dis.			S_e
		S_1	S_2	S_3	S_4	S_5	S_6	S_7	S_8	S_9	
42	300	272	286	277	246	251	244	236	222	232	226
56	331	309	300	302	275	267	262	257	242	245	242
70	285	262	249	269	233	213	235	212	198	217	203
84	257	242	221	236	203	208	189	188	190	172	176
98	273	229	253	250	214	205	202	193	184	188	180
112	308	280	266	271	228	224	215	197	195	202	190
126	297	256	253	268	207	203	214	190	180	182	172
140	264	226	224	237	189	184	175	155	164	164	151
154	251	219	210	222	171	177	168	148	156	152	143

TABLE II-20 COD Concentrations for Rubber Cell Samples at Different Distances

T_s	S_i	2 m Dis.			5 m Dis.			8 m Dis.			S_e
		S_1	S_2	S_3	S_4	S_5	S_6	S_7	S_8	S_9	
0	344	333	324	318	293	308	299	289	290	276	278
14	325	301	318	299	280	289	275	269	259	276	262
28	368	348	356	335	325	310	316	282	313	310	294
42	300	264	290	292	251	249	265	242	237	250	235
56	331	306	317	301	273	271	288	266	259	261	253
70	285	268	263	261	239	239	227	220	227	216	213
84	257	238	246	227	209	201	214	201	191	196	188
98	273	259	255	237	203	225	223	199	197	210	192
112	308	282	269	283	233	229	241	210	221	215	203
126	297	260	274	258	223	217	220	199	197	210	188
140	264	233	229	240	195	186	201	184	169	181	165
154	251	216	222	228	190	174	187	169	161	174	155

TABLE II-21 TSS Concentrations for Plastic Cell Samples at Different Distances

T_s	S_i	2 m Dis.			5 m Dis.			8 m Dis.			S_e
		S_1	S_2	S_3	S_4	S_5	S_6	S_7	S_8	S_9	
0	150	118	124	120	85	82	85	72	68	70	66
14	148	121	122	113	82	84	80	68	71	69	64

TABLE II-21 (Continued)

T_s	S_i	2 m Dis.			5 m Dis.			8 m Dis.			S_e
		S_1	S_2	S_3	S_4	S_5	S_6	S_7	S_8	S_9	
28	162	134	127	129	92	87	88	77	71	74	70
42	180	140	147	146	102	96	97	84	79	83	76
56	146	118	110	120	77	80	80	68	63	64	61
70	143	118	109	115	78	73	80	65	61	62	59
84	156	129	121	122	85	85	79	66	69	69	64
98	166	136	128	129	91	86	87	73	68	75	67
112	172	139	132	138	89	94	90	75	76	71	69
126	138	109	106	112	75	69	72	59	58	60	55
140	144	114	117	111	76	73	76	63	60	63	57
154	155	122	120	124	81	80	82	66	68	64	61

TABLE II-22 TSS Concentrations for Gravel Cell Samples at Different Distances

T_s	S_i	2 m Dis.			5 m Dis.			8 m Dis.			S_e
		S_1	S_2	S_3	S_4	S_5	S_6	S_7	S_8	S_9	
0	150	127	135	128	98	105	106	89	94	92	86
14	148	120	130	131	100	99	104	91	89	90	84
28	162	133	145	139	111	112	107	100	92	99	91
42	180	151	160	151	125	115	123	107	103	111	100
56	146	130	129	116	98	94	102	84	84	89	80
70	143	120	126	121	93	93	99	82	85	81	77
84	156	136	129	134	102	106	101	86	93	92	83
98	166	145	138	140	110	106	111	94	93	98	88
112	172	142	150	146	114	109	112	99	100	95	91
126	138	114	117	120	87	92	90	77	76	81	72
140	144	125	122	119	94	92	93	82	79	83	75
154	155	129	132	132	98	98	104	89	85	87	80

TABLE II-23 TSS Concentrations for Rubber Cell Samples at Different Distances

T_s	S_i	2 m Dis.			5 m Dis.			8 m Dis.			S_e
		S_1	S_2	S_3	S_4	S_5	S_6	S_7	S_8	S_9	
0	150	133	130	127	103	109	106	95	97	93	91
14	148	125	128	131	108	99	105	96	91	92	88

TABLE II-23 (Continued)

T_s	S_i	2 m Dis.				5 m Dis.			8 m Dis.		S_e
		S_1	S_2	S_3	S_4	S_5	S_6	S_7	S_8	S_9	
28	162	145	141	134	115	109	115	98	103	103	96
42	180	157	158	150	127	128	120	113	105	115	105
56	146	131	125	123	101	99	103	91	87	89	85
70	143	127	122	120	99	95	100	88	89	83	82
84	156	129	136	136	110	103	105	94	92	96	89
98	166	144	148	134	109	115	115	100	96	101	93
112	172	149	141	151	118	110	121	106	97	103	96
126	138	118	120	116	95	91	93	83	79	83	77
140	144	124	125	120	97	99	95	83	87	85	80
154	155	132	129	135	108	101	103	94	89	90	85

For the intermediate sampling points for each cell (S_1 to S_9), the average value for S_1, S_2, and S_3 represents pollutant concentration at 2 m from inlet; the average value for S_4, S_5, and S_6 represents pollutant concentration at 5 m; and the average value for S_7, S_8, and S_9 gives pollutant concentration at 8 m. The effluent concentration was taken at a distance of 10 m from cell inlet (outlet).

T_s = time from start of sampling
S_i = influent water sample
S_e = effluent water samples
S_1, S_2, S_3 = sampling point at distance of 2 m from inlet (right, middle, and left, respectively)
S_4, S_5, S_6 = sampling point at distance of 5 m from inlet (right, middle, and left, respectively)
S_7, S_8, S_9 = sampling point at distance of 8 m from inlet (right, middle, and left, respectively)

4. Additional Parameters

The properties of hydrogen ion and water temperature were monitored through this study as they are important ones in treatment performance of constructed wetland systems. Tables II-24 to II-29 illustrate the values of these properties in both set up and steady stages at influent and effluents points for plastic, gravel, and rubber cells.

Tables of Water Samples Analysis

TABLE II-24 Temperature and pH for Influent and Effluent Points (Set Up Stage)

Date	Run No.	T_s	Influent pH	Tem.	Rubber pH	Tem.	Gravel pH	Tem.	Plastic pH	Tem.
31/10/2009	1	0	6.67	24.5	7.35	24.3	6.98	24.2	7.48	24.3
14/11/2009	2	14	6.72	23.4	7.39	23.2	7.10	23.2	7.39	23.3
28/11/2009	3	28	6.66	23.4	7.42	23.2	6.90	23.1	7.26	23.2
12/12/2009	4	42	6.81	20.6	7.45	20.5	7.25	20.4	7.35	20.4
26/12/2009	5	56	6.84	20.3	7.44	20.2	7.03	20.1	7.25	20.1
09/01/2010	6	70	6.64	18.8	7.25	18.7	7.03	18.5	7.05	18.6
23/01/2010	7	84	6.86	18.5	7.23	18.3	6.90	18.2	7.31	18.3
06/02/2010	8	98	6.71	19.3	7.24	19.1	6.80	19.1	7.25	19.1
20/02/2010	9	112	6.68	19.6	7.39	19.4	7.04	19.2	7.55	19.3
06/03/2010	10	126	6.79	21.5	7.43	21.3	6.94	21.5	7.33	21.3
20/03/2010	11	140	6.73	22.6	7.25	22.3	7.11	22.4	7.35	22.4
03/04/2010	12	154	6.85	23.4	7.31	23.2	7.13	23.2	7.37	23.3
The average value			6.75	21.3	7.35	21.1	7.02	21.1	7.33	21.1

TABLE II-25 Temperature and pH for Inlet and Outlet (Cycle No. 1 – Steady Stage)

Date	Run No.	T_s	Inlet (C_i) pH	Tem.	Rubber (C_o) pH	Tem.	Gravel (C_o) pH	Tem.	Plastic (C_o) pH	Tem.
17/04/2010	1	168	6.65	22.7	7.41	22.7	7.12	22.6	7.42	22.5
24/04/2010	2	175	6.69	22.9	7.42	22.7	7.10	22.7	7.40	22.5
01/05/2010	3	182	6.75	22.0	7.53	21.7	7.12	21.7	7.28	21.6
08/05/2010	4	189	6.78	24.1	7.42	23.7	7.24	23.8	7.26	23.9
15/05/2010	5	196	6.82	24.7	7.23	24.4	7.11	24.4	7.38	24.3
The average value			6.74	23.3	7.40	23.0	7.14	23.0	7.35	23.0

TABLE II-26 Temperature and pH for Inlet and Outlet (Cycle No. 2 – Steady Stage)

Date	Run No.	T_s	Inlet (C_i) pH	Tem.	Rubber (C_o) pH	Tem.	Gravel (C_o) pH	Tem.	Plastic (C_o) pH	Tem.
22/05/2010	6	203	6.83	24.9	7.41	24.5	7.33	24.6	7.37	24.6
29/05/2010	7	210	6.80	24.4	7.46	24.2	7.31	24.3	7.37	24.2
05/06/2010	8	217	6.70	26.0	7.20	25.8	7.50	25.9	7.30	25.8
12/06/2010	9	224	6.77	25.4	7.33	25.3	7.31	25.4	7.09	25.2
19/06/2010	10	231	7.42	25.5	7.36	25.2	7.46	25.3	7.42	25.3
The average value			6.90	25.2	7.35	25.0	7.38	25.1	7.31	25.0

TABLE II-27 Temperature and pH for Inlet and Outlet (Cycle No. 3 – Steady Stage)

Date	Run No.	T_s	Inlet (C_i) pH	Tem.	Rubber (C_o) pH	Tem.	Gravel (C_o) pH	Tem.	Plastic (C_o) pH	Tem.
26/06/2010	11	238	7.26	25.4	7.48	25.2	7.48	25.2	7.57	25.1
03/07/2010	12	245	6.84	26.3	7.55	26.2	7.18	26.4	7.55	26.4
10/07/2010	13	252	6.76	27.0	7.48	26.8	7.26	26.7	7.40	26.9
17/07/2010	14	259	6.81	26.5	7.34	26.3	7.30	26.2	7.31	26.5
24/07/2010	15	266	6.74	27.7	7.51	27.3	7.31	27.4	7.30	27.3
The average value			6.88	26.6	7.47	26.4	7.31	26.4	7.43	26.4

TABLE II-28 Temperature and pH for Inlet and Outlet (Cycle No. 4 – Steady Stage)

Date	Run No.	T_s	Inlet (C_i) pH	Tem.	Rubber (C_o) pH	Tem.	Gravel (C_o) pH	Tem.	Plastic (C_o) pH	Tem.
31/07/2010	16	273	7.05	27.9	7.36	27.8	7.40	27.9	7.41	27.8
07/08/2010	17	280	6.88	28.0	7.48	27.8	7.14	27.7	7.32	27.9
14/08/2010	18	287	6.64	28.5	7.39	28.4	7.43	28.5	7.39	28.3
21/08/2010	19	294	6.69	28.4	7.42	28.2	7.59	28.3	7.38	28.3
28/08/2010	20	301	7.15	29.3	7.58	29.2	7.51	29.2	7.36	29.1
The average value			6.88	28.4	7.45	28.3	7.41	28.3	7.37	28.3

TABLE II-29 Temperature and pH for Inlet and Outlet (Cycle No. 5 – Steady Stage)

Date	Run No.	T_s	Inlet (C_i) pH	Tem.	Rubber (C_o) pH	Tem.	Gravel (C_o) pH	Tem.	Plastic (C_o) pH	Tem.
18/09/2010	21	322	6.53	28.1	7.52	27.9	7.10	28.0	7.34	28.0
25/09/2010	22	329	6.42	28.2	7.43	28.1	7.09	28.0	7.36	28.1
02/10/2010	23	336	6.84	25.9	7.45	25.8	7.16	25.7	7.41	25.8
09/10/2010	24	343	7.19	24.5	7.50	24.3	7.32	24.4	7.35	24.3
16/10/2010	25	350	7.23	24.7	7.3	24.6	7.48	24.7	7.45	24.5
The average value			6.84	26.3	7.44	26.1	7.23	26.2	7.38	26.1

Wastewater inlet pH values throughout the wetland cells operation varied between 6.74 and 6.90 with an average value of 6.83 and at the outlets ranged from 7.02 to 7.47 with an average value of 7.34. The mean min. and max. temperature during the study period were 21.2 and 28.3°C.

5. MPN Index

Table II-30 gives the MPN index (Most Probable Number) and 95% confidence limits for various combinations of positive results when five tubes are used per dilution (Fecal Coliform computation by multiple-tube fermentation) (APHA, 1998).

TABLE II-30 MPN Index and 95% Confidence Limits for Fecal Coliforms Computation

Combination of Positives	MPN Index	95% Confidence Limits Lower	95% Confidence Limits Upper	Combination of Positives	MPN Index	95% Confidence Limits Lower	95% Confidence Limits Upper
0–0–0	<0.18	–	0.68	4–0–3	2.50	0.98	7.00
0–0–1	0.18	0.009	0.68	4–1–0	1.70	0.60	4.00
0–1–0	0.18	0.009	0.69	4–1–1	2.10	0.68	4.20
0–1–1	0.36	0.070	1.00	4–1–2	2.60	0.98	7.00
0–2–0	0.37	0.070	1.00	4–1–3	3.10	1.00	7.00
0–2–1	0.55	0.180	1.50	4–2–0	2.20	0.68	5.00
0–3–0	0.56	0.180	1.50	4–2–1	2.60	0.98	7.00
1–0–0	0.20	0.010	1.00	4–2–2	3.20	1.00	7.00
1–0–1	0.40	0.070	1.00	4–2–3	3.80	1.40	10.0
1–0–2	0.60	0.180	1.50	4–3–0	2.70	0.99	7.00
1–1–0	0.40	0.071	1.20	4–3–1	3.30	1.00	7.00
1–1–1	0.61	0.180	1.50	4–3–2	3.90	1.40	10.0
1–1–2	0.81	0.340	2.20	4–4–0	3.40	1.40	10.0
1–2–0	0.61	0.180	1.50	4–4–1	4.00	1.40	10.0
1–2–1	0.82	0.340	2.20	4–4–2	4.70	1.50	12.0
1–3–0	0.83	0.340	2.20	4–5–0	4.10	1.40	10.0
1–3–1	1.00	0.350	2.20	4–5–1	4.80	1.50	12.0
1–4–0	1.00	0.350	2.20	5–0–0	2.30	0.68	7.00
2–0–0	0.45	0.079	1.50	5–0–1	3.10	1.00	7.00
2–0–1	0.68	0.180	1.50	5–0–2	4.30	1.40	10.0
2–0–2	0.91	0.340	2.20	5–0–3	5.80	2.20	15.0
2–1–0	0.68	0.180	1.70	5–1–1	3.30	1.00	10.0
2–1–1	0.92	0.340	2.20	5–1–2	4.60	1.40	12.0
2–1–2	1.20	0.410	2.60	5–1–3	6.30	2.20	15.0

TABLE II-30 (Continued)

Combination of Positives	MPN Index	95% Confidence Limits Lower	95% Confidence Limits Upper	Combination of Positives	MPN Index	95% Confidence Limits Lower	95% Confidence Limits Upper
2–2–0	0.93	0.340	2.20	5–1–3	8.40	3.40	22.0
2–2–1	1.20	0.410	2.60	5–2–0	4.90	1.50	15.0
2–2–2	1.40	0.590	3.60	5–2–1	7.00	2.20	17.0
2–3–0	1.20	0.410	2.60	5–2–2	9.40	3.40	23.0
2–3–1	1.40	0.590	3.60	5–2–3	12.0	3.60	25.0
2–4–0	1.50	0.590	3.60	5–2–4	15.0	5.80	40.0
3–0–0	0.78	0.210	2.20	5–3–1	7.90	2.20	22.0
3–0–1	1.10	0.350	2.30	5–3–2	11.0	3.40	25.0
3–0–2	1.30	0.560	3.50	5–3–3	14.0	5.20	40.0
3–1–0	1.10	0.350	2.60	5–3–3	17.0	7.00	40.0
3–1–1	1.40	0.560	3.60	5–3–4	21.0	7.00	40.0
3–1–2	1.70	0.600	3.60	5–4–0	13.0	3.60	40.0
3–2–0	1.40	0.570	3.60	5–4–1	17.0	5.80	40.0
3–2–1	1.70	0.680	4.00	5–4–2	22.0	7.00	44.0
3–2–2	2.00	0.680	4.00	5–4–3	28.0	10.0	71.0
3–3–0	1.70	0.680	4.00	5–4–4	35.0	10.0	71.0
3–3–1	2.10	0.680	4.00	5–4–5	43.0	15.0	110
3–3–2	2.40	0.980	7.00	5–5–0	24.0	7.00	71.0
3–4–0	2.10	0.680	4.00	5–5–1	35.0	10.0	110
3–4–1	2.40	0.680	7.00	5–5–2	54.0	15.0	170
3–5–0	2.50	0.980	7.00	5–5–3	92.0	22.0	260
4–0–0	1.30	0.410	3.50	5–5–4	160	40.0	460
4–0–1	1.70	0.590	3.60	5–5–5	> 160	70.0	–
4–0–2	2.10	0.680	4.00				

6. Table of Standard Values

There are some limits for draining and disposing of certain substances in the irrigation drains. These limits are summarized in Law No. 48 of 1982, which concerning with the protection of the River Nile. The amounts of drained substances indicated hereunder shall not exceed the limits indicated next to each of them.

Table II-31 contains the standard limit of wastewaters for parameters under study in irrigation drains (NAWQAM, 2002).

TABLE II-31 Limit Values Specified in Current Legislation

Wastewater Quality Parameters	Limit Values
Iron (Fe)	1.5 mg/l
Cadmium (Cd)	0.05 mg/l
Ammonia (NH_3)	3.0 mg/l
Temperature (T_e)	35°C
Phosphate (PO_4^{-3})	5.0 mg/l
Focal Coliform (FC)	< 5000 MPN/100 ml
Chemical Oxygen Demand (COD)	80 mg/l
Biochemical Oxygen Demand (BOD)	60 mg/l
Total Suspended Solids (TSS)	50 mg/l
Dissolved Oxygen (DO)	> 4.0 mg/l
Hydrogen Ion (pH)	6: 9
Manganese (Mn)	1.0 mg/l
Lead (Pb)	0.5 mg/l
Zinc (Zn)	5.0 mg/l

APPENDIX III

ANNs PROGRAMS AND TABLES OF DATA AND RESULTS

1. ARTIFICIAL NEURAL NETWORKS PROGRAMS

The investigated pollutants in this study are represented by artificial neural networks programs. Three general programs were designed, one for the set up stage and two programs for the steady stage. In ANNs programs, every pollutant was written as "POL."

A. Set up Stage

In the set up stage, three pollutants were studied (BOD, COD, and TSS). Different hidden layers were tested and the best one which gave the least square error was selected. The following program (ANNs Program No. 1) is fit for the above mentioned three parameters with the change of the axes limits and number of hidden layers.

ANNs Program No. 1

```
%netPOLSETUP=init(netPOLSETUP)
netPOLSETUP=newff(POL_SETUP_INPUT,    POL_SETUP_OUTPUT,
[5 4], {'tansig' 'tansig' 'purelin'}, 'trainlm')
% =================== Network Parameters ===============
netPOLSETUP.trainparam.show=10;
netPOLSETUP.trainparam.epochs=1500;
netPOLSETUP.trainparam.goal=0;
netPOLSETUP.trainparam.max_fail=20;
netPOLSETUP.trainparam.lr=0.3;
netPOLSETUP.trainparam.mu=1;
netPOLSETUP.trainparam.mu_dec=0.3;
netPOLSETUP.trainparam.mu_inc=1.3;
```

```
netPOLSETUP.trainparam.mem_reduc=1;
% ================= Divide Function =================
trainInd=1:120;
valInd=1:120;
testInd=1:120; [trainPOL_SETUP_INPUT, valPOL_SETUP_INPUT, testPOL_SETUP_INPUT]=divideind(POL_SETUP_INPUT, trainInd, valInd, testInd);
[trainPOL_SETUP_OUTPUT, valPOL_SETUP_OUTPUT, testPOL_SETUP_OUTPUT]=divideind(POL_SETUP_OUTPUT, trainInd, valInd, testInd);
% =============== Network Training ===================
[netPOLSETUP, tr]=train(netPOLSETUP, POL_SETUP_INPUT, POL_SETUP_OUTPUT)
% ============== Network Simulation ==================
a=sim(netPOLSETUP, POL_SETUP_INPUT)
x=1:120;
plot(x, a, x, POL_SETUP_OUTPUT)
axis([0 130 0 375])
xlabel('Pattern number')
ylabel('NN output')
POL_SETUP_ERROR=a-POL_SETUP_OUTPUT;
grid on
figure
plot(x, POL_SETUP_ERROR)
axis([0 130 −20 20])
xlabel('Pattern number')
ylabel('Error')
grid on
POL_TEST_SETUP_OUTPUT=sim(netPOLSETUP, POL_SETUP_INPUT_TEST)
figure
x1=1:24
plot(x1, POL_TEST_SETUP_OUTPUT, x1, POL_SETUP_OUTPUT_TEST_Target)
axis([0 25 0 375])
xlabel('Pattern number')
ylabel('Output')
grid on
```

```
figure
Error_of_test=(POL_SETUP_OUTPUT_TEST_Target-POL_TEST_
SETUP_OUTPUT)
   plot(x1, Error_of_test)
   axis([0 25 −10 10])
   xlabel('Pattern number')
   ylabel('Error')
   grid on
   INPUT_LAYER_WEIGHTS_MATRIX= netPOLSETUP.iw
   HIDDEN_LAYER_WEIGHTS_MATRIX= netPOLSETUP.Lw
   BIASES_VECTOR= netPOLSETUP.b
```

B. Steady Stage

In steady stage, 12 parameters were studied. Two programs were designed to represent this stage, one for DOD, COD, TSS (ANNs Program No. 2); and the other for the residual parameters (ANNs Program No. 3). These two programs are suitable for the above-mentioned parameters with the change of the axes limits and number of hidden layers. The two programs are given below:

ANNs Program No. 2

```
%netPOLSTEADY=init(netPOLSTEADY)
   netPOLSTEADY=newff(POL_STEADY_INPUT,    POL_STEADY_
OUTPUT, [6 5], {'tansig' 'tansig"purelin'}, 'trainlm')
   % =============== Network Parameters ==================
   netPOLSTEADY.trainparam.show=10;
   netPOLSTEADY.trainparam.epochs=1500;
   netPOLSTEADY.trainparam.goal=0;
   netPOLSTEADY.trainparam.max_fail=20;
   netPOLSTEADY.trainparam.lr=0.3;
   netPOLSTEADY.trainparam.mu=1;
   netPOLSTEADY.trainparam.mu_dec=0.3;
   netPOLSTEADY.trainparam.mu_inc=1.3;
   netPOLSTEADY.trainparam.mem_reduc=1;
   % ================= Divide Function ==================
   trainInd=1:240;
   valInd=1:240;
```

```
testInd=1:240;
[trainPOL_STEADY_INPUT, valPOL_STEADY_INPUT, testPOL_STEADY_INPUT]=divideind(POL_STEADY_INPUT, trainInd, valInd, testInd);
[trainPOL_STEADY_OUTPUT, valPOL_STEADY_OUTPUT, testPOL_STEADY_OUTPUT]=divideind(POL_STEADY_OUTPUT, trainInd, valInd, testInd);
% =============== Network Training ===================
[netPOLSTEADY, tr]=train(netPOLSTEADY, POL_STEADY_INPUT, POL_STEADY_OUTPUT)
% =============== Network Simulation ===================
a=sim(netPOLSTEADY, POL_STEADY_INPUT)
x=1:240;
plot(x, a, x, POL_STEADY_OUTPUT)
axis([0 250 0 325])
xlabel('Pattern number')
ylabel('NN output')
POL_STEADY_ERROR=a-POL_STEADY_OUTPUT;
grid on
figure
plot(x, POL_STEADY_ERROR)
axis([0 250 −20 20])
xlabel('Pattern number')
ylabel('Error')
grid on
POL_TEST_STEADY_OUTPUT=sim(netPOLSTEADY, POL_STEADY_INPUT_TEST)
figure
x1=1:60
plot(x1, POL_TEST_STEADY_OUTPUT, x1, POL_STEADY_OUTPUT_TEST_Target)
axis([0 60 0 325])
xlabel('Pattern number')
ylabel('Output')
grid on
figure
Error_of_test=(POL_STEADY_OUTPUT_TEST_Target-POL_TEST_STEADY_OUTPUT)
```

```
plot(x1, Error_of_test)
axis([0 60 −15 15])
xlabel('Pattern number')
ylabel('Error')
grid on
INPUT_LAYER_WEIGHTS_MATRIX= netPOLSETUP.iw
HIDDEN_LAYER_WEIGHTS_MATRIX= netPOLSETUP.Lw
BIASES_VECTOR= netPOLSETUP.b
```

ANNs Program No. 3

```
%netPOLSTEADY=init(netPOLSTEADY)
netPOLSTEADY=newff(POL_STEADY_INPUT,  POL_STEADY_OUTPUT, [4 3], {'tansig' 'tansig' 'purelin'}, 'trainlm')
% ============== Network Parameters ==================
netPOLSTEADY.trainparam.show=10;
netPOLSTEADY.trainparam.epochs=1500;
netPOLSTEADY.trainparam.goal=0;
netPOLSTEADY.trainparam.max_fail=20;
netPOLSTEADY.trainparam.lr=0.3;
netPOLSTEADY.trainparam.mu=1;
netPOLSTEADY.trainparam.mu_dec=0.3;
netPOLSTEADY.trainparam.mu_inc=1.3;
netPOLSTEADY.trainparam.mem_reduc=1;
% ================= Divide Function ==================
trainInd=1:60;
valInd=1:60;
testInd=1:60;
[trainPOL_STEADY_INPUT, valPOL_STEADY_INPUT, testPOL_STEADY_INPUT]=divideind(POL_STEADY_INPUT, trainInd, valInd, testInd);
[trainAmm_STEADY_OUTPUT, valPOL_STEADY_OUTPUT, testAmm_STEADY_OUTPUT]=divideind(POL_STEADY_OUTPUT, trainInd, valInd, testInd);
% =============== Network Training ==================
[netPOLSTEADY,tr]=train(netPOLSTEADY,POL_STEADY_INPUT, POL_STEADY_OUTPUT)
% ============== Network Simulation ==================
a=sim(netPOLSTEADY, POL_STEADY_INPUT)
```

x=1:60;
plot(*x*, a, *x*, POL_STEADY_OUTPUT)
axis([0 60 0 30])
xlabel('Pattern number')
ylabel('NN output')
POL_STEADY_ERROR=a–POL_STEADY_OUTPUT;
grid on
figure
plot(*x*, POL_STEADY_ERROR)
axis([0 60 −3 3])
xlabel('Pattern number')
ylabel('Error')
grid on
POL_TEST_STEADY_OUTPUT=sim(netPOLSTEADY, POL_STEADY_INPUT_TEST)
figure
x1=1:15
plot(x1,POL_TEST_STEADY_OUTPUT,x1,POL_STEADY_OUTPUT_TEST_Target)
axis([0 15 0 30])
xlabel('Pattern number')
ylabel('Output')
grid on
figure
Error_of_test=(POL_STEADY_OUTPUT_TEST_Target-POL_TEST_STEADY_OUTPUT)
plot(x1, Error_of_test)
axis([0 15 −3 3])
xlabel('Pattern number')
ylabel('Error')
grid on
INPUT_LAYER_WEIGHTS_MATRIX= netPOLSETUP.iw
HIDDEN_LAYER_WEIGHTS_MATRIX= netPOLSETUP.Lw
BIASES_VECTOR= netPOLSETUP.b

2. TABLES OF MODELS DATA

The following tables give the whole data used in the artificial neural networks and SPSS analysis mentioned in chapter eight:

Appendices

- Tables III-1 to III-3 represent data for BOD, COD, and TSS (Set Up Stage).
- Tables III-4 to III-6 represent data for BOD, COD, and TSS (steady stage).
- Tables III-7 to III-9 represent data for NH_3, PO_4, FC, and DO.
- Table III-10 to III-12 represent data for heavy metals.

TABLE III-1 BOD, COD, and TSS Data (Rubber Cell – Set Up Stage)

Basin	v (m/d)	T_o (d)	BOD (mg/l) C_i	BOD (mg/l) C_o	COD (mg/l) C_i	COD (mg/l) C_o	TSS (mg/l) C_i	TSS (mg/l) C_o
First Basin 2 m × 2 m $A_s = 178$ m² & $q = 1.695$ m/d	15.36	56	220	207	344	325	150	130
	15.47	70	211	198	325	306	148	128
	15.64	84	232	217	368	346	162	140
	15.82	98	198	185	300	282	180	155
	15.97	112	205	190	331	308	146	126
	16.01	126	191	176	285	264	143	123
	16.07	140	172	158	257	237	156	134
	16.11	154	180	164	273	250	166	142
	16.16	168	194	174	308	278	172	147
	16.22	182	184	163	297	264	138	118
	16.28	196	174	154	264	234	144	123
	16.32	210	168	148	251	222	155	132
Second Basin 2 m × 5 m $A_s = 569$ m² & $q = 0.678$ m/d	14.48	56	220	190	344	300	150	106
	14.54	70	211	181	325	281	148	104
	14.65	84	232	198	368	317	162	113
	14.79	98	198	167	300	255	180	125
	14.90	112	205	170	331	277	146	101
	14.93	126	191	156	285	235	143	98
	14.96	140	172	138	257	208	156	106
	14.99	154	180	142	273	217	166	113
	15.05	168	194	146	308	234	172	116
	15.08	182	184	135	297	220	138	93
	15.14	196	174	127	264	194	144	97
	15.17	210	168	122	251	184	155	104

TABLE III-1 (Continued)

Basin	v (m/d)	T_o (d)	BOD (mg/l) C_i	BOD (mg/l) C_o	COD (mg/l) C_i	COD (mg/l) C_o	TSS (mg/l) C_i	TSS (mg/l) C_o
Third Basin 2 m × 8 m $A_s = 960$ m² & $q = 0.424$ m/d	13.80	56	220	183	344	285	150	95
	13.85	70	211	175	325	268	148	93
	13.96	84	232	191	368	302	162	101
	14.09	98	198	161	300	243	180	111
	14.20	112	205	163	331	262	146	89
	14.23	126	191	149	285	221	143	87
	14.26	140	172	132	257	196	156	94
	14.29	154	180	134	273	202	166	99
	14.34	168	194	136	308	215	172	102
	14.37	182	184	126	297	202	138	82
	14.43	196	174	118	264	178	144	85
	14.45	210	168	113	251	168	155	91
Last Basin 2 m × 10 m $A_s = 1138$ m² & $q = 0.339$ m/d	13.71	56	220	177	344	278	150	91
	13.78	70	211	169	325	262	148	88
	13.90	84	232	184	368	294	162	96
	14.04	98	198	154	300	235	180	105
	14.16	112	205	156	331	253	146	85
	14.19	126	191	142	285	213	143	82
	14.23	140	172	125	257	188	156	89
	14.26	154	180	126	273	192	166	93
	14.31	168	194	127	308	203	172	96
	14.35	182	184	116	297	188	138	77
	14.40	196	174	108	264	165	144	80
	14.43	210	168	103	251	155	155	85

TABLE III-2 BOD, COD, and TSS Data (Gravel Cell – Set Up Stage)

Basin	v (m/d)	T_o (d)	BOD (mg/l) C_i	BOD (mg/l) C_o	COD (mg/l) C_i	COD (mg/l) C_o	TSS (mg/l) C_i	TSS (mg/l) C_o
First Basin 2 m × 2 m $A_s = 210$ m² & $q = 1.668$ m/d	17.21	56	220	205	344	323	150	130
	17.36	70	211	196	325	304	148	127
	17.57	84	232	215	368	344	162	139
	17.80	98	198	182	300	278	180	154
	17.96	112	205	187	331	304	146	125

TABLE III-2 (Continued)

Basin	v (m/d)	T_o (d)	BOD (mg/l) C_i	BOD (mg/l) C_o	COD (mg/l) C_i	COD (mg/l) C_o	TSS (mg/l) C_i	TSS (mg/l) C_o
	18.01	126	191	173	285	260	143	122
	18.09	140	172	155	257	233	156	133
	18.14	154	180	160	273	244	166	141
	18.19	168	194	170	308	272	172	146
	18.27	182	184	159	297	259	138	117
	18.32	196	174	150	264	229	144	122
	18.37	210	168	144	251	217	155	131
Second Basin 2 m × 5 m $A_s = 741$ m² & $q = 0.667$ m/d	19.98	56	220	186	344	295	150	103
	20.09	70	211	176	325	275	148	101
	20.30	84	232	192	368	309	162	110
	20.58	98	198	161	300	247	180	121
	20.69	112	205	164	331	268	146	98
	20.75	126	191	150	285	227	143	95
	20.81	140	172	132	257	200	156	103
	20.86	154	180	135	273	207	166	109
	20.92	168	194	138	308	222	172	112
	20.98	182	184	127	297	208	138	90
	21.04	196	174	119	264	183	144	93
	21.10	210	168	114	251	172	155	100
Third Basin 2 m × 8 m $A_s = 1272$ m² & $q = 0.417$ m/d	19.04	56	220	178	344	276	150	92
	19.14	70	211	168	325	256	148	90
	19.35	84	232	183	368	288	162	97
	19.61	98	198	153	300	230	180	107
	19.72	112	205	155	331	248	146	86
	19.77	126	191	141	285	209	143	83
	19.82	140	172	124	257	183	156	90
	19.88	154	180	125	273	188	166	95
	19.93	168	194	126	308	198	172	98
	19.99	182	184	115	297	184	138	78
	20.05	196	174	107	264	161	144	81
	20.10	210	168	103	251	152	155	87

TABLE III-2 (Continued)

Basin	v (m/d)	T_o (d)	BOD (mg/l) C_i	BOD (mg/l) C_o	COD (mg/l) C_i	COD (mg/l) C_o	TSS (mg/l) C_i	TSS (mg/l) C_o
Last Basin	17.47	56	220	173	344	272	150	86
2 m × 10 m	17.59	70	211	163	325	253	148	84
$A_s = 1482$	17.79	84	232	178	368	285	162	91
m² & $q =$	18.03	98	198	148	300	226	180	100
0.334 m/d	18.15	112	205	149	331	242	146	80
	18.20	126	191	135	285	203	143	77
	18.26	140	172	117	257	176	156	83
	18.31	154	180	118	273	180	166	88
	18.36	168	194	119	308	190	172	91
	18.43	182	184	106	297	172	138	72
	18.48	196	174	99	264	151	144	75
	18.53	210	168	95	251	143	155	80

TABLE III-3 BOD, COD, and TSS Data (Plastic Cell – Set Up Stage)

Basin	V (m/d)	T_o (d)	BOD (mg/l) C_i	BOD (mg/l) C_o	COD (mg/l) C_i	COD (mg/l) C_o	TSS (mg/l) C_i	TSS (mg/l) C_o
First Basin	11.75	56	220	202	344	314	150	121
2 m × 2 m	11.83	70	211	193	325	296	148	119
$A_s = 285$ m² &	11.93	84	232	211	368	333	162	130
$q = 1.594$ m/d	12.05	98	198	179	300	270	180	144
	12.14	112	205	184	331	295	146	116
	12.17	126	191	170	285	253	143	114
	12.20	140	172	152	257	226	156	124
	12.23	154	180	157	273	237	166	131
	12.26	168	194	165	308	261	172	136
	12.31	182	184	154	297	247	138	109
	12.33	196	174	145	264	219	144	114
	12.36	210	168	140	251	208	155	122
Second Basin	8.85	56	220	183	344	289	150	84
2 m × 5 m	8.88	70	211	174	325	271	148	82
$A_s = 1133$ m² &	8.92	84	232	187	368	300	162	89
$q = 0.638$ m/d	8.99	98	198	156	300	239	180	98
	9.03	112	205	159	331	259	146	79
	9.05	126	191	144	285	217	143	77

Appendices

TABLE III-3 (Continued)

Basin	V (m/d)	T_o (d)	BOD (mg/l) C_i	BOD (mg/l) C_o	COD (mg/l) C_i	COD (mg/l) C_o	TSS (mg/l) C_i	TSS (mg/l) C_o
	9.06	140	172	128	257	193	156	83
	9.07	154	180	130	273	199	166	88
	9.09	168	194	131	308	210	172	91
	9.11	182	184	119	297	194	138	72
	9.13	196	174	112	264	172	144	75
	9.15	210	168	108	251	163	155	81
Third Basin 2 m × 8 m A_s = 1980 m² & q = 0.399 m/d	8.43	56	220	175	344	270	150	70
	8.46	70	211	166	325	252	148	69
	8.50	84	232	178	368	279	162	74
	8.57	98	198	148	300	222	180	82
	8.61	112	205	149	331	237	146	65
	8.62	126	191	135	285	199	143	63
	8.63	140	172	118	257	174	156	68
	8.64	154	180	119	273	178	166	72
	8.66	168	194	117	308	183	172	74
	8.68	182	184	105	297	167	138	59
	8.70	196	174	99	264	148	144	62
	8.72	210	168	95	251	140	155	66
Last Basin 2 m × 10 m A_s = 2265 m² & q = 0.319 m/d	9.02	56	220	167	344	263	150	66
	9.06	70	211	157	325	244	148	64
	9.12	84	232	168	368	269	162	70
	9.20	98	198	139	300	213	180	76
	9.25	112	205	139	331	226	146	61
	9.26	126	191	125	285	188	143	59
	9.28	140	172	110	257	166	156	64
	9.29	154	180	109	273	167	166	67
	9.32	168	194	106	308	170	172	69
	9.34	182	184	95	297	155	138	55
	9.36	196	174	89	264	136	144	57
	9.38	210	168	85	251	128	155	61

TABLE III-4A BOD, COD, and TSS Data (Rubber Cell – Steady Stage)

Basin	q & v (m/d)	BOD (mg/l) C_i	BOD (mg/l) C_o	COD (mg/l) C_i	COD (mg/l) C_o	TSS (mg/l) C_i	TSS (mg/l) C_o
First Basin 2 m × 2 m (A_s = 178 m²)	q = 1.280 v = 12.33	190	164	297	257	147	123
		185	159	285	246	140	117
		182	156	289	249	135	113
		193	166	292	253	153	128
		197	170	318	276	149	125
	q = 0.871 v = 8.39	200	165	303	251	131	105
		203	168	322	268	139	112
		209	173	337	280	170	138
		188	155	285	236	151	122
		194	160	290	240	141	114
	q = 0.599 v = 5.77	180	143	286	228	162	124
		192	153	291	233	148	113
		198	158	319	256	165	127
		187	149	279	223	160	123
		182	145	272	217	144	110
	q = 0.424 v = 4.09	178	139	266	209	177	131
		162	126	242	189	169	125
		176	138	267	210	164	121
		199	156	316	249	159	117
		189	148	305	240	173	128
	q = 0.316 v = 3.04	183	142	273	213	171	122
		177	138	268	210	176	127
		193	150	306	239	180	129
		187	146	302	237	165	118
		179	139	271	212	168	120
Second Basin 2 m × 5 m (A_s = 569 m²)	q = 0.512 v = 10.91	190	131	297	207	147	93
		185	126	285	196	140	88
		182	123	289	197	135	85
		193	132	292	202	153	98
		197	136	318	221	149	96

TABLE III-4A (Continued)

Basin	q & v (m/d)	BOD (mg/l) C_i	C_o	COD (mg/l) C_i	C_o	TSS (mg/l) C_i	C_o
	q = 0.348 v = 7.42	200	124	303	190	131	73
		203	126	322	202	139	79
		209	131	337	213	170	98
		188	116	285	177	151	86
		194	120	290	181	141	79
	q = 0.240 v = 5.11	180	99	286	159	162	79
		192	106	291	162	148	72
		198	110	319	179	165	82
		187	104	279	157	160	79
		182	100	272	151	144	69
	q = 0.170 v = 3.62	178	92	266	139	177	73
		162	84	242	127	169	69
		176	91	267	139	164	67
		199	104	316	167	159	65
		189	98	305	160	173	72
	q = 0.126 v = 2.69	183	91	273	137	171	62
		177	87	268	133	176	65
		193	96	306	154	180	66
		187	93	302	151	165	59
		179	88	271	135	168	60

TABLE III-4B BOD, COD, and TSS Data (Rubber Cell – Steady Stage)

Basin	q & v (m/d)	BOD (mg/l) C_i	C_o	COD (mg/l) C_i	C_o	TSS (mg/l) C_i	C_o
Third Basin 2 m × 8 m (A_s = 960 m²)	q = 0.320 v = 10.62	190	114	297	177	147	81
		185	110	285	168	140	76
		182	107	289	169	135	74
		193	116	292	175	153	86
		197	119	318	191	149	84

TABLE III-4B (Continued)

Basin	q & v (m/d)	BOD (mg/l) C_i	BOD (mg/l) C_o	COD (mg/l) C_i	COD (mg/l) C_o	TSS (mg/l) C_i	TSS (mg/l) C_o
	$q = 0.218$ $v = 7.22$	200	104	303	157	131	58
		203	106	322	167	139	63
		209	110	337	177	170	79
		188	96	285	145	151	69
		194	100	290	149	141	63
	$q = 0.150$ $v = 4.97$	180	76	286	120	162	59
		192	82	291	124	148	53
		198	86	319	138	165	62
		187	81	279	120	160	59
		182	77	272	114	144	51
	$q = 0.106$ $v = 3.52$	178	69	266	103	177	46
		162	62	242	92	169	43
		176	68	267	103	164	41
		199	78	316	123	159	40
		189	73	305	117	173	45
	$q = 0.079$ $v = 2.62$	183	65	273	97	171	37
		177	63	268	95	176	40
		193	70	306	111	180	40
		187	67	302	108	165	35
		179	63	271	95	168	35
Last Basin $2 \text{ m} \times 10 \text{ m}$ ($A_s = 1138 \text{ m}^2$)	$q = 0.256$ $v = 10.91$	190	105	297	165	147	73
		185	101	285	156	140	68
		182	98	289	157	135	66
		193	105	292	160	153	77
		197	109	318	177	149	76
	$q = 0.174$ $v = 7.42$	200	92	303	140	131	50
		203	94	322	150	139	55
		209	98	337	159	170	70
		188	85	285	130	151	61
		194	88	290	132	141	55

TABLE III-4B (Continued)

Basin	q & v (m/d)	BOD (mg/l) C_i	C_o	COD (mg/l) C_i	C_o	TSS (mg/l) C_i	C_o
	q = 0.120	180	63	286	101	162	50
	v = 5.11	192	68	291	104	148	45
		198	72	319	117	165	53
		187	67	279	101	160	50
		182	64	272	96	144	43
	q = 0.085	178	54	266	81	177	33
	v = 3.62	162	49	242	74	169	31
		176	54	267	82	164	29
		199	62	316	99	159	28
		189	58	305	94	173	33
	q = 0.063	183	50	273	75	171	24
	v = 2.69	177	48	268	73	176	26
		193	54	306	86	180	26
		187	52	302	84	165	22
		179	48	271	73	168	22

TABLE III-5A BOD, COD, and TSS Data (Gravel Cell – Steady Stage)

Basin	q & v (m/d)	BOD (mg/l) C_i	C_o	COD (mg/l) C_i	C_o	TSS (mg/l) C_i	C_o
First Basin 2 m × 2 m (A_s = 210 m²)	q = 1.259 v = 13.88	190	162	297	255	147	119
		185	157	285	243	140	113
		182	154	289	246	135	109
		193	164	292	250	153	125
		197	168	318	273	149	122
	q = 0.857 v = 9.44	200	163	303	249	131	104
		203	166	322	266	139	110
		209	172	337	280	170	136
		188	153	285	234	151	120
		194	158	290	238	141	112

TABLE III-5A (Continued)

Basin	q & v (m/d)	BOD (mg/l) C_i	BOD (mg/l) C_o	COD (mg/l) C_i	COD (mg/l) C_o	TSS (mg/l) C_i	TSS (mg/l) C_o
	q = 0.589 v = 6.49	180	141	286	226	162	122
		192	151	291	231	148	111
		198	156	319	254	165	125
		187	148	279	223	160	121
		182	143	272	215	144	108
	q = 0.417 v = 4.60	178	137	266	206	177	128
		162	125	242	188	169	122
		176	136	267	208	164	118
		199	154	316	246	159	115
		189	146	305	237	173	126
	q = 0.311 v = 3.42	183	140	273	211	171	121
		177	136	268	208	176	125
		193	148	306	237	180	127
		187	144	302	234	165	116
		179	137	271	209	168	118
Second Basin 2 m × 5 m (A_s = 741 m²)	q = 0.504 v = 13.99	190	122	297	193	147	85
		185	119	285	186	140	79
		182	116	289	187	135	77
		193	124	292	190	153	89
		197	127	318	208	149	87
	q = 0.343 v = 9.52	200	115	303	177	131	69
		203	117	322	188	139	74
		209	122	337	200	170	92
		188	107	285	164	151	81
		194	111	290	168	141	75
	q = 0.236 v = 6.55	180	89	286	143	162	71
		192	96	291	147	148	64
		198	100	319	164	165	73
		187	94	279	142	160	71
		182	90	272	136	144	62

TABLE III-5A (Continued)

Basin	q & v (m/d)	BOD (mg/l) C_i	C_o	COD (mg/l) C_i	C_o	TSS (mg/l) C_i	C_o
	q = 0.167 v = 4.64	178	82	266	124	177	66
		162	74	242	112	169	63
		176	81	267	124	164	60
		199	93	316	150	159	59
		189	88	305	144	173	65
	q = 0.124 v = 3.45	183	81	273	123	171	57
		177	78	268	120	176	59
		193	87	306	140	180	60
		187	84	302	137	165	54
		179	78	271	120	168	55

TABLE III-5B BOD, COD, and TSS Data (Gravel Cell – Steady Stage)

Basin	q & v (m/d)	BOD (mg/l) C_i	C_o	COD (mg/l) C_i	C_o	TSS (mg/l) C_i	C_o
Third Basin 2 m × 8 m (A_s = 1272 m²)	q = 0.315 v = 14.03	190	106	297	164	147	72
		185	102	285	155	140	67
		182	100	289	157	135	65
		193	107	292	161	153	76
		197	110	318	176	149	75
	q = 0.214 v = 9.54	200	94	303	141	131	54
		203	96	322	151	139	59
		209	100	337	160	170	73
		188	87	285	131	151	65
		194	90	290	133	141	59
	q = 0.147 v = 6.57	180	66	286	104	162	52
		192	72	291	108	148	47
		198	76	319	121	165	54
		187	71	279	105	160	52
		182	68	272	101	144	45

TABLE III-5B (Continued)

Basin	q & v (m/d)	BOD (mg/l) C_i	C_o	COD (mg/l) C_i	C_o	TSS (mg/l) C_i	C_o
	q = 0.104	178	58	266	86	177	41
	v = 4.65	162	52	242	77	169	38
		176	58	267	87	164	36
		199	67	316	105	159	36
		189	63	305	101	173	41
	q = 0.078	183	56	273	83	171	30
	v = 3.46	177	54	268	81	176	32
		193	60	306	94	180	32
		187	58	302	93	165	28
		179	53	271	80	168	28
Last Basin 2 m × 10 m (A_s = 1482 m^2)	q = 0.252 v = 13.99	190	97	297	153	147	66
		185	93	285	144	140	61
		182	91	289	146	135	60
		193	97	292	148	153	70
		197	101	318	164	149	69
	q = 0.171	200	84	303	128	131	48
	v = 9.52	203	86	322	138	139	53
		209	90	337	146	170	66
		188	78	285	119	151	58
		194	80	290	120	141	53
	q = 0.118	180	55	286	88	162	44
	v = 6.55	192	60	291	92	148	39
		198	64	319	104	165	46
		187	60	279	90	160	44
		182	56	272	84	144	37
	q = 0.083	178	46	266	69	177	29
	v = 4.64	162	41	242	62	169	27
		176	46	267	70	164	25
		199	54	316	86	159	25
		189	50	305	81	173	29
	q = 0.062	183	43	273	65	171	19
	v = 3.45	177	41	268	63	176	21
		193	46	306	74	180	20
		187	45	302	73	165	18
		179	40	271	61	168	17

TABLE III-6A BOD, COD, and TSS Data (Plastic Cell – Steady Stage)

Basin	q & v (m/d)	BOD (mg/l) C_i	BOD (mg/l) C_o	COD (mg/l) C_i	COD (mg/l) C_o	TSS (mg/l) C_i	TSS (mg/l) C_o
First Basin 2 m × 2 m (A_s = 285 m²)	$q = 1.204$ $v = 9.35$	190	156	297	243	147	112
		185	152	285	233	140	106
		182	149	289	235	135	103
		193	159	292	240	153	117
		197	162	318	260	149	114
	$q = 0.819$ $v = 6.36$	200	158	303	238	131	94
		203	161	322	254	139	100
		209	165	337	265	170	124
		188	148	285	223	151	109
		194	153	290	227	141	101
	$q = 0.563$ $v = 4.37$	180	137	286	216	162	111
		192	146	291	220	148	101
		198	151	319	242	165	114
		187	142	279	211	160	110
		182	138	272	205	144	98
	$q = 0.399$ $v = 3.10$	178	133	266	198	177	121
		162	121	242	180	169	115
		176	131	267	198	164	112
		199	149	316	235	159	108
		189	141	305	226	173	118
	$q = 0.297$ $v = 2.31$	183	135	273	201	171	115
		177	130	268	196	176	119
		193	142	306	224	180	121
		187	138	302	222	165	111
		179	131	271	198	168	112
Second Basin 2 m × 5 m (A_s = 1133 m²)	$q = 0.481$ $v = 7.09$	190	111	297	175	147	71
		185	107	285	166	140	66
		182	105	289	169	135	65
		193	113	292	173	153	75
		197	116	318	189	149	74

TABLE III-6A (Continued)

	q = 0.328 v = 4.82	200	103	303	158	131	48
		203	106	322	170	139	52
		209	109	337	178	170	65
		188	96	285	147	151	57
		194	99	290	149	141	52
	q = 0.225 v = 3.32	180	80	286	128	162	52
		192	86	291	132	148	47
		198	89	319	145	165	54
		187	84	279	127	160	52
		182	81	272	122	144	45
	q = 0.160 v = 2.35	178	73	266	110	177	51
		162	66	242	100	169	48
		176	73	267	112	164	46
		199	83	316	133	159	45
		189	78	305	127	173	50
	q = 0.119 v = 1.75	183	70	273	106	171	44
		177	68	268	104	176	46
		193	75	306	120	180	47
		187	73	302	119	165	42
		179	68	271	104	168	43

TABLE III-6B BOD, COD, and TSS Data (Plastic Cell – Steady Stage)

Basin	q & v (m/d)	BOD (mg/l) C_i	C_o	COD (mg/l) C_i	C_o	TSS (mg/l) C_i	C_o
Third Basin 2 m × 8 m (A_s = 1980 m²)	q = 0.301 v = 6.69	190	94	297	145	147	56
		185	90	285	137	140	52
		182	88	289	138	135	51
		193	95	292	142	153	59
		197	98	318	156	149	59
	q = 0.205 v = 4.55	200	83	303	124	131	34
		203	85	322	133	139	37
		209	88	337	140	170	48
		188	76	285	114	151	42
		194	79	290	117	141	37

TABLE III-6B (Continued)

Basin	q & v (m/d)	BOD (mg/l) C_i	BOD (mg/l) C_o	COD (mg/l) C_i	COD (mg/l) C_o	TSS (mg/l) C_i	TSS (mg/l) C_o
	$q = 0.141$ $v = 3.13$	180	58	286	91	162	30
		192	63	291	94	148	26
		198	66	319	105	165	32
		187	61	279	90	160	30
		182	59	272	87	144	25
	$q = 0.100$ $v = 2.22$	178	50	266	74	177	30
		162	45	242	66	169	28
		176	50	267	75	164	26
		199	58	316	91	159	26
		189	54	305	86	173	30
	$q = 0.074$ $v = 1.65$	183	46	273	68	171	20
		177	44	268	66	176	22
		193	49	306	77	180	21
		187	48	302	76	165	19
		179	44	271	66	168	19
Last Basin 2 m × 10 m ($A_s = 2265$ m^2)	$q = 0.241$ $v = 7.09$	190	83	297	131	147	50
		185	79	285	123	140	45
		182	77	289	123	135	45
		193	84	292	128	153	53
		197	86	318	140	149	53
	$q = 0.164$ $v = 4.82$	200	70	303	107	131	25
		203	72	322	115	139	28
		209	75	337	122	170	37
		188	64	285	98	151	32
		194	67	290	101	141	28
	$q = 0.113$ $v = 3.32$	180	44	286	71	162	21
		192	48	291	73	148	19
		198	51	319	83	165	23
		187	47	279	71	160	22
		182	45	272	68	144	17

TABLE III-6B (Continued)

Basin	q & v (m/d)	BOD (mg/l) C_i	C_o	COD (mg/l) C_i	C_o	TSS (mg/l) C_i	C_o
	q = 0.080	178	36	266	54	177	17
	v = 2.35	162	32	242	48	169	16
		176	36	267	55	164	14
		199	42	316	67	159	14
		189	39	305	64	173	17
	q = 0.059	183	30	273	45	171	10
	v = 1.75	177	28	268	43	176	12
		193	32	306	51	180	11
		187	31	302	50	165	10
		179	28	271	43	168	9

TABLE III-7 Ammonia, Phosphate, F.C., and DO Data (Rubber Cell – Steady Stage)

Basin	q & v (m/d)	Ammonia C_i	C_o	Phosphate C_i	C_o	Fecal Coliform C_i	C_o	DO C_o
Last Basin	q = 0.256	19.43	10.29	3.24	1.72	210,000	6100	1.7
2 m × 10 m	v = 10.91	16.76	9.38	2.96	1.54	240,000	6300	1.7
(A_s = 1138 m²)		26.16	13.82	2.77	1.42	210,000	6000	1.9
		22.44	12.50	3.54	1.88	250,000	6300	2.1
		15.28	8.22	3.32	1.75	240,000	6100	1.8
	q = 0.174	17.79	8.75	2.85	1.17	240,000	4800	2.9
	v = 7.42	18.48	9.32	2.71	1.12	260,000	4900	2.8
		20.87	10.25	2.53	1.03	220,000	4700	3.0
		24.86	12.46	3.11	1.26	250,000	4800	3.1
		23.55	11.14	3.26	1.36	270,000	4900	3.0
	q = 0.120	16.35	7.16	1.89	0.72	230,000	3900	3.3
	v = 5.11	23.74	11.06	2.05	0.76	210,000	3900	3.5
		18.64	8.51	2.33	0.85	200,000	3800	3.4
		15.92	6.78	2.14	0.77	210,000	3800	3.4
		21.31	9.25	2.57	0.99	260,000	4000	3.4
	q = 0.085	15.73	6.68	2.70	0.90	220,000	3200	3.8
	v = 3.62	19.92	8.32	2.94	0.97	250,000	3200	3.8
		25.38	11.31	3.05	0.98	260,000	3300	3.9

TABLE III-7 (Continued)

Basin	q & v (m/d)	Ammonia C_i	C_o	Phosphate C_i	C_o	Fecal Coliform C_i	C_o	DO C_o
		16.11	6.75	3.22	1.10	220,000	3100	3.7
		20.37	8.46	3.16	1.10	210,000	3100	3.9
	$q = 0.063$ $v = 2.69$	20.22	7.44	3.15	0.84	270,000	2400	4.1
		18.57	6.70	2.80	0.76	210,000	2300	4.2
		15.96	6.38	2.66	0.72	240,000	2400	4.1
		21.83	8.07	2.89	0.79	200,000	2200	4.2
		19.45	7.35	2.97	0.83	220,000	2300	4.2

TABLE III-8 Ammonia, Phosphate, F.C., and DO Data (Gravel Cell – Steady Stage)

Basin	q & v (m/d)	Ammonia C_i	C_o	Phosphate C_i	C_o	Fecal Coliform C_i	C_o	DO C_o
Last Basin 2 m × 10 m ($A_s = 1482$ m²)	$q = 0.252$ $v = 13.99$	19.43	9.86	3.24	1.42	210,000	5600	2.8
		16.76	8.99	2.96	1.27	240,000	5800	2.5
		26.16	13.25	2.77	1.18	210,000	5800	2.4
		22.44	12.23	3.54	1.55	250,000	6000	2.9
		15.28	7.73	3.32	1.44	240,000	6000	2.5
	$q = 0.171$ $v = 9.52$	17.79	8.21	2.85	0.98	240,000	4500	3.4
		18.48	8.36	2.71	0.92	260,000	4700	3.3
		20.87	9.23	2.53	0.86	220,000	4500	3.5
		24.86	11.92	3.11	1.09	250,000	4600	3.4
		23.55	10.63	3.26	1.09	270,000	4700	3.3
	$q = 0.118$ $v = 6.55$	16.35	6.85	1.89	0.60	230,000	3700	4.0
		23.74	9.85	2.05	0.64	210,000	3600	4.1
		18.64	8.03	2.33	0.72	200,000	3600	4.1
		15.92	6.39	2.14	0.69	210,000	3700	4.2
		21.31	8.76	2.57	0.79	260,000	3800	4.0
	$q = 0.083$ $v = 4.64$	15.73	6.30	2.70	0.79	220,000	2800	4.2
		19.92	8.00	2.94	0.86	250,000	3100	4.3
		25.38	9.98	3.05	0.87	260,000	3100	4.2
		16.11	6.50	3.22	0.99	220,000	2700	4.4
		20.37	7.99	3.16	0.95	210,000	2700	4.3

TABLE III-8 (Continued)

Basin	q & v (m/d)	Ammonia C_i	C_o	Phosphate C_i	C_o	Fecal Coliform C_i	C_o	DO C_o
	q = 0.062	20.22	6.05	3.15	0.77	270,000	2200	4.5
	v = 3.45	18.57	6.29	2.80	0.67	210,000	2100	4.6
		15.96	5.19	2.66	0.63	240,000	2300	4.5
		21.83	7.32	2.89	0.72	200,000	2100	4.7
		19.45	6.17	2.97	0.72	220,000	2100	4.5

TABLE III-9 Ammonia, Phosphate, F.C., and DO Data (Plastic Cell – Steady Stage)

Basin	q & v (m/d)	Ammonia C_i	C_o	Phosphate C_i	C_o	Fecal Coliform C_i	C_o	DO C_o
Last Basin	q = 0.241	19.43	8.19	3.24	0.99	210,000	5400	3.0
2 m × 10 m	v = 7.09	16.76	7.72	2.96	0.93	240,000	5500	3.1
(A_s = 2265 m²)		26.16	11.09	2.77	0.86	210,000	5500	3.3
		22.44	9.36	3.54	1.13	250,000	5600	2.9
		15.28	6.52	3.32	0.99	240,000	5400	3.4
	q = 0.164	17.79	7.10	2.85	0.78	240,000	4300	3.8
	v = 4.82	18.48	7.66	2.71	0.73	260,000	4500	3.9
		20.87	8.64	2.53	0.70	220,000	4100	3.7
		24.86	10.45	3.11	0.85	250,000	4300	3.8
		23.55	10.00	3.26	0.87	270,000	4500	3.7
	q = 0.113	16.35	5.82	1.89	0.44	230,000	3500	4.2
	v = 3.32	23.74	8.33	2.05	0.49	210,000	3400	4.3
		18.64	6.44	2.33	0.54	200,000	3400	4.4
		15.92	5.75	2.14	0.48	210,000	3500	4.3
		21.31	7.55	2.57	0.58	260,000	3600	4.3
	q = 0.080	15.73	4.25	2.7	0.48	220,000	2600	4.6
	v = 2.35	19.92	5.56	2.94	0.51	250,000	2700	4.7
		25.38	7.18	3.05	0.52	260,000	2700	4.6
		16.11	4.45	3.22	0.59	220,000	2600	4.5
		20.37	5.29	3.16	0.58	210,000	2500	4.4
	q = 0.059	20.22	4.44	3.15	0.32	270,000	2000	5.2
	v = 1.75	18.57	3.52	2.8	0.33	210,000	1700	5.1
		15.96	3.65	2.66	0.31	240,000	2000	4.9
		21.83	4.99	2.89	0.34	200,000	1700	4.9
		19.45	4.03	2.97	0.34	220,000	1800	5.0

Appendices

TABLE III-10 Heavy Metals Data (Rubber Cell – Steady Stage)

Cell No.	q & v (m/d)	Zinc C_i	Zinc C_o	Iron C_i	Iron C_o	Manganese C_i	Manganese C_o	Lead C_i	Lead C_o	Cadmium C_i	Cadmium C_o
Last Basin 2 m × 10 m ($A_s = 1138$ m^2)	$q = 0.256$ $v = 10.91$	1.55	0.96	0.72	0.58	0.41	0.34	0.040	0.032	0.00173	0.00142
		1.85	1.13	0.82	0.65	0.27	0.22	0.043	0.035	0.00252	0.00206
		1.66	1.00	0.63	0.50	0.22	0.18	0.061	0.049	0.00217	0.00178
		2.01	1.20	0.65	0.52	0.30	0.25	0.031	0.024	0.00306	0.00251
		2.16	1.31	0.95	0.76	0.28	0.23	0.021	0.017	0.00211	0.00175
	$q = 0.174$ $v = 7.42$	1.42	0.80	1.12	0.74	0.15	0.11	0.033	0.024	0.00191	0.00153
		1.41	0.79	1.05	0.69	0.48	0.36	0.039	0.028	0.00223	0.00178
		1.73	0.96	0.96	0.63	0.26	0.19	0.025	0.018	0.00298	0.00237
		1.91	1.05	0.84	0.56	0.42	0.32	0.027	0.019	0.00212	0.00167
		1.74	0.95	0.79	0.53	0.50	0.38	0.048	0.035	0.00233	0.00182
	$q = 0.120$ $v = 5.11$	1.28	0.66	0.85	0.46	0.36	0.21	0.036	0.018	0.00196	0.00149
		1.38	0.70	0.77	0.42	0.45	0.27	0.044	0.022	0.00185	0.00139
		1.57	0.79	0.64	0.34	0.29	0.17	0.027	0.014	0.00267	0.00200
		1.66	0.83	0.66	0.35	0.38	0.22	0.064	0.032	0.00284	0.00215
		1.79	0.89	0.87	0.47	0.17	0.10	0.047	0.023	0.00290	0.00224

TABLE III-10 (Continued)

Cell No.	q & v (m/d)	Zinc C_i	Zinc C_o	Iron C_i	Iron C_o	Manganese C_i	Manganese C_o	Lead C_i	Lead C_o	Cadmium C_i	Cadmium C_o
	q = 0.085 v = 3.62	1.43	0.53	1.19	0.58	0.28	0.14	0.027	0.012	0.00255	0.00170
		1.85	0.72	0.89	0.43	0.23	0.11	0.071	0.031	0.00273	0.00183
		1.79	0.69	0.64	0.31	0.24	0.12	0.061	0.029	0.00261	0.00172
		1.92	0.72	0.97	0.46	0.34	0.17	0.055	0.025	0.00235	0.00154
		1.64	0.64	0.83	0.40	0.37	0.18	0.043	0.020	0.00247	0.00167
	q = 0.063 v = 2.69	1.50	0.45	0.98	0.36	0.26	0.10	0.072	0.031	0.00175	0.00112
		1.69	0.54	0.76	0.28	0.25	0.10	0.045	0.019	0.00130	0.00082
		1.44	0.46	0.74	0.27	0.31	0.12	0.061	0.026	0.00126	0.00080
		1.38	0.43	0.89	0.31	0.41	0.16	0.032	0.014	0.00182	0.00116
		1.87	0.59	0.88	0.32	0.44	0.17	0.066	0.028	0.00241	0.00152

Appendices

TABLE III-11 Heavy Metals Data (Gravel Cell – Steady Stage)

Cell No.	q & v (m/d)	Zinc C_i	Zinc C_o	Iron C_i	Iron C_o	Manganese C_i	Manganese C_o	Lead C_i	Lead C_o	Cadmium C_i	Cadmium C_o
Last Basin 2 m × 10 m (A_s = 1482 m²)	q = 0.252 v = 13.99	1.55	0.88	0.72	0.50	0.41	0.32	0.040	0.029	0.00173	0.00132
		1.85	1.03	0.82	0.57	0.27	0.21	0.043	0.031	0.00252	0.00191
		1.66	0.94	0.63	0.43	0.22	0.17	0.061	0.044	0.00217	0.00164
		2.01	1.15	0.65	0.46	0.30	0.24	0.031	0.023	0.00306	0.00237
		2.16	1.20	0.95	0.67	0.28	0.22	0.021	0.015	0.00211	0.00164
	q = 0.171 v = 9.52	1.42	0.70	1.12	0.72	0.15	0.11	0.033	0.022	0.00191	0.00138
		1.41	0.70	1.05	0.66	0.48	0.33	0.039	0.026	0.00223	0.00160
		1.73	0.84	0.96	0.60	0.26	0.18	0.025	0.017	0.00298	0.00212
		1.91	0.91	0.84	0.53	0.42	0.30	0.027	0.018	0.00212	0.00152
		1.74	0.83	0.79	0.50	0.50	0.35	0.048	0.032	0.00233	0.00167
	q = 0.118 v = 6.55	1.28	0.56	0.85	0.37	0.36	0.16	0.036	0.017	0.00196	0.00140
		1.38	0.58	0.77	0.34	0.45	0.21	0.044	0.020	0.00185	0.00130
		1.57	0.65	0.64	0.27	0.29	0.13	0.027	0.012	0.00267	0.00187
		1.66	0.70	0.66	0.28	0.38	0.17	0.064	0.029	0.00284	0.00201
		1.79	0.78	0.87	0.38	0.17	0.08	0.047	0.021	0.00290	0.00202
	q = 0.083 v = 4.64	1.43	0.47	1.19	0.49	0.28	0.12	0.027	0.011	0.00255	0.00161
		1.85	0.65	0.89	0.38	0.23	0.10	0.071	0.030	0.00273	0.00170
		1.79	0.63	0.64	0.27	0.24	0.10	0.061	0.026	0.00261	0.00162

TABLE III-11 (Continued)

Cell No.	q & v (m/d)	Zinc C_i	Zinc C_o	Iron C_i	Iron C_o	Manganese C_i	Manganese C_o	Lead C_i	Lead C_o	Cadmium C_i	Cadmium C_o
		1.92	0.66	0.97	0.41	0.34	0.15	0.055	0.022	0.00235	0.00152
		1.64	0.56	0.83	0.33	0.37	0.16	0.043	0.017	0.00247	0.00159
	$q = 0.062$ $v = 3.45$	1.50	0.36	0.98	0.22	0.26	0.07	0.072	0.027	0.00175	0.00087
		1.69	0.42	0.76	0.18	0.25	0.07	0.045	0.017	0.00130	0.00064
		1.44	0.36	0.74	0.18	0.31	0.09	0.061	0.023	0.00126	0.00062
		1.38	0.36	0.89	0.22	0.41	0.12	0.032	0.012	0.00182	0.00089
		1.87	0.48	0.88	0.22	0.44	0.12	0.066	0.025	0.00241	0.00116

TABLE III-12 Heavy Metals Data (Plastic Cell – Steady Stage)

Cell No.	Q & v (m/d)	Zinc C_i	Zinc C_o	Iron C_i	Iron C_o	Manganese C_i	Manganese C_o	Lead C_i	Lead C_o	Cadmium C_i	Cadmium C_o
Last Basin 2 m × 10 m ($A_s = 2265$ m^2)	$q = 0.241$ $v = 7.09$	1.55	0.76	0.72	0.47	0.41	0.28	0.040	0.024	0.00173	0.00122
		1.85	0.90	0.82	0.54	0.27	0.18	0.043	0.026	0.00252	0.00175
		1.66	0.82	0.63	0.41	0.22	0.15	0.061	0.037	0.00217	0.00154
		2.01	0.98	0.65	0.42	0.30	0.21	0.031	0.018	0.00306	0.00219
		2.16	1.03	0.95	0.61	0.28	0.19	0.021	0.012	0.00211	0.00150
	$q = 0.164$ $v = 4.82$	1.42	0.53	1.12	0.51	0.15	0.07	0.033	0.017	0.00191	0.00130
		1.41	0.51	1.05	0.47	0.48	0.24	0.039	0.020	0.00223	0.00149

Appendices

TABLE III-12 (Continued)

Cell No.	Q & v (m/d)	Zinc C_i	Zinc C_o	Iron C_i	Iron C_o	Manganese C_i	Manganese C_o	Lead C_i	Lead C_o	Cadmium C_i	Cadmium C_o
		1.73	0.62	0.96	0.43	0.26	0.13	0.025	0.013	0.00298	0.00198
		1.91	0.71	0.84	0.39	0.42	0.20	0.027	0.014	0.00212	0.00146
		1.74	0.66	0.79	0.36	0.50	0.23	0.048	0.024	0.00233	0.00160
	$q = 0.113$ $v = 3.32$	1.28	0.41	0.85	0.26	0.36	0.13	0.036	0.015	0.00196	0.00131
		1.38	0.44	0.77	0.23	0.45	0.17	0.044	0.018	0.00185	0.00122
		1.57	0.49	0.64	0.19	0.29	0.10	0.027	0.011	0.00267	0.00174
		1.66	0.50	0.66	0.21	0.38	0.13	0.064	0.026	0.00284	0.00185
		1.79	0.55	0.87	0.28	0.17	0.06	0.047	0.019	0.00290	0.00191
	$q = 0.080$ $v = 2.35$	1.43	0.41	1.19	0.33	0.28	0.09	0.027	0.010	0.00255	0.00145
		1.85	0.52	0.89	0.24	0.23	0.07	0.071	0.027	0.00273	0.00153
		1.79	0.49	0.64	0.17	0.24	0.08	0.061	0.023	0.00261	0.00146
		1.92	0.52	0.97	0.28	0.34	0.11	0.055	0.022	0.00235	0.00130
		1.64	0.46	0.83	0.22	0.37	0.12	0.043	0.016	0.00247	0.00141
	$q = 0.059$ $v = 1.75$	1.50	0.28	0.98	0.19	0.26	0.06	0.072	0.024	0.00175	0.00078
		1.69	0.32	0.76	0.15	0.25	0.05	0.045	0.015	0.00130	0.00057
		1.44	0.27	0.74	0.15	0.31	0.06	0.061	0.020	0.00126	0.00055
		1.38	0.24	0.89	0.19	0.41	0.09	0.032	0.011	0.00182	0.00079
		1.87	0.32	0.88	0.19	0.44	0.10	0.066	0.021	0.00241	0.00111

3. ANN WEIGHTS MATRICES AND BIASES VECTORS

Once the desired performance was achieved the weights and biases of the trained artificial neural network were recorded. Three weights matrices and three biases vectors were obtained for each network. Tables III-13 to III-27 demonstrate the input layer weights matrix (from input to first hidden layer), the first hidden layer weights matrix (from first to second hidden layer), and the second hidden layer weights matrix (from second hidden layer to the output layer). Also these tables show the biases vectors for the selected networks.

TABLE III-13 Weights Matrices and Biases Vectors for BOD (Set-Up Stage)

Input Layer Weights Matrix						Biases V.
0.01457	−0.09568	−0.00991	0.13851	−0.00770		−2.91278
−0.88484	0.13511	−0.89381	−0.15297	1.21617		0.28141
−0.98724	0.15300	−0.23739	0.05405	0.88495		−1.28056
−4.46708	0.10168	−0.18168	−0.10951	2.20198		0.39175
0.11368	−0.90367	−0.03867	0.47142	−0.10385		2.20107

First Hidden Layer Weights Matrix				1.84377
−3.65348	−0.01673	−1.84143		3.38190
−0.28342	−2.77303	2.15049		−3.67731
−4.04718	0.17798	−3.84123		
−5.54623	−0.13020	4.09077		4.35437
−9.94267	−0.02178	−2.68480		

SHL WM	
−0.17822	
−4.61961	Network
1.23700	5–5–3–1

Appendices

TABLE III-14 Weights Matrices and Biases Vectors for COD (Set-Up Stage)

Input Layer Weights Matrix					Biases V.
0.10709	−0.04205	−0.13399	0.27791		−0.00493
0.15423	0.45311	−1.59861	−0.07426		0.74575
−0.04768	−0.00666	−0.27104	−0.15592		−0.48817
−0.00101	−0.01969	0.01981	0.05175		0.66071
0.09118	−0.08362	−0.26368	0.82406		

					−0.89689
First Hidden Layer Weights Matrix					−1.03883
4.72659	0.32656	−2.29674			−0.03004
0.11277	2.88498	−1.38519			
0.75364	0.80233	−0.70758			0.64219
−1.26737	−1.62590	1.32845			

SHL WM					
1.29002					
−1.76899					Network
−2.91563					5–4–3–1

TABLE III-15 Weights Matrices and Biases Vectors for TSS (Set-Up Stage)

Input Layer Weights Matrix					Biases V.
0.09663	0.08887	−0.28535	0.21119	0.34465	2.42190
0.53839	−0.00901	0.36072	−0.17290	0.24502	0.14932
−0.22025	−0.20611	3.39456	−0.16869	1.76799	−0.93389
2.23709	−0.02612	2.10123	−0.18155	−0.75190	0.30287
0.34575	−0.04132	−0.26905	−0.18817	−1.70615	1.96071

First Hidden Layer Weights Matrix					2.27691
−0.61083	−0.09737	−0.07118			0.60480
−4.71648	−1.30144	−0.39155			−2.17093
−0.29108	0.03095	2.43901			
1.40064	−0.60671	−2.09188			2.19294
−0.04315	0.07709	1.64752			

SHL WM					
−2.58118					
−0.93186					Network
−0.00137					5–5–3–1

TABLE III-16 Weights Matrices and Biases Vectors for BOD (Steady Stage)

Input Layer Weights Matrix					Biases V.
0.09336	−0.15973	−0.06184	−0.20766	0.04324	−2.35915
−2.68110	−0.10513	5.47613	−0.03370	−4.07115	0.69107
1.44060	0.07541	−0.74022	2.87472	−0.50181	4.23106
1.56861	−0.22630	−2.13745	1.02476	0.85907	2.78776
					−3.67932

First Hidden Layer Weights Matrix				Biases V.
0.45786	−3.86982	1.36473	−4.02603	0.12051
−1.27386	−5.76130	−0.14082	−1.14345	2.52572
0.05808	−0.98664	1.11305	4.38468	−1.31069
−0.08711	−3.26327	0.38571	1.86280	−9.29525
−1.52599	−8.03160	−2.11955	−1.97920	
				−0.17113

SHL WM	
1.98711	
0.35848	
−1.33002	**Network**
0.17552	4–5–4–1

TABLE III-17 Weights Matrices and Biases Vectors for COD (Steady Stage)

Input Layer Weights Matrix								Biases V.
0.12670	−0.06027	0.02123	0.48226	−0.18026	0.07036	−0.31125		−0.30740
2.19328	0.12923	−0.15447	1.09481	−2.73125	−12.73110	4.16938		−0.13426
−3.51169	−0.04241	0.12246	3.84660	3.97933	0.75703	0.26023		0.22033
−1.11716	−0.32494	0.35672	−1.30648	2.03425	1.47309	−0.89154		1.97631
								−0.13320

First Hidden Layer Weights Matrix						−12.66068
5.92949	0.35754	−1.09604	−0.66018	−4.43266		3.75866
−4.20651	−2.76939	0.48785	2.48572	9.27350		
3.37572	−2.61515	−1.55823	0.62303	−4.22482		−3.81454
4.16971	0.17891	2.54887	2.59192	1.83161		−0.86222
2.72266	0.25103	−0.97542	−0.80416	2.94304		4.09403
1.92611	−0.31020	4.64017	1.10441	5.13426		0.48398
4.95556	0.18322	1.32074	1.33751	−3.18085		0.59740

TABLE III-17 (Continued)

SHL WM		1.64001
0.06636		
3.06743		
2.93315		
−3.15153		**Network**
−0.12638		4–7–5–1

TABLE III-18 Weights Matrices and Biases Vectors for TSS (Steady Stage)

Input Layer Weights Matrix						Biases V.
0.29551	−0.27770	0.00543	0.00717	0.01330	−0.10353	−1.90732
0.13996	0.10706	1.11939	6.43810	5.59278	−3.50783	−1.39889
−2.06317	−0.08497	−0.20882	−2.66984	−1.40916	−2.36668	1.58819
0.41837	0.28595	0.42737	−2.94655	0.40282	7.01688	4.44598

	4.22361
First Hidden Layer Weights Matrix	3.39591

−2.87108	3.04954	−5.93848	0.93290	4.31506
2.43879	2.48732	0.62115	−1.56676	2.72618
−1.73939	1.20727	1.48170	−0.29593	1.26467
0.03263	−1.81537	0.19146	0.27196	6.55801
0.19186	−2.96976	−0.75334	1.06264	−1.79744
0.11015	1.35877	−4.92757	−1.05656	−5.68244

−0.02615
1.80048
−0.56365
−1.71703
7.85999

SHL WM		0.08194
−1.09035		
−0.15869		
0.21067		
0.90402		**Network**
−0.11579		4–6–5–1

TABLE III-19 Weights Matrices and Biases Vectors for Ammonia (Steady Stage)

Input Layer Weights Matrix	
−0.11034	0.11223
−0.61912	0.89009
0.14241	−0.19692
−0.23426	0.35583

First Hidden Layer Weights Matrix	
5.39589	1.70212
3.60643	−2.74440

SHL WM
−4.913946
2.9853105

Biases Vectors
−0.05794
0.22868

0.27105
3.06748

0.28279

Network
4–2–2–1

TABLE III-20 Weights Matrices and Biases Vectors for Phosphate (Steady Stage)

Input Layer Weights Matrix			
1.28022	−1.95700	0.96344	0.59619
1.03193	1.02570	1.34579	0.03706
−0.72367	1.09446	−0.73787	−1.44469
−0.08549	2.55089	−0.17168	0.89522

First Hidden Layer Weights Matrix			
0.91183	−0.02355	0.50785	
1.01276	0.48180	−0.71153	
−2.84519	0.67134	−0.83600	
0.40784	0.26980	1.13288	

SHL WM
0.81261
2.79705
1.10135

Biases V.
−0.50042
3.70385
1.23222
−2.91256

−2.34402
0.51951
1.83613

−1.13627

Network
4–4–3–1

TABLE III-21 Weights Matrices and Biases Vectors for DO (Steady Stage)

Input Layer Weights Matrix				Biases V.
−1.13226	0.98168	1.01305	2.79332	2.56920
1.15403	1.53945	−0.67906	−0.96378	−1.73417
0.42860	−1.19304	−1.77128	−1.39794	1.28160
				1.18632

First Hidden Layer Weights Matrix				
−1.25043	0.27912	1.17913	1.48041	2.07144
−0.45782	1.41549	0.74040	1.39990	−1.28739
1.38441	−1.09053	0.54490	0.69844	1.15940
0.28440	0.83345	2.31386	−1.95550	1.73844

SHL WM		
0.10230		−0.65655
−0.95419		
−1.06111		**Network**
1.23029		3–4–4–1

TABLE III-22 Weights Matrices and Biases Vectors for FC (Steady Stage)

Input Layer Weights Matrix			Biases Vectors
4.33014	8.00547	−0.01180	1.63942
−2.31254	−2.77707	−0.23348	−2.53148
4.03479	−2.72758	0.01394	−0.64973
−1.64934	−1.95998	0.02352	
			−0.27137

First Hidden Layer Weights Matrix			
			−0.28601
0.80882	0.00195	−0.09609	−4.87545
−3.74042	−0.06789	−0.40500	
−0.57959	−3.32224	−4.85311	−4.15717

SHL WM	
−0.16224	
7.22693	**Network**
2.45122	4–3–3–1

TABLE III-23 Weights Matrices and Biases Vectors for Zinc (Steady Stage)

Input Layer Weights Matrix				Biases V.
1.81915	1.40886	0.39607	−2.95577	−1.91192
1.62515	−3.37961	−7.18130	−6.64934	−0.63125
−1.43447	10.09722	−1.03702	9.00874	−8.34397
−0.96658	9.96257	−0.78184	−1.99936	−7.45295

First Hidden Layer Weights Matrix				
				−15.92006
−3.55096	−1.74408	−12.27970		2.87349
−3.95523	−0.34921	−16.05991		0.85692
12.37950	7.22397	2.03484		
14.84162	10.35660	14.74638		5.54812

SHL WM		Biases
−4.01443		
−5.71467		**Network**
−0.58685		4–4–3–1

TABLE III-24 Weights Matrices and Biases Vectors for Iron (Steady Stage)

Input Layer Weights Matrix				Biases V.
0.06037	−0.04208	−0.16270	−0.16829	−0.10025
0.26168	−1.72910	0.50102	3.34077	−0.99551
0.20030	−1.61298	1.98860	2.90975	3.89960
−0.33304	2.48019	3.46288	−0.27718	4.22161

First Hidden Layer Weights Matrix				
				1.79738
−1.55706	−3.54246	−0.70134		−0.28469
0.67358	−2.11413	2.44333		−1.39878
0.96189	−3.31323	1.24838		
0.50376	3.06926	−1.06473		−1.24658

SHL WM		
0.47920		
−2.91597		**Network**
−2.23044		4–4–3–1

Appendices

TABLE III-25 Weights Matrices and Biases Vectors for Manganese (Steady Stage)

Input Layer Weights Matrix				Biases V.
−0.17624	0.28760	−0.11003	−0.28050	2.30063
−3.36501	1.77790	2.01908	−0.31942	−1.43955
1.02639	−0.54126	−0.02095	0.07490	1.65240
1.17893	−0.48646	−0.28364	0.04710	−0.72992

First Hidden Layer Weights Matrix				
				1.88140
−1.58172	0.19631	1.29065	−0.87305	−0.92172
−0.78467	0.65206	1.71123	−0.78961	−0.81592
0.96319	−3.31592	−0.22093	0.89046	−2.32133
−0.27868	−0.06635	−1.39168	1.31446	

	−0.39346

SHL WM
−0.49347
−1.48316
2.05082
0.01162

Network
4–4–4–1

TABLE III-26 Weights Matrices and Biases Vectors for Lead (Steady Stage)

Input Layer Weights Matrix			Biases Vectors
−0.05549	−0.38745	−0.63179	−0.74848
−1.47854	−0.32583	−0.04322	−0.65011
−0.32674	−1.58738	−0.02243	−0.51758
−0.55412	0.75321	0.06728	

	−1.80515

First Hidden Layer Weights Matrix		
0.44462	0.31032	1.12979
−0.14586	−0.26683	
1.83599	1.38586	−0.40972

SHL WM
−0.85219
−1.61651

Network
4–3–2–1

TABLE III-27 Weights Matrices and Biases Vectors for Cadmium (Steady Stage)

Input Layer Weights Matrix				Biases V.
1.10073	0.41377	0.14719	−0.08589	−2.51671
−0.93654	2.18817	0.02647	4.96939	−1.56809
−0.74939	−3.64881	−0.02343	0.38446	−0.03534
−0.51778	0.98490	−0.03052	1.97255	6.43234

First Hidden Layer Weights Matrix			
			0.50780
0.39727	−0.54292	−1.37419	−1.00655
−0.59266	0.83153	1.67863	−1.32107
1.16077	1.90060	2.39856	
−0.01161	0.84041	−2.55618	1.04481

SHL WM	
2.20023	
1.55943	Network
1.60812	4–4–3–1

4. LIST OF MATLAB ORDERS

Various operators and commands which have been used in Matlab, the names of parameters and their description are given below (Demuth et al., 2009):

A. Functions

Disp	Displays a network's properties
Init	Returns neural network net with weight and bias values updated according to the network initialization function
Initzero	Zero weight and bias initialization function
Sim	Simulates neural network
Train	Train neural network
Newff	Creates feed-forward back propagation network
Mse	A network performance function. It measures the network's performance according to the mean of squared errors.

Divideint	Used to separate input and target vectors into three sets:
Plotperform	Plotperform(tr) plots the training, validation, and test performances given the training record TR returned by the function train
plottrainstate	Plottrainstate(tr) plots the training state from a training record TR returned by train.
removerows	Process matrices by removing rows with specified indices
Gensim	Creates a Simulink system containing a block that simulates neural network net.
Trainlm	A network training function that updates weight and bias values according to Marquardt-Levenberg optimization.
Tansig	Hyperbolic tangent sigmoid transfer function
Logsig	Log-sigmoid transfer function
Purelin	Linear transfer function

B. Training Parameters

net.trainparam.delt_dec	Value of decrease in learning rate (adaptive learning rate)
net.trainparam.delt_inc	Value of decrease in learning rate (adaptive learning rate)
net.trainparam.epochs	Value of maximum number of epochs
net.trainparam.goal	Value of minimum performance error
net.trainparam.lr	Value of learning rate
net.trainparam.show	Value of epochs after which error is displayed

APPENDIX IV

TABLES OF STATISTICAL ANALYSIS

This appendix highlights the tables of the statistical analysis for the field and models outputs (standard deviation, normality, one way ANOVA, and Chi-square tests).

1. MEAN, RANGE, AND S.D. FOR SET UP AND STEADY STAGES

Table IV-1 gives the values of mean, range, and standard deviation (S.D.) of the period between 31/10/2009 and 3/4/2010 (set up stage) for inlet and outlet concentrations (10 m from inlet). While Tables IV-2 to IV-7 present these values for the period between 17/4 and 17/10/2010 (steady stage).

TABLE IV-1 Mean, Range, and S.D. for Set Up Stage at Cell Inlet and Outlet

Pollutant	Value	C_i (mg/l)	C_{or} (mg/l)	C_{og} (mg/l)	C_{op} (mg/l)
BOD	Mean	194.1	140.6	133.3	124.1
	Range	168–232	103–184	95–178	85–168
	S.D.	20.02	27.23	28.63	29.63
COD	Mean	300.3	218.8	207.8	193.8
	Range	251–368	155–294	143–285	128–269
	S.D.	36.58	45.05	47.15	48.28
TSS	Mean	155.0	88.9	83.9	64.1
	Range	138–180	77–105	72–100	55–76
	S.D.	12.77	7.87	7.87	5.96

C_i = influent concentration;
C_{or} = effluent concentration at the end of rubber cell;
C_{og} = effluent concentration at the end of gravel cell;
C_{op} = effluent concentration at the end of plastic cell.

TABLE IV-2 Mean, Range, and S.D. for BOD & COD at Cell Inlet and Outlet – Steady

Cycle No.	Value	Biochemical Oxygen Demand (mg/l)					Chemical Oxygen Demand (mg/l)			
		C_i	C_{or}	C_{og}	C_{op}	C_i	C_{or}	C_{og}	C_{op}
1	Mean	189.4	103.6	95.8	81.8	296.2	163.0	151.0	129.0
	Range	182–197	98–109	91–101	77–86	285–318	156–177	144–164	123–140
	S.D.	6.02	4.22	3.90	3.70	12.95	8.57	8.00	7.04
2	Mean	198.8	91.4	83.6	69.6	307.4	142.2	130.2	108.6
	Range	188–209	85–98	78–90	64–75	285–337	130–159	119–146	98–122
	S.D.	8.11	5.08	4.77	4.28	21.87	12.26	11.67	9.91
3	Mean	187.8	66.8	59.0	47.0	289.4	103.8	91.6	73.2
	Range	180–198	63–72	55–64	44–51	272–319	96–117	84–104	68–83
	S.D.	7.36	3.56	3.61	2.74	18.04	7.92	7.54	5.76
4	Mean	180.8	55.4	47.4	37.0	279.2	86.0	73.6	57.6
	Range	162–199	49–62	41–54	32–42	242–316	74–99	62–86	48–67
	S.D.	13.99	4.88	4.88	3.74	30.52	10.22	9.71	7.77
5	Mean	183.8	50.4	43.0	29.8	284.0	78.2	67.2	46.4
	Range	177–193	48–54	40–46	28–32	268–306	73–86	61–74	43–51
	S.D.	6.42	2.61	2.55	1.79	18.40	6.30	5.93	3.85

TABLE IV-3 Mean, Range, and S.D. for TSS & NH_3 at Cell Inlet and Outlet – Steady

Cycle No.	Value	Total Suspended Solids (mg/l)					Ammonia (mg/l)			
		C_i	C_{or}	C_{og}	C_{op}	C_i	C_{or}	C_{og}	C_{op}
1	Mean	144.8	72.0	65.2	49.2	20.01	10.84	10.41	8.58
	Range	135–153	66–77	61–70	45–53	15.3–26.2	8.2–13.8	7.7–13.3	6.5–11.1
	S.D.	7.22	4.85	4.55	4.02	4.39	2.29	2.28	1.74
2	Mean	146.4	58.2	55.6	30.0	21.11	10.38	9.67	8.77
	Range	131–170	50–70	48–66	25–37	17.8–24.9	8.8–12.5	8.2–11.9	7.1–10.5
	S.D.	14.99	7.66	6.80	4.64	3.08	1.47	1.58	1.45
3	Mean	155.8	48.2	42.0	20.4	19.19	8.55	7.98	6.78
	Range	144–165	43–53	37–46	17–23	15.9–23.7	6.8–11.1	6.4–9.9	5.8–8.3
	S.D.	9.23	4.09	3.81	2.41	3.33	1.72	1.41	1.13
4	Mean	168.4	30.8	27.0	15.6	19.50	8.30	7.75	5.35
	Range	159–177	28–33	25–29	14–17	15.7–25.4	6.7–11.3	6.3–10.0	4.3–7.2
	S.D.	7.13	2.28	2.00	1.52	3.91	1.88	1.48	1.16
5	Mean	172.0	24.0	19.0	10.4	19.21	7.19	6.20	4.13
	Range	165–180	22–26	17–21	9–12	16.0–21.8	6.4–8.1	5.2–7.3	3.5–5.0
	S.D.	6.04	2.00	1.58	1.14	2.17	0.66	0.76	0.60

TABLE IV-4 Mean, Range, and S.D. for DO & FC at Cell Inlet and Outlet – Steady

Cycle No.	Value	C_{or}	Dissolved Oxygen (mg/l) C_{og}	C_{op}	C_i	Fecal Coliforms (1000 MPN/100 ml) C_{or}	C_{og}	C_{op}
1	Mean	1.84	2.62	3.14	230	6.16	5.84	5.48
	Range	1.7–2.1	2.4–2.9	2.9–3.4	210–250	6.0–6.3	5.6–6.0	5.4–5.6
	S.D.	0.17	0.22	0.21	18.71	0.13	0.17	0.08
2	Mean	2.96	3.38	3.78	248	4.82	4.60	4.34
	Range	2.8–3.1	3.3–3.5	3.7–3.9	220–270	4.7–4.9	4.5–4.7	4.1–4.5
	S.D.	0.11	0.08	0.08	19.24	0.08	0.10	0.17
3	Mean	3.40	4.08	4.30	222	3.88	3.68	3.48
	Range	3.3–3.5	4.0–4.2	4.2–4.4	200–260	3.8–4.0	3.6–3.8	3.4–3.6
	S.D.	0.07	0.08	0.07	23.88	0.08	0.08	0.08
4	Mean	3.82	4.28	4.56	232	3.18	2.88	2.62
	Range	3.7–3.9	4.2–4.4	4.4–4.7	210–260	3.1–3.3	2.7–3.1	2.5–2.7
	S.D.	0.08	0.08	0.11	21.68	0.08	0.21	0.08
5	Mean	4.16	4.56	5.02	228	2.32	2.16	1.84
	Range	4.1–4.2	4.5–4.7	4.9–5.2	200–270	2.2–2.4	2.1–2.3	1.7–2.0
	S.D.	0.05	0.09	0.13	27.75	0.08	0.09	0.15

TABLE IV-5 Mean, Range, and S.D. for PO_4 & Zn at Cell Inlet and Outlet – Steady

Cycle No.	Value	Phosphate (mg/l)				Zinc Element (mg/l)			
		C_i	C_{or}	C_{og}	C_{op}	C_i	C_{or}	C_{og}	C_{op}
1	Mean	3.17	1.66	1.37	0.98	1.85	1.12	1.04	0.90
	Range	2.77–3.54	1.42–1.88	1.18–1.55	0.86–1.13	1.55–2.16	0.96–1.31	0.88–1.20	0.76–1.03
	S.D.	0.30	0.18	0.15	0.10	0.25	0.14	0.14	0.11
2	Mean	2.89	1.19	0.99	0.79	1.64	0.91	0.80	0.60
	Range	2.53–3.26	1.03–1.36	0.86–1.09	0.70–0.87	1.41–1.91	0.79–1.05	0.70–0.91	0.51–0.71
	S.D.	0.30	0.13	0.10	0.07	0.22	0.11	0.09	0.09
3	Mean	2.20	0.82	0.69	0.51	1.54	0.77	0.65	0.48
	Range	1.89–2.57	0.72–0.99	0.60–0.79	0.44–0.58	1.28–1.79	0.66–0.89	0.56–0.78	0.41–0.55
	S.D.	0.26	0.11	0.07	0.05	0.21	0.09	0.09	0.05
4	Mean	3.01	1.01	0.89	0.54	1.73	0.66	0.59	0.48
	Range	2.70–3.22	0.90–1.10	0.79–0.99	0.48–0.59	1.43–1.92	0.53–0.72	0.47–0.66	0.41–0.52
	S.D.	0.21	0.09	0.08	0.05	0.20	0.08	0.08	0.05
5	Mean	2.89	0.79	0.70	0.33	1.58	0.49	0.40	0.29
	Range	2.66–3.15	0.72–0.84	0.63–0.77	0.31–0.34	1.38–1.87	0.43–0.59	0.36–0.48	0.24–0.32
	S.D.	0.18	0.05	0.05	0.01	0.20	0.07	0.05	0.03

TABLE IV-6 Mean, Range, and S.D. for Fe & Mn at Cell Inlet and Outlet – Steady

Cycle No.	Value	Iron Element (mg/l)				Manganese Element (mg/l)			
		C_i	C_{or}	C_{og}	C_{op}	C_i	C_{or}	C_{og}	C_{op}
1	Mean	0.75	0.60	0.53	0.49	0.30	0.24	0.23	0.20
	Range	0.63–0.95	0.50–0.76	0.43–0.67	0.41–0.61	0.22–0.41	0.18–0.34	0.17–0.32	0.15–0.28
	S.D.	0.13	0.11	0.10	0.08	0.07	0.06	0.06	0.05
2	Mean	0.95	0.63	0.60	0.43	0.36	0.27	0.25	0.17
	Range	0.79–1.12	0.53–0.74	0.50–0.72	0.36–0.51	0.15–0.50	0.11–0.38	0.11–0.35	0.07–0.24
	S.D.	0.14	0.09	0.09	0.06	0.15	0.12	0.10	0.07
3	Mean	0.76	0.41	0.33	0.23	0.33	0.19	0.15	0.12
	Range	0.64–0.87	0.34–0.47	0.27–0.38	0.19–0.28	0.17–0.45	0.10–0.27	0.08–0.21	0.06–0.17
	S.D.	0.11	0.06	0.05	0.04	0.11	0.06	0.05	0.04
4	Mean	0.90	0.44	0.37	0.25	0.29	0.14	0.13	0.10
	Range	0.64–1.19	0.31–0.58	0.27–0.49	0.17–0.33	0.23–0.37	0.11–0.18	0.10–0.16	0.07–0.12
	S.D.	0.20	0.10	0.08	0.06	0.06	0.03	0.03	0.02
5	Mean	0.85	0.31	0.20	0.17	0.33	0.13	0.09	0.07
	Range	0.74–0.98	0.27–0.36	0.18–0.22	0.15–0.19	0.25–0.44	0.10–0.17	0.07–0.12	0.05–0.10
	S.D.	0.10	0.04	0.02	0.02	0.09	0.03	0.03	0.02

TABLE IV-7 Mean, Range, and S.D. for Pb & Cd at Cell Inlet and Outlet – Steady

Cycle No.	Value	Lead Element (µg/l)					Cadmium Element (µg/l)			
		C_i	C_{or}	C_{og}	C_{op}	C_i	C_{or}	C_{og}	C_{op}
1	Mean	39.2	31.4	28.4	23.4	2.32	1.90	1.78	1.64
	Range	21–61	17–49	15–44	12–37	1.73–3.06	1.42–2.51	1.32–2.37	1.22–2.19
	S.D.	14.19	12.10	10.71	9.37	0.50	0.41	0.39	0.36
2	Mean	34.4	24.8	23.0	17.6	2.31	1.83	1.66	1.57
	Range	25–48	18–35	17–32	13–24	1.91–2.98	1.53–2.37	1.38–2.12	1.30–1.98
	S.D.	9.37	6.98	6.16	4.51	0.40	0.32	0.28	0.26
3	Mean	43.6	21.8	19.8	17.8	2.44	1.85	1.72	1.61
	Range	27–64	14–32	12–29	11–26	1.85–2.90	1.39–2.24	1.30–2.02	1.22–1.91
	S.D.	13.79	6.72	6.22	5.54	0.50	0.39	0.34	0.32
4	Mean	51.4	23.4	21.2	19.6	2.54	1.69	1.61	1.43
	Range	27–71	12–31	11–30	10–27	2.35–2.73	1.54–1.83	1.52–1.70	1.30–1.53
	S.D.	16.99	7.64	7.46	6.66	0.14	0.10	0.06	0.08
5	Mean	55.2	23.6	20.8	18.2	1.71	1.08	0.84	0.76
	Range	32–72	14–31	12–27	11–24	1.26–2.41	0.8–1.52	0.62–1.16	0.55–1.11
	S.D.	16.39	6.95	6.18	5.17	0.47	0.29	0.22	0.23

2. NORMALITY TEST FOR FIELD AND MODELS OUTPUTS

Tables IV-8 to IV-10 present the significant test values to show that each group of the data follows the normal distribution, for both set up and steady stages parameters of the media under study. While Tables IV-11 to IV-13 present these values for experimental, ANNs, and SPSS results for the verification processes.

TABLE IV-8 Normality Test for Set Up and Steady Parameters (Rubber Cell)

Stage	Parameter	Kolmogorov-Smirnov			Test
		Statistic	D. F.	Sig.	
Set-up	BOD	0.191	12	0.200*	Normal Dis.
	COD	0.141	12	0.200*	Normal Dis.
	TSS	0.107	12	0.200*	Normal Dis.
Steady	BOD	0.162	25	0.091	Normal Dis.
	COD	0.181	25	0.034	Abnormal Dis.
	TSS	0.170	25	0.059	Normal Dis.
	Ammonia	0.123	25	0.200*	Normal Dis.
	Phosphate	0.149	25	0.158	Normal Dis.
	DO	0.138	25	0.200*	Normal Dis.
	FC	0.122	25	0.200*	Normal Dis.
	Zinc	0.099	25	0.200*	Normal Dis.
	Iron	0.110	25	0.200*	Normal Dis.
	Manganese	0.139	25	0.200*	Normal Dis.
	Lead	0.086	25	0.200*	Normal Dis.
	Cadmium	0.117	25	0.200*	Normal Dis.

TABLE IV-9 Normality Test for Set Up and Steady Parameters (Gravel Cell)

Stage	Parameter	Kolmogorov-Smirnov			Test
		Statistic	D. F.	Sig.	
Set-up	BOD	0.192	12	0.200*	Normal Dis.
	COD	0.147	12	0.200*	Normal Dis.
	TSS	0.107	12	0.200*	Normal Dis.
Steady	BOD	0.166	25	0.072	Normal Dis.
	COD	0.183	25	0.031	Abnormal Dis.

TABLE IV-9 (Continued)

Stage	Parameter	Kolmogorov-Smirnov			Test
		Statistic	**D. F.**	**Sig.**	
	TSS	0.161	25	0.092	Normal Dis.
	Ammonia	0.108	25	0.200*	Normal Dis.
	Phosphate	0.146	25	0.179	Normal Dis.
	DO	0.217	25	0.004	Abnormal Dis.
	FC	0.110	25	0.200*	Normal Dis.
	Zinc	0.133	25	0.200*	Normal Dis.
	Iron	0.107	25	0.200*	Normal Dis.
	Manganese	0.146	25	0.180	Normal Dis.
	Lead	0.092	25	0.200*	Normal Dis.
	Cadmium	0.140	25	0.200*	Normal Dis.

TABLE IV-10 Normality Test for Set Up and Steady Parameters (Plastic Cell)

Stage	Parameter	Kolmogorov-Smirnov			Test
		Statistic	**D. F.**	**Sig.**	
Set-up	BOD	0.183	12	0.200*	Normal Dis.
	COD	0.189	12	0.200*	Normal Dis.
	TSS	0.114	12	0.200*	Normal Dis.
Steady	BOD	0.158	25	0.108	Normal Dis.
	COD	0.181	25	0.035	Abnormal Dis.
	TSS	0.159	25	0.102	Normal Dis.
	Ammonia	0.100	25	0.200*	Normal Dis.
	Phosphate	0.161	25	0.092	Normal Dis.
	DO	0.142	25	0.200*	Normal Dis.
	FC	0.143	25	0.200*	Normal Dis.
	Zinc	0.179	25	0.037	Abnormal Dis.
	Iron	0.164	25	0.083	Normal Dis.
	Manganese	0.152	25	0.138	Normal Dis.
	Lead	0.073	25	0.200*	Normal Dis.
	Cadmium	0.131	25	0.200*	Normal Dis.

TABLE IV-11 Normality Test for Set Up and Steady Parameters (Exp. Outputs).

Stage	Parameter	Kolmogorov-Smirnov Statistic	D. F.	Sig.	Test
Set-up	BOD	0.124	24	0.200*	Normal Dis.
	COD	0.110	24	0.200*	Normal Dis.
	TSS	0.146	24	0.200*	Normal Dis.
Steady	BOD	0.123	60	0.025	Abnormal Dis.
	COD	0.092	60	0.200*	Normal Dis.
	TSS	0.114	60	0.049	Abnormal Dis.
	Ammonia	0.153	15	0.200*	Normal Dis.
	Phosphate	0.152	15	0.200*	Normal Dis.
	DO	0.125	15	0.200*	Normal Dis.
	FC	0.121	15	0.200*	Normal Dis.
	Zinc	0.218	15	0.054	Normal Dis.
	Iron	0.105	15	0.200*	Normal Dis.
	Manganese	0.164	15	0.200*	Normal Dis.
	Lead	0.154	15	0.200*	Normal Dis.
	Cadmium	0.155	15	0.200*	Normal Dis.

TABLE IV-12 Normality Test for Set Up and Steady Parameters (ANNs Outputs)

Stage	Parameter	Kolmogorov-Smirnov Statistic	D. F.	Sig.	Test
Set-up	BOD	0.114	24	0.200*	Normal Dis.
	COD	0.096	24	0.200*	Normal Dis.
	TSS	0.140	24	0.200*	Normal Dis.
Steady	BOD	0.119	60	0.025	Abnormal Dis.
	COD	0.094	60	0.200*	Normal Dis.
	TSS	0.116	60	0.042	Abnormal Dis.
	Ammonia	0.162	15	0.200*	Normal Dis.
	Phosphate	0.158	15	0.200*	Normal Dis.
	DO	0.132	15	0.200*	Normal Dis.
	FC	0.115	15	0.200*	Normal Dis.
	Zinc	0.188	15	0.163	Normal Dis.
	Iron	0.143	15	0.200*	Normal Dis.

Appendices

TABLE IV-12 (Continued)

Stage	Parameter	Kolmogorov-Smirnov			Test
		Statistic	D. F.	Sig.	
	Manganese	0.176	15	0.200*	Normal Dis.
	Lead	0.126	15	0.200*	Normal Dis.
	Cadmium	0.171	15	0.200*	Normal Dis.

TABLE IV-13 Normality Test for Set Up and Steady Parameters (SPSS Outputs)

Stage	Parameter	Kolmogorov-Smirnov			Test
		Statistic	D. F.	Sig.	
Set-up	BOD	0.113	24	0.200*	Normal Dis.
	COD	0.088	24	0.200*	Normal Dis.
	TSS	0.161	24	0.107	Normal Dis.
Steady	BOD	0.105	60	0.097	Normal Dis.
	COD	0.081	60	0.200*	Normal Dis.
	TSS	0.110	60	0.066	Normal Dis.
	Ammonia	0.162	15	0.200*	Normal Dis.
	Phosphate	0.115	15	0.200*	Normal Dis.
	DO	0.104	15	0.200*	Normal Dis.
	FC	0.117	15	0.200*	Normal Dis.
	Zinc	0.179	15	0.200*	Normal Dis.
	Iron	0.126	15	0.200*	Normal Dis.
	Manganese	0.132	15	0.200*	Normal Dis.
	Lead	0.122	15	0.200*	Normal Dis.
	Cadmium	0.130	15	0.200*	Normal Dis.

* This is a lower bound of the true significance.

3. ONE WAY ANOVA TEST FOR FIELD AND MODELS OUTPUTS

The following Tables IV-14 to IV-27 exhibit comparisons between results using the one way analysis of variance at 5% level:

- Table IV-14 and IV-21 for the outputs of the used media (set up and steady stages).
- Table IV-22 and IV-27 for the outputs of experimental and both ANNs and SPSS outputs, also the outputs of SPSS and ANNs (verification processes).

TABLE IV-14 ANOVA Test for Rubber, Gravel, and Plastic Cells (Set-Up Stage)

Pollutant	Source	Sum of Squares	D. F.	Mean Square	F-ratio	Sig.
BOD	Bet. Groups	1641.500	2	820.750	1.010	0.375
	Within Groups	26828.500	33	812.985		Non-significant
	Total	28470.000	35			
COD	Bet. Groups	3792.056	2	1896.028	0.864	0.431
	Within Groups	72414.167	33	2194.369		Non-significant
	Total	76206.222	35			
TSS	Bet. Groups	4140.222	2	2070.111	38.975	0.000
	Within Groups	1752.750	33	53.114		Significant
	Total	5892.972	35			

TABLE IV-15 ANOVA Test for Rubber, Gravel, and Plastic Cells (Steady)

Pollutant	Source	Sum of Squares	D. F.	Mean Square	F-ratio	Sig.
BOD	Bet. Groups	5345.387	2	2672.693	6.043	0.004
	Within Groups	31845.760	72	442.302		Significant
	Total	37191.147	74			
COD	Abnormal distribution for rubber, gravel, and plastic groups					
	Using chi-square test for significant difference					
TSS	Bet. Groups	6365.120	2	3182.560	11.050	0.000
	Within Groups	20736.960	72	288.013		Significant
	Total	27102.080	74			

TABLE IV-15 ANOVA Test for Rubber, Gravel, and Plastic Cells (Steady)

Parameter	Source	Sum of Squares	D. F.	Mean Square	F-ratio	Sig.
NH_3	Bet. Groups	72.589	2	36.295	8.099	0.001
	Within Groups	322.671	72	4.482		Significant
	Total	395.260	74			

TABLE IV-15 (Continued)

Parameter	Source	Sum of Squares	D. F.	Mean Square	F-ratio	Sig.
PO_4	Bet. Groups	2.792	2	1.396	16.870	0.000
	Within Groups	5.958	72	0.083		Significant
	Total	8.750	74			
DO	Abnormal distribution for gravel group					
	Using chi-square test for significant difference					
FC	Bet. Groups	3386666.667	2	1693333.333	0.957	0.389
	Within Groups	1.274E08	72	1769544.444		Non-Significant
	Total	1.308E08	74			

TABLE IV-15 ANOVA Test for Rubber, Gravel, and Plastic Cells (Steady)

Pollutant	Source	Sum of Squares	D. F.	Mean Square	F-ratio	Sig.
Zn	Abnormal distribution for plastic group					
	Using chi-square test for significant difference					
Fe	Bet. Groups	0.327	2	0.163	7.554	0.001
	Within Groups	1.558	72	0.022		Significant
	Total	1.884	74			
Mn	Bet. Groups	0.053	2	0.027	4.412	0.016
	Within Groups	0.435	72	0.006		Significant
	Total	0.488	74			
Pb	Bet. Groups	0.000	2	0.000	3.667	0.030
	Within Groups	0.004	72	0.000		Significant
	Total	0.004	74			
Cd	Bet. Groups	0.000	2	0.000	2.579	0.083
	Within Groups	0.000	72	0.000		Non-Significant
	Total	0.000	74			

TABLE IV-16 ANOVA Test for Rubber and Gravel Cells (Set-Up Stage)

Pollutant	Source	Sum of Squares	D. F.	Mean Square	F-ratio	Sig.
BOD	Bet. Groups	315.375	1	315.375	0.404	0.532
	Within Groups	17171.583	22	780.527		Non-significant
	Total	17486.958	23			
COD	Bet. Groups	737.042	1	737.042	0.347	0.562
	Within Groups	46777.917	22	2126.269		Non-significant
	Total	47514.958	23			
TSS	Bet. Groups	150.000	1	150.000	2.423	0.134
	Within Groups	1361.833	22	61.902		Non-significant
	Total	1511.833	23			

TABLE IV-17A ANOVA Test for Rubber and Gravel Cells (Steady Stage)

Pollutant	Source	Sum of Squares	D. F.	Mean Square	F-ratio	Sig.
BOD	Bet. Groups	752.720	1	752.720	1.647	0.206
	Within Groups	21942.800	48	457.142		Non-significant
	Total	22695.520	49			
COD	Abnormal distribution for rubber and gravel groups					
	Using chi-square test for significant difference					
TSS	Bet. Groups	297.680	1	297.680	0.900	0.348
	Within Groups	15882.320	48	330.882		Non-significant
	Total	16180.000	49			

TABLE IV-17B ANOVA Test for Rubber and Gravel Cells (Steady Stage)

Parameter	Source	Sum of Squares	D. F.	Mean Square	F-ratio	Sig.
NH_3	Bet. Groups	5.294	1	5.294	1.216	0.276
	Within Groups	208.959	48	4.353		Non-significant
	Total	214.254	49			
PO_4	Bet. Groups	0.339	1	0.339	3.570	0.065
	Within Groups	4.564	48	0.095		Non-significant
	Total	4.904	49			

Appendices

TABLE IV-17B (Continued)

Parameter	Source	Sum of Squares	D. F.	Mean Square	F-ratio	Sig.
DO	Abnormal distribution for gravel group					
	Using chi-square test for significant difference					
FC	Bet. Groups	720000.000	1	720000.000	0.400	0.530
	Within Groups	8.640E07	48	1800100.000		Non-Significant
	Total	8.712E07	49			

TABLE IV-17C ANOVA Test for Rubber and Gravel Cells (Steady Stage)

Pollutant	Source	Sum of Squares	D. F.	Mean Square	F-ratio	Sig.
Zn	Bet. Groups	0.114	1	0.114	2.045	0.159
	Within Groups	2.682	48	0.056		Non-significant
	Total	2.796	49			
Fe	Bet. Groups	0.061	1	0.061	2.603	0.113
	Within Groups	1.116	48	0.023		Non-significant
	Total	1.177	49			
Mn	Bet. Groups	0.008	1	0.008	1.176	0.284
	Within Groups	0.334	48	0.007		Non-significant
	Total	0.343	49			
Pb	Bet. Groups	0.000	1	0.000	1.099	0.300
	Within Groups	0.003	48	0.000		Non-significant
	Total	0.003	49			
Cd	Bet. Groups	0.000	1	0.000	1.584	0.214
	Within Groups	0.000	48	0.000		Non-Significant
	Total	0.000	49			

TABLE IV-18 ANOVA Test for Rubber and Plastic Cells (Set-Up Stage)

Pollutant	Source	Sum of Squares	D. F.	Mean Square	F-ratio	Sig.
BOD	Bet. Groups	1633.500	1	1633.500	2.017	0.170
	Within Groups	17813.833	22	809.720		Non-significant
	Total	19447.333	23			

TABLE IV-18 (Continued)

Pollutant	Source	Sum of Squares	D.F.	Mean Square	F-ratio	Sig.
COD	Bet. Groups	3775.042	1	3775.042	1.732	0.202
	Within Groups	47957.917	22	2179.905		Non-significant
	Total	51732.958	23			
TSS	Bet. Groups	3700.167	1	3700.167	75.948	0.000
	Within Groups	1071.833	22	48.720		Significant
	Total	4772.000	23			

TABLE IV-19A ANOVA Test for Rubber and Plastic Cells (Steady Stage)

Pollutant	Source	Sum of Squares	D.F.	Mean Square	F-ratio	Sig.
BOD	Bet. Groups	5242.880	1	5242.880	12.033	0.001
	Within Groups	20913.200	48	435.692		Significant
	Total	26156.080	49			
COD	Abnormal distribution for rubber and plastic groups					
	Using chi-square test for significant difference					
TSS	Bet. Groups	5788.880	1	5788.880	21.374	0.000
	Within Groups	13000.400	48	270.842		Significant
	Total	18789.280	49			

TABLE IV-19B ANOVA Test for Rubber and Plastic Cells (Steady Stage)

Parameter	Source	Sum of Squares	D.F.	Mean Square	F-ratio	Sig.
NH_3	Bet. Groups	68.141	1	68.141	15.046	0.000
	Within Groups	217.387	48	4.529		Significant
	Total	285.528	49			
PO_4	Bet. Groups	2.714	1	2.714	30.884	0.000
	Within Groups	4.219	48	0.088		Significant
	Total	6.933	49			

TABLE IV-19B (Continued)

Parameter	Source	Sum of Squares	D. F.	Mean Square	F-ratio	Sig.
DO	Bet. Groups	10.672	1	10.672	18.752	0.000
	Within Groups	27.318	48	0.569		Significant
	Total	37.990	49			
FC	Bet. Groups	3380000.000	1	3380000.000	1.902	0.174
	Within Groups	8.529E07	48	1776933.333		Non-Significant
	Total	8.867E07	49			

TABLE IV-19C ANOVA Test for Rubber and Plastic Cells (Steady Stage)

Pollutant	Source	Sum of Squares	D. F.	Mean Square	F-ratio	Sig.
Zn	Abnormal distribution for plastic group					
	Using chi-square test for significant difference					
Fe	Bet. Groups	0.325	1	0.325	16.496	0.000
	Within Groups	0.945	48	0.020		Significant
	Total	1.270	49			
Mn	Bet. Groups	0.052	1	0.052	9.368	0.004
	Within Groups	0.269	48	0.006		Significant
	Total	0.321	49			
Pb	Bet. Groups	0.090	1	0.090	12.920	0.001
	Within Groups	0.334	48	0.007		Significant
	Total	0.423	49			
Cd	Bet. Groups	0.000	1	0.000	5.282	0.026
	Within Groups	0.000	48	0.000		Significant
	Total	0.000	49			

TABLE IV-20 ANOVA Test for Gravel and Plastic Cells (Set-Up Stage)

Pollutant	Source	Sum of Squares	D. F.	Mean Square	F-ratio	Sig.
BOD	Bet. Groups	513.375	1	513.375	0.605	0.445
	Within Groups	18671.583	22	848.708		Non-significant
	Total	19184.958	23			
COD	Bet. Groups	1176.000	1	1176.000	0.516	0.480
	Within Groups	50092.500	22	2276.932		Non-significant
	Total	51268.500	23			
TSS	Bet. Groups	2360.167	1	2360.167	48.444	0.000
	Within Groups	1071.833	22	48.720		Significant
	Total	3432.000	23			

TABLE IV-21A ANOVA Test for Gravel and Plastic Cells (Steady Stage)

Pollutant	Source	Sum of Squares	D. F.	Mean Square	F-ratio	Sig.
BOD	Bet. Groups	2022.480	1	2022.480	4.659	0.036
	Within Groups	20835.520	48	434.073		Significant
	Total	22858.000	49			
COD	Abnormal distribution for gravel and plastic groups					
	Using chi-square test for significant difference					
TSS	Bet. Groups	3461.120	1	3461.120	13.194	0.001
	Within Groups	12591.200	48	262.317		Significant
	Total	16052.320	49			

TABLE IV-21B ANOVA Test for Gravel and Plastic Cells (Steady Stage)

Parameter	Source	Sum of Squares	D. F.	Mean Square	F-ratio	Sig.
NH_3	Bet. Groups	35.448	1	35.448	7.770	0.008
	Within Groups	218.995	48	4.562		Significant
	Total	254.444	49			
PO_4	Bet. Groups	1.134	1	1.134	17.374	0.000
	Within Groups	3.133	48	0.065		Significant
	Total	4.267	49			

Appendices

TABLE IV-21B (Continued)

Parameter	Source	Sum of Squares	D. F.	Mean Square	F-ratio	Sig.
DO	Abnormal distribution for gravel group					
	Using chi-square test for significant difference					
FC	Bet. Groups	980000.000	1	980000.000	0.566	0.456
	Within Groups	8.312E07	48	1731600.000		Non-Significant
	Total	8.410E07	49			

TABLE IV-21C ANOVA Test for Gravel and Plastic Cells (Steady Stage)

Pollutant	Source	Sum of Squares	D. F.	Mean Square	F-ratio	Sig.
Zn	Abnormal distribution for plastic group					
	Using chi-square test for significant difference					
Fe	Bet. Groups	0.105	1	0.105	4.778	0.034
	Within Groups	1.054	48	0.022		Significant
	Total	1.159	49			
Mn	Bet. Groups	0.019	1	0.019	3.468	0.069
	Within Groups	0.266	48	0.006		Non-significant
	Total	0.285	49			
Pb	Bet. Groups	0.000	1	0.000	2.837	0.099
	Within Groups	0.002	48	0.000		Non-significant
	Total	0.002	49			
Cd	Bet. Groups	0.000	1	0.000	0.977	0.328
	Within Groups	0.000	48	0.000		Non-Significant
	Total	0.000	49			

TABLE IV-22 ANOVA Test for Exp. and ANNs Outputs (Set-Up Stage)

Pollutant	Source	Sum of Squares	D. F.	Mean Square	F-ratio	Sig.
BOD	Bet. Groups	1.277	1	1.277	0.002	0.966
	Within Groups	31082.218	46	675.700		Non-significant
	Total	31083.495	47			

TABLE IV-22 (Continued)

Pollutant	Source	Sum of Squares	D. F.	Mean Square	F-ratio	Sig.
COD	Bet. Groups	0.048	1	0.048	0.000	0.997
	Within Groups	113873.820	46	2475.518		Non-significant
	Total	113873.868	47			
TSS	Bet. Groups	0.394	1	0.394	0.001	0.980
	Within Groups	28025.099	46	609.241		Non-significant
	Total	28025.494	47			

TABLE IV-23A ANOVA Test for Exp. and ANNs Outputs (Steady Stage)

Pollutant	Source	Sum of Squares	D. F.	Mean Square	F-ratio	Sig.
BOD	Abnormal distribution for exp. and ANNs groups					
	Using chi-square test for significant difference					
COD	Bet. Groups	0.724	1	0.724	0.000	0.989
	Within Groups	443819.886	118	3761.185		Non-significant
	Total	443820.610	119			
TSS	Abnormal distribution for exp. and ANNs groups					
	Using chi-square test for significant difference					

TABLE IV-23B ANOVA Test for Exp. and ANNs Outputs (Steady Stage)

Parameter	Source	Sum of Squares	D. F.	Mean Square	F-ratio	Sig.
NH_3	Bet. Groups	0.021	1	0.021	0.004	0.947
	Within Groups	134.785	28	4.814		Non-significant
	Total	134.806	29			
PO_4	Bet. Groups	0.000	1	0.000	0.003	0.958
	Within Groups	2.677	28	0.096		Non-significant
	Total	2.677	29			

Appendices

TABLE IV-23B (Continued)

Parameter	Source	Sum of Squares	D. F.	Mean Square	F-ratio	Sig.
DO	Bet. Groups	0.001	1	0.001	0.001	0.975
	Within Groups	20.330	28	0.726		Non-significant
	Total	20.331	29			
FC	Bet. Groups	974.700	1	974.700	0.001	0.982
	Within Groups	5.275E07	28	1883755.986		Non-Significant
	Total	5.275E07	29			

TABLE IV-23C ANOVA Test for Exp. and ANNs Outputs (Steady Stage)

Pollutant	Source	Sum of Squares	D. F.	Mean Square	F-ratio	Sig.
Zn	Bet. Groups	0.000	1	0.000	0.002	0.968
	Within Groups	2.121	28	0.076		Non-significant
	Total	2.121	29			
Fe	Bet. Groups	0.000	1	0.000	0.005	0.943
	Within Groups	1.022	28	0.037		Non-significant
	Total	1.022	29			
Mn	Bet. Groups	0.000	1	0.000	0.004	0.951
	Within Groups	0.213	28	0.008		Non-significant
	Total	0.213	29			
Pb	Bet. Groups	0.000	1	0.000	0.086	0.771
	Within Groups	0.001	28	0.000		Non-significant
	Total	0.001	29			
Cd	Bet. Groups	0.000	1	0.000	0.033	0.857
	Within Groups	0.000	28	0.000		Non-Significant
	Total	0.000	29			

TABLE IV-24 ANOVA Test for Exp. and SPSS Outputs (Set-Up Stage)

Pollutant	Source	Sum of Squares	D. F.	Mean Square	F-ratio	Sig.
BOD	Bet. Groups	0.054	1	0.054	0.000	0.993
	Within Groups	33028.359	46	718.008		Non-significant
	Total	33028.413	47			
COD	Bet. Groups	16.591	1	16.591	0.007	0.934
	Within Groups	111218.933	46	2417.803		Non-significant
	Total	111235.524	47			
TSS	Bet. Groups	4.496	1	4.496	0.008	0.931
	Within Groups	27551.799	46	598.952		Non-significant
	Total	27556.294	47			

TABLE IV-25A ANOVA Test for Exp. and SPSS Outputs (Steady Stage)

Pollutant	Source	Sum of Squares	D. F.	Mean Square	F-ratio	Sig.
BOD	Abnormal distribution for exp. group					
	Using chi-square test for significant difference					
COD	Bet. Groups	8.976	1	8.976	0.003	0.960
	Within Groups	421222.396	118	3569.681		Non-significant
	Total	421231.372	119			
TSS	Abnormal distribution for exp. group					
	Using chi-square test for significant difference					

TABLE IV-25B ANOVA Test for Exp. and SPSS Outputs (Steady Stage)

Parameter	Source	Sum of Squares	D. F.	Mean Square	F-ratio	Sig.
NH_3	Bet. Groups	0.160	1	0.160	0.035	0.852
	Within Groups	126.031	28	4.501		Non-significant
	Total	126.191	29			

TABLE IV-25B (Continued)

Parameter	Source	Sum of Squares	D. F.	Mean Square	F-ratio	Sig.
PO$_4$	Bet. Groups	0.006	1	0.006	0.064	0.802
	Within Groups	2.793	28	0.100		Non-significant
	Total	2.799	29			
DO	Bet. Groups	0.010	1	0.010	0.014	0.906
	Within Groups	20.520	28	0.733		Non-significant
	Total	20.530	29			
FC	Bet. Groups	6931.200	1	6931.200	0.004	0.952
	Within Groups	5.300E07	28	1892962.057		Non-Significant
	Total	5.301E07	29			

TABLE IV-25C ANOVA Test for Exp. and SPSS Outputs (Steady Stage)

Pollutant	Source	Sum of Squares	D. F.	Mean Square	F-ratio	Sig.
Zn	Bet. Groups	0.015	1	0.015	0.182	0.673
	Within Groups	2.299	28	0.082		Non-significant
	Total	2.313	29			
Fe	Bet. Groups	0.005	1	0.005	0.162	0.691
	Within Groups	0.925	28	0.033		Non-significant
	Total	0.930	29			
Mn	Bet. Groups	0.000	1	0.000	0.000	1.000
	Within Groups	0.201	28	0.007		Non-significant
	Total	0.201	29			
Pb	Bet. Groups	0.000	1	0.000	0.018	0.895
	Within Groups	0.001	28	0.000		Non-significant
	Total	0.001	29			
Cd	Bet. Groups	0.000	1	0.000	0.038	0.847
	Within Groups	0.000	28	0.000		Non-Significant
	Total	0.000	29			

TABLE IV-26 ANOVA Test for SPSS and ANNs Outputs (Set-Up Stage)

Pollutant	Source	Sum of Squares	D. F.	Mean Square	F-ratio	Sig.
BOD	Bet. Groups	0.806	1	0.806	0.001	0.973
	Within Groups	32400.660	46	704.362		Non-significant
	Total	32401.466	47			
COD	Bet. Groups	18.426	1	18.426	0.008	0.931
	Within Groups	112587.503	46	2447.554		Non-significant
	Total	112605.929	47			
TSS	Bet. Groups	7.553	1	7.553	0.012	0.912
	Within Groups	27908.981	46	606.717		Non-significant
	Total	27916.534	47			

TABLE IV-27A ANOVA Test for SPSS and ANNs Outputs (Steady Stage)

Pollutant	Source	Sum of Squares	D. F.	Mean Square	F-ratio	Sig.
BOD	Abnormal distribution for ANNs group						
	Using chi-square test for significant difference						
COD	Bet. Groups	4.602	1	4.602	0.001	0.972
	Within Groups	427900.815	118	3626.278		Non-significant
	Total	427905.417	119			
TSS	Abnormal distribution for ANNs group						
	Using chi-square test for significant difference						

TABLE IV-27B ANOVA Test for SPSS and ANNs Outputs (Steady Stage)

Parameter	Source	Sum of Squares	D. F.	Mean Square	F-ratio	Sig.
NH_3	Bet. Groups	0.298	1	0.298	0.064	0.803
	Within Groups	131.263	28	4.688		Non-significant
	Total	131.561	29			
PO_4	Bet. Groups	0.004	1	0.004	0.039	0.845
	Within Groups	2.907	28	0.104		Non-significant
	Total	2.911	29			

TABLE IV-27B (Continued)

Parameter	Source	Sum of Squares	D. F.	Mean Square	F-ratio	Sig.
DO	Bet. Groups	0.006	1	0.006	0.008	0.931
	Within Groups	20.535	28	0.733		Non-significant
	Total	20.541	29			
FC	Bet. Groups	2707.500	1	2707.500	0.001	0.970
	Within Groups	5.267E07	28	1881003.757		Non-Significant
	Total	5.267E07	29			

TABLE IV-27C ANOVA Test for SPSS and ANNs Outputs (Steady Stage)

Pollutant	Source	Sum of Squares	D. F.	Mean Square	F-ratio	Sig.
Zn	Bet. Groups	0.018	1	0.018	0.221	0.642
	Within Groups	2.256	28	0.081		Non-significant
	Total	2.274	29			
Fe	Bet. Groups	0.004	1	0.004	0.111	0.741
	Within Groups	0.885	28	0.032		Non-significant
	Total	0.889	29			
Mn	Bet. Groups	0.000	1	0.000	0.004	0.949
	Within Groups	0.197	28	0.007		Non-significant
	Total	0.197	29			
Pb	Bet. Groups	0.000	1	0.000	0.025	0.876
	Within Groups	0.001	28	0.000		Non-significant
	Total	0.001	29			
Cd	Bet. Groups	0.000	1	0.000	0.000	0.999
	Within Groups	0.000	28	0.000		Non-Significant
	Total	0.000	29			

4. ONE WAY ANOVA TEST FOR FIELD AND MODELS OUTPUTS

The following Tables IV-28 to IV-34 present comparisons between results using the Chi-square test for abnormal distribution of data groups at 5% level:

- Table IV-28 and IV-31 for the outputs of the used media.
- Table IV-32 and IV-34 for the outputs of experimental and both ANNs and SPSS outputs, also the outputs of ANNs and SPSS (verification processes).

TABLE IV-28 Chi-Square Test for Rubber, Gravel, and Plastic Cells (Steady Stage)

Stage	Parameter	Kruskal Wallis			Test
		Chi-Square	D. F.	Sig.	
Steady	COD	11.162	2	0.004	Significant
	DO	15.556	2	0.000	Significant
	Zinc	12.405	2	0.002	Significant

TABLE IV-29 Chi-Square Test for Rubber and Gravel Cells (Steady Stage)

Stage	Parameter	Kruskal Wallis			Test
		Chi-Square	D. F.	Sig.	
Steady	COD	2.233	1	0.135	Non-significant
	DO	6.144	1	0.013	Significant

TABLE IV-30 Chi-Square Test for Rubber and Plastic Cells (Steady Stage)

Stage	Parameter	Kruskal Wallis			Test
		Chi-Square	D. F.	Sig.	
Steady	COD	9.886	1	0.002	Significant
	Zinc	11.403	1	0.001	Significant

TABLE IV-31 Chi-Square Test for Gravel, and Plastic Cells (Steady Stage)

Stage	Parameter	Kruskal Wallis			Test
		Chi-Square	D. F.	Sig.	
Steady	COD	4.598	1	0.032	Significant
	DO	3.231	1	0.072	Non-significant
	Zinc	5.026	1	0.025	Significant

TABLE IV-32 Chi-Square Test for Exp. and ANNs Outputs (Steady Stage)

Stage	Parameter	Kruskal Wallis			Test
		Chi-Square	D. F.	Sig.	
Steady	BOD	0.000	1	0.992	Non-significant
	TSS	0.002	1	0.962	Non-significant

TABLE IV-33 Chi-Square Test for Exp. and SPSS Outputs (Steady Stage)

Stage	Parameter	Kruskal Wallis			Test
		Chi-Square	D. F.	Sig.	
Steady	BOD	0.022	1	0.883	Non-significant
	TSS	0.005	1	0.941	Non-significant

TABLE IV-34 Chi-Square Test for SPSS and ANNs Outputs (Steady Stage)

Stage	Parameter	Kruskal Wallis			Test
		Chi-Square	D. F.	Sig.	
Steady	BOD	0.017	1	0.898	Non-significant
	TSS	0.022	1	0.883	Non-significant

APPENDIX V

WETLAND GLOSSARY

Glossary of Wetland Terminology

The definitions and purposes are specific to this study but have been conformed to common usage as much as possible (USDA, 2002).

Abiotic	Nonliving (usually refers to substances or environmental factors).
Absorption	The passing of nutrient material into the body of a plant or animal.
Activated Sludge	A method of biological wastewater treatment involving aeration with flocculating microbial biomass.
Adaptation	Behavioral feature of an organism that increases its chance of survival.
Adsorption	The adherence of a gas, liquid or dissolved chemical to the surface of a solid, such as a sediment particle.
Aerobic	Occurring in the presence of free oxygen, either as a gas in the atmosphere or dissolved in water.
Agricultural Waste	Waste normally associated with the production and processing of food and fiber on farms, feedlots, ranches, and forests that may include animal manure, crop and food processing residue, agricultural chemicals, and animal carcasses.
Algae	Simple photosynthetic rootless plants. They typically grow in sunlit waters in proportion to the amount of available nutrients.

Anaerobic	Occurring in conditions devoid of oxygen.
Ammonification	The first stage of nitrogen transformation by reduction to ammoniacal nitrogen.
Aquatic Plants	Plant that grow in standing water, usually submerged, or with floating leaves; or emergent, rooted beneath the water surface, but growing above it.
Aspect Ratio	The ratio of length to width of a constructed wetland.
Autotrophic	The production of organic carbon from inorganic substances.
Bauxite	A naturally occurring mixture of minerals rich in hydrated aluminum oxides and ferric oxides and low in alkali metals.
Biochemical Oxygen Demand	A measure of the amount of oxygen consumed by aerobic bacteria during the decomposition of organic matter. An indicator of the concentration of organic material in water.
Biofilm	An organic layer typically composed of algae, microfauna and bacteria, which adsorbs small particles and nutrients during the water treatment process.
Biomass	The total mass of living organisms in a designated area at any given time.
Biota	Community of living organisms.
Bog	A wetland ecosystem that is highly acidic and has an accumulation of decomposed plants known as peat.
Burnt Oil Shale	The waste product from heating oil shale to produce mineral oil.
Cation Exchange	The interchange between a cation in solution and another cation in the boundary layer surface of negatively charged material, such as organic matter.
Chemical Oxygen Demand	A measure of the amount of oxygen consumed by a chemical in water. An indicator of the concentration of organic material.

Appendices

Coefficient of Determination	A common measure of the goodness of fit in regression analysis between one or more independent variables and a single dependent variable. Coefficient of determination equals the square of the correlation coefficient.
Constructed Wetland	A wetland constructed for the purpose of pollution control and wastewater treatment. The flow rate, detention time and other factors are controlled in order to enhance the removal and/or decomposition of contaminants.
Contaminant	Harmful substance deposited in the air or water or land.
Denitrification	A process that transforms nitrate or nitrite into nitrogen gas in the absence of freely available oxygen.
Detention Time (Retention Time – Residence Time)	The average period of time that wastewater stays in a treatment system. Detention times vary for different types of wastewater treatment systems and can range from hours to weeks.
Dewatered Alum Sludge	A widely generated by-product of drinking water treatment plants using aluminum salts as coagulants.
Diffusion	The process by which matter is transported from one part of a system to another as the result of random molecular movement; movement is from areas of high concentration to areas of low concentration influenced by temperature and the nature of the medium.
Dispersion	Mechanical mixing of a dissolved chemical as it flows through a solution. Dispersion causes chemicals to spread away from the straight pathway into a wider path. Diffusion is a special case of dispersion.
Dissolved Oxygen	The amount of free oxygen in water. Low DO levels can indicate pollution.
Ecosystem	A network of plants and animals that live together and depend on each other for survival.

Effluent	The liquid that flows out of a treatment system.
Emergent Plants	Aquatic plants rooted in the support medium with much of their green parts above the surface of the water.
Escherichia Coli	A common bacterium that normally inhabits the intestinal tracts of humans and animals. Used as an indicator of the presence of human sewage or animal manure in water.
Evapo-transpiration	Moisture lost through passive evaporation, and active transpiration, the water that plants lose when converting food to energy.
Fecal Coliform	Bacteria found in the intestinal tract of warm-blooded animals, used as an indicator of the presence of human sewage or animal manure in water.
Fen	Lowland covered wholly or partially with water but producing sedge, coarse grasses, or other aquatic plants.
First Order Reaction	A reaction where the rate of removal or production of a particular component is directly proportional to its available concentration.
Fixation	The conversion of nitrogen gas to ammonia by a select group of bacteria.
Fly Ash	An inorganic waste product from coal combustion, consisting mainly of spherical glassy particles of silica (SiO_2), alumina (Al_2O_3), and iron oxides. The material was furnace bottom ash which has coarser particle size and, hence higher permeability than fine fly ash.
Freeboard	Vertical distance between the maximum water level in a basin and the top of the side walls of the basin.
Generation Time	The time required for a bacterial population to double.
Habitat	The environment in which an organism lives.
Heavy Metals	Metal elements and their derivatives including zinc, lead, copper, iron, mercury, cadmium, cobalt, manganese, and nickel.

Heterotrophic Organism	A nonphotosynthetic organism using organic carbon as an energy source.
Hydraulic Conductivity	The ability to support medium to conduct fluid through the interstices between particles which make up the medium.
Hydraulic Load	Influent discharge into a constructed wetland.
Hydrogen Ion	A measure of the hydrogen ion concentration in solution, indicating the presence of acidic, neutral or alkaline conditions. It has a logarithmic scale of 0 to 14. The pH of natural, unpolluted water is generally between 6 and 9, lower or higher pH can indicate pollution and can be harmful to human or aquatic life.
Hydrology	The study of water and how it moves across and under the land.
Hydroponics	The cultivation of plants in a liquid film of nutrients.
Infiltration	The process of water percolation into soil or substrate.
Influent	The liquid that flows into a treatment system.
Ion	An electrically charged atom, radical, or molecule formed by the loss or gain of one or more electrons.
Ion Exchange	Movement of chemical ions between sites.
Immersed	Growing under water.
Impervious Surface	A surface that acts as a barrier to the downward movement of water into the soil.
Kjeldahl Nitrogen	Nitrogen in the form of organic proteins or their decomposition product ammonia, as measured by the Kjeldahl method.
Loading	The quantity of a substance entering the environment, such as the quantity of a nutrient to a constructed wetland.
Leachate	Water that collects contaminants, pesticides or fertilizers; for example, the liquid formed when water soaks into and through a landfill.
Litter	A layer of dead vegetation which serves as an important site for microorganisms.

Lagoon	a shallow lake or pond, especially one connected with a larger body of water.
Limestone	Composed largely of calcium carbonate (CaCO3) in the form of calcite.
Light Expanded Clay Aggregates	Formed by expanding special clay minerals at high temperature producing lightweight ceramic pebbles (ρ = 300: 400 kg/m3).
Macrophyte	Plants that can be seen with the naked eye. Often used to describe large aquatic plants.
Marsh	An environment where terrestrial and aquatic habitats overlap; a wetland dominated by grasses.
Medium	Soil, gravel or other material used in a constructed wetland in which to grow plants and support microorganisms.
Micro-organism	An organism that is not visible with the naked eye.
Mineralization	The decomposition of organic matter into its inorganic constituents.
Most Probable Number	A statistical determination of the number of bacteria per weight or volume of sample. It is based on the fact that the greater the number of bacteria in a sample, the more dilution is needed to reduce the density to the point at which no bacteria are left to grow in a dilution series.
Nitrification	The oxidation of ammoniac nitrogen to nitrite and nitrate by autotrophic bacteria under aerobic conditions.
Nitrogen	When high amounts of nutrient are present, this can create an increase of bacteria and algae in water, leading to the depletion of oxygen. Present in various forms (NO_2, NO_3, NO, NH_4, NH_3) in wastewater.
Nutrients	A substance that provides the nourishment needed for the survival of an organism.
Organic loading	The strength of wastewater measured by its amount of organic matter able to be consumed by biochemical processes.

Organic matter	Mass of matter that contains living organisms or nonliving material derived from organisms.
Oxidation	The addition of oxygen, removal of hydrogen, or the removal of electrons from an element or compound. In the environment, organic matter is oxidized to more stable substances.
Oxidation Pond (Stabilization or Facultative Pond)	A man made body of water in which algae and bacteria are involved in the decomposition of waste. These ponds can be aerobic, anaerobic, or include both aerobic and anaerobic conditions.
Oxidation-Reduction Potential	The electric potential required to transfer electrons from one compound or element (the oxidant) to another compound (the reductant).
Oxidized	In this case, combined with oxygen- specifically when organic material, which is mostly carbon, is decomposed by microorganisms into carbon dioxide (and other compounds).
Pathogen	Micro-organisms that can cause disease in other organisms or in humans, animals, and plants. They may be bacteria, viruses, fungi, or parasites.
Pattern	Point which called by artificial neural networks program.
Peat	Organic material (leaves, bark, nuts) that has decayed partially. It is dark brown with identifiable plant parts, and can be found in peat-lands and bogs.
Permeable	Allowing movement of liquids and gasses.
Phosphorus	A nutrient, present in wastewater, which acts as a fertilizer in surface waters, causing algal blooming and eutrophication of the water body.
Photosynthesis	The biological synthesis of organic matter from inorganic matter in the presence of sunlight and chlorophyll. Photosynthesis process is used to build up plant tissue from carbon dioxide and water.
Polishing	General term used to describe the final treatment of wastewater prior to discharge.

Pollution	Waste, often made by humans, that damages the water, the air, and the soil.
Pond	A body of standing water smaller than a lake, often artificially formed.
Porosity	A measure of the total void space in a media.
Precipitation	The process by which chemicals move out of solution to be deposited in solid form; also rain, snow, sleet or hail.
Pretreatment	Treatment of wastewater to reduce the concentrations of solids and other constituents before discharge to a facility for further management.
Reduction	Chemical changes resulting from the absence of oxygen, or a decrease in positive valence, or an increase in negative valence by the gaining of electrons.
Reuse of Wastewater	The beneficial use of treated wastewater; for example, for irrigation purposes or groundwater recharge.
Rhizome	A root-like stem that produces roots from the lower surface and leaves, and stems from the upper surface.
Rhizosphere	The soil surrounding and directly influenced by plant roots.
Root	The part of a plant, usually below the ground, that holds the plant in position, draws water and nutrients from the soil, stores food, and is typically nongreen.
Root Zone Method	A wastewater treatment method using higher aquatic plants with well-developed root systems. Developed in the 1960 s by the Max Planck Institute in Germany, these systems were early prototypes to today's constructed wetlands.
Secondary Treatment	The biological and chemical process of reducing organic matter in effluent from primary treatment systems.

Seepage	The loss of water by percolation into the soil from a canal, ditch, lateral, watercourse, reservoir, storage facility, or other body of water, or from a field.
Sewage	Wastewater flows originating from human domestic activities.
Shale	An argillaceous rock, derived from the lower limestone group of the carboniferous system, which is highly fissile and which splits readily into very thin laminae.
Short Circuiting	When water finds a more direct course from inlet to outlet than was intended. This is generally undesirable because it may result in short contact, reaction, or settling time in comparison with the presumed detention times.
Sludge	The accumulation of solids resulting from chemical coagulation, flocculation, and sedimentation after water or wastewater treatment.
Sorption	The removal of an ion or molecule from solution by adsorption and absorption.
Stem	Main axis of a plant typically above the soil surface, having leaves or scales, and a characteristic arrangement of the vascular tissue.
Storm Water Run-off	The pulse of surface water following a rainstorm. This water carries sediment, gas, oil, animal feces and other waste from the watershed to receiving waters.
Submersed Plants	Plants' growing with their root, stems, and leaves completely under the surface of the water.
Substrate (Media)	A supporting surface on which organisms grow.
Subsurface Flow Wetlands	A type of constructed wetland in which primarily treated waste flows through deep gravel and/or other porous medium planted with wetland vegetation. The water is not exposed to the air, thereby avoiding problems with odor and direct contact.

Surface Flow Wetlands	A constructed wetland consisting of shallow earthen basin planted with rooted, emergent vegetation in which contaminated water is treated by higher aquatic plants and associated organisms.
Tertiary Wastewater Treatment	The treatment of effluent beyond that of primary and secondary methods. This process may use flocculation basins, clarifiers, filters, chlorine basins, ozone or ultraviolet radiation processes, or constructed wetlands.
Total Suspended Solids	The suspended matter in wastewater capable of being removed by filtration.
Treatment	Chemical, biological, or mechanical procedures applied to sources of contamination to remove, reduce, or neutralize contaminants.
Volatilization	Loss of gaseous components, such as ammonium nitrogen.
Wastewater	Contaminated water containing various products from domestic, industrial, and agricultural wastes.
Water Budget	An accounting of inflow, outflow and storage changes of water in a system.
Water Quality	The excellence of water in comparison with its intended uses.
Wetlands	Land areas that is saturated with water and which contains plants and animals that are adapted to living on, near, or in water. Wetlands have special hydric soils and are usually located between a body of water and land.
Zeolite	A hydrated aluminum-silicate mineral with the aluminum and silicate polyhedral linked by the sharing of oxygen atoms.

INDEX

A

Abiotic process, 27, 40, 47, 48, 641
Acetone (ketone class), 60
Actinomycetes decomposers, 57
Activated sludge, 641
Actual velocity, 123–125, 219, 274, 358, 430, 521
   water, 19, 114, 123–226
Adsorption, 8, 16, 31, 35, 40, 48, 53, 56, 105, 107, 641, 649
Aerial plant tissue, 37
Aerobic, 9, 13, 37, 48, 50–52, 641, 642, 646, 647
Agricultural
   chemicals, 641
   waste, 641
Agricultural wastewater, 27, 35, 67
Algae, 36, 48, 53, 641, 642, 646, 647
Algorithm, 121–123, 359, 360
Aliphatic class (1-butanol), 60
Alum adding unit, 71
Aluminum (Al), 35, 52
Alyamani/sen formula, 24
Ammonia (NH), 2, 13, 34, 35, 39, 49, 51, 59, 60, 91, 99, 106, 118, 126, 127, 130, 143, 158, 230, 287–294, 348, 349, 388, 390, 392–395, 398, 405, 447, 448, 452, 455, 492, 518, 521, 522, 560, 644, 645
   concentration, 164, 165, 287
   inlet–outlet relationships, 287–289
   neural network training tool, 393
   removal efficiency TR impact, 289–292
Ammoniac nitrogen, 35
Ammonification, 48, 57, 642
Ammonium chloride, 103
Ammonium-nitrogen, 47
Anaerobic, 59, 642
   microbial methane formation, 51
   sludge, 16
Animal manure, 641, 644

ANOVA Test, 624–637
Aquatic organisms, 100
Aquatic plants, 36, 48, 53, 57, 642, 644, 646, 648, 650
Aromatic class (tetrahydrofuran), 60
Arrhenius equation, 44
Arsenic, 52, 53
Artificial neural networks (ANNs), 2–4, 64, 65, 111, 112, 119, 121–127, 357, 358, 360, 361, 366, 374, 376, 377, 388–393, 407, 410, 411, 416, 430, 434, 435, 438, 441, 445, 447, 452, 455, 457, 461, 467, 470–472, 477–480, 488–514, 522–526, 573, 575, 577, 620, 622, 623, 631–633, 636–639, 647
   modeling, 118
      set up stage, 358, 470
      steady stage, 374, 388, 407, 479, 492, 500
   training algorithms, 121, 122
   types, 119–121
ANNs structure, 358–363, 375, 376, 389–392, 407
Aspect ratio, 63, 74, 76, 526, 642
Assimilation, 31
Atmospheric
   carbon dioxide, 39
   oxygen, 48
Atomic absorption spectrometer (AAS), 109, 169
Atomic
   number, 109
   weight, 109
Autotrophic bacteria, 57, 642, 646
Average
   negative, 478, 488, 492, 500
   percentage errors, 471, 488, 492, 500
   positive, 488, 492, 500
   removal (k), 340
Axes limits, 388, 407, 573, 575

B

Back propagation (BP), 121, 122
Bacteria, 11, 27, 36, 39, 49, 53, 55, 60, 71, 107, 136, 200, 228, 275, 301, 644
  wetlands, 57, 58
Bauxite, 55, 56, 642
Bias initialization function, 610
Biases vectors, 359, 375, 389, 407, 602
Biochemical
  activities, 102
  degradation, 8
  transformations, 49
Biochemical oxygen demand (BOD), 2, 3, 12, 29, 34, 35, 43, 44, 48, 50, 57, 62, 64–67, 91, 98, 99, 103, 104, 108, 117, 125–130, 136, 138–140, 146–153, 180–182, 184–200, 213, 229–248, 267, 341–348, 357–388, 430–445, 470, 472, 477–482, 488–490, 516, 517, 520–526, 557, 559–562, 571, 573, 579, 584, 591, 602, 604, 614
  concentrations, 561, 562
  removal efficiency, 186, 187, 192, 194–196, 198, 199, 233–235, 242–248, 516, 517
  time effect treatment efficiency, 198–200
  TR impact removal efficiency, 191–197
  treatment distance effect, 182–187
  treatment Q impact, 187–189
Biodegradable substances, 103
Bio-degradation, 91, 517
Biological
  processes, 2, 15, 48, 51, 108, 357
  productivity, 15
  purification, 9
  uptake, 107
Biofilm bacteria, 11, 25, 32, 33, 35, 37, 39, 57, 60, 69, 70, 80, 91, 136–138, 181, 187, 193, 195, 200, 203, 210, 213, 219, 226, 228, 229, 241, 244, 263, 274, 275, 293, 358, 374, 516, 517, 525, 642
Biomass, 9, 36, 53, 54, 57, 63, 641, 642
Biomat, 25, 49
Biota, 642
Bog, 642
Breyer formula, 23
Bulk porosity, 525
Bulrush (*Scirpus*), 36
Burnt oil shale, 55, 642

C

Cadmium (Cd), 35, 52, 109, 143, 169, 177, 179, 180, 229, 230, 334, 335, 337–340, 349, 352, 354, 407, 409–417, 429, 458, 461, 464, 467, 519, 521, 524, 571, 610
  concentration, 179, 180, 335
  neural network training tool, 416
  removal efficiency, 338, 339
Calcite, 49, 55, 646
Calcium (Ca), 35
Calibration process, 4, 41, 85, 111, 126, 127, 367, 377, 392, 430, 434, 438, 441, 445, 447, 452, 457, 461, 477, 488, 490, 492, 495, 496, 498, 500, 504, 510, 514, 521, 523, 525
Carbon
  cycle, 53, 54
  fixation, 15
Carbon dioxide, 15, 39, 57, 153, 647
Carbonates, 49, 52
Cation exchange, 642
Cattail (*Typha*), 11, 36, 58
Cells construction stages, 76
  final, 76
  initial, 76
  transitional, 76
Cell
  depth, 76
  inlet, 76, 112, 124, 125, 138, 144, 152, 185, 199, 200, 202–204, 213, 216–230, 233, 234, 252, 268, 566
  length, 181, 182, 189, 233
  length, 76
  width, 76
Central nervous system, 118
Chemical
  mass balances, 38
  oxidant, 104
  oxidation, 40
  oxygen demand (COD), 2, 3, 12, 29, 34, 35, 43, 44, 50, 57, 65–67, 91, 98, 99, 104, 117, 125–130, 136, 139–141,

Index

146, 147, 153–158, 180, 181, 200–213, 229, 248–267, 341–348, 357–491, 516, 517, 520–526, 557–560, 563, 564, 571, 573, 575, 579, 584, 591, 603, 614
  precipitation, 40
  reaction, 50, 108
  reduction, 40
Chi-square test, 127, 514, 638, 639
Chromium, 53
Clogging, 24, 26, 32, 33, 37, 67, 77, 131, 515, 524, 525
Closed-loop feed back type system, 121
Coagulants, 643
Coarse gravel, 48, 73, 77, 80–82, 90–94, 113, 125, 131, 132, 143, 189, 191, 192, 241, 264, 515, 516, 540, 554, 557
Coastal protection, 15
COD
  concentrations, 563, 564
  TR impact removal efficiency, 207–213
  treatment Q impact, 203–205
Coefficient discharge (Cd), 85
Coliforms removal, 44, 305–307, 518
Coliforms, 2, 35, 44, 58, 67, 71, 91, 99, 107, 118, 126, 127, 130, 143, 158, 169, 180, 230, 301–307, 348, 349, 388, 390, 392–395, 398, 405, 447, 448, 452, 455, 471, 492, 518–521, 524, 560, 644
Coliphages removal, 58
Concentration, 46, 139–143, 148–173, 185, 188, 202, 216, 219, 230–233, 248, 249, 265, 267, 283, 286, 308
Conductivity, 21–27, 36, 47, 52, 56, 62, 63, 105, 526, 645
Constructed wetlands (CWs), 1–15, 20, 26, 27, 30–32, 34–38, 40–61, 64–67, 69, 74, 88, 91, 116, 119, 129, 133, 143, 181, 230, 357, 374, 522, 525–527, 566, 642, 645–650
Constructed wetlands implementation, 61
  HSSF wetlands
    construction, 63
    design, 62, 63
    operation and maintenance, 63, 64
      maintenance, 64
      operation, 64

  wetland planning principals, 61, 62
Constructed wetlands models, 42
  reaction rate models, 43
    first order, 44, 45
    Monod-type, 45
    plug flow dispersion (PFD), 46, 47
    relaxed TIS concentration, 46
    tanks in series (TIS), 45, 46
    zero order, 43, 44
  regression equations, 42, 43
  thumb rules, 42
Constructed wetlands nutrient cycles, 53
  bacteria in wetlands, 57, 58
  carbon, 53, 54
  Cyperusalternatifolius wetland plant, 65
  nitrogen, 54, 55
  phosphorous, 55, 56
Convective gas transport processes, 36
Conventional
  sewage technology, 8
  system, 73, 74
  treatment, 8, 13, 58, 59, 60
  wastewater, 1
Convergence diagrams, 361, 376, 392, 410
Copper, 35, 52, 53
Corrugated plastic media, 136
Cycle discharge, 88
Cylindrical buckets, 88

D

Dakahlia governorate, 14, 70
Danish sands, 56
Dead zones, 527
Decomposers, 57
Decomposition, 8, 53, 57, 106, 108, 642, 643, 645–647
Definition of wetlands, 8
Denitrification, 8, 13, 31, 36, 48, 51, 54, 55, 57, 108, 643
Denitrifiers, 49
Dependent variables, 119
Detention time, 643, 649
Deterioration, 64, 74, 293

Determination coefficient (R^2), 132, 185, 203, 216, 219, 233, 252, 268, 282, 307, 327, 337, 430, 438, 439, 448, 458
Di-potassium hydrogen,
 orthophosphate unhydrous, 103
  di-sodium hydrogen, 103
Die-off, 518
 fecal coliforms, 518
  treatment performance of, 518
Diffusion, 39, 643
Diffusive flow, 8
Discharge, 2–4, 29, 49, 52, 56, 71, 73, 74, 82–85, 88, 112, 115, 123, 125, 133, 134, 144, 187, 191, 192, 230–233, 237, 238, 243–249, 252–254, 259–272, 277, 279–284, 288, 289, 293–295, 299–302, 307, 308, 313–315, 319–321, 326–334, 339–341, 515, 519, 525, 557, 645, 647, 648
 cycles, 115, 247
 variation, 230
Dispersion, 42, 47, 643
Dissolution, 102
Dissolved oxygen (DO), 2, 35, 48, 50, 57, 102, 103, 129, 130, 158, 165, 230, 282–287, 357, 388–392, 396, 398, 400–406, 447–449, 451, 452, 454–457, 492, 494, 497–499, 518, 522–525, 571, 579, 607, 616, 643
Distilled water, 101, 103, 104
Divideint, 611
DO treatment,
 Q impact, 284, 285
Domestic wastewater, 2, 14, 16, 28, 29, 34, 35, 51, 57, 58, 66
Donor, 50, 51
Drainage hole edge, 89
Duck potato arrowhead (*Sagittaria*), 36
Dupuit-Forcheimer equation, 21

E

Effluent concentrations, 125, 136, 143, 264, 294, 308, 309, 317, 322, 329, 335
 TSS concentrations, 219
Egypt, 1, 2, 5, 7, 13–15, 29, 67, 340
Egyptian
 pounds (LE), 73, 74

rural areas, 2
Egypt wetlands, 13
 constructed wetlands, 14, 15
 natural wetlands, 14
Electric conductivity, 47, 52
Electrons, 50, 51, 645, 647, 648
Emergent
 aquatic plants, 48
 plants, 2, 644
 vegetation, 11, 56, 650
Empty wetland velocity, 19
Equality diagram, 368, 369, 372, 373, 384, 385, 387, 388, 401, 402, 405–407, 422–424, 427–429, 433, 434, 437, 438, 442–446
Equilibrium, 92
Ergun equation, 22
Escherichia coli (*E. coli*), 107, 644
Evaporative loss, 51
Evapo-transpiration, 18, 20, 40, 644
Experimental
 data, 4, 65, 70, 111, 117, 118, 125, 129, 340, 430, 438, 447, 457, 470, 477, 479, 488, 492, 500, 511, 521, 525
 facilities, 3, 93
 measurements, 125
 procedures, 70, 93
Experimental procedures, 93
 media arrangement, 94
 wetland cells construction, 93
Exponential function, 185, 189, 196, 202, 206, 211, 216, 222, 226, 233, 237, 250, 255, 267, 273, 288

F

Facultative pond (FP), 58
Fecal coliforms (FC), 2, 30, 35, 58, 66, 91, 99 104, 107, 129, 130, 158, 170, 171, 230, 301, 306, 307, 348, 349, 357, 388, 390, 392, 396, 398–407, 447–450, 452, 454–457, 492–500, 518, 520, 523–525, 561, 560, 571, 579, 616, 644
Feed-forward neural networks (FFNNs), 65, 119, 610
Fen, 644
Ferric chloride, 103
Ferric oxides, 642

Index 655

Fescue (monoculture systems), 58
Filtering effect, 37
Filtralite, 34, 35
Filtration, 8, 9, 13, 16, 31, 48, 58, 105, 650
Fine gravel (FG), 34, 82
First hidden layer (FHL), 359
First order, 2, 65, 116, 340
Floodwater storage, 15
Flow rate, 17, 18, 19, 20, 44, 62, 64, 67, 83, 84, 99, 125, 133, 643
Fly ash, 56, 644
Fonthill sand, 34
Freeboard, 76, 644

G

Gas adherence, 641
Gaseous oxygen, 165
Generation time, 644
Gensim, 611
Giant cutgrass (*Zizaniopsis*), 36
Glass-fiber filter, 105
Gradation, 88, 91
Graphical method, 117, 118
Gravel, 511, 515–527, 541, 554, 555, 557, 560, 566
  bed wetlands, 11, 22
  cells, 173, 196, 226, 298
  media, 13, 19, 21, 35, 49, 66, 91, 94, 131, 132, 137, 142, 143, 188, 189, 192, 196, 205, 218, 219, 237, 241, 247, 253, 259, 271, 272, 279, 288, 300, 322, 335, 515–518, 525, 526, 541

H

Harvesting, 3, 56, 64, 143
Hazen formula, 23
Heavy metals, 2, 35, 53, 99, 109, 118, 126, 127, 130, 143, 169, 180, 230, 349, 354, 355, 407, 409–411
Heavy metals treatment concentrations, 169
  cadmium element, 177–180
  iron element, 173
  lead element, 177
  manganese element, 173–176
  zinc element, 169–173
Helophytes, 33, 36
Herbicide, 16
Heterogeneous distributions, 23
Heterotrophic
  bacteria, 57
  organism hydraulic conductivity, 645
Hidden layers, 119, 120, 122, 358, 359, 361, 370, 375, 388, 389, 407, 573, 575
Homogenous inflow distribution tool, 77
Horizontal flow, 11, 13, 54, 55
Horizontal subsurface flow (HSSF), 1, 3–5, 12–26, 34–37, 42–44, 49, 53–56, 58, 59, 61–63, 66–69, 74, 76, 91, 119, 129, 133, 181, 374, 515, 522, 524, 526, 527
  bed media clogging, 24–26
  constructed wetlands physical model, 74
    inlet zone configurations, 76–78
    outlet zone configurations, 78–80
    treatment media, 80–82
    wetland treatment cells, 74–76
  porous media flow, 20–22
  porous media hydraulic conductivity, 22–24
    Alyamani and Sen formula, 24
    Breyer formula, 23
    Hazen formula, 22, 23
    Kozeny-Carman formula, 23
    Slitcher formula, 23
    Terzaghi formula, 24
    USBR formula, 24
  wetland hydrology, 17
    hydraulic loading rate, 17
    hydraulic retention time, 18, 19
    mean water depth, 17
    overall water mass balances, 19, 20
    wetland cell water velocity, 19
  wetlands hydraulics, 20
    long-term, 25, 26
    short-term, 24, 25
Hybrid systems, 10, 13, 57
Hydrated aluminum oxides, 642
Hydraulic
  application rate, 17

conductivity (HC), 21–27, 36, 56, 62, 63, 105, 527
gradient, 21
loading rate, 17, 34, 43, 62, 112, 117, 123, 125, 133, 134, 182, 187, 188, 189, 204, 205, 218, 219, 237–241, 257, 258, 275, 285, 291, 297, 298, 304, 309, 311, 317, 324, 331, 337, 341, 349, 358, 375, 389, 407, 439, 526
parameter, 46, 112, 144, 525
retention time, 3, 4, 18, 45, 112, 128, 134–136, 143, 144, 182, 192, 196, 207, 211, 222, 259, 279, 286, 292, 299, 306, 312, 318, 325, 326, 332, 338, 520, 557
Hydrogen, 49, 571, 645
ion concentration (pH), 29, 34, 47, 49, 53, 56, 57, 66, 99, 100, 101, 102, 130, 567, 568, 571, 645
ion, 47, 100, 101, 130, 566, 645
sulfide, 39
Hydrologic condition, 8, 17
Hydrology, 645
Hydroponics, 14, 645
Hyperbolic design equation, 65

I

Impervious surface, 645
Implementation, 7
Incubation, 50, 103
Independent variables, 119, 430, 470, 643
Inflow measurement, 83
Infiltration, 20, 52, 74, 645
Influent concentrations, 10, 15, 26, 40, 43, 44, 62, 63, 67, 73, 99, 114, 117, 122, 123, 125, 136–139, 143, 146, 153, 158, 165, 169, 173, 177, 182, 185, 189, 202, 206, 207, 219, 222, 230, 231, 235, 237, 238, 248, 252, 255, 264, 282, 287–289, 294–297, 301, 302, 308, 314, 321, 322, 327–329, 334, 335, 341, 349, 358, 388, 389, 522, 526, 560, 566
Initial
porosity, 88, 91
stage, 76
Inlet–outlet
BOD concentrations, 189–191

COD concentrations, 205–207
TSS concentrations, 219–222
Inlet zone, 25, 26, 78, 82, 205
configurations, 76, 77
media, 77
Input
concentration, 124, 213, 439, 448
layer, 119
vectors, 122
Insoluble
chemical precipitates, 25, 26
sulfides metals, 52
Interconnections of, neural network, 118
Intermittent feeding strategy, 133
Interpolation, 113, 135, 136
Intestinal tracts, 644
Iron concentration, 173–175, 315
Iron (Fe), 35, 51–53, 109, 143, 169, 173–175, 229, 230, 314–320, 349, 354, 407, 409, 410–413, 415, 416, 417, 429, 458, 461, 464, 467, 518, 520, 571
neural network training tool, 413
removal efficiency, 320

J

Juvenile fish, 106

K

Ketone, 60
Kinetic energy, 9
Kjeldahl nitrogen, 645
Kozeny-Carman formula, 23

L

Lagoon, 646
Lake manzala, 14, 15
Landfill leachate, 35, 50, 53
Leachate, 12, 28, 29, 35, 50, 53, 645
treatment, 29
Lead (Pb), 35, 52, 53, 109, 143, 169, 177, 178, 229, 230, 327, 328, 329, 331, 332, 333, 349, 351, 354, 407, 409, 410, 411, 412, 414, 415, 417, 429, 458, 461, 464, 467, 519, 521, 571
Lead concentration, 177, 178, 328
Least square method, 126
Levees, 63

Index

Light
  attenuation, 37
  expanded, 646
Lignite pyrolysis wastewater, 55
Limestone, 49, 55, 56, 646, 649
  dissolution, 49
Linear
  interpolation process, 113
  regression, 126, 361, 430
  transfer function, 359
Litter, 40, 48, 49, 645
Logarithmic
  equations, 188, 204, 219, 240, 247, 257, 264, 275, 282, 285, 287, 289, 294, 297, 300, 304, 307, 309, 312, 313, 317, 319, 320, 322, 325, 327, 329, 331, 333, 337, 339, 340
  function, 188, 204, 219, 240, 247, 257, 264, 275, 282, 285, 287, 289, 294, 297, 300, 304, 307, 309, 312, 313, 317, 319, 320, 322, 325, 327, 329, 331, 333, 337, 339, 340, 519
Logsig, 611
  transfer functions, 358
Longitudinal distance, 21, 153, 182, 213, 264
Long-term clogging, 25

M

Macrophytes, 34, 36, 39, 66
Manganese (Mn), 52, 53, 109, 143, 169, 173, 175, 176, 229, 230, 321, 322, 324–327, 349, 354, 407, 409–417, 429, 458, 461, 464, 467, 519, 521, 571, 609
  concentration, 175, 176, 321
  neural network training tool, 414
Marquardt-Levenberg (ML) algorithm, 121, 122, 360
Marsh, 9, 646
Mass removal rate (JA), 38
Matlab neural network toolbox, 359, 388
Maturation stage, 91
Mean
  square error, 122, 359, 360, 361, 376, 392, 410, 430, 447, 458, 521, 522
  water depth, 17
Measuring arrangements, 82, 83
  inflow measurement, 83–87

media porosity measurement, 88–91
media surface area estimation, 91
  gravel media, 91
  plastic media, 92, 93
  rubber media, 92
mean square error (MSE), 360, 362, 365, 366, 376, 381, 382, 390, 397, 398, 417–419
set-up stages water discharges, 88
Media
  arrangement, 94, 95
  cells, 82, 113, 234, 264
  clogging, 64
  pores, 49, 136, 191
  porosity measurement, 88
  porosity, 2, 19, 62, 83, 88, 89, 109, 113, 114, 125, 136, 181, 207, 229, 374, 542–551
  removal efficiencies, 114
  surface area, 2–5, 82, 91, 123, 125, 247, 275, 358, 521
  voids, 89, 91
Medium gravel, 82
Mercury, 52
Metabolism, 108
Metal binding, 49
Metals removal, 52, 519
Meteorological parameters, 65
Methane, 39
Microbes, 25, 36, 57
Microbial, 9, 11, 25, 27, 28, 38–40, 48–51, 53, 57, 58, 518, 641
  biofilm, 25, 49
  biomat formation, 25
  conversion, 53
  decomposition, 53
  degradation, 518
  growth, 50, 57
  metabolism, 9, 57
  mineralization, 51
  rock-reed filters, 11
Micro-
  biota, 36
  fauna, 642
  organism, 16, 27, 31, 36, 39, 40, 50, 57, 58, 103, 108, 645–647
Mineral solids, 63
Mineralization, 646

Mitigation, 16
Molecular movement, 643
Momentum factor (Mu), 359
Monoculture systems planted, 58
Monod-type model, 45
Most probable number (MPN), 30, 108, 169, 170, 171, 301, 302, 304, 349, 394, 405, 448, 455, 471, 518, 569
Multilayer feed forward neural network (MFFNN), 120, 121, 359, 375
Multilayer perceptron (MLP), 64, 119
Multiple regression analysis (MRA), 64
Municipal wastewater, 1, 2, 11, 28, 30, 50, 60, 525

N

Natural
 gravel, 2, 81
 wetland, 9, 10, 19, 61
Nessler reagent, 106
Network
 calibration process, 363–369, 376–378, 392–397, 411–414, 430–434, 439–441, 448
 generalization, 122
 training tool, 394, 415
 validation process, 370–374, 378–388, 397, 414, 434, 441, 452, 464
Neural networks, 3, 111, 118–120, 123, 126, 358, 359, 392, 411
 hidden layers, 119
 input layer, 119
 output layer, 119
Neurons, 118–120, 358, 359, 361, 370, 375, 389, 407, 521
Nickel, 52, 53
Nile
 delta, 29, 30
 valley, 29, 30
Nitrification, 8, 31, 37, 48, 51, 54, 57, 108, 646
Nitrifiers, 49
Nitrite, 51, 643, 646
Nitrobacter, 54
Nitrogen (N), 10, 13, 28, 34, 35, 48, 49, 54, 55, 59, 65, 106, 642–644, 646, 650
 cycle, 54, 106
Nitrogenous oxygen demand, 48

Nitrosomones, 54
Nitrous oxide, 39
Non-vegetated cell, 67
Nonhalogenated polar organic solvents, 60
Nonlinear regression, 2, 111, 126, 430, 438, 447
Nonparametric test, 511
Normality test, 511
Nutrients, 28, 53, 646

O

Odor elimination, 13
Optimum model structure, 360, 361, 373, 374, 377, 390–392, 409, 411
Organic
 chemicals, 49
 loading, 646
 material, 25, 26, 50, 57, 104, 153, 642, 647
 matter, 25, 28, 35, 47, 50, 103, 104, 153, 526, 642, 646–648
 particulates, 53
Outlet
 dissolved oxygen, 167–169
 DO concentration, 282–284
 structures, 11, 64
Output layer, 119
Overlapping, 198, 212, 227, 526
 reduction potential, 47, 50, 647
Oxidation, 8, 40, 47, 50–52, 646, 647
Oxygen, 2, 9, 11, 12, 15, 22, 36, 37, 47–50, 53–59, 66, 99, 102, 103, 125, 127, 143, 153, 158, 165, 229, 230, 282, 284–286, 388–395, 398, 405, 447, 448, 452, 455, 492, 518, 519, 524, 560, 641–643, 646–650

P

Pathogens, 9, 16, 57, 58, 647
Pathogenic
 bacteria, 39
 organisms, 28
Peat, 647
Pebble media, 65
Percentage error, 366, 377, 388, 434, 477, 478–480, 489–491, 496–499, 505–509

Index

Perdition, 16
Perforated media buckets, 89
Periodic table, 109
Permeable soils, 62, 647
Petroleum hydrocarbons, 16
Phenol, 65
Phosphate (PO), 103, 107, 166, 167, 229, 295, 298, 300, 388, 394
   concentration, 166, 167, 295
   inlet–outlet concentrations, 294, 295
   removal efficiency TR impact, 298, 299
   treatment Q impact, 295–298
Phosphorous (P), 28, 34, 35, 48, 53, 56, 55, 59, 165, 647
   cycle, 55
Photo-plankton, 48
Photometer, 105
Photosynthesis, 39, 647
Phragmites Australis, 35, 94, 129, 515
Physical model hydraulic representation, 112
   actual water velocity, 114
   hydraulic loading rate, 112
   hydraulic retention time, 112, 113
   media removal efficiencies, 114
      set up stage, 114, 115
      steady stage, 115
   pollutant removal efficiency, 113, 114
Phyto
   accumulation, 40
   degradation, 40
   stabilization, 40
   voiatilization, 40
Plant uptake, 8, 9, 40, 48, 53, 54
Plastic
   cell notch, 87
   cells, 140–143, 151, 152, 156–158, 162, 163, 193, 208, 222, 230, 248, 265, 269, 561–564, 591, 621, 624, 625, 627–630, 638, 639
   media, 76, 80, 90–93, 96, 109, 123, 131, 132, 136–139, 142, 152, 189, 192, 200, 201, 205, 208, 213, 220, 228, 236, 237, 241, 244, 252, 259, 271, 272, 279, 288, 293, 294, 300,
313, 320, 326, 332, 339, 348, 355, 511, 515–520, 524–527
   pieces, 35, 516, 523
   pipes, 2, 5, 80, 81, 94, 129
Plotperform, 611
Plottrainstate, 611
Plug flow
   dispersion (PFD), 46
   hydraulics, 63
   kinetics, 116, 340
Pollutant
   removal efficiency, 113, 244
   removal rate, 517, 526
Pollution
   interception, 15
   reduction, 8
Polyculture systems, 58
Polyethylene bottles, 98, 99
Polynomial function, 516, 517
Polyvinyl chloride (PVC), 65, 71, 77, 78, 88, 89, 93
Porosity, 2, 3, 4, 18, 19, 22–24, 32, 33, 56, 62, 65, 78, 82, 83, 88–93, 99, 109, 112–114, 125, 129–136, 143, 144, 180, 181, 192, 199, 207, 208, 219, 221, 223, 228, 229, 241, 242, 259, 276, 355, 358, 374, 515–517, 523–526, 540, 554, 557
   measurements, 130
   values, 131, 135, 143, 241, 515
Porous media, 2, 11, 20, 32, 33, 65
Potassium di hydrogen orthophosphate, 103
Potential energy, 9
Power function, 187, 203, 217, 235, 244, 252, 261, 268, 279, 292, 299, 306
Predation, 518
Pretreatment media, 88, 648
Principal component analysis (PCA), 65
Printed screen, 362–366, 378–382, 393–398, 412–419
Protozoan parasites, 58
Pumps, 63, 64, 70, 71
Purelin, 573, 575, 577, 611

Q

Quality parameter, 67, 99
Queenston shale, 34

R

Radial basis function network (RBFN), 64, 65
Radial basis networks, 119
Rate constant, 46, 47, 116
Raw wastewater, 29, 30
Reaction rate models, 42, 43
Reciprocating flow constructed wetland (RFCW), 15
Reduction, 8, 15, 26, 34, 35, 38, 40, 43, 46, 47, 50–53, 63, 66, 136, 182, 195, 201, 213, 231, 239, 249, 516, 642, 648
Reed (*Phragmites*), 11, 14, 36, 96, 129, 143, 199, 213, 228, 515, 517
  bed establishment, 94, 97
  roots, 199, 213, 228, 517
Regression equation, 4, 42, 43, 111, 126, 128, 358, 430, 432, 434, 438, 439, 447, 448, 457, 458, 514
Removal efficiency (RE), 4, 35, 37, 66, 107, 113–115, 137–143, 148–167, 170, 171, 175–180, 187, 196, 199, 203, 211, 218, 235, 244, 247, 261, 262, 279, 294, 301, 306–308, 311–314, 318–321, 324–329, 331–334, 337–340, 516–520, 526, 557, 560, 313, 325, 326, 333, 334, 338, 339
Removal rate constants, 2, 4, 46, 111, 117, 118, 128, 341–355, 519–521, 526
  calculation method, 117, 118
  graphical method, 117
Removerows, 611
Residual parameters, 3, 575
Restopper, 102
Retention time, 3, 4, 18, 19, 34, 35, 41, 45, 53, 65, 73, 83, 112, 117, 125, 128, 130, 133–136, 143, 144, 182, 191–196, 200, 207, 209–213, 222–230, 242–244, 248, 253, 254, 259, 260, 264, 272, 276–279, 282, 285–287, 289–294, 298–301, 305–308, 311–314, 318, 319, 321, 324–327, 329–334, 338–341, 349, 515, 517, 519, 520, 525, 526, 555, 557
Reynolds number, 21
Rhizodegfadation, 40
Rhizome, 13, 36, 37, 63, 73 96
Rhizosphere, 36, 51, 53, 55, 66

Root zone method (RZM), 11, 36, 48, 52, 648
Rubber
  cell notch, 85, 87
  cells, 114, 115, 173, 181, 184–187, 198, 199, 202, 204, 211, 215, 216, 231, 233, 238, 240, 244, 247, 249, 257, 260, 265, 267, 269, 272, 275, 284, 286, 289, 291, 293, 297, 299–304, 307, 309, 313, 317, 320, 322, 326, 329, 331, 332, 335, 338, 341–343, 348, 349, 521, 560, 566
  media, 76, 80, 92, 94, 96, 109, 117, 131, 132, 135, 136, 138, 142, 182, 192, 195, 204, 205, 213, 219, 221, 228, 231, 235, 237, 241, 244, 254, 263, 272, 275, 278, 287–289, 294–302, 305, 308–315, 320–324, 326, 328, 329, 332, 334, 337, 339, 348, 349, 355, 515–526, 555, 557, 560
Rules of thumb, 42
Rush (*Juncus*), 36

S

Samaha, 14, 69–71, 73, 74, 76, 77, 80, 83, 93, 133, 525, 526
  plant, 14, 74, 93
  sampling period, 98, 100, 108, 129
  waste water treatment plant, 109
Sampling, 3, 70, 76, 78, 83, 93, 96–100, 108, 109, 112, 113, 129, 135, 144, 146, 181, 182, 192, 196, 198, 200, 211, 213, 226, 230, 517, 521, 557, 560, 566
Sanitary drainage company, 74
Sanitation facilities, 71
Scirpusmaritimus, 35
Second hidden layer (SHL), 359
Second order polynomial, 227, 517
Sediment
  accumulation, 91, 526
  particle, 641
Sedimentation, , 9, 15, 31, 37, 40, 48, 54, 64, 107, 649
Seepage, 10, 11, 18, 74, 93 ,649
Selenium, 53
Set up stage, 3, 4, 70, 88, 91, 98, 112, 114, 122, 123, 125, 127–129, 131, 133, 135, 136, 142, 143, 182, 186, 189–193,

Index

199, 200, 205, 207, 210, 213, 217, 220, 226, 228, 241, 358, 359, 362–366, 370, 371, 430, 435, 470, 515–517, 522, 525, 555, 557, 560, 573, 613
Settling, 16, 40
Set-up
  operation, 98
  pollutants treatment, 136–138
    biochemical oxygen demand, 138, 139
    chemical oxygen demand, 139–141
    total suspended solids, 142,
  sampling, 98
  stage data, 122
Sewage, 649
  network, 29, 30
Shale, 34, 55, 56, 642, 649
Short
  circuiting, 649
  term clogging, 24, 25
Shredded tires, 2, 5, 34, 76, 80, 81, 92, 526
Slitcher formula, 23
Sludge, 641, 643, 649
Sodium thiosulfate, 99, 103
Solar radiation, 9, 65
Solid surface vicinity, 39
Solution
  acidity, 49
  alkalinity, 49
Sorption, 40, 649
Spectro photometer, 106, 107
Stability, 70, 181, 392
Stabilization
  conditions, 143
  phase, 63
Stage pollutants, 430, 471, 557, 558
Standard deviation ($S.D.$), 136, 143, 613
Start-up phase, 63
Statistical
  analysis, 2, 4, 67, 111, 126, 357, 613
  modeling, 3, 514
  tests, 511
Steady stage, 3, 4, 70, 88, 99, 100, 109, 115–117, 125, 127–131, 143, 144, 146, 153, 158, 165, 169, 173, 177, 180, 229, 230, 232–234, 236, 242, 250–252, 255, 259, 266, 267, 271, 273, 282, 286–289, 291, 294, 301, 305, 326, 334, 340, 355, 358, 374–376, 378–382, 394, 397, 398, 410, 411, 417–419, 438, 441, 445, 457, 470, 488, 522, 523, 557, 560, 573, 575, 579, 613
  sampling, 99
Steadystate conditions, 98
Stochastic package for social science (SPSS), 2, 3, 4, 66, 67, 111, 112, 126–128, 357, 358, 430–472, 477–514, 524–526, 578, 620, 623, 634–639
Stochastic package, 111, 358
Storm water, 649
Submersed plants, 649
Substrate, 649
Subsurface flow (SSF), 7, 10, 11, 12, 15, 16, 19, 20, 21, 28, 30, 32, 34, 48, 50, 52, 53, 55, 56, 59, 66, 69, 70, 80, 112, 116, 340, 357
  constructed wetlands components, 26, 27
    functions, 36, 37
    plant communities, 36
    sources of pollutants, 27–30
  constructed wetland media, 30–36
    characteristics, 32, 33
    configuration, 33
    functions, 31
    particle size, 31, 32
    types, 33–35
Suction pump, 106
Sulfide compounds, 52
Sulfide oxidation, 51
Sulfuric acid, 103, 105
Sulphate reduction, 51
Surface area, 17, 20, 22, 32, 33, 37, 50, 56, 60, 62, 91–93, 112, 123–125, 133–136, 187, 203, 218–221, 238, 244, 263, 264, 275, 279, 282, 293, 358, 430, 439, 516, 523, 524
Surface flow (SF), 10, 11, 15, 50, 59, 60, 650
Suspended particulate matter, 16, 28
Suspended solids (SS), 2, 9, 13, 25, 26, 40, 47, 49, 52, 58–60, 71, 105, 115, 116, 143, 158, 230, 269, 270–273, 560
Synthetic wastewater, 66

T

Tanks-in-series (TIS), 41, 42, 45, 46
Tannery wastewater, 34
Tansig, 611
Target vectors, 122, 360, 611
Temperature, 51, 108, 567, 568
Tennessee valley authority (TVA), 26
Terzaghi formula, 24
Tetrahydrofuran, 60
Theoretical discharges, 85–87
   gravel cell notch, 86
   plastic cell notch, 87
Third order polynomial, 198, 212, 517
Tire chips, 524, 525
Total coliforms (TC), 107
Total dissolved solids (TDS), 29, 52
Total kjeldahl nitrogen (TKN), 34, 66
Total phosphorus (TP), 30, 34, 35, 57, 66
Total suspended solids (TSS), 2, 3, 13, 26, 29, 34, 35, 43, 44, 48, 57, 58, 62, 66, 91, 98, 99, 105, 116, 117, 125–130, 136, 142, 143, 146, 147, 159–163, 180, 181, 213, 215–230, 264–282, 341–348, 357–388, 430–447, 470, 477–480, 485–491, 516, 517, 520–526, 557–560, 564, 565, 571–575, 579, 584, 591, 615
   concentrations, 564, 565
   removal efficiency, 213, 217, 218, 222–228, 264, 268, 276–282, 516, 517
   time effect treatment efficiency, 226–228
   treatment distance effect, 213–218
   treatment Q impact, 218, 219
Training
   parameters, 611
   performance, 367, 368, 382, 383, 399, 400, 419–421, 431, 432, 440, 441, 448, 449
   vectors, 361
Trainlm, 361, 376, 392, 410, 573, 575, 577, 611
Transition habitat, 9
Transitional stage, 76
Treatment concentrations, 158
   ammonia (NH3), 158–165
   dissolved oxygen (DO), 165–169
   fecal coliforms (FC), 169
   phosphate (PO4), 165
Treatment, 7, 34, 37, 48, 67, 73, 129, 138, 192, 229, 648, 650
   ammonia, 287
   cadmium, 334
   cells, 2, 15, 73, 74, 76, 78, 79, 82, 83, 88, 94, 109, 130, 133
   concentrations, 146
      biochemical oxygen demand (BOD), 146–152
      chemical oxygen demand (COD), 153–158
      total suspended solids (TSS), 158
   distance effect, 200–203
   efficiency, 35, 69, 74, 181, 185, 200, 228, 234, 248, 355, 517, 526, 527
   fecal coliforms, 301
   functions, 91
   iron, 314
   lead, 327
   length, 186, 187, 203, 217, 230–235, 249, 251, 264, 268
   manganese, 321
   media, 2, 80, 83, 109, 190, 526
   performance, 10, 15, 64, 67, 74, 97, 112, 114, 125, 130, 181, 210, 218, 220, 226, 230, 234, 236, 239, 256, 261, 279, 284, 286, 293, 297, 299, 302, 309, 317, 319, 332, 334, 339, 375, 516, 518, 526, 566
      chemical mass balances, 38
      internal hydraulics models, 40, 41
      pollutants removal processes, 38–40
   phosphate, 294
   zinc, 308
Typha species, 34
Typhalatifolia wetland plant, 34, 65

U

Ultraviolet radiation, 39, 650
Uniformity (U), 19, 23, 47, 610
   coefficient, 23, 24
Uranium, 53
U.S. Bureau of Reclamation (USBR), 24

Index

V

Validation, 4, 112, 126, 127, 361, 370, 371, 376, 378, 392, 397, 410, 415, 430, 434, 435, 438, 441, 445, 447, 452, 455, 457, 464, 467, 470, 479, 488–492, 495, 497–500, 510, 514, 525, 611
  performance, 370, 371, 385, 386, 403, 404, 424–426, 435, 436, 443, 444, 453, 454, 465–467
  process, 127, 371, 434, 435, 455, 470, 479, 492, 500
Vegetated cells, 67
Vegetation, 37
Vegetative submerged bed systems, 11
Velocity (v), 19, 89, 114, 123, 125, 358, 375, 389, 407, 430, 439
Vertical flow (VF), 11–13, 15, 29, 37
Volatilization, 40, 48, 51, 53, 650
Volumetric removal rate constant (k), 117, 343–349, 352–354

W

Warm-blooded animals, 107
Wastewater treatment plant (WWTP), 69, 70, 80
Water
  budget, 650
  diffusion, 48
  level control system (WLCS), 78, 79
  pollution, 102, 158
  quality, 2, 8, 14, 16, 51, 61, 63, 67, 99, 109, 526
  quality analysis, 100
    ammonia exists, 106
    biochemical oxygen demand, 103, 104
    chemical oxygen demand, 104, 105
    dissolved oxygen, 102, 103
    fecal coliforms, 107, 108
    heavy metals, 109
    hydrogen ion, 100–102
    phosphate, 107
    total suspended solids, 105, 106
    water temperature, , 108
  sampling, 70, 98–100, 181
    set-up stage, 98, 99
    steady stage, 99, 100
  velocity (v), 19
Weight matrixes, 359, 375, 389, 407
Wetland, 7, 8, 9, 16, 18, 31, 37, 39, 42–45, 48, 59, 60, 69, 649, 650
  bed, 25, 26, 49, 56, 61, 63, 97, 136
  cell, 2, 3, 14, 15, 19, 35, 48, 63, 64, 70, 73, 74, 77–79, 85, 88, 93, 94, 98, 99, 112–115, 122–125, 129, 130, 133–135, 144, 158, 165, 169, 177, 181, 187, 191, 192, 203, 207, 218, 219, 222, 230, 233, 238, 241, 244, 256, 261, 267, 272, 273, 277, 285, 287, 292, 306, 307, 314, 317, 319, 320, 327, 329, 333, 334, 340, 388, 511, 525, 540, 557, 560, 568
    water velocity, 19
  components, 3, 64
  definitions, 67
  functions, 15, 67
    biological productivity, 15
    carbon dioxide balance, 15
    carbon fixation, 15
    coastal protection, 15
    floodwater storage, 15
    groundwater recharge, 15
    pollution interception, 15
    wildlife habitat, 15
  hydraulics, 3, 67
  hydrology, 17
  inlet bed, 24
  internal hydraulics, 41
  modeling, 67
  planning basic principals, 61
  sediments, 49
  systems, 1, 3, 9, 10, 28, 30, 36, 37, 42, 51, 56–59, 61, 108, 146, 181, 374, 517, 520, 526, 527, 566
  treatment cells, 73, 74
  types, 3, 9, 37, 67
    constructed, 10–13
    natural, 9, 10
    subsurface flow, 11–13
    surface flow, 11
  values, 67
  vegetation, 36
  water parameters, 47

biochemical oxygen demand, 50
chemical oxygen demand, 50
dissolved oxygen, 47–49
electric conductivity, 52
hydrogen ion concentration, 49
metals removal, 52, 53
oxidation-reduction potential, 50, 51
total dissolved solids, 52
total suspended solids, 49, 50
water temperature, 51, 52

Wildlife habitat, 15, 60

Z

Zeolite, 35, 55, 65, 650
Zero
   order model, 43
   weight, 610
Zinc (Zn), 35, 52, 53, 109, 143, 169, 171–173, 229, 230, 308, 309, 311–314, 349, 354, 407, 409–417, 429, 458, 461, 464, 467, 518, 520, 571, 608
   neural network training tool, 412

Printed in the United States
By Bookmasters